Agricultural and Food Electroanalysis

Agricultural and Food Electroanalysis

Edited by

ALBERTO ESCARPA
MARÍA CRISTINA GONZÁLEZ
MIGUEL ÁNGEL LÓPEZ

Analytical Chemistry,
Physical Chemistry and Chemical Engineering Department,
University of Alcala,
Spain

WILEY

This edition first published 2015

Registered office
John Wiley & Sons Ltd, The Atrium, Southern Gate, Chichester, West Sussex, PO19 8SQ, United Kingdom

For details of our global editorial offices, for customer services and for information about how to apply for permission to reuse the copyright material in this book please see our website at www.wiley.com.

Library of Congress Cataloging-in-Publication Data

Agricultural and food electroanalysis / edited by Alberto Escarpa, Maria Cristina Gonzalez and Miguel Angel Lopez.
pages cm
Includes bibliographical references and index.
ISBN 978-1-119-96186-4 (cloth)
1. Food–Analysis. 2. Electrochemical analysis. I. Escarpa Miguel, Alberto, 1968- editor. II. Gonzalez Martin, Maria Cristina, editor. III. Lopez Gil, Miguel Angel, editor.
TX545.A39 2015
664′.07–dc23

2015016626

A catalogue record for this book is available from the British Library.

ISBN: 9781119961864
Cover Image: pawel.gaul/iStockphoto

Set in 10/12pt TimesLTStd by SPi Global, Chennai, India
Printed and bound in Malaysia by Vivar Printing Sdn Bhd

1 2015

At the time we were editing this book, Professor Mascini passed away. Probably, one of the latest excellent contributions done in his vast successful career is found in this book. In memoriam, editors would like to dedicate him these words as proof of his valuable contribution to the field of electrochemical biosensors in food analysis.

Contents

List of Contributors

Lúcio Angnes, Instituto de Química, Universidade de São Paulo, Brazil

Alberto Sánchez Arribas, Departamento de Química Analítica y Análisis Instrumental, Universidad Autónoma de Madrid, Spain

Craig E. Banks, School of Chemistry and the Environment, Division of Chemistry and Environmental Science, Manchester Metropolitan University, UK

Mónica Moreno Barambio , Departamento de Química Analítica y Análisis Instrumental, Universidad Autónoma de Madrid, Spain

Simona Benedetti, Dipartimento di Scienze per gli Alimenti, la nutrizione e l'ambiente Università degli Studi di Milano, Italy

Erika Bustos, Centro de Investigación y Desarrollo Tecnológico en Electroquímica, Parque Tecnológico Querétaro Sanfandila, México

Gang Chen, School of Pharmacy, Fudan University, China

Wendell K. T. Coltro, Institute of Chemistry, Federal University of Goias, Brazil

Stella M. Cosio, Dipartimento di Scienze e Tecnologie Alimentari e Microbiologiche, Università degli Studi di Milano, Italy

Agustín Costa-García, Departamento de Química Física y Analítica, Facultad de Química, Universidad de Oviedo, Spain

Alberto Escarpa, Departamento de Química Analítica, Química Física e Ingeniería Química, Universidad de Alcalá, Spain

Fabiana Silva Felix, Instituto de Química, Universidade de São Paulo, Brazil

M. Teresa Fernández-Abedul, Departamento de Química Física y Analítica, Facultad de Química, Universidad de Oviedo, Spain

Carlos D. Garcia, Department of Chemistry, The University of Texas at San Antonio, USA

Luis A. Godínez, Centro de Investigación y Desarrollo Tecnológico en Electroquímica, Parque Tecnológico Querétaro Sanfandila, México

María Cristina González, Departamento de Química Analítica, Química Física e Ingeniería Química, Universidad de Alcalá, Spain

Araceli González-Cortés, Departamento de Química Analítica, Facultad de Ciencias Química, Universidad Complutense de Madrid, Spain

M. Begoña González-García , Departamento de Química Física y Analítica, Facultad de Química, Universidad de Oviedo, Spain

Miguel A. López, Departamento de Química Analítica, Química Física e Ingeniería Química, Universidad de Alcalá, Spain

Saverio Mannino, Dipartimento di Scienze e Tecnologie Alimentari e Microbiologiche, Università degli Studi di Milano, Italy

Juan Manríquez , Centro de Investigación y Desarrollo Tecnológico en Electroquímica, Parque Tecnológico Querétaro Sanfandila, México

Marco Mascini, Dipartimento di Chimica Ugo Schiff, Università degli Studi di Firenze, Italy

Sandra Mendoza, Departamento de Investigación y Posgrado en Alimentos, Universidad Autónoma de Querétaro, México

Solomon Lemma Mengistu, Free University of Bolzano, Italy

Arben Merkoçi, Institució Catalana de Recerca i Estudis Avançats (ICREA), Spain; Nanobioelectronics and Biosensors Group, Universitat Autònoma de Barcelona, Spain

Jonathan P. Metters, School of Chemistry and the Environment, Division of Chemistry and Environmental Science, Manchester Metropolitan University, UK

Maria F. Mora, Jet Propulsion Laboratory, California Institute of Technology, USA

Geza Nagy, General and Physical Chemistry, Faculty of Science, University of Pécs, Hungary

Lívia Nagy, Szentagothai Research Centre, University of Pécs, Hungary

Ilaria Palchetti, Dipartimento di Chimica Ugo Schiff, Università degli Studi di Firenze, Italy

Briza Pérez-López, Nanobioelectronics and Biosensors Group, Universitat Autònoma de Barcelona, Spain; LEITAT Technological Center, Spain

José M. Pingarrón, Department of Analytical Chemistry, Faculty of Chemistry, University Complutense of Madrid, Spain

Manuel Chicharro Santamaría, Departamento de Química Analítica y Análisis Instrumental, Universidad Autónoma de Madrid, Spain

Matteo Scampicchio, Free University of Bolzano, Italy

Joseph Wang, Departments of Nanoengineering, University of California, USA

Paloma Yáñez-Sedeño, Department of Analytical Chemistry, Faculty of Chemistry, University Complutense of Madrid, Spain

Preface

This pioneer book *Agricultural and Food Electroanalysis* provides a description and rationale use of modern electroanalytical techniques, strategies, and approaches in the exciting field of agricultural and food analysis. Electrochemical techniques offer very valuable features such as very good sensitivity, tunable selectivity, low cost, simple use, inherent miniaturization, high compatibility with modern technologies required from microfabrication techniques to build "lab-on-a-chip" devices, high compatibility with surface modification employing biological reagents as well as exciting nanomaterials such as nanoparticles, nanotubes, and nanowires. Without any question, with the incursion of advanced approaches such as screen printed technology, biosensors, microchips, and nanotechnology, among others, electroanalysis is living a truly *Renaissance* and new frontiers have been clearly opened in the last years.

This book is divided in three parts and contains 16 chapters written by truly well-recognized experts in the field.

The first chapter, *Electroanalysis and Food Analysis,* has the important role to introduce the readers in the whole book where the adequacy of electroanalysis to agricultural and food analysis is exposed.

Following this initial introductory chapter, the book is structured in three parts. The first part discusses different *Electroanalytical Techniques in Batch and Continuous Systems* as highly remarkable tools in the agricultural and food field. In this sense, Chapter 2 deeply explores the sweep potential electroanalytical techniques, while Chapter 3 allows the readers to obtain fundamental information on voltammetric techniques coupled to flow systems which could proportionate faster analysis, reproducible results, high sensitivity, with additional advantages such as the requirement of less sample, and the use of simpler instrumentation.

Separation techniques coupled to electrochemical detectors are also studied in this section. Chapter 4 deals with the design and integration of electrochemical detectors within the HPLC separation system and their compatibility and compromise with the chromatographic conditions necessary to achieve the optimum resolution of the analytes in agricultural and food field. Chapter 5 introduces the key strategies in capillary electrophoresis using electrochemical detection for separating and detecting a variety of constituents in foods and agricultural products. That includes the commonly used separation modes of capillary electrophoresis, its coupling with electrochemical detection, and its application in agricultural and food analysis.

The largest part of the book, organized in the following nine chapters, is dedicated to the *Electrochemical Sensing in Food Analysis*. Chapter 6 introduces microelectrodes and microelectrode arrays which can be used for both fundamental and applied electrochemistry. Different approaches to fabricate such transducers are critically overviewed with special emphasis on the requirement of sensors that can be used at the site of sampling, being cost-effective and reproducible. Besides, the importance of potentiometric sensors and electrochemical biosensing approaches is deeply studied in Chapters 7–11.

Electrochemical transducers combined with an enzyme as a biochemical component constitute the largest category of biosensors, thus becoming an important tool for the detection of highly concern analytes in agricultural and food monitoring. This matter is considered in Chapter 8. The design, chemical construction, and application in agricultural and food electroanalysis of the further most important biosensing approaches such as immunosensors and genosensors are studied in Chapter 9 and Chapter 10, respectively. Additionally, Chapter 11 discusses the recent trends that have led to powerful nanomaterial-based electrochemical biosensing devices and examines the related prospects and challenges suggesting considerable promise for diverse applications in the food and agricultural field.

The next two chapters of this book section address the novel micro- and nanotechnologies impact in the field. Electroanalysis on board of microfluidics and lab-on-a-chip platforms is studied in Chapter 12 and selected nanoelectrochemistry applications for food analysis are covered in Chapter 13. To conclude this part, Chapter 14 deals with the principles and food applications using electrochemical impedance spectroscopy.

The book finishes with two chapters configuring the third part regarding *Industrial Implications: Electroanalysis in Food Process Control* (Chapter 15) and *Instrumental Aspects of Food Analysis by Electrochemical Methods* (Chapter 16). Unlike traditional chemical analysis, performed in well-equipped laboratories with the aim to identify and quantify small amounts of analytes, the goal of process analytical chemistry is to supply quantitative and qualitative information about a chemical process that can be used not only to monitor and control the process, but also to optimize its efficient use of energy, time, and raw materials. In addition, it is possible to simultaneously minimize plant effluent release and to improve product quality. These important concepts adapted to food and agricultural electroanalysis are focused in Chapter 15. In contrast, despite the clear advantages associated with electrochemical detectors, the training of personnel and somehow the limited availability of commercial instruments has traditionally limited the development of electrochemical methods applied to agricultural and food-related samples. In the light of such considerations, Chapter 16 aims to provide a brief overview of the key instrumental aspects linked to agricultural and food electroanalysis.

In sum, and in editor's opinion, this valuable text offers a comprehensive vision about different electrochemical techniques following their basic principles, instrumentation, and main applications in the field of food and agricultural analysis. Also, editors hope that the critical and attractive vision of the book will help readers to get introduced in the exciting area of agricultural and food electroanalysis.

Finally, the editors would like to thank to all the authors for their excellent contributions to this pioneer book in the agricultural and food analysis field.

Alberto Escarpa, María Cristina González, and Miguel Ángel López
Editors

1

Electroanalysis and Food Analysis

Paloma Yáñez-Sedeño and José M. Pingarrón
Department of Analytical Chemistry, Faculty of Chemistry, University Complutense of Madrid, 28040, Madrid, Spain

1.1 Introduction and Adequacy of Electroanalysis for Food Analysis

Electroanalysis is a powerful analytical tool for food analysis. Since the early times of polarography and potentiometry until the current developments of chemical sensors, biosensors and lab-on-a-chip (LOC) devices involving electrochemical detection principles and many other electroanalytical methodologies have demonstrated their usefulness to accomplish the requirements imposed by the food industry for the analytical monitoring and control of raw materials and foodstuffs. This is particularly true in the last decades where impressing advances exhibited in automation, miniaturization, and easy handling of electroanalytical devices including both the corresponding instrumentation and the electrodes employed as electrochemical transducers, have led to user-friendly methods of analysis that are competitive against other well-established analytical techniques such as chromatography and spectroscopy. This, together with the inherent affordable costs of electroanalytical approaches and the superior sensitivity that can be achieved using modern voltammetric techniques or coupling with amplification response methodologies involving nanomaterials, makes modern electroanalysis a more than suitable strategy to face up to the increasingly demanding requirements of food industry to ensure food quality, control, and safety [1].

The development and innovation in food industry rely basically on the concepts of food safety and food quality. However, lately, products as foodstuffs supplemented with compounds such as omega-3 acids, vitamins, fiber, and so on, which confer them particular properties sought by specific strata of society, have burst into the modern food industry providing products with a high added value. Obviously, as the food chain is increasingly

Agricultural and Food Electroanalysis, First Edition.
Edited by Alberto Escarpa, María Cristina González and Miguel Ángel López.

complex, there is a high demand for the development of efficient traceability systems which are able to guarantee the firmness of the whole chain. These systems should possess high sensitivity, ability to be implemented rapidly, and permit automatic screening.

Food quality can be understood as a set of factors which are able to differentiate food products according to their organoleptic characteristics, composition, and functional properties. An increased regulatory action together with an increased consumer demand for information have led to the extensive labeling of major and minor constituents of the foodstuffs. The scientific evaluation of the food freshness is another important task concerning food quality assessment. A list of the most current compounds to be analyzed for food quality assessment can be found in [2]. Moreover, continuous monitoring of food industrial processes allow real-time detection of possible errors in the chain production as well as taking decisions to rectify such errors in an immediate manner. The assessment of food safety is the other key axe for the modern food industry. In general, one can speak about food contamination when dealing with harmful substances or microorganisms that are not intentionally added to the food. Contaminants may enter the food chain during growth, cultivation, or preparation, accumulate in food during storage, form in the food through the interaction of chemical components or may be concentrated from the natural components of the food [3]. However, chemicals are also added during food processing in the form of additives. At present, pathogen microorganisms, pesticides, animal-drug residues, and antimicrobial drug resistance are the main concerns for food safety. Food regulatory agencies have established control programs, such as the HACCP (Hazard Analysis Critical Control Point) program, to avoid the entering of these substances into the food chain [4].

Electroanalysis has played a relevant role in food quality and food safety assessment and, in the last few years, it is increasingly significant due to the combination of sensors and biosensors technology, even in a disposable manner, with efficient electrochemical transduction techniques allowing the implementation of rapid and reliable detection methods. To provide an overview of the state of art in the use of electrochemical techniques in the field of agricultural and food analysis, we discuss in this chapter some examples on the latest advances in this field, pointing out on relevant methodologies related to the measurement techniques, including the development of electrochemical sensors and biosensors for food components, and the use of nanostructured electrodes.

1.2 Methodologies Related to Measurement Techniques

1.2.1 Continuous Detection Methods

In general, the application of electrochemical techniques for the detection of analytes in a continuous mode has demonstrated to be able to improve the sensitivity and selectivity of well-established analytical methods. Electrochemical detection has shown to be appropriate to be combined with high-performance liquid chromatography (HPLC), flow injection analysis, capillary electrophoresis, or microfluidics-based methodologies. There are numerous examples regarding food analysis where the improvements achieved using electrochemical detection techniques can be illustrated. With respect to liquid chromatography, methods are still being developed for detecting healthy food components. Representative examples are the simultaneous determination of hydroxy polymethoxy-flavones in citrus products and orange juice [5], and a very recent method for the determination of phenolics in olive oil [6]. In both cases, HPLC with coulometric detection at multichannel CoulArray detector was used which enabled a high sensitivity to be obtained. Moreover, methodologies developed for the detection of drugs and pesticide residues using electrochemical detection can be found in the recent literature. As examples, the efficient separation and sensitive determination of sulfonamides in shrimps using a monolithic column and amperometric detection at a boron-doped diamond electrode [7], and the detection of carbamate pesticides in fruits and vegetables using an acetylcholinesterase biosensor where the enzyme was immobilized on a polyaniline–carbon nanotubes composite electrode [8], can be cited.

Flow-injection methods with electrochemical detection continue to attract great attention in the field of food analysis due to the inherent simplicity of these approaches and the good analytical performance provided. A recent and interesting application involves a single-line flow injection system combined with multiple pulse amperometric detection with a boron-doped diamond electrode for the simultaneous determination of two pairs of food colorants: tartrazine (TT) and sunset yellow (SY) (TT–SY) or brilliant blue (BB) and SY (BB–SY) in sports drink beverages, gelatin, and powdered juice. A dual-potential waveform was applied to the electrode for both colorants in each pair to be determined with detection limits ranging between 0.80 and 3.5 μM [9]. Batch injection analysis (BIA) combined with electrochemical detection has also been applied in this field taking advantage of the versatility, reproducibility, high analytical frequency, sensitivity, portability, and sample size provided by this combination [10]. Using these systems, precise sample plugs are directly injected onto the working electrode surface which is immersed in a large-volume blank solution, and the electrochemical responses are recorded directly. For example, amperometric detection at a Prussian Blue-modified graphite-composite electrode was recently described for determining H_2O_2 in high- and low-fat milk samples [11]. In this method, an electronic micropipette injected 100-μl aliquots of 10-fold diluted samples directly onto the modified electrode immersed in the BIA cell (Figure 1.1). The detection limit was low (10 μM), and good recovery values were achieved for spiked samples.

Within this family of continuous methodologies, it is also important to mention the sequential injection lab-on-valve (SI-LOV) technique, which allows increasing sampling capacity and the automation of the analytical methods [12]. In a recent article, an SI-LOV system was used for the sensitive determination of hypoxanthine [13]. As one of the purine bases, hypoxanthine is produced during the degradation process of fresh meat and fish, so that the content of this compound can be envisaged as a valuable indicator of food freshness [14]. In the cited work, a Fe_3O_4/multiwalled carbon nanotubes (MWCNTs)/β-cyclodextrin (β-CD) (Fe_3O_4/MWCNTs/β-CD) modified electrode was employed to measure the electrochemical oxidation of hypoxanthine. A diagram of the SI-LOV system used is depicted in Figure 1.2. After aspiration of 500 μl phosphate buffer solution (PBS) into the holding coil, various microvolumes of carrier, air, sample solution, and PBS, were aspirated and transferred into electrochemical flow cell (EFC) for the analyte accumulation on the modified electrode at 0.1 V. Then, the stripping voltammogram of hypoxanthine was recorded. Under the optimized conditions, a linear dependence between log I_p *vs.* log [hypoxanthine] was found in the 5.0×10^{-8}–1.0×10^{-5} range, and the method was applied in determining hypoxanthine in meat samples.

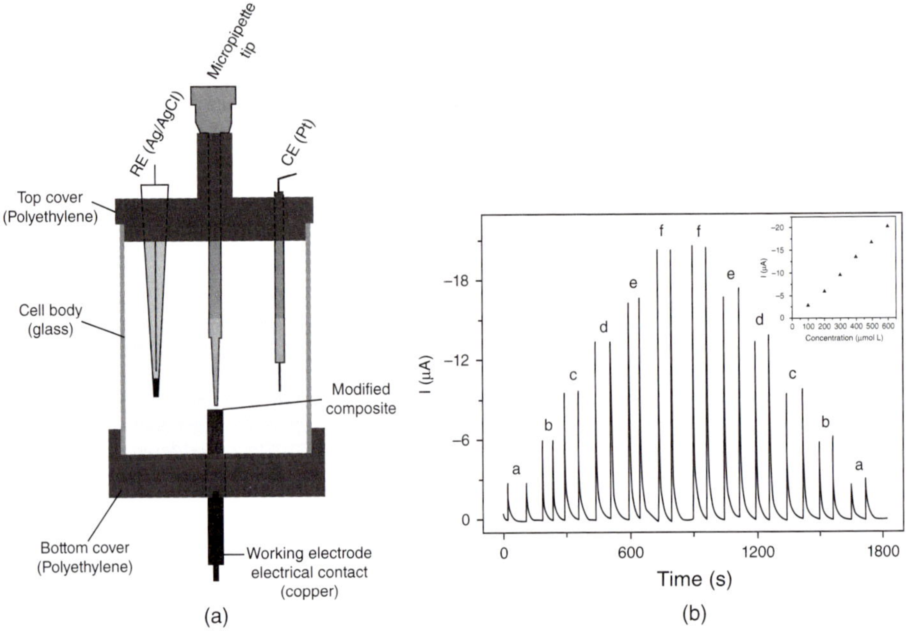

Figure 1.1 *(a) Schematic diagram of the batch injection cell containing the three-electrode system. (b) BIA amperometric responses of PB-modified graphite composite electrode for 100–600 μmol/l H_2O_2. Reproduced from Ref. [11] with permission from Elsevier*

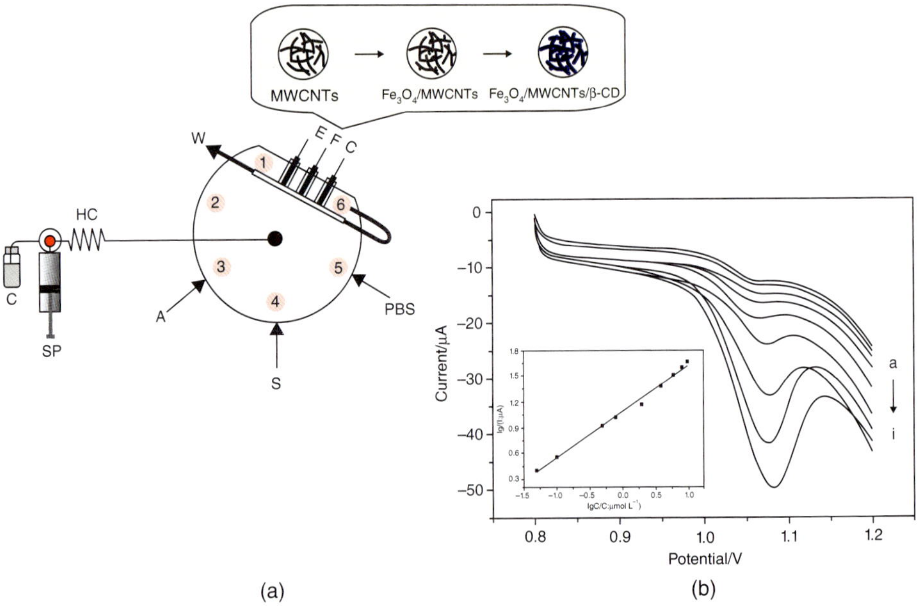

Figure 1.2 *(a) Schematic diagram of SI-LOV manifold for hypoxanthine analysis: C, carrier (H_2O); SP, syringe pump; HC, holding coil; W, waste; A, air; S, sample; PBS, phosphate buffer solution, EFC, electrochemical flow cell (internal volume 200 μl). (b) Stripping voltammograms for (a–i) 0.05–10 mmol/l hypoxanthine. Inset: log I_p vs. log hypoxanthine concentration calibration plot. Reproduced from Ref. [13] with permission from Elsevier*

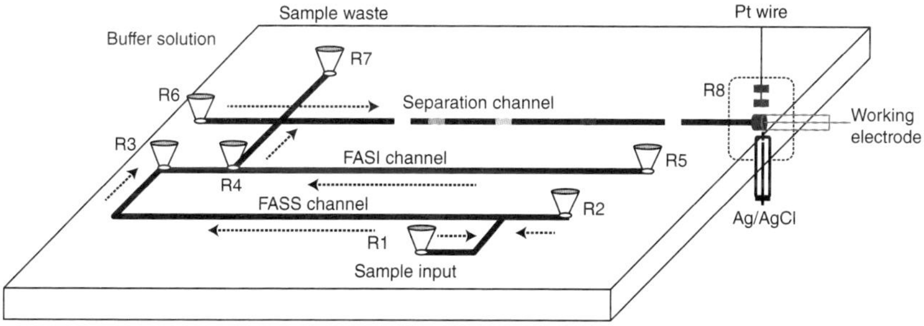

Figure 1.3 *Scheme of the microfluidic device for sulfonamides separation and detection. Reproduced from Ref. [16] with permission from Elsevier*

During the past 20 years, a great progress has been made in the development of miniaturized systems for chemical analysis. In this context, microfluidics has attracted special interest because of offering remarkable sensitivity, inherent miniaturization, low cost, portability, compatibility with mass fabrication, and on-site analysis. The manipulation of small amounts of fluids through microchannels can be combined with miniaturization technologies for developing "LOC" devices, which intend the integration of different steps involved in the analytical process. When a separation step is required, capillary electrophoresis (CE) has demonstrated to be very appropriate for miniaturized technology. A representative example of application of these systems to agricultural and food analysis is the determination of phenolic compounds (tyrosol, hydroxytyrosol, and oleuropein glucoside) in olive oil using glass microchip electrophoresis with end-channel amperometric detection at a 100-μm gold wire working electrode [15]. Other recent application makes use of a microfluidic device for the simultaneous detection of five sulfonamides in meat involving preconcentration of the analytes in the microfluidic device (Figure 1.3) followed by their electrokinetic separation and amperometric detection at Al_2O_3–gold nanoparticles (AuNPs)-modified carbon paste electrodes. A linear range between 0.01 and 2,025 pM, and detection limits between 0.91 and 2.21 fM were obtained [16].

Capillary electrophoresis with amperometric detection was also used for the determination of four electroactive preservatives (methylparaben, ethylparaben, propylparaben, and butylparaben) and, indirectly, two nonelectroactive preservatives (potassium sorbate and sodium lactate) in various types of foodstuffs [17]. Moreover, high-performance micellar electrokinetic capillary chromatography with amperometric detection (MECC–AD) has been also employed for the fast determination of melamine (2,4,6-triamino-s-triazine), which was occasionally used to increase the apparent protein content of milk products [18].

1.2.2 Stripping Analysis

Mercury film electrode has been widely used in stripping voltammetry for a long time owing to its excellent electrochemical behavior. However, the toxicity of mercury has led to the development and use of the environmentally friendly bismuth film electrode (BiFE). This electrode is characterized by the simple preparation process, large enough accessible potential window, high sensitivity, well-defined, and undistorted stripping signal, as well as good resolution of neighboring peaks. In addition, BiFE is less sensitive to the presence of dissolved oxygen than MFE [19]. An interesting example of recent applications of stripping methodologies with BiFE in food analysis is the determination of azo-compounds used as dyes for food, beverages, and textile industry coloring, which represent a human hazard because their degradation products, including amines, are carcinogenic. A bismuth/poly(*p*-aminobenzene sulfonic acid) (*p*-ABSA) composite film-coated glassy carbon electrode (GCE) (Bi/poly(*p*-ABSA)/GCE), prepared by depositing bismuth on the poly(*p*-ABSA) modified electrode at −0.9 V, was used in this method. Azo-compounds such as 1-(2-pyridylazo)-2-naphthol (PAN), 4-(2-pyridylazo)-resorcinol (PAR), and azobenzene were determined by differential pulse voltammetry in orange and lemon beverages [20]. Other metallic films have also been employed for the preparation of modified electrodes to be applied in stripping methodologies for food analysis. For example, adsorptive stripping voltammetry (AdSV) with an *in situ* plated lead film GCE was used for the determination of vitamin B1 (thiamine) in juices. Thiamine was preconcentrated at −1.25 V and then electrochemically reduced by scanning the potential from −1.25 to −1.55 V using square wave voltammetry (SWV). A range of linearity between 0.0133 and 0.265 mg/l thiamine was reported with this methodology [21].

Despite the attempts to substitute mercury electrodes by other greener film electrodes, some recent applications involving Hg electrodes still appear in the literature. As an example, a method for the determination of the antibiotic ceftiofur (CF) in milk has been reported implying the adsorptive accumulation of the drug on a hanging mercury-drop electrode (HMDE) and reductive SWV. CF is a widely used broad-spectrum third-generation cephalosporin, which is approved for the treatment of infections in cattle, swine, sheep, goats, turkeys, and chickens. By application of the AdSV methodology, a linear calibration plot between 52.4 and 524 ng/ml, which allowed testing of the established tolerance level of 100 ng/ml for CF residues in bovine milk, was achieved. The method was applied to determine CF in spiked milk samples [22]. HMDE and SWV were also used for trace determination of azoxystrobin [methyl (E)-2-{2-[6-(2-cyanophenoxy)-pytimidin-4-yloxy]phenyl}-3-methoxyacrylate] and dimoxy-strobin [(E)-*o*-(2,5-dimethylphenoxymethyl)-2-methoxyimino-*N*-methyl phenylacetamide] in potatoes, grapes, and grape juice. These compounds are synthetic pesticides, the strobilurins, which are derived from the natural occurring *β*-methoxyacrylates. In this method, limits of quantification as low as 119 µg/l in grape juice and 45 µg/kg in potatoes and grapes were found using deposition potential and deposition time values of −300 mV and 30 s, respectively [23]. Differential pulse stripping voltammetry at an HMDE was also used for the simultaneous determination of tetracycline antibiotics in spiked animal feed and fresh fish muscle dosed with the drugs. The voltammograms from the drug mixture produced complex, overlapping profiles, and chemometrics methods were applied for calibration modeling. The analytical linear ranges were within 0.02–0.18 µg/ml and the corresponding limit of detections (LODs) were within 3–5 µg/l [24].

However, metal traces in foods were also determined by stripping analysis. For example, in a recent method, the sequential voltammetric determination of Hg(II) and Cu(II) at a gold electrode, and of Cu(II), Pb(II), Cd(II), Zn(II) at an HMDE by SW anodic stripping voltammetry in matrices involved in food chain as whole meal, wheat, and maize meal, was proposed. The supporting electrolyte was 0.01 mol/l EDTA-Na_2 + 0.06 mol/l NaCl + 2.0 mol/l $HClO_4$, and the analytical procedure was validated by the analysis of standard reference materials [25].

1.2.3 Potentiometry and Chronopotentiometry

Potentiometric detection using ion selective electrodes (ISEs) has had wide application in the field of food analysis since long time ago due to the fair selectivity, wide linear dynamic range, low cost, and automation ability of the derived methods [26]. Advances ISEs, developed in the last years, have also found application in this area. For example, solid-contact ion-selective platforms based on GCEs coated with electropolymerized polyaniline (PANI) and tetrasubstituted thiacalix[4]arene ionophores were reported for the discrimination of the brands of apple juices and herbal liqueurs. The samples were diluted and spiked with Fe^{3+}, and the variation of the signal from this ion, which is related to its reactivity with the organic ligands, was monitored. The method was also applied to the determination of antioxidants (ascorbic, malic, oxalic acids, hydroquinone, and quercetin) in the range from 5.0×10^{-6} to 1.0×10^{-2} M [27]. A potentiometric fumarate (FUM) ion selective electrode (Pt/Hg/Hg^{2+} FUM/graphite) has been recently developed for the determination of the acidulant additive fumaric acid in powdered foods such as gelatin, instant pudding, and ice cream. The achieved sensitivity was (-29.2 ± 0.6) mV/decade over a concentration range between 7.5×10^{-7} and 1.0×10^{-2} M [26]. A graphite carbon electrode coupled with a flow system was also used for the potentiometric determination of citrate in fruit juices and an isotonic drink. The fundamentals of this procedure involve the ion-exchange adsorption process of citrate on the electrode surface and the subsequent potential change. The electrode exhibits a linear response, with a slope of -29.0 ± 1.0 mV/decade, in a 0.07–7.0 mM concentration range with an LOD of 3.0 μM [28].

Chronopotentiometric analytical methods have also found application in the field of food analysis. In this technique, the oxidation or reduction of species at a constant current is carried out, and the transition time is measured as the quantitative characteristic [29]. In this context, recent chronopotentiometric methods have been reported for histamine determination in foodstuffs. One of these methods was based on the oxidation of the amine at a planar gold disc electrode in the presence of electrogenerated chlorine which facilitates charge transfer between the analyte and the electrode surface. Well-defined signals were observed at +1.15 V in hydrochloric acid medium, giving rise to a linear calibration plot in the 2–100 mg/l concentration range with an LOD of 0.27 mg/l histamine. The method was applied to the determination of histamine in fermented sausages [30]. The same authors used a mercury film electrode to develop a chronopotentiometric method for histamine in cheese [29].

1.2.4 Electronic Tongues

Electronic tongues are multisensor systems with marked mix-response, capable of giving a wide and complete response toward the analyzed species [31]. Advanced mathematical procedures for signal processing based on pattern recognition and/or multivariate analysis, able to extract meaningful data from the complex readings, are usually needed in their applications [32]. Electronic tongues involving arrays of electrodes suitable for voltammetric experiments have been applied in food quality studies, namely in wines [33], milk [34], and fruit juices [35]. An illustrative recent example is the development of an electronic tongue that combined non-noble metals (Ni, Co, and Cu) and noble metals (Au, Pt, Rh, Ir, and Ag) for the determination of chloride, nitrate, and nitrite in minced meat. A rational design of the waveform used by the electronic tongue and multivariate analysis including cross validation and partial least square (PLS) to build suitable management and prediction models for the analysis were reported [36]. Another electronic tongue has been developed for the simultaneous determination of the ethanol acetaldehyde, diacetyl, lactic acid, acetic acid, and citric acid content in probiotic fermented milk. The sensor array comprised of seven nonspecific, cross-sensitive sensors coupled with a reference Ag/AgCl electrode. Samples of plain, strawberry, apple-pear, and forest-fruit flavored probiotic fermented milk were analyzed and the results were used for the development of neural network models for rapid estimation of the aroma compounds content in probiotic fermented milk [37].

Monitoring of biotechnological processes, including fermentation, by determination of physicochemical parameters allows the suitable control of the process ongoing including the detection of possible relevant perturbations. A sensor array composed of potentiometric and voltammetric sensors (Figure 1.4a) was proposed as an efficient tool to control the production process of beer. The sensor array consisted of 10 miniaturized ion-selective electrodes and silicon based three-electrode voltammetric transducers. The obtained results were processed using PLSs and PLS-DA (partial least squares-discriminant analysis). The samples originated from batch of homemade beer fermentation and from two stages of the process: fermentation reaction and maturation of beer [38]. Also, a bioelectronic tongue has been constructed for the estimation of the polyphenol content in wine. The approach involved an array of four voltammetric enzyme biosensors (Figure 1.4b) using epoxy–graphite composites and a chemometric processing tool, which is able to interpret the chemical signals and extract meaningful data from the complex readings. One blank

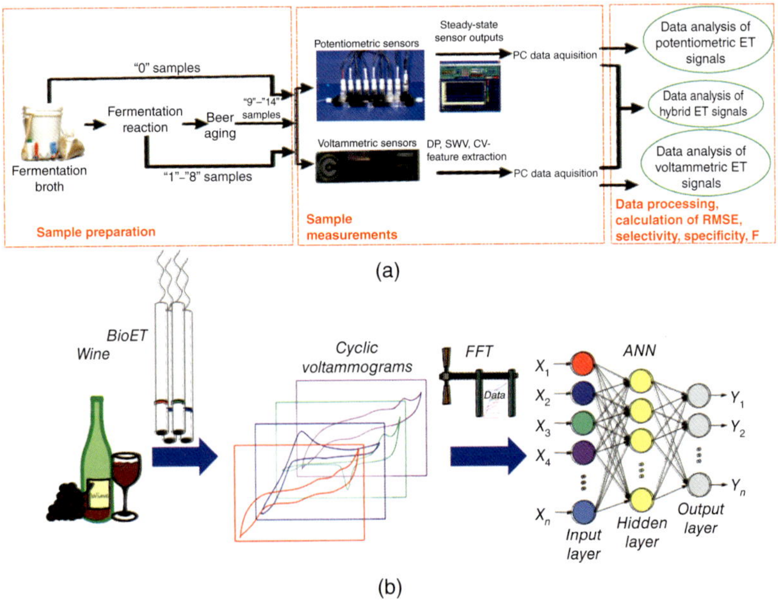

Figure 1.4 Schemes of the experimental setups of: (a) a hybrid electronic tongue for the control of the beer production process. Reproduced from Ref. [38] with permission from Elsevier. (b) A bioelectronic tongue for quantification of polyphenols in wine. Reproduced from Ref. [31] with permission from Elsevier

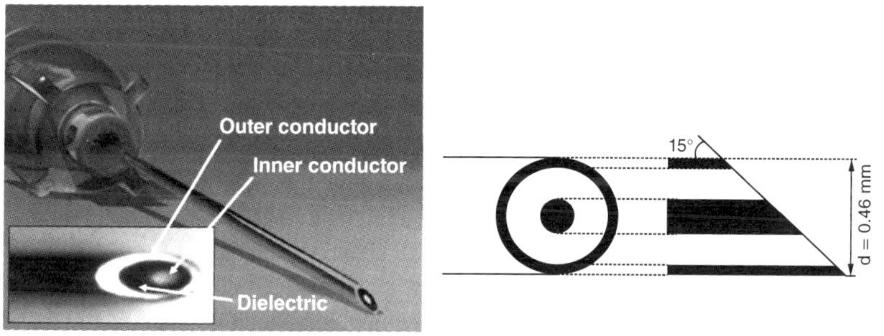

Figure 1.5 *Coaxial needle electrode used for the analysis of food samples by impedance measurements. Reproduced from Ref. [42] with permission from Elsevier*

electrode and the other three bulk modified with tyrosinase and laccase on one side, and copper nanoparticles on the other side, were used in order to incorporate differentiated or catalytic response to different polyphenols present in wine. The obtained voltammetric responses were preprocessed employing the fast Fourier transform (FFT) whereas the obtained coefficients fed an artificial neural network (ANN) model that accomplished the quantification of total polyphenol content [31].

1.2.5 Impedance Spectroscopy

Various applications of impedance spectroscopy have been described in food engineering, such as monitorization of yogurt processing [39], salt and moisture measurement in salmon fillets [40], and testing of meat quality [41]. A relevant example is the construction of a low-cost and nondestructive system to evaluate the salt levels in food based on a punctual measurement of the impedance in the samples. A coaxial electrode, consisted of an isolated wire inserted into a hollow needle, facilitating the placement inside the food sample, was used (Figure 1.5). Furthermore, the impedance modulus and phase values obtained for each frequency were processed using PLS in order to estimate and predict the salt content in minced pork meat [42].

1.3 Electrochemical Sensors and Biosensors for Food Components

Food analysis is a major application area of electrochemical sensors and biosensors. One of the key aspects driving their dramatic development in recent years has been, among others, the use of nanomaterials. As it is outlined below, nanoscience and nanotechnology has strongly influenced the design and construction of recent electrochemical sensors and biosensors paving the way for nanostructured electrode surfaces which are able to improve the quality of the electrochemical response and allowing the efficient immobilization of biomolecules.

1.3.1 Molecularly Imprinted Electrodes

Apart from ISEs, there are very few examples of electrochemical sensors, which are able to provide specific responses to particular analytes without the support of biological recognition elements. Among these, devices using molecularly imprinting polymers (MIPs) have found a certain degree of popularity lately. As it is well known, molecularly imprinting is as a synthetic methodology intended to create tailor-made binding sites with the memory of shape, size, and functional groups toward the template molecules [43]. Various designs of molecularly imprinted electrodes, mainly based on the preparation of electropolymerized imprinted films have been applied to food analysis. For example, a computer-assisted design of MIPs was used to screen functional monomers and solvents for cyanazyne, a triazine herbicide, as the template molecule. The obtained MIP, embedded in a carbon paste electrode, functioned as a preconcentrator agent and selective recognition element for cyanazine determination by using cathodic stripping voltammetry. The efficiency of the prepared MIP electrode for the extraction of the CZ in tomato, onion, lettuce, and rice, spiked with cyanazyne, was demonstrated [44]. A molecularly imprinted electrochemical impedimetric sensor was constructed for the selective determination of melamine. The sensor consisted of a gold electrode modified by electrochemical polymerization of 2-mercaptobenzimidazole (2-MBI) using cyclic voltammetry (CV) in the presence of the template molecule melamine. A linear response was obtained between 1.0×10^{-8} and 5.0×10^{-5} M, with a detection limit of 3.0×10^{-9} M and the method was applied to the analysis of spiked liquid, powder milk and yogurt samples [45] (Figure 1.6).

1.3.2 Enzyme Biosensors

Twenty years ago, Professor Saverio Mannino and Professor Joseph Wang, published a review article entitled "Electrochemical Methods for Food and Drink Analysis." In this

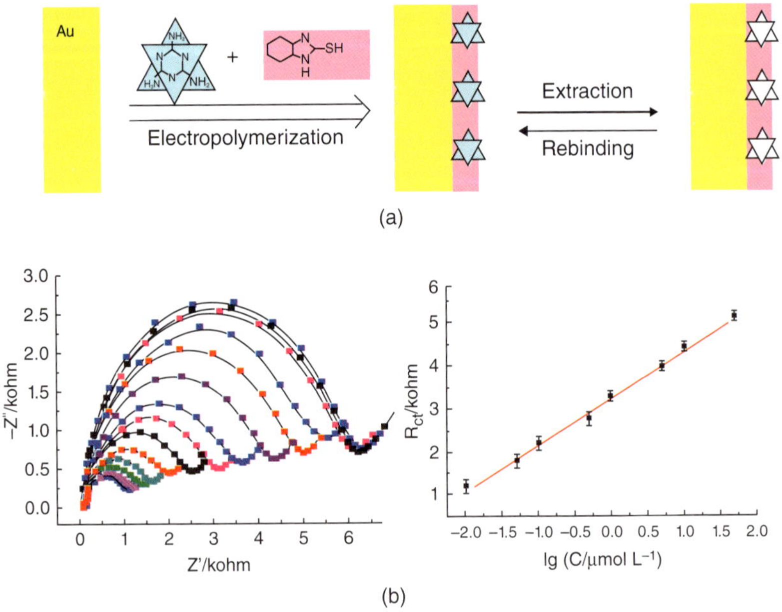

Figure 1.6 (a) Scheme of the sensor preparation and (b) electrochemical impedance responses for 0–0.001 M melamine and calibration plot. Reproduced from Ref. [45] with permission from Elsevier

review they had stated that: "Many target analytes in the food and beverage industries are primary substrates of oxidase or dehydrogenase enzymes, and can thus be easily sensed with amperometric transducers. Other biological entities should play a growing role in the future" [46]. In fact, at present, a great variety of enzyme electrode-based biosensors are available to detect many analytes of interest in the food industry. Coupling of the inherent specificity of enzymes with the advantageous properties of electrochemical transducers, particularly, their high sensitivity, low cost, and simplicity of the methodologies still remains the main attraction for the application of these analytical tools in food industries. The most studied analytes include glucose, fructose, and other carbohydrates, cholesterol, and ethanol. However, recently, enzymatic biosensors have been designed and applied to the determination of other less common target analytes in food samples. For example, an amperometric and impedimetric biosensor for detecting trimethylamine (TMA), which represents a good parameter for estimating fish freshness, has been developed. A conducting polypyrrole substituted with ferrocenyl was used to immobilize flavin-containing monooxygenase 3 (FMO3) enzyme which catalyzed the monooxygenation TMA to trimethylamine N oxide (TMO). The evaluation of fish freshness was realized by measuring the extract sample from horse-mackerel fish [47]. An electrochemical microdevice was also fabricated for on-site determination of rapid freshness of rice by monitoring the activity of peroxidase using hydrogen peroxide as enzyme substrate and hydroquinone as redox mediator for amperometric detection. Peroxidase is one of the key enzymes involved in the elimination of reactive oxygen species and is directly related to the deterioration of rice grains. The rate of current change originating from benzoquinone, a reduction product of the mediated enzymatic reaction, on a gold electrode, depended on the rice freshness, reflecting the remaining activity of peroxide which is more rapid with fresh rice (Figure 1.7). The device with multiple sensing sites demonstrated the possibility to identify a mixture of grains [48].

As it was mentioned above, the use of nanomaterials has attracted a great attention in the development of a new generation of biosensors. The unique physical, chemical, and electronic properties of nanomaterials make possible the intimate attachment of enzymes providing enhanced electroanalytical performance. Among the most widely used nanomaterials, gold nanoparticles have been found to be particularly useful due to the easy synthesis and functionalization, high chemical stability, low inherent toxicity (biocompatibility), and tunable optical and electronic properties [49]. Many interesting examples of biosensors involving the use of gold nanostructures as an appropriate support to immobilize enzymes can be found in the recent literature. For example, an electrochemical biosensor for the determination of total cholesterol was fabricated by co-immobilizing three enzymes, cholesterol oxidase (ChOx), cholesterol esterase (ChE), and horseradish peroxidase (HRP), on nanoporous gold networks which were directly grown on a titanium substrate (Ti/NPAu/ChOx–HRP–ChE). The developed biosensor possessed high selectivity and high sensitivity ($29.33\,\mu A\,mM^{-1}\,cm^{-2}$), as well as a wide linear range up to $300\,mg\,dl^{-1}$. The fabricated biosensor was validated by analyzing margarine, butter, and fish oil samples [49].

Nanostructured enzyme electrochemical biosensors were also recently prepared for the determination of polyphenols, the most abundant antioxidants present in diet. One configuration involved immobilization of laccase onto nickel nanoparticles (NiNPs) decorated-carboxylated multiwalled carbon nanotubes (cMWCNTs)/PANI and was applied to the

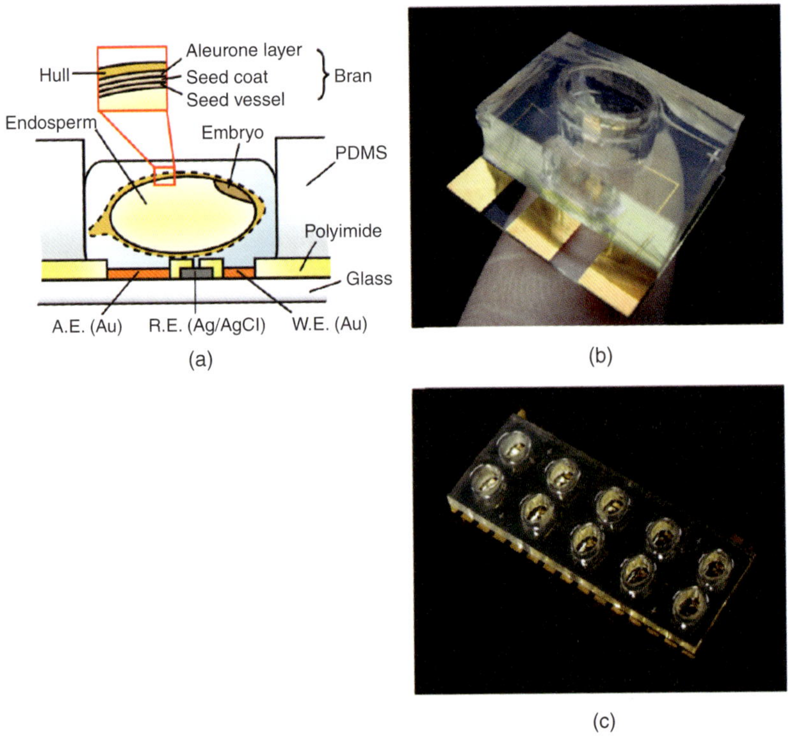

Figure 1.7 *(a) Cross-section of the electrochemical microdevice for the determination of rice freshness, and a rice grain. (b and c) Images of the devices for 1 and 10 rice grains. Reproduced from Ref. [48] with permission from Elsevier*

determination of total polyphenols in fruit The enzyme electrode could be used up to 200 times over a period of four months when it was stored at 4°C [50]. Another example relies on laccase immobilization onto nanocomposites of silver nanoparticles and zinc oxide nanoparticles which were electrochemically deposited onto gold electrodes. The method was applied to the determination of total phenolic compounds in wine and a linear range for guaiacol was reported in the 0.1–500 μM concentration range [51].

1.3.3 Affinity Biosensors

The applications of immunosensors and genosensors in the field of food analysis are diverse, the more recent encompassing allergen detection and monitoring of species associated with celiac disease, toxins, microorganisms, genetically modified organisms (GMOs), and pesticide or drug residues, among others. The detection of okadaic acid (OA) is a challenging and important issue for shellfish industries worldwide. This compound is a lipophilic marine biotoxin produced by Dinophysis and Prorocentrum dinoflagellates. OA intoxication is considered as the most of concern diarrhetic shellfish poisoning (DSP) for human health [52]. An electrochemical immunosensor was developed recently for the determination of OA based on an indirect competitive immunoassay format under

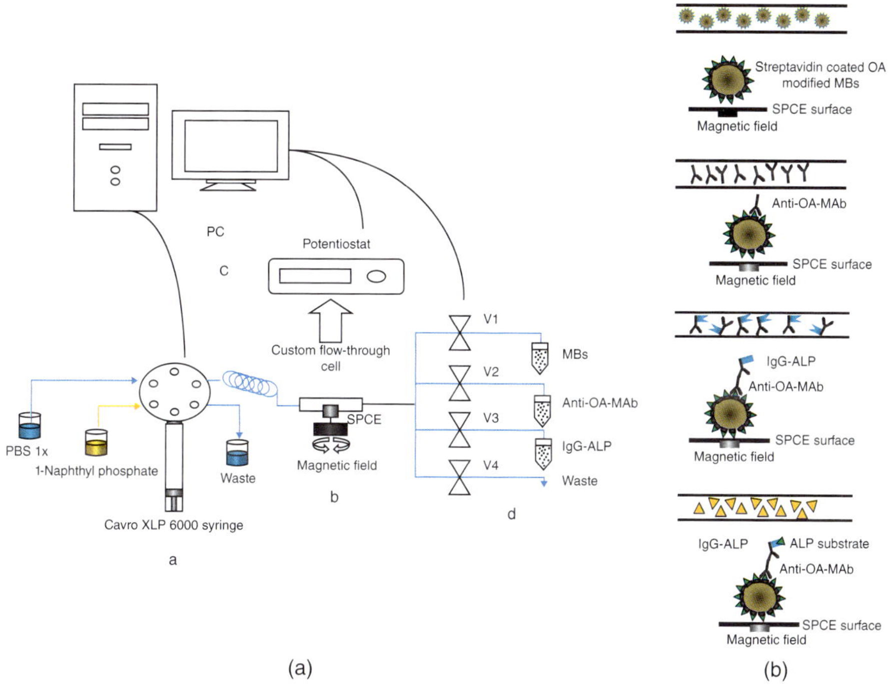

Figure 1.8 *Schematic diagram of the apparatus used for continuous flow automated okadaic acid analysis. A: (a) Syringe pump; (b) electromagnetic device; (c) recorder device; and (d) selection valves. B: Schematic representation of the rationale relying the flow-based immunoassay. Reproduced from Ref. [52] with permission from Elsevier*

automated flow conditions. The biosensor was fabricated by injecting OA modified magnetic beads onto screen printed carbon electrode in the flow system (Figure 1.8). The OA present in the sample competed with the immobilized OA to bind with anti-okadaic antibody, and the secondary alkaline phosphatase-labeled antibody was used to perform electrochemical detection. The method was applied to the analysis of mussels [52].

A disposable amperometric immunosensor was also developed for sensitive detection of chlorpyrifos-methyl, a persistent insecticide widely used in farming and horticulture. The device was fabricated by modification of the screen-printed carbon electrodes with nanocomposites made by skillful doping of bovine serum albumin conjugated chlorpyrifos-methyl and platinum colloid into silica sol–gel. The immobilization of BSA–Ag conjugates on the nanocomposite retained their immunoactivities, which allowed the immobilized BSA–Ag to effectively capture unbound Ab-HRP in the detection solution. A linear response to chlorpyrifos-methyl concentration ranging from 0.4 to 20 ng/ml was reported and good results were obtained for the detection of the insecticide in treated soil and grape samples [53].

Gliadin, a constituent of the cereal protein gluten, is responsible for the intolerance generated in celiac disease. Control of gliadin contents in gluten-free foods is needed to prevent the inappropriate immune response in patients. Various immunosensors have been reported in the literature with this objective. As an example, an electrochemical magneto

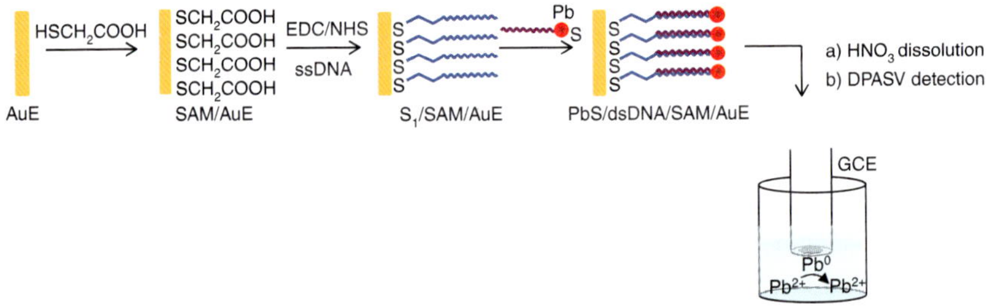

Figure 1.9 *Scheme of the preparation of an electrochemical DNA biosensor using PbS nanoparticle labels*

immunosensor was described relying on the immobilization of gliadin onto tosyl-activated magnetic beads. A direct competitive format providing a detection limit in the order of $\mu g\, l^{-1}$ was used to determine gliadin in spiked gluten-free foodstuffs (skimmed milk and beer) obtaining excellent recoveries [54].

Regarding DNA sensors, a few examples can be considered to demonstrate their usefulness in the field of food analysis. One concerns microorganisms as target analytes and involved the development of a biosensor based on target-induced aptamer displacement for direct detection of *Escherichia coli O111.* The aptamer was immobilized on a gold electrode through hybridization with a capture probe anchored to the electrode surface *via* Au-thiol binding. In the presence of *E. coli O111*, the aptamer was dissociated from the capture probe–aptamer duplex due to the stronger interaction between the aptamer and *E. coli O111.* The consequent single-strand capture probe could be hybridized with biotinylated detection probe and tagged with streptavidin–alkaline phosphatase, producing sensitive enzyme-catalyzed electrochemical responses. Using this strategy, a detection limit of 305 CFU ml^{-1} in milk within 3.5 h was obtained [55].

Electrochemical DNA biosensors are also appropriate for the detection of GMOs due to their well-known advantages of simplicity, low cost, and ease of use. In a recent article, a DNA biosensor was developed for determining the CaMV 35S sequence, which was derived from the cauliflower mosaic virus from *Agrobacterium tumefaciens,* and is often used as an insert and indicator of GMOs. The DNA biosensor design involved the synthesis of mercaptoacetic acid-modified lead sulfide nanoparticles which were used as specific ssDNA sequence labels. The target ssDNA was covalently fixed on a SAM-modified gold electrode (Figure 1.9), and the hybridization process was monitored by dissolution of PbS nanoparticles anchored on the hybrids by oxidation with nitric acid and determination of lead ion by differential pulse anodic stripping voltammetry. The CaMV 35S target sequence was detected with a detection limit of 4.38×10^{-12} mol l^{-1} [56].

1.4 Nanomaterials for Electrochemical Analysis of Food

We have already commented that the application of nanotechnology has highly improved the designs and the way of preparing platforms for electrochemical analysis. In addition to the examples already described, other representative configurations should be mentioned in this section because of their relevance, as they are representative examples of the great usefulness of nanomaterials in the field of food analysis.

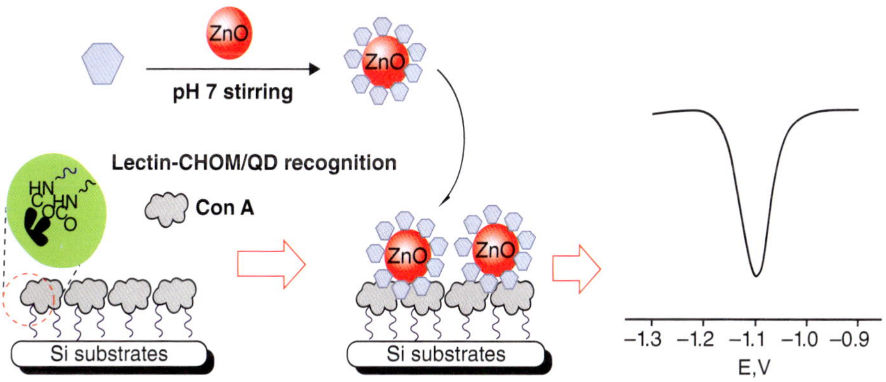

Figure 1.10 *Scheme of the preparation of bioconjugates for the electrochemical detection of chicken ovomucoid allergen (CHOM). Reproduced from Ref. [57] with permission from Elsevier*

For example, the quantum dot (QD) labeled methodology, which has been profusely applied in the construction of optical sensors, has been also used for the preparation of bioconjugates that can be electrochemically detected. The high surface-to-volume ratio of QDs combined with stripping voltammetry led to signal amplification allowing highly sensitive detection. An illustrative example is the development of a concanavalin A-based sensor for the direct electrochemical detection of chicken ovomucoid allergen. The biomimetic and high isoelectric point ZnO QDs material was used for self-assembling of the analyte through electrostatic interaction between negatively charged protein and positively charged QDs surface (Figure 1.10). Concanavalin A was employed as a recognition element for ZnO QDs-labeled allergen, and the detection process was monitored by stripping SWV of captured QDs [57].

Glucose monitoring has great interests in food industry. Numerous glucose oxidase-based enzyme electrochemical biosensors have been developed for this purpose. However, the immobilization of the enzyme on the surface of transducers can complicate the preparation of biosensors and affect their final performance. Moreover, instability of glucose oxidase humpers the use of these biosensors for continuous monitoring in production or transformation processes such as fermentation. Nonenzymatic glucose sensors, which are able to determine glucose in a direct manner, constitute an attractive alternative to overcome these problems. Direct electrochemical oxidation of glucose is readily noticeable onto Ni, NiO, or $Ni(OH)_2^-$ modified electrodes due to the redox couple of $Ni(OH)_2$/NiOOH formed on the electrode surface on alkaline medium. For example, nano NiO-modified carbon paste electrodes were employed to determine glucose in real juice samples [58].

Lately, graphene-based electrochemical platforms are being increasingly used for numerous applications. Although there are still few examples in the area of food analysis, it can be mentioned that some interesting biosensor configurations involving the nanomaterial modification by covalent functionalization have appeared in the recent literature. For example, a label-free voltammetric immunosensor for the sensitive detection of β-lactoglobulin in milk using graphene-modified screen-printed electrodes was described. Graphene was derivatized by electrografting of 4-nitrophenyl diazonium cations followed by electrochemical reduction and activation with glutaraldehyde for the covalent attachment of β-lactoglobulin antibodies. The analyte determination was performed by monitoring the decrease in the differential pulse voltammetric current from $[Fe(CN)_6]^{3-/4-}$. A detection limit of 0.85 pg/ml and a dynamic range from 1 pg/ml to 100 ng/ml β-lactoglobulin were achieved, and the method was validated by application to the analysis of various foodstuffs containing milk derivatives and comparison with commercial ELISA protocols.

1.5 Future Trends

Throughout this chapter, different recent examples on application of electroanalytical techniques and methodologies in the field of food analysis have been commented. They are nice examples of the great potentiality of modern electroanalysis in this important technological and industrial area not only for the detection and quantification of different species at very low concentrations, taking advantage of the great sensitivity and low limits of detection that contemporary electroanalytical techniques allow, but also for the monitoring and control of food components with high precision and accuracy. These inherent excellent performance of electroanalysis together with the low cost of the required instrumentation and maintenance, constitute synergistic strengths that it is expected to contribute for offering new, affordable, and attractive alternative methods suitable to be implemented in quality control laboratories in the near future. Furthermore, it is most likely that new applications of biosensors using both enzymatic designs and affinity designs will appear and the continued use of new generation nanomaterials will contribute in providing amplified and enhanced electroanalytical responses as well as improving the immobilization ability of the modified electrode surfaces. Regarding relevant target analytes, it is foreseen that new electroanalytical methods will be devoted to the detection of allergens, drug residues, and toxins, as well as to the determination of components of interest that are present in functional foods and their relationship with deficiency diseases or obesity, among others.

Acknowledgments

Financial support from the Ministerio de Ciencia e Innovación (Projects CTQ2012-35041 and CTQ2012-34238), and Comunidad de Madrid S2013/MT-3029 (NANOAVAnSENS) is gratefully acknowledged.

References

1. Escarpa, A. (2012) Food electroanalysis: sense and simplicity. *Chem. Rec.*, **12**, 72–91.
2. Serra, B., Reviejo, A.J. and Pingarrón, J.M. (2007) in Electrochemical Sensor Analysis, Wilson & Wilson's Comprehensive Analytical Chemistry, Chapter 13 (eds S. Alegret and A. Merkoçi), Elsevier, Amsterdam, pp. 257–258.
3. Pividori, M.I. and Alegret, S. (2007) in Electrochemical Sensor Analysis, Wilson & Wilson's Comprehensive Analytical Chemistry, Chapter 22 (eds S. Alegret and A. Merkoçi), Elsevier, Amsterdam, p. 467.

4. Rooney, R. and Wall, P.G. (2003) Food safety, in Encyclopedia of Food Science and Nutrition (eds B. Caballero, L. Trugo and P.M. Finglas), Academic Press, New York, pp. 27–30.
5. Dong, P., Qiu, P., Zhu, Y. *et al.* (2010) Simultaneous determination of four 5-hydroxy polymethoxyflavones by reversed-phase high performance liquid chromatography with electrochemical detection. *J. Chromatogr. A*, **1217**, 642–647.
6. Bayram, B., Ozcelik, B., Schultheiss, G. *et al.* (2013) A validated method for the determination of selected phenolics in olive oil using high-performance liquid chromatography with coulometric electrochemical detection and a fused-core column. *Food Chem.*, **138**, 1663–1669.
7. Sangjarusvichai, H., Dungchaia, W., Siangproh, W. and Chailapakula, O. (2009) Rapid separation and highly sensitive detection methodology for sulfonamides in shrimp using a monolithic column coupled with BDD amperometric detection. *Talanta*, **79**, 1036–1041.
8. Cesarino, I., Moraes, F.C., Lanza, M.R.V. and Machado, S.A.S. (2012) Electrochemical detection of carbamate pesticides in fruit and vegetables with a biosensor based on acetylcholinesterase immobilised on a composite of polyaniline-carbon nanotubes. *Food Chem.*, **135**, 873–879.
9. Medeiros, R.A., Lourencao, B.C., Rocha-Filho, R.C. and Fatibello-Filho, O. (2012) Flow injection simultaneous determination of synthetic colorants in food using multiple pulse amperometric detection with a boron-doped diamond electrode. *Talanta*, **99**, 883–889.
10. Quintino, M.S.M. and Angnes, L. (2004) Batch injection analysis: an almost unexplored powerful tool. *Electroanalysis*, **16**, 513–523.
11. Silva, R.A.B., Montes, R.H.O., Richter, E.M. and Muñoz, R.A.A. (2012) Rapid and selective determination of hydrogen peroxide residues in milk by batch injection analysis with amperometric detection. *Food Chem.*, **133**, 200–204.
12. de Castro, M.D.L., Ruiz-Jimenez, J. and Perez-Serradilla, J.A. (2008) Lab-on-valve: A useful tool in biochemical analysis. *TrAC, Trends Anal. Chem.*, **27**, 118–126.
13. Wang, Y., Wang, L., Tian, T. *et al.* (2012) A highly sensitive and automated method for the determination of hypoxanthine based on lab-on-valve approach using Fe_3O_4/MWCNTs/β-CD modified electrode. *Talanta*, **99**, 840–845.
14. Hernández-Cázares, A.S., Aristoy, M.C. and Toldrá, F. (2010) Hypoxanthine-based enzymatic sensor for determination of pork meat freshness. *Food Chem.*, **123**, 949–954.
15. Godoy-Caballero, M.P., Acedo-Valenzuela, M.I., Galeano-Díaz, T. *et al.* (2012) Microchip electrophoresis with amperometric detection for a novel determination of phenolic compounds in olive oil. *Analyst*, **137**, 5153–5160.
16. Won, S.Y., Chandra, P., Hee, T.S. and Shim, Y.B. (2013) Simultaneous detection of antibacterial sulfonamides in a microfluidic device with amperometry. *Biosens. Bioelectron.*, **39**, 204–209.
17. Wang, W., Wang, Y., Zhang, J. *et al.* (2010) Simultaneous determination of electroactive and non-electroactive food preservatives by novel capillary electrophoresis with amperometric detection. *Anal. Chim. Acta*, **678**, 39–43.

18. Wang, J., Jiang, L., Chu, Q. and Ye, J. (2010) Residue analysis of melamine in milk products by micellar electrokinetic capillary chromatography with amperometric detection. *Food Chem.*, **121**, 215–219.
19. Wang, J. (2005) Stripping analysis at bismuth electrodes: a review. *Electroanalysis*, **17**, 1341–1346.
20. Xie, X.Y., Luo, H.Q. and Li, N.B. (2010) Determination of azo compounds by differential pulse voltammetry at a bismuth/poly(*p*-aminobenzene sulfonic acid) film electrode and application for detection in food stuffs. *J. Electroanal. Chem.*, **639**, 175–180.
21. Tyszczuk-Rotko, K. (2012) New voltammetric procedure for determination of thiamine in commercially available juices and pharmaceutical formulation using a lead film electrode. *Food Chem.*, **134**, 1239–1243.
22. Barbosa, A.M.J., de Araújo, T.A., Trinidade, M.A.G. and Ferreira, V.S. (2011) A simple method for electroanalytical determination of ceftiofur in UHT milk samples using square-wave voltammetry technique. *J. Appl. Electrochem.*, **41**, 249–255.
23. Pacheco, W.F., Doyle, A., Duarte, D.R.A. *et al.* (2010) Square-wave adsorptive stripping voltammetry for trace determination of dimoxystrobin and azoxystrobin in potatoes and grapes. *Food Anal. Methods*, **3**, 205–210.
24. Ni, Y., Li, S. and Kokot, S. (2011) Simultaneous voltammetric analysis of tetracycline antibiotics in foods. *Food Chem.*, **124**, 1157–1163.
25. Locatelli, C. and Melucci, D. (2012) Voltammetric determination of ultra-trace total mercury and toxic metals in meals. *Food Chem.*, **130**, 460–466.
26. Oppermann Santini, A., Redigolo Pezza, H. and Pezza, L. (2012) Development of a sensitive potentiometric sensor for determination of fumaric acid in powdered food products. *Food Chem.*, **134**, 483–487.
27. Evtugyn, G.A., Belyakova, S.V., Shamagsumova, R.V. *et al.* (2010) Discrimination of apple juice and herbal liqueur brands with solid-state electrodes covered with polyaniline and thiacalixarenes. *Talanta*, **82**, 613–619.
28. Araujo, C.L., Melo, E.I. and Coelho, N.M.M. (2011) Potentiometric detection of citrate in beverages using a graphite carbon electrode. *Talanta*, **84**, 1169–1173.
29. Švarc-Gajić, J. and Stojanović, Z. (2011) Determination of histamine in cheese by chronopotentiometry on a thin film mercury electrode. *Food Chem.*, **124**, 1172–1176.
30. Stojanovic, Z.S. and Svarc-Gajic, J.V. (2011) A simple and rapid method for histamine determination in fermented sausages by mediated chronopotentiometry. *Food Control*, **22**, 2013–2019.
31. Cetó, X., Céspedes, F. and del Valle, M. (2012) Bioelectronic tongue for the quantification of total polyphenol content in wine. *Talanta*, **99**, 544–551.
32. del Valle, M. (2010) Electronic tongues employing electrochemical sensors. *Electroanalysis*, **22**, 1539–1555.
33. Pigani, L., Foca, G., Ulrici, A. *et al.* (2009) Classification of red wines by chemometric analysis of voltammetric signals from PEDOT-modified electrodes. *Anal. Chim. Acta*, **643**, 67–73.
34. Dias, L.A., Peres, A.M., Veloso, A.C.A. *et al.* (2009) An electronic tongue taste evaluation: identification of goat milk adulteration with bovine milk. *Sens. Actuators, B*, **136**, 209–217.

35. Gutés, A., Ibáñez, A.B., del Valle, M. and Céspedes, F. (2006) Automated SIA e-Tongue employing a voltammetric biosensor array for the simultaneous determination of glucose and ascorbic acid. *Electroanalysis*, **18**, 82–88.
36. Campos, I., Masot, R., Alcañiz, M. *et al.* (2010) Accurate concentration determination of anions nitrate, nitrite and chloride in minced meat using a voltammetric electronic tongue. *Sens. Actuators, B*, **149**, 71–78.
37. Hruskar, M., Major, N., Krpan, M. and Vahcic, N. (2010) Simultaneous determination of fermented milk aroma compounds by a potentiometric sensor array. *Talanta*, **82**, 1292–1297.
38. Kutyła-Olesiuk, A., Zaborowski, M., Prokaryn, P. and Ciosek, P. (2012) Monitoring of beer fermentation based on hybrid electronic tongue. *Bioelectrochemistry*, **87**, 104–113.
39. Kitamura, Y., Toyoda, K. and Park, B. (2000) Electric impedance spectroscopy for yogurt processing. *Food Sci. Technol. Res.*, **6**, 310–313.
40. Chevalier, D., Ossart, F. and Ghommids, C. (2006) Development of a non-destructive salt and moisture measurement method in salmon (Salmo salar) fillets using impedance technology. *Food Control*, **17**, 342–347.
41. Damez, J.L., Clerjon, S., Abouelkaram, S. and Lepetit, J. (2008) Beef meat electrical impedance spectroscopy and anisotropy sensing for non-invasive early assessment of meat ageing. *J. Food Eng.*, **85**, 116–122.
42. Masot, R., Alcañiz, M., Fuentes, A. *et al.* (2010) Design of a low-cost non-destructive system for punctual measurements of salt levels in food products using impedance spectroscopy. *Sens. Actuators, A*, **158**, 217–223.
43. Chen, L., Xu, S. and Li, J. (2011) Recent advances in molecular imprinting technology: current status, challenges and highlighted applications. *Chem. Soc. Rev.*, **40**, 2922–2942.
44. Gholivand, M.B., Torkashvand, M. and Malekzadeh, G. (2012) Fabrication of an electrochemical sensor based on computationally designed molecularly imprinted polymers for determination of cyanazine in food samples. *Anal. Chim. Acta*, **713**, 36–44.
45. Wu, B., Wang, Z., Zhao, D. and Lu, X. (2012) A novel molecularly imprinted impedimetric sensor for melamine determination. *Talanta*, **101**, 374–381.
46. Mannino, S. and Wang, J. (1992) Electrochemical methods for food and drink analysis. *Electroanalysis*, **4**, 835–840.
47. Bourigua, S., El Ichi, S., Youssoufi, H.K. *et al.* (2011) Electrochemical sensing of trimethylamine based on polypyrrole–flavin-containing monooxygenase (FMO3) and ferrocene as redox probe for evaluation of fish freshness. *Biosens. Bioelectron.*, **28**, 105–111.
48. Koyachi, E., Kojima, K., Qiu, X. *et al.* (2013) Electrochemical microdevice for on-site determination of rice freshness. *Biosens. Bioelectron.*, **42**, 640–645.
49. Ahmadalinezhad, A. and Chen, A. (2011) High-performance electrochemical biosensor for the detection of total cholesterol. *Biosens. Bioelectron.*, **26**, 4508–4513.
50. Chawla, S., Rawal, R., Sharma, S. and Pundir, C.S. (2012) An amperometric biosensor based on laccase immobilized onto nickel nanoparticles/carboxylated multiwalled carbon nanotubes/polyaniline modified gold electrode for determination of phenolic content in fruit juices. *Biochem. Eng. J.*, **68**, 76–84.

51. Chawla, S., Rawal, R., Kumar, D. and Pundir, C.S. (2012) Amperometric determination of total phenolic content in wine by laccase immobilized onto silver nanoparticles/zinc oxide nanoparticles modified gold electrode. *Anal. Biochem.*, **430**, 16–23.
52. Dominguez, R.B., Hayat, A., Sassolas, A. *et al.* (2012) Automated flow-through amperometric immunosensor for highly sensitive and on-line detection of okadaic acid in mussel sample. *Talanta*, **99**, 232–237.
53. Wei, W., Zong, X., Wang, X. *et al.* (2012) A disposable amperometric immunosensor for chlorpyrifos-methyl based on immunogen/platinum doped silica sol–gel film modified screen-printed carbon electrode. *Food Chem.*, **135**, 888–892.
54. Laube, T., Kergaravat, S.V., Fabiano, S.N. *et al.* (2011) Magneto immunosensor for gliadin detection in gluten-free foodstuff: towards food safety for celiac patients. *Biosens. Bioelectron.*, **27**, 46–52.
55. Luo, C., Lei, Y., Yan, L. *et al.* (2012) A rapid and sensitive aptamer-based electrochemical biosensor for direct detection of *Escherichia coli O111. Electroanalysis*, **24**, 1186–1191.
56. Sun, W., Zhong, J., Qin, P. and Jiao, K. (2008) Electrochemical biosensor for the detection of cauliflower mosaic virus 35 S gene sequences using lead sulfide nanoparticles as oligonucleotide labels. *Anal. Biochem.*, **377**, 115–119.
57. Yang, C., Gu, B., Xu, C. and Xu, X. (2011) Self-assembled ZnO quantum dot bioconjugates for direct electrochemical determination of allergen. *J. Electroanal. Chem.*, **660**, 97–100.
58. Mu, Y., Jia, D., He, Y. *et al.* (2011) Nano nickel oxide modified non-enzymatic glucose sensors with enhanced sensitivity through an electrochemical process strategy at high potential. *Biosens. Bioelectron.*, **26**, 2948–2952.

Part I

Electroanalytical Techniques in Batch and Continuous Systems in Food Analysis

2

Voltammetric Techniques

Sandra Mendoza[1], *Erika Bustos*[2], *Juan Manríquez*[2] *and Luis A. Godínez*[2]
[1]*Departamento de Investigación y Posgrado en Alimentos, Universidad Autónoma de Querétaro, Cerro de las Campanas s/n, Querétaro, QRO, 76010, México*
[2]*Department of Science, Centro de Investigación y Desarrollo Tecnológico en Electroquímica, S. C., Parque Tecnológico Querétaro s/n, Sanfandila, Pedro Escobedo, QRO 76703, México*

2.1 Introduction

Measuring the quality, safety, nutrition, and stability of food products is a concern of analytical chemistry. Although thin layer chromatography (TLC)-densitometry, high performance liquid chromatography coupled to mass spectrometry (HPLC-MS), gas chromatography coupled to mass spectrometry (GC-MS), capillary electrophoresis, and spectroscopy among others, have proven to be reliable methods for the identification and quantification of several components in food products, there is still a need for fast, accurate, reproducible, and sensitive analytical techniques. In this regard, conventional methods suffer from some disadvantages, such as high cost, long analysis time, sample pretreatment requirements, and in some cases, low sensitivity and selectivity. Compared with other methods, electrochemical methods are characterized by simplicity, high sensitivity, good stability, low-cost instrumentation, and small sample requirements. In this chapter, a brief review of sweep potential electrochemical techniques as well as some relevant literature concerning identification and quantification of food components will be addressed.

2.2 An Overview of Sweep Potential Electrochemical Techniques

Potential sweep electrochemical techniques involve the interfacial potential scan of a selected potential window and the corresponding measurement of the current responses,

Agricultural and Food Electroanalysis, First Edition.
Edited by Alberto Escarpa, María Cristina González and Miguel Ángel López.

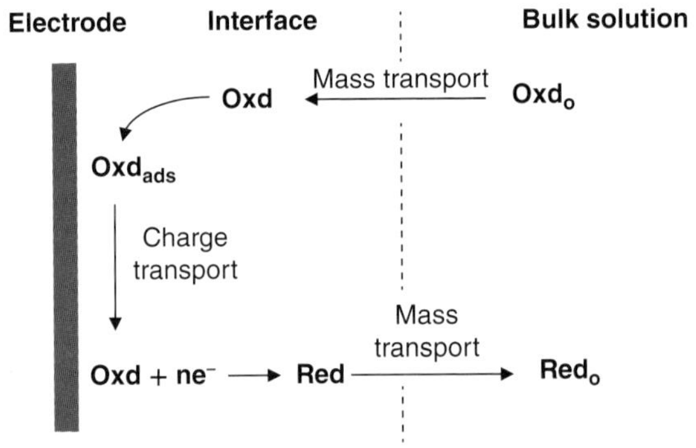

Figure 2.1 Electrochemical reaction events

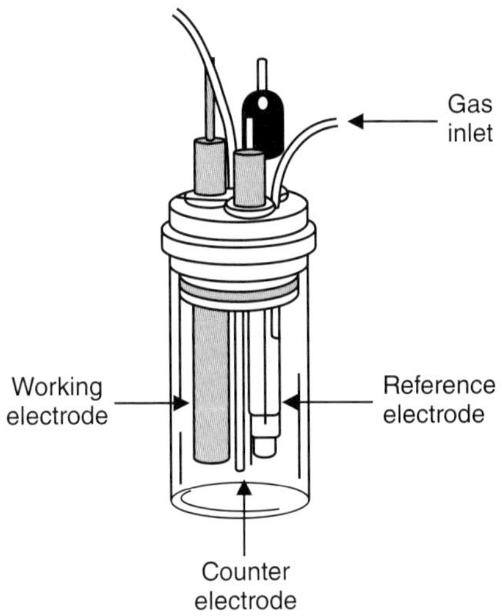

Figure 2.2 Scheme of a typical three-electrode cell used in voltammetry techniques, where the gas inlet is used for bubbling the electrolyte solution with an inert gas

which are produced from electron transfer events between the electroactive species and the working-electrode material (Figure 2.1). Since this phenomenon involves oxidation or reduction reactions of the electroactive species, the measured currents can be considered proportional to the concentrations.

The instrumentation for carrying out potential sweep techniques typically involves a three-electrode cell (Figure 2.2), where working- and counter electrodes are immersed in an electrolyte solution containing the electroactive species. A reference electrode however, is used to maintain a controlled potential at the working electrode by means of a potentiostat. It is also important to point out that the electrolytic solutions are commonly bubbled with an inert gas (such as nitrogen or argon) in order to remove oxygen, which could react in the potential window under study and interfere with the relevant current responses [1–3].

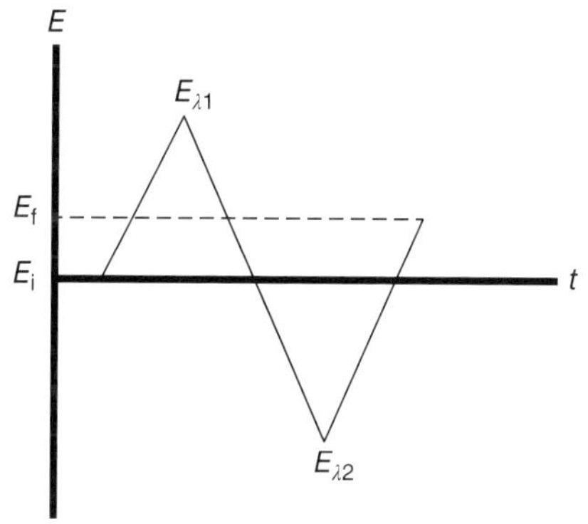

Figure 2.3 *CV potential vs. time perturbation wave-form*

2.2.1 Linear Sweep Voltammetry/Cyclic Voltammetry

Linear sweep voltammetry (LSV) and cyclic voltammetry (CV) are the most commonly used potential sweep techniques. The perturbation signal of the working electrode (Figure 2.3) is based on a linear potential wave-form, where the potential E is changed with the time (t) in a linear fashion. The rate of potential change is known as the *scan rate* v $(= dE/dt)$, which typically varies between 1 mVs^{-1} and 1 Vs^{-1} [3, 4].

In LSV, the explored potential region is scanned from the initial potential (E_i) to the final potential (E_f). CV however, is an extension of LSV since the direction of the potential scan is reversed at the end of the first scan (which then becomes the switching potential, $E_{\lambda1}$), and the final potential becomes the starting potential. In some experiments, the perturbation can be stopped at $E_f = E_i$, or instead, the potential can be scanned past this potential to a predefined second switching potential value ($E_{\lambda2}$), where the direction of the potential scan is reversed again. The potential can therefore be cycled between the two switching potentials for as many cycles as needed before the experiment is ended at the E_f.

The current response, I, in voltammetry experiments corresponds to the addition of charging (I_c) and faradaic (I_f) current contributions, which in turn are associated with the interfacial charging and the electron transfer events that take place at the electrode surface (see Equation 2.1),

$$I = I_c + I_F = vC_d + I_F \tag{2.1}$$

while $I_c \propto v$ (the interfacial capacitance, C_d, is the proportionality constant), $I_F \propto C_s v^{1/2}$, where C_s corresponds to the concentration of the electroactive species in the electrolytic solution. In practical terms however, the capacitive contribution in Equation 2.1 is eliminated by subtraction of the blank response (electrolytic solution without electroactive species). In this way, detection limits for LSV and CV have been reported to be around 10^{-5} M [5].

The most common response of an LSV experiment is a peak-shaped curve (Figure 2.4) from which it can be observed that the current starts to increase at the activation potential (E_a). This potential value corresponds to the energy onset at which the electrochemical reaction takes place. As the potential moves from E_a, a continuous depletion of electroactive species near the electrode surface occurs, reaching a point at the peak potential (E_p), in

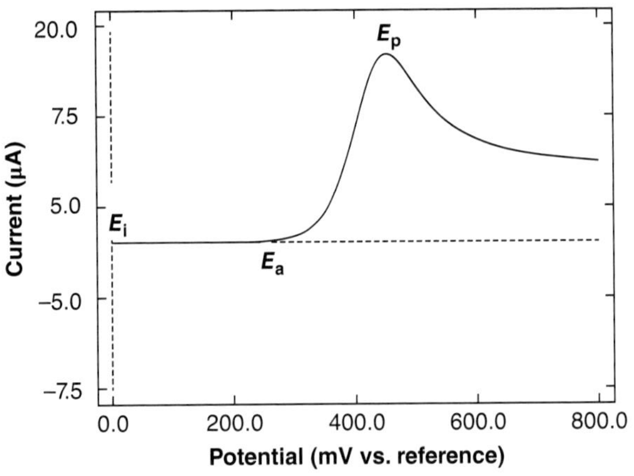

Figure 2.4 *Typical LSV experimental response*

which the electrochemically reactive species has been completely transformed. Beyond this potential value, the current response is characterized by a current decreasing curve that reflects the mass transport control of the electrochemical reaction.

During the reverse potential sweep in CV experiments on the other hand, the product of the initial oxidation or reduction of the electroactive species is complementary reduced or oxidized (Figure 2.5). In this context, E_{pa} and E_{pc} correspond to the anodic and cathodic potential peaks, respectively [1, 6, 7]. It is important to note that for reactions characterized by fast electron transfer events relative to the time-scale of the potential sweep rate (*reversible reactions,* where the difference $E_{pa} - E_{pc}$ is independent of v), the formal thermodynamic electrode potential $E^{\circ\prime}$ can be readily estimated using Equation 2.2.

$$E^{\circ\prime} = \frac{1}{2}(E_{pa} + E_{pc}) \tag{2.2}$$

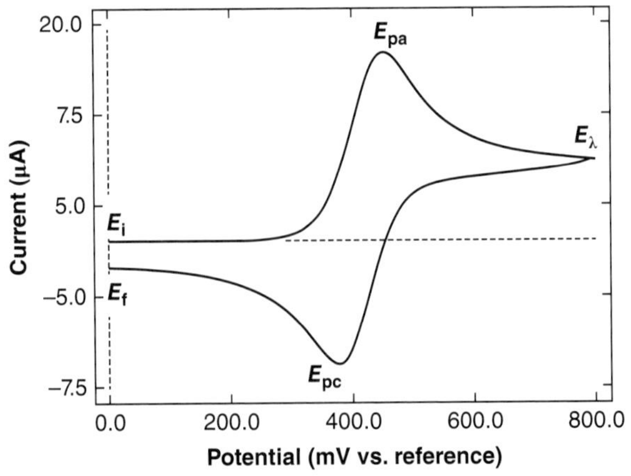

Figure 2.5 *Typical CV experimental response*

2.2.2 Pulse Voltammetry Techniques

Pulse voltammetry techniques are characterized by a succession of potential steps. During the sequential potential steps, the rates of current decay of the capacitive (I_c) and the faradaic currents (I_F) are essentially different (specifically, while I_c in Equation 2.1 decays exponentially with time, I_F decreases as a function of $t^{-\frac{1}{2}}$, characteristic of a diffusion-controlled electrochemical reaction). In this way, the rate of decay of I_c is significantly faster than that of I_F, and thus I_c is negligible at a time of $\sim 5R_uC_d$ after the potential step is imposed (where R_uC_d is the time constant, τ_{cell}, for the electrochemical cell having values from microseconds to milliseconds, and R_u is the uncompensated resistance between reference and working electrodes). Consequently, I_F is the main contribution to the measured current I when its value is measured at the end of a potential step. The detection limits of these techniques therefore fall around 10^{-7} M making them suitable for quantitative analysis.

The most important parameters for pulse voltammetric techniques are defined as follows. *Pulse amplitude* is the height of the potential pulse, which may or may not be constant depending on the technique. *Pulse width* is the duration of the potential pulse. *Sample period* is the time of the pulse at which the current is measured. A number of different pulse techniques are available in commercial potentiostats, which essentially differ in their potential step wave-forms and the number of sampling points [1].

2.2.3 Normal Pulse Voltammetry

The potential wave-form for normal pulse voltammetry (NPV) is shown in Figure 2.6. This typical perturbation consists of a series of pulses of increasing potential amplitude that ends at the initial potential value. The sigmoidal shape response for NPV technique (Figure 2.7) resembles that of a typical I–E curve obtained in a classical polarography experiment [2, 7].

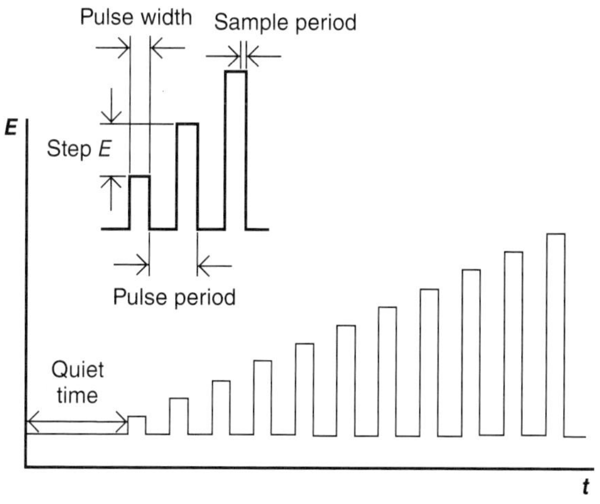

Figure 2.6 Potential perturbation wave-form in a NPV experiment

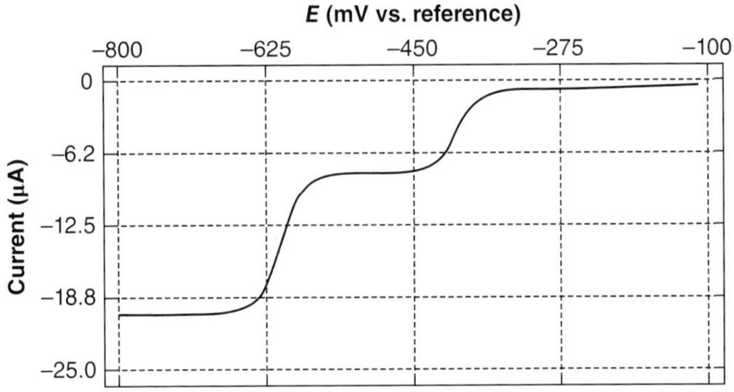

Figure 2.7 Sigmoidal-type response obtained from a NPV experiment

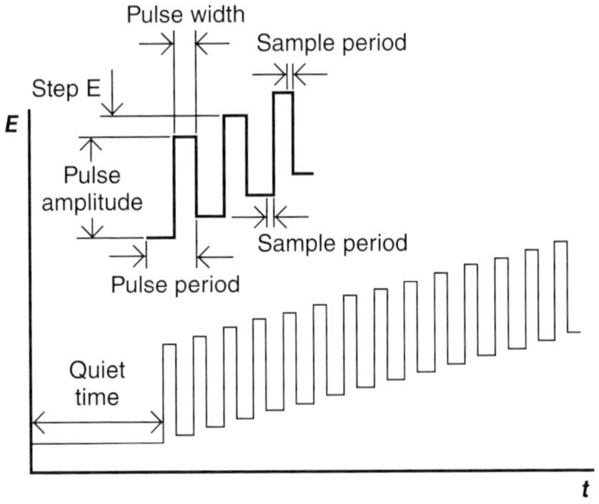

Figure 2.8 Potential wave form used as perturbation in a DPV experiment

2.2.4 Differential Pulse Voltammetry

The potential wave-form for differential pulse voltammetry (DPV) is shown in Figure 2.8. The perturbation consists of a series of pulses having constant amplitude, superimposed at the same time upon a staircase wave-form. In contrast with NPV, the current is sampled twice at each pulse period (at the beginning and at the end of the pulse). The difference between these two current values is recorded and displayed as function of the applied potential, *E*, as shown in Figure 2.9, where it is observed that the current function response is characterized by symmetric peaks [4, 8].

2.2.5 Square Wave Voltammetry

The excitation potential wave-form for square wave voltammetry (SWV) is shown in Figure 2.10. The perturbation consists of a square wave having constant amplitude,

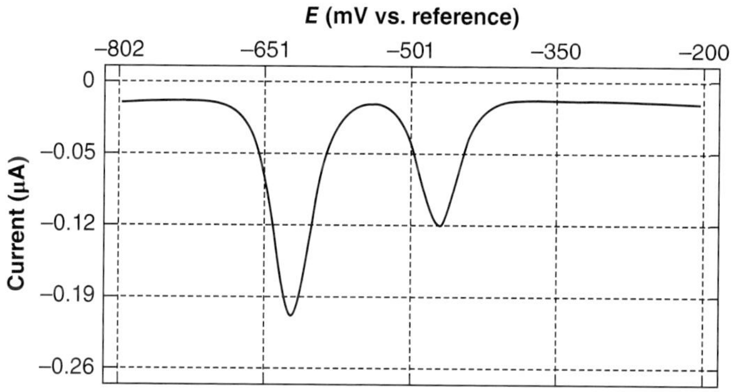

Figure 2.9 Response from a DPV experiment showing symmetric peaks

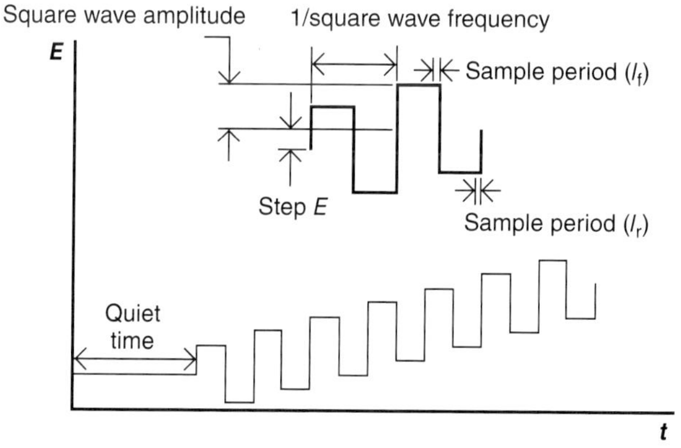

Figure 2.10 Perturbation potential wave-form in an SWV experiment

superimposed to a staircase wave form. The current is measured at the end of each forward half-cycle (I_f) and at the end of each reverse half-cycle (I_r). The difference between both current values ($I_f - I_r$) is displayed as a function of the applied potential E as shown in Figure 2.11. Therefore, the effect of the charging current is notably decreased in SWV since any residual charging current is completely removed [1, 4].

2.2.6 Stripping Voltammetry

The first stage in stripping electroanalytical methods is the accumulation of the electroactive substance either on the surface or in the bulk of a liquid electrode (mercury electrodes). The second stage consists of electrode polarization, obtaining cathodic or anodic voltammograms (stripping voltammetry (SV)) that give information on the nature and concentration of electroactive analytes [7].

In this context, anodic stripping voltammetry (ASV) is a very sensitive electrochemical technique mainly used for the analysis of trace concentrations of metallic species in solution, although detection of some organic species has also been carried out. Detection limits

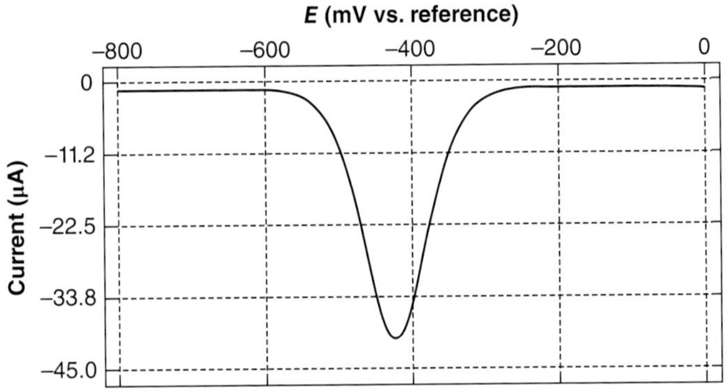

Figure 2.11 Response of an SWV experiment where residual charging currents are removed

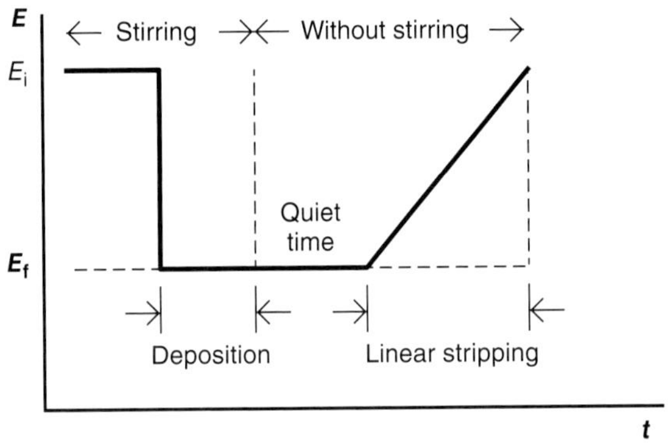

Figure 2.12 Potential sequence used as perturbation in an ASV experiment

for metal ions at ~10^{-11} M (or sub-parts per billion) concentrations have been reported. As shown in Figure 2.12, there are three sequential steps in a typical ASV experiment: deposition or preconcentration, quiet time, and linear stripping. During the deposition step, the electroactive species are reduced to their zero-valence form under stirring (forced convection) of the relevant electrolytic solutions. Later, nonstirring is for a quiet time in order to eliminate convection transport effects. Finally, the previously reduced species are selectively oxidized during an anodic linear sweep and the current is recorded as function of the applied potential (Figure 2.13) without stirring the electrolytic solution [1].

Cathodic stripping voltammetry (CSV) is the inverse of ASV. Mercury electrodes can be used to detect and quantify species that form sparingly soluble mercury salts such as halide ions, sulfide, cyanide, thiols, and penicillins. CSV is not restricted to mercury electrodes. Any electrode material that, on oxidizing, forms sparingly soluble salts will work, such as silver electrodes for the determination of halide ions. Another possibility is that the ion of interest oxidizes to form a precipitate on the electrode surface.

Potentiometric stripping analysis (PSA) on the other hand, is very similar to ASV in that the accumulation stage is the same. The difference is that in the stripping step, an oxidant in solution, or alternatively an anodic current, causes the oxidation of the reduced analyte. In mixtures, the stripping potential identifies the analyte.

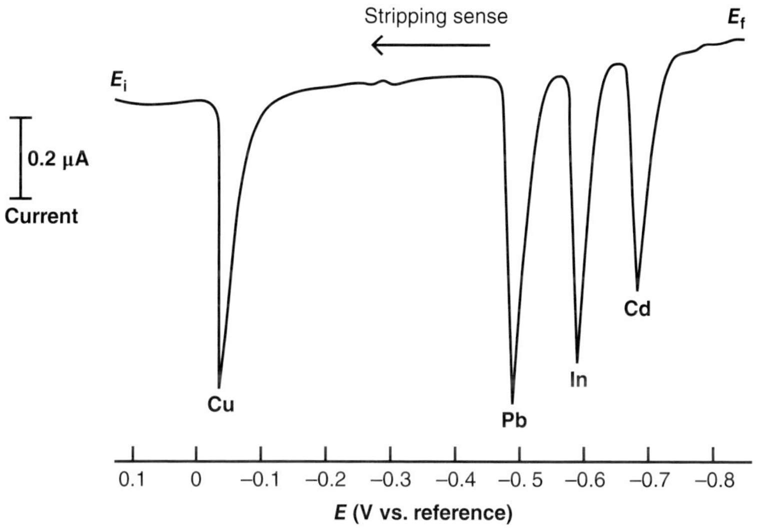

Figure 2.13 Response from an ASV experiment, where metallic Cu, Pb, In, and Cd were detected

Adsorptive stripping voltammetry (AdSV) is based on the accumulated adsorption of a complex between the analyte and a complexing agent at the electrode surface, often at an applied potential chosen so as to help the adsorption process. Either the metal ion or the ligand can be measured in the presence of an excess of the other component in bulk solution. Usually, the determination step involves reduction of the metal ion from the adsorbed complex. Adsorption follows an isotherm, usually of Langmuir type [3].

2.3 Applications of Voltammetric Techniques in Food Analysis

2.3.1 Food Contaminants: Heavy Metals, Pesticides, and Toxic Substances

Food contaminants are any harmful substances unintentionally added to food, as a result of processing procedures, as well as environmental pollution and natural sources. Tables 2.1 and 2.2 summarize the recent work devoted to the identification of heavy metals, pesticides, and toxic substances through the use of the above discussed electrochemical techniques. As can be observed in Table 2.1, differential pulse stripping voltammetric (DPSV) techniques are the most popular to determine heavy metal content such as lead, cadmium, and mercury in food matrices. Square wave stripping voltammetry (SWSV) has also been carried out to analyze lead, cadmium, and arsenic in wholemeal, wheat, and maize meals [12]. Considering the inconvenience of the toxicity of the hanging drop mercury electrode, there are reported alternative DPASV methodologies where a ring gold disk electrode and a Nafion-coated bismuth film electrode were used, respectively [11, 14]. In this regard, some reviews have discussed the use of SV techniques as an analytical tool in food analysis [23, 24].

Pesticides such as fungicides, herbicides, and insecticides are also found as contaminants in food products. Alghamdi for instance, describes data on the use of AdSV in analyzing

Table 2.1 *Applications of voltammetric techniques in the determination of toxic metals in food samples*

Analyte	Sample	Electrochemical technique	Working electrode	Pretreatment of sample	Supporting electrolyte	Studied concentration/ detection limit	References
Pb (II)	Rice, flour, soya, sugar, water	AdCSV	HMDE	Microwave acid digestion	0.01 M ammonia buffer solution, pH 8.5, 8.0 mM carbidopa	Pb: 0.1–210 nM, DL: 0.09 nM and Zn: 0.2–170 nM, DL: 0.15 nM	[9]
Pb (II)	Milk	DPCSV, DPASV	HMDE	Acid digestion	0.1 M HCl	185–8.7 μg l^{-1}	[10]
Hg (II)	Refined beet sugar	DPASV	RGDE	None	0.448 g KCl/0.372 g, Na_2EDTA/22 ml of $HClO_4$	–	[11]
Pb(II), Cd(II), As(III),	Wholemeal, wheat, and maize meal	SWASV, SWCSV	HMDE	Acid digestion	0.5 M HCl for Pb(II), Cd(II), and 0.5 M HCl/0.8 mM $CuCl_2$/0.04 mM N_2H_4 2HCl for As(III)	DL < 10^3 mg l^{-1}	[12]
Pb(II), Cd(II)	Rice, soya, and sugar	AdDPCSV	HMDE	Acid digestion	Borate buffer, pH 8.0, 2–1.0 × 10^{-4} M mercaptobenzothiazole	Pb: 0.5–70 ng ml^{-1}, DL: 0.017 ng ml^{-1} and Cd: 0.2–30 ng ml^{-1}, DL: 0.01 ng ml^{-1}	[13]
Pb(II), Cd(II)	Vegetables	DPASV	Nafion-coated bismuth film electrode	Acid/dry-ashing digestion	0.1 M acetate buffer solution, pH 4.5	DL: 0.17 μg l^{-1}	[14]
Cd(II), Pb(II)	Commercial broiler poultry feeds	DPASV	HMDE	Acid digestion	Acetate buffer, pH 5	DL: 0.35 μg kg^{-1}, Cd and DL: 0.68 μg kg^{-1}, Pb	[15]

HMDE = Hanging mercury drop electrode; RGDE = Rotating gold disk electrode; and DL = detection limit.

Table 2.2 *Applications of voltammetric techniques in the determination of some toxic substances in food samples*

Analyte	Sample	Electrochemical technique	Working electrode	Pretreatment of sample	Supporting electrolyte	Studied concentration/ detection limit (DL)	References
Acrylamide	Potato chips	CV, OSWV	Carbon hemoglobin (Hb)	Water extraction	0.05 M NaBr, 0.2 M acetate buffer, pH 4.8; and 0.05 M NaBr, 0.01 M phosphate buffer, pH 7.4.	DL: 1.2×10^{-10} M	[16]
Aflatoxins B1 (AFB1) and B2 (AFB2)	Groundnut	AdCSV	HMDE	Solvent extraction	3.0 M KCl	AFB1: 0.4–40 ng ml^{-1}, DL: 0.15 ng ml^{-1} and AFB2: 0.2–70 ng ml^{-1}, DL: 0.10 ng ml^{-1}	[17]
Azinphos-methyl and parathion-methyl	Honey	AdDPSV	HMDE	Solvent: buffer extraction	0.05 M Britton–Robinson buffer, pH 7 and 10	DL: 51.71 μg kg^{-1}, MeP and DL: 65.87 μg kg^{-1}, AzMet	[18]
Furan	Coffee	SWV	Pt electrodes	None	Acetonitrile with tetrabutylammonium, perchlorate	1.5×10^{-5}–1×10^{-3}M, DL: 1.2×10^{-5}M	[19]
Furfural (F) and 5-hydroxymethyl-2-furaldehyde (HMF)	Food products pharmaceutical syrups	SWV and Ad-SWSV	Static mercury drop electrode	*In situ* derivatization with Girard's reagent T	NH_3–NH_4Cl buffer, pH 9.5	DL: 400 ng g^{-1}, F and DL: 10 ng g^{-1}, HMF	[20]
Oxytetracycline	Milk and egg	SWV	Carbon fiber microelectrodes	Ethyl acetate extraction	5.0×10^{-4}M TBAP solution in acetonitrile	1.0×10^{-6}–1×10^{-4}M, DL: 2.9×10^{-7}M	[21]
Paraquat	Natural water, food, and beverages	Multiple SWV	Au-microelectrode	None	0.1 M Na_2SO_4	5.0×10^{-7}–1.04×10^{-5} M, DL: 0.044 μg l^{-1}	[22]

HMF: Hydroxymethyl-2-furaldehyde.

fenthion, pyridafenthion, phosalone, and carbophos in olive oil, wine, potato, and tomato respectively, as well as the use of ASV for measuring ziram, dinoseb from rice, and apple juice, respectively [23]. Table 2.2, on the other hand, describes relevant information regarding the quantification of paraquat and parathion in beverages and honey, respectively [18, 22].

Regarding the quantification of toxic substances in food samples, SWV has been used to detect furan, a carcinogen compound formed by heat treatment in canned, jarred, and roasted foods, in coffee samples, and oxytetracycline, an antibiotic used for the prevention and treatment of animal diseases, in milk samples [19, 21]. An adsorptive cathodic stripping voltammetry (AdCSV) methodology has been proposed to determine the content of aflatoxin B1 and B2 in groundnut samples. These toxins are produced by fungi and can be found in cereals, peanuts, and dried fruits. They are responsible for carcinogenic, mutagenic, teratogenic, and immunosuppressive effects. Furthermore, CV and Osteryoung SWV (OSWV) on a carbon paste electrode modified with hemoglobin have been successfully used to evaluate the content of acrylamide in potato chips [16]. Acrylamide is a compound produced when starch-rich food is heated to temperatures above 120 °C, such as fried or roasted potatoes, cookies, bread, and coffee. It can generate neurotoxicity, neuropathy, and it is considered as a possible carcinogenic. Figure 2.14, for instance, shows the OSWV response of an aqueous extract of potato crisps in the absence and in the presence of additions of acrylamide. It can be observed that the peak current values decrease as the acrylamide concentration increases. As in the case of some other reported processes, an adduct is formed. In this case, the hemoglobin–acrylamide adduct produces structural changes of the hemoglobin therefore decreasing the current of the hemoglobin redox reaction values.

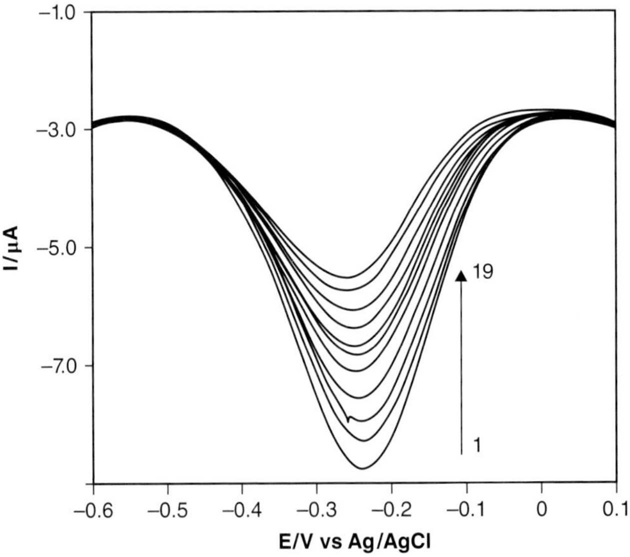

Figure 2.14 *OSWV response of carbon paste electrodes modified with Hb-DDAB liposomes toward acrylamide in the presence of water extract from the potato crisp. Reproduced from Ref. [16] with permission from Elsevier*

2.3.2 Trace Essential Elements

Trace elements and minerals are essential in biological processes for human normal growth and development; however, high concentrations, accumulation, or low intake may lead to deficiencies and/or diseases. Among the most important elements found in food, Cu, Fe, Mn, Zn, Co, and Se stand out. The conventional method to quantify trace metals in these samples is flame atomic absorption spectrometry. Table 2.3 shows some recent literature regarding quantification of trace elements and minerals by electrochemical techniques in food samples.

The concentration of Cu is usually low in food samples, therefore developing sensitive and selective methods for Cu quantification are required. Among the electrochemical techniques, stripping analysis following adsorption accumulation of Cu chelates has demonstrated high sensitivity; however after some time of accumulation, metal and chelates may strongly adsorb on the electrode surface. Most of the authors used the standard addition method not only to minimize the food matrix effects, but also to improve the resolution of the voltammetric technique. An ASV procedure at a platinum rotatory disk electrode without the need of Cu-complex formation has been proposed to determine the Cu content of tequila samples [26].

DPSV procedures have been performed to quantify Se, Zn, Ni, and Co. Taking advantage of the suitability and versatility of the electrochemical techniques, several authors have proposed methodologies to perform simultaneous determination of metals in food samples. For example, a sequential determination of Cu(II), Pb(II), Cd(II), Zn(II), As(III), Se(IV), Mn (II), and Fe(III) has been performed by diverse SWV techniques on standard reference and market food samples of wholemeal, wheat, and rice flour [12]. Besides the detection of the metals, some details of the analytical method were addressed suggesting the use of the peak area instead of the peak current to achieve lower detection limits. Figure 2.15 shows the SW anodic stripping voltammogram of the digested sample of wholemeal where Cu, Pb, Cd, and Zn were identified.

2.3.3 Food Additives

In food industry, several substances are used as preservatives, nutritional, flavoring, coloring, and texturizing agents. However, despite the benefits attributed to food additives, consumption of concentrations of these compounds above recommended levels are related to toxic effects and therefore, it is important to develop analytical strategies that can offer high sensitivity and selectivity.

For example, sulfite has been used as food preservative mainly to inhibit microbial growth and prevent discoloration associated with nonenzymatic browning reactions. It can be found in dried fruits and vegetables, pickles, coconut water, vinegar, juices, wine, and others. However, it has been related to hypersensitivity producing nausea, diarrhea, gastric irritation, nettle rash, or swelling and the aggravation of asthmatic conditions. The official methods to determine sulfites in foods and beverages include titration, photometry, and iodometry, techniques that require long analysis time and sample pretreatment. Recently however, alternative electrochemical methods have been proposed where the quantitative reduction of bisulfite and sulfur dioxide formed after the acidification of samples containing sulfite is determined (Table 2.4) [43, 45].

Table 2.3 *Applications of voltammetric techniques in the determination of trace elements and minerals in food samples*

Analyte	Sample	Electrochemical technique	Working electrode	Pretreatment of sample	Supporting electrolyte	Studied concentration/ detection limit (DL)	References
Al (III)	Flour, tea, bread cookies, and rice	ASV	HMDE	Acid digestion	Cupferron formed complex in 0.4 M $(NH_4)_2SO_4$ solution	4.0×10^{-8}–2.0×10^{-9} g ml^{-1} DL: 8×10^{-10} g ml^{-1}	[25]
Cu (II)	Tequila	ASV	Pt rotating disk	None	0.2 M HCl	–	[26]
Cu (II)	Rice and vegetables	ASV	HMDE	Microwave acid digestion	0.05 M PBS, pH 2.5 and 2-(2,3,5-triazolylazo)-5-dimethylaninobenzonic acid	9.5×10^{-9}–4.7×10^{-7}M, DL: 9.5×10^{-9}M	[27]
Cu (II)	Olive oil	Ad-SWSV	HMDE	Extraction with hot concentrated HCl	7.86×10^{-5}M DCDT, 2.4% ethanol, and 1 M HCl	0–35 ng ml^{-1}, DL: 0.49 ng ml^{-1}	[28]
Cu (II)	Water, tomato, blood, rice, and tea	AdSV	HMDE	Acid digestion	0.02 M Britton–Robinson buffer, pH 9.7, and 2.5×10^{-6}M thiosemi-carbazide	0.01–90.0 ng ml^{-1}, DL: 0.007 ng ml^{-1}	[29]
Cu (II)	–	ASV	HMDE	Acid digestion	3-Aminophthalhydrazide (luminol)	0.5–105.0 ng ml^{-1}, DL: 0.04 ng ml^{-1}	[30]
Cu(II), Zn(II), Se(IV), Mn(II), Fe(III)	Wholemeal, wheat, and maize meal	SWASV, SWCSV, SWV	HMDE	Acid digestion	0.5 M HCl, Cu, and Zn, 0.5 M HCl/0.8 mM $CuCl_2$/0.04 mM N_2H_4, 2HCl, Se and ammonia-ammonium chloride buffer, pH 8.8, Mn and Fe	DL < 10^3 mg l^{-1}	[12]

Ni (II), Co (II)	Refined beet sugar	DPCSV	HMDE	None	0.1 M ammonia buffer, pH 9.2 and 0.1 M DMG	–	[31]
Se (IV)	Garlic	CSV	HMDE	Acid digestion	0.1 M HCl	2.0×10^{-8}–6.0×10^{-7} M	[32]
Se (IV)	Milk	DPCSV, DPASV	HMDE	Acid digestion	0.1 M HCl	75–1.2 μg l^{-1}	[10]
Zn (II)	Vegetables	DPASV	Nafion-coated bismuth film electrode	Acid/dry-ashing digestion	0.1 M acetate buffer solution, pH 4.5	DL: 0.30 μgl^{-1}	[14]
Zn (II), Cu (II)	Commercial broiler poultry feeds	DPASV	HMDE	Acid digestion	Acetate buffer, pH 5	DL: 0.69 μg kg^{-1}, Zn and DL: 0.24 μg kg^{-1}, Cu	[15]
Se (IV)	Rice	DPCSV	HMDE	Acid/dry-ashing digestion	Cu(II) solution (1020 mg/l), 0.1 M HCl, 0.6 M MgCl2	0.15–8 ppb, DL: 0.07 ppb	[33]
Zn (II), Cu (II)	Fish	DPASV	HMDE	Lyophilized, milled and decomposed in an oxygen plasma ashing chamber	$HClO_4$	–	[34]

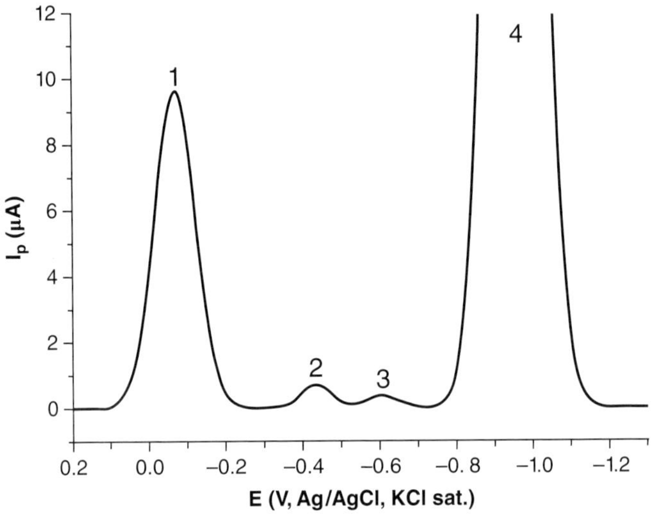

Figure 2.15 *SW anodic stripping voltammogram of the solution obtained from the digestion of Wholemeal BCR-CRM 189 standard reference material. Element peaks: Cu(II) [1]; Pb(II) [2]; Cd(II)[3]; and Zn(II) [4]. Reproduced from Ref. [12] with permission from John Wiley & Sons*

Color is a quality parameter that impacts the consumer preference; therefore food dyes must guarantees stability during food processing resulting in intensive and permanent color of food products. Due to the adverse effects produced by synthetic dyes, the addition of natural dyes in food products is preferred. However, most dyes from natural sources do not demonstrate enough stability through food processing. An example of synthetic dye family is azo-dyes, which are metabolized to aromatic amines that are responsible for adult headaches and hyperactive syndrome in children. In this regard, many analytical methods have been explored to determine the content of synthetic dyes in food such as chromatographic (TLC, HPTLC, and HPLC) and spectrophotometric methods. For these techniques, pretreatment is necessary to extract the dye from food matrices. Table 2.4 summarizes the information from recent literature regarding the quantification of synthetic dyes in food samples. Electrochemical techniques such as CV, DPV, and adsorptive square wave stripping voltammetry (AdSWSV) have been employed to quantify azo compounds, amaranth, carmine, ponceau 4R, sunset yellow, and tartrazine without pretreatment of food samples. Noteworthy, glassy carbon electrodes modified with bismuth/poly (*p*-aminobenzene sulfonic acid) [36], alternated layers of iron(III) tetra-(*N*-methyl-4-pyridyl)-porphyrin (FeT4MPyP) and copper tetrasulfonated phthalocyanine (CuTSPc) [39], acetylene black nanoparticles [40], carbon nanotubes [41, 43], magnetic Fe_3O_4 and gold nanoparticles [42, 44] have been successfully used to increase sensitivity.

Other widely used type of food additives is the synthetic antioxidant group in which compounds such as butylated hydroxyanisole (BHA), butylated hydroxytoluene (BHT), tertiary butyl hydroquinone (TBHQ), 2,4,5-trihydroxybutyrophenone (THBP), di-tertbutyl-4-hydroxymethylphenol (IONOX-100), propyl gallate (PG), octyl gallate (OG), nordihydroguaiaretic acid (NDGA), 4-hexylresorcinol (4HR), 2-naphthol (2NL), 4-phenylphenol (OPP), and 2,4-dichlorophenoxyacetic acid (2,4-DA) are included.

Table 2.4 Applications of voltammetric techniques in the determination of food additives

Analyte	Sample	Electrochemical technique	Working electrode	Pretreatment of sample	Supporting electrolyte	Studied concentration/ detection limit (DL)	References
Amaranth	Soft drinks	AdSWSV	HMDE	None	Carbonate buffer, pH 10	1×10^{-8}–1×10^{-7}M, DL: 1.7×10^{-9} M	[35]
Azo compunds	Beverages	DPV	Modified GCE	None	Acetate buffer, pH 4.7	DL: 4.1×10^{-8}M, PAN, and DL: 3.8×10^{-8} M, PAR	[36]
BHA and BHT	Mayonnaise	SWV	MCCE	Solvent extraction	Potassium nitrate/10% ethanol (v/v) at pH 6.7	3.4×10^{-7}–4.1×10^{-5} M, DL: 7.2×10^{-8}M, BHA and DL: 9.3×10^{-8}M, BHT	[37]
Carmine (E120)	Ice cream and soft drinks	AdSWSV	HMDE	None	Acetate buffer, pH 3	5×10^{-8}–1.25×10^{-7}M, DL: 1.43×10^{-9} M	[38]
Nitrite	Sausage and vegetables	DPV	Modified GCE	Precipitation of proteins	0.1 M phosphate buffer, pH 7	0.5–7.5 μM, DL: 0.1 μM	[39]
Ponceau 4R Tartrazine	Soft drinks	AdASV	Modified-GCE	None	Phosphate buffer, pH 7	Ponceat 4R, 0.05–4 mg kg^{-1}, DL: 0.03 mg kg^{-1} and Tartrazine 0.15–18 mg kg^{-1}, DL: 0.1 mg kg^{-1}	[40]
Quinoline yellow	Soft drinks	DPV	MWNT/GCE	None	Phosphate buffer, pH 8	0.75–20 mg l^{-1}, DL: 0.5 mg l^{-1}	[41]
Sudan I	Chili, ketchup, strawberry, tomato sauces, duck egg yolk	CV and DPV	Modified GCE	Solvent extraction	0.1 M PBS, pH 6.0	0.01–1 and 1–20 μM, DL: 0.001 μM	[42]
Sulfite	Vinegar, pickle and coconut water, shredded coconut	SWV	GCE/PAH/MWCNT	Adjusting pH	Acetate buffer, pH 4.6	DL: 4.2 mM	[43]
Sunset yellow and Tartrazine	Soft drinks	CV and DPV	Modified CPE	None	PBS, pH 4.0	0.05–1.6 μM, DL: 3.0×10^{-8} M, SY and DL: 2.0×10^{-9} M, Tz	[44]

GCE = Glassy carbon electrode; MWCNT = Multiwall carbon nanotubes; CPE = Carbon paste electrode.

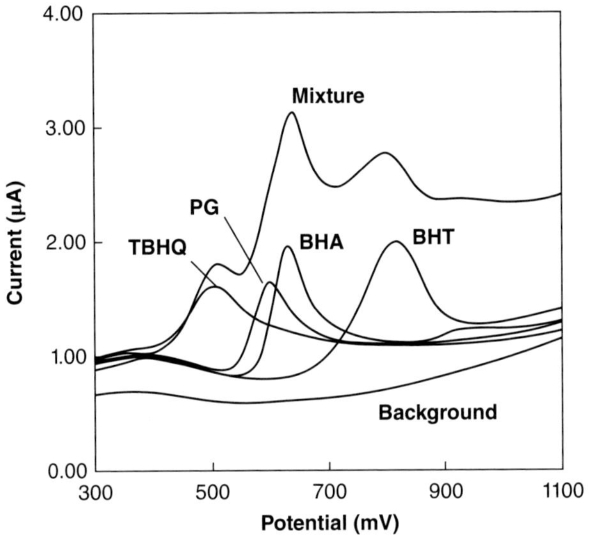

Figure 2.16 LSV of 5.0 mg l^{-1} of BHA, BHT, PG, TBHQ, and their mixture at a glassy carbon electrode in 0.1 M of perchloric acid containing 1% methanol. Reproduced from Ref. [46] with permission from Elsevier

Although their protecting effect on the product quality during food distribution is clear, antioxidant excess in food products might result in toxicities and thus endanger human health. For instance, the voltammetric response of BHA, BHT, PG, and TBHQ, at a glassy carbon electrode, was investigated [46]. Figure 2.16 shows the LSV responses of the individual antioxidants and their mixture. The four compounds showed well-defined oxidation waves, while the mixture displayed overlapped signals and therefore the use of chemometric approaches, such as classical least squares, partial least squares, and principal component regression were needed for quantification purposes. Furthermore, an electrochemical method for the simultaneous determination of BHA and BHT in food samples using SWV at a carbon composite electrode modified with copper(II) phosphate immobilized in a polyester resin was developed (Figure 2.17) [37]. Compared with the previously described work, this modified electrode allowed the detection of BHA at lower potential values with a peak separation of about 430 mV between the oxidation potentials of BHA and BHT in binary mixtures. Moreover, the detection limits were improved from 1.05×10^{-6}, 6.8×10^{-7} to 7.2×10^{-8}, 9.3×10^{-8} M for BHA and BHT, respectively.

2.3.4 Nutraceuticals: Phenolic Acids, Flavonoids, and Others

Functional foods represent one of the most interesting areas of research and innovation in food industry. Actually, the role of a balanced nutrition is not only to ensure an adequate intake of nutrients to efficiently enable all metabolic processes but also to contribute to sustained human health. Plant-based foods, such as fruits, vegetables, and whole grains, which contain significant amounts of bioactive phytochemicals also known as nutraceuticals, may provide desirable health benefits beyond basic nutrition to significantly reduce the incidence of chronic diseases, such as cancer, cardiovascular diseases, and

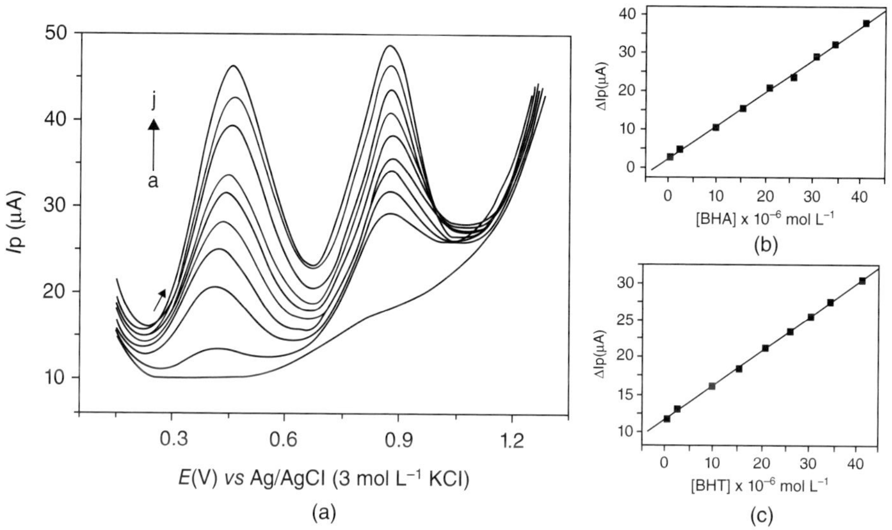

Figure 2.17 SW voltammograms ($f = 50\,Hz$; $a = 70\,mV$; and $\Delta E_s = 5\,mV$) acquired on MCCE–$Cu_3(PO_4)_2$-Poly for BHA and BHT (a) in 0.1 M KNO_3/10% ethanol (v/v) (pH 6.7). (a) Supporting electrolyte; (b)–(j) 3.40×10^{-7} M to 4.12×10^{-5} M. Analytical curves for BHA (b), and BHT (c). Reproduced from Ref. [37] with permission from Elsevier

other aging-related pathologies. The pleiotropic, additive, antagonic, and synergistic effects of nutraceutical compounds present in the diet are responsible for the observed biological effects, and therefore a reliable identification and accurate quantification of these compounds in food is of paramount importance.

Examples of polyphenols quantification including phenolic acids and flavonoids in food samples are listed in Table 2.5. Several sweep potential electrochemical techniques such as CV, AdSV, and SWV combined with a variety of working electrodes have been used to quantify these organic compounds. For example, ellagic acid, a dimeric derivative of gallic acid found in vegetables, fruits, and some nuts has important antioxidant, antiviral, antimutagenic, and anticarcinogenic properties. For its determination in food samples, HPLC, spectroscopic, electrophoretic, and polarographic methods are used. Since it is usually present as ellagitannin, time consuming pretreatment of samples is needed. Recently, the employment of CV, DPV, and SWV techniques allowed the study of the electrochemical behavior of ellagic acid at a glassy carbon electrode [56]. Figure 2.18 shows the ellagic acid CV response where two oxidation peaks can be observed. The same electroactive species shows in the DPV and SWV corresponding responses only one oxidation wave due to the short time scale of these measurements. For ellagic acid, SWV showed a slightly better response than that obtained for DPV. Possible interference in the determination of ellagic content with these methodologies could be ascorbic acid, ferulic acid, and quercetin. Deconvolution procedures of the SWV responses of ellagic acid, in the presence of quercetin, have been suggested for quantification purposes [50].

Capsaicin is another nutraceutic compound valued not only for its sensory attributes of pungency, aroma, and color but also for its biological properties such as analgesic, antioxidant, antibacterial, and anticarcinogenic, among others. CV and AdSV techniques have been proposed for identification and quantification [47, 48].

Table 2.5 *Applications of voltammetric techniques in the determination of nutraceuticals in food samples*

Analyte	Sample	Electro-chemical technique	Working electrode	Pretreatment of sample	Supporting electrolyte	Studied concentration/ detection limit (DL)	References
Capsaicin	Hot pepper sauce	AdSV, CV	MWCNT-BPPGE, MWCNT-SPE	Solvent extraction	0.05 M Britton–Robinson buffer pH 1.0	0.5–15 μM, DL: 0.31 μM and 0.5–35 μM, DL: 0.45 μM	[47]
Capsaicin	Commercial pepper products	AdSV	Boron-doped diamond electrode	Solvent extraction	Britton–Robinson buffer, pH 1.0–9.0 and sodium dodecylsulfate	0.16–20 mM, DL: 0.034 mM	[48]
Catechin	Green and black tea	SWV	GCE	None	pH 2	DL: 0.04 μM, EGCG and DL: 2.48 μM, EGC	[49]
Ellagic acid	Strawberry, raspberry, blackberry	SWV	GCE and paraffin-impregnated graphite rod	Acid hydrolysis	0.1 M KNO_3, pH 2	5×10^{-7}–8×10^{-6} M, DL: 1.35×10^{-7} M	[50]
Naringin	Grapefruit juice	DPCSV	HMDE	None	Phosphate buffer, pH 11/acetonitrile	1.72×10^{-7}–6.88×10^{-5} M, DL: 55 nM	[51]
Polyphenols	Green, oolong and black tea and an instant coffee	CV	GCE	None	Phosphate buffer, pH 7.0	–	[52]
Polyphenol and sulfur dioxide	Wine	CV	GCE	None	pH 3.3 model wine solution	–	[53]
Rutin	Soybean cultivars	AdCSV	HMDE	Solvent extraction	phosphate buffer pH 6 and 1.6×10^{-6} M Cu(II)	2.0×10^{-7}–1.4×10^{-6} M, DL: 7.0×10^{-9} M	[54]
Trans-resveratrol	Red wine	AdSWSV	GCE	Solvent extraction followed by solid phase separation	0.10 M perchloric medium	5.0–35 ng ml^{-1}, DL: 4.2 ng ml^{-1}	[55]

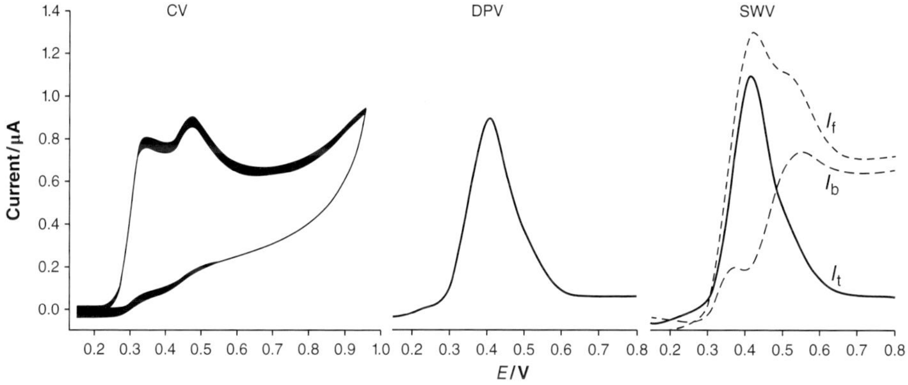

Figure 2.18 *Background corrected CV (scan rate 50 mVs^{-1}), DPV (amplitude 50 mV, potential increment 2 mV, modulation time 0.05 s, and interval time 0.1 s), and SWV (f = 10 Hz, a = 50 mV, and ΔE_s = 2 mV) responses of 1.5 × 10^{-5}M EA in 0.2 M acetic/acetate buffer of pH 5.5. I_t, I_f, and I_b, correspond to total, forward and backward currents, respectively. Reproduced from Ref. [56] with permission from Elsevier*

Numerous epidemiological and clinical studies have linked the high intake of dietary antioxidants, to lower incidence of chronic diseases such as diabetes, atherosclerosis, rheumatoid, arthritis, neurodegenerative, and coronary diseases and cancer and thus mortality rates [57–59]. The potential physiological benefits of dietary antioxidants have led to a dramatic growth of publications regarding the antioxidant capacity evaluation of extracts, pure compounds, and food products. Among the most popular methods to determine antioxidant capacity, 2,2′-azinobis-3-ethyl benzothiazoline-6-sulfonic acid (ABTS), 1,1-diphenyl-2-picrylhydrazyl (DPPH), oxygen radical absorbance capacity (ORAC), and ferric reducing antioxidant power (FRAP) assays stand out. However, electroanalytical techniques are emerging as a rapid, affordable, and reliable approach to evaluate the antioxidant activity of complex biological and food samples [60, 61]. The most employed electrochemical techniques to evaluate the antioxidant capacity are CV and DPV. Their inherent analytical potential is based on the fact that the oxidation potential is conceptually directly related to the antioxidant capacity. Therefore, it is expected that the lower the oxidation potential obtained during the analysis of a target sample extract, the stronger the antioxidant capacity. Moreover, the amperometric current and/or charge measured under fixed oxidation conditions can be related to the extension of the antioxidant capacity as well as to the estimation of the total content of redox compound families in the sample. The "electrochemical index" (EI) concept, defined as the total amount of antioxidants obtained by selective oxidation of them at different target potentials has been introduced [62]. In this context, the antioxidant capacity of dark fruit juices [63], cane and palm sugar [64], seaweeds [65], and wines [66] has been evaluated using CV. Furthermore, DPV and SWV have been performed to determine the antioxidant capacity of red wines and fruit juices, respectively [67, 68]. Significant linear correlation was observed between the voltammperometric measurements and the results obtained from conventional spectrophotometric techniques. Furthermore, the electrochemical techniques along with multivariate data analysis methods have been performed for classification purposes of edible vegetables oils [69] and tea liquor [70].

2.4 Concluding Remarks

The voltammetric electrochemical techniques discussed in this chapter are widely used in food sample analysis. While linear sweep and CV are preferred for exploring electron transfer reaction mechanisms, pulse techniques are used to achieve better detection and quantification limits of organic and inorganic species in food analysis. Together, these techniques can provide not only quantitative data but also important kinetic and thermodynamic information regarding the electron transfer processes of the complex food matrix.

Abbreviations

ABTS 2,2′-azinobis-3-ethyl benzothiazoline-6-sulfonic acid; ASV anodic stripping voltammetry; AdSV adsorptive stripping voltammetry; AdASV adsorptive anodic stripping voltammetry; AdCSV adsorptive cathodic stripping voltammetry; AdDPSV adsorptive differential pulse stripping voltammetry; AdDPCSV adsorptive differential pulse cathodic stripping voltammetry; AdSWSV adsorptive square wave stripping voltammetry; CSV cathodic stripping voltammetry; CV cyclic voltammetry; DPPH 1,1-diphenyl-2-picrylhydrazyl; DPSV differential pulse stripping voltammetry; DPASV differential pulse anodic stripping voltammetry; DPCSV differential pulse cathodic stripping voltammetry; FRAP ferric reducing antioxidant power; LSV linear sweep voltammetry; NPV normal pulse voltammetry; ORAC oxygen radical absorbance capacity; OSWV Osteryoung square wave voltammetry; SV stripping voltammetry; SWV square wave voltammetry; SWSV square wave stripping voltammetry; SWASV square wave anodic stripping voltammetry; SWCSV square wave cathodic stripping voltammetry.

References

1. Scholz, F. (2002) *Electroanalytical Methods: Guide to Experiments and Applications*, Springer-Verlag, Berlin, p. 331.
2. Bard, A.J. and Faulkner, L.R. (2001) *Electrochemical Methods: Fundamentals and Applications*, John Wiley & Sons, Inc., New York, p. 174.
3. Brett, C.M.A. and Oliveira, A.M. (1998) *Electroanalysis*, Oxford University Press, New York.
4. Bard, A.J. and Rubinstein, I. (1999) *Electroanalytical Chemistry: A Series of Advances*, Marcel Dekker, Inc..
5. Fleury, M.B. and Largeron, M. (1999) *Techniques Electrochimiques D'Analyse*, Polytechnica, Paris.
6. Gosser, D.K. (1993) *Cyclic Voltammetry: Simulation and Analysis of Reaction Mechanisms*, VCH, New York.
7. Galus, Z., Chalmers, R.A. and Bryce, W.A.J. (1994) *Fundamentals of Electrochemical Analysis*, Polish Scientific Publishers.
8. Ivaska, A., Lewenstam, A. and Sara, R. (1990) *Contemporary Electroanalytical Chemistry*, Plenum Press, New York and London.
9. Gholivand, M.B., Ahmadi, F. and Sohrabi, A. (2007) Adsorptive stripping voltammetric determination of ultra trace of zinc and lead with carbidopa as complexing agent in food and water samples. *Electroanalysis*, **19**, 2465–2471.

10. Inam, R. and Somer, G. (2000) A direct method for the determination of selenium and lead in cow's milk by differential pulse stripping voltammetry. *Food Chem.*, **69**, 345–350.
11. Sancho, D., Debán, L., Campos, I. *et al.* (2001) Determination of mercury in refined beet sugar by anodic stripping voltammetry without sample pretreatment. *Food Chem.*, **74**, 527–531.
12. Locatelli, C. (2004) Heavy metals in matrices of food interest: sequential voltammetric determination at trace and ultratrace level of copper, lead, cadmium, zinc, arsenic, selenium, manganese an iron in meals. *Electroanalysis*, **16**, 1478–1486.
13. Abbasi, S., Khodarahmiyan, K. and Abbasi, F. (2011) Simultaneous determination of ultra trace amounts of lead and cadmium in food samples by adsorptive stripping voltammetry. *Food Chem.*, **128**, 254–257.
14. Xu, H., Zeng, L., Huang, D. *et al.* (2008) A nafion-coated bismuth film electrode for the determination of heavy metals in vegetable using differential pulse anodic stripping voltammetry: an alternative to mercury-based electrodes. *Food Chem.*, **109**, 834–839.
15. Mahesar, S., Sherazi, S., Niaz, A. *et al.* (2010) Simultaneous assessment of zinc, cadmium, lead and copper in poultry feeds by differential pulse anodic stripping voltammetry. *Food Chem. Toxicol.*, **48**, 2357–2360.
16. Stobiecka, A., Radecka, H. and Radecki, J. (2007) Novel voltammetric biosensor for determining acrylamide in food samples. *Biosens. Bioelectron.*, **22**, 2165–2170.
17. Hajian, R. and Ensafi, A. (2009) Determination of aflatoxins B1 and B2 by adsorptive cathodic stripping voltammetry in groundnut. *Food Chem.*, **115**, 1034–1037.
18. Tsiafoulis, C. and Nanos, C. (2010) Determination of azinphos-methyl and parathion-methyl in honey by stripping voltammetry. *Electrochim. Acta*, **56**, 566–574.
19. Falciola, M., Pifferi, V., Possenti, M.L. and Carrara, V. (2012) Square wave voltammetric detection of furan on platinum and platinum-based screen printed electrodes. *J. Electroanal. Chem.*, **664**, 100–104.
20. Shamsipur, M., Belgi, A., Teymoun, M. *et al.* (2010) Electrocatalytic application of Girard's reagent T to simultaneous determination of furaldehydes in pharmaceutical and food matrices by highly sensitive voltammetric methods. *Electroanalysis*, **22**, 1314–1322.
21. Agüí, L., Guzman, A., Pedrero, M. *et al.* (2003) Voltammetric and flow injection determination of oxytetracycline residues in food samples using carbon fiber microelectrodes. *Electroanalysis*, **15**, 601–607.
22. de Souza, D., Machado, S. and Pires, R. (2006) Multiple square wave voltammetry for analytical determination of paraquat in natural water, food, and beverages using microelectrodes. *Talanta*, **69**, 1200–1207.
23. Alghamdi, A. (2010) Applications of stripping voltammetric techniques in food analysis. *Arabian J. Chem.*, **3**, 1–7.
24. Brainina, K., Malakhova, N. and Stojko, N. (2000) Stripping voltammetry in environmental and food analysis. *Fresenius J. Anal. Chem.*, **368**, 307–325.
25. Qiong, L., Lirong, W., Danli, X. and Guanghan, L. (2006) Analytical, nutritional and clinical methods determination of trace aluminum in foods by stripping voltammetry. *Food Chem.*, **97**, 176–180.
26. Carreon-Alvarez, A., Casillas, N., Ibanez, J. *et al.* (2008) Determination of Cu in tequila by anodic stripping voltammetry. *Anal. Lett.*, **41**, 469–477.

27. Lua, G., Longb, D., Lia, D. *et al.* (2004) Determination of copper in food by stripping voltammetry using Cu(II)–[2-(2,3,5-triazolylazo)-5-dimethylaninobenzonic] complex. *Food Chem.*, **84**, 319–322.
28. Galeano, T., Guiberteau, A., López-Soto, M. and Ortiz, J. (2006) Determination of copper with 5,5-dimethylcyclohexane-1,2,3-trione 1,2-dioxime 3-thiosemicarbazone in olive oils by adsorptive stripping square wave voltammetry. *Food Chem.*, **96**, 156–162.
29. Abbasi, S., Khani, H. and Tabaraki, R. (2010) Determination of ultra trace levels of copper in food samples by a highly sensitive adsorptive stripping voltammetric method. *Food Chem.*, **123**, 507–512.
30. Abbasi, S., Bahiraei, A. and Abbasai, F. (2011) A highly sensitive method for simultaneous determination of ultra trace levels of copper and cadmium in food and water samples with luminol as a chelating agent by adsorptive stripping voltammetry. *Food Chem.*, **129**, 1274–1280.
31. Sancho, D., Debán, L., Campos, I. *et al.* (2000) Determination of nickel and cobalt in refined beet sugar by adsorptive cathodic stripping voltammetry without sample pretreatment. *Food Chem.*, **71**, 139–145.
32. Inam, R. and Somer, G. (1999) Determination of selenium in garlic by cathodic stripping voltammetry. *Food Chem.*, **66**, 381–385.
33. Panigati, M., Falciola, L., Mussini, P. *et al.* (2007) Determination of selenium in Italian rices by differential pulse cathodic stripping voltammetry. *Food Chem.*, **105**, 1091–1098.
34. Celik, U. and Oehlenschläger, J. (2004) Determination of zinc and copper in fish samples collected from Northeast Atlantic by DPSAV. *Food Chem.*, **87**, 343–347.
35. Alghamdi, A. (2005) A square-wave adsorptive stripping voltammetric method for the determination of amaranth, a food additive dye. *J. AOAC Int.*, **88**, 788–793.
36. Ying, X., Qun, H. and Bing, N. (2010) Determination of azo compounds by differential pulse voltammetry at a bismuth/poly(*p*-aminobenzene sulfonic acid) film electrode and application for detection in food stuffs. *J. Electroanal. Chem.*, **639**, 175–180.
37. Garcia, K. and Fatibello-Filho, O. (2010) Simultaneous determination of butylated hydroxyanisole (BHA) and butylated hydroxytoluene (BHT) in food samples using a carbon composite electrode modified with $Cu_3(PO_4)_2$ immobilized in polyester resin. *Talanta*, **81**, 1102–1108.
38. Alghamdi, A., Alshammery, H., Abdalla, M. and Alghamdi, A. (2009) Determination of carmine food dye (E120) in foodstuffs by stripping voltammetry. *J. AOAC Int.*, **92**, 1454–1459.
39. Santos, W., Lima, P., Tanaka, A. *et al.* (2009) Determination of nitrite in food samples by anodic voltammetry using a modified electrode. *Food Chem.*, **113**, 1206–1211.
40. Yang, X., Quin, H., Gao, M. and Zhang, H. (2011) Simultaneous detection of Ponceat 4R and tartrazine in food using adsorptive stripping voltammetry on an acetylene black nanoparticle-modified electrode. *J. AOAC Int.*, **91**, 2821–2825.
41. Zhao, J., Zhand, Y., Wu, K. *et al.* (2011) Electrochemical sensor for hazardous food colourant quinolone yellow based on carbon nanotube-modified electrode. *Food Chem.*, **128**, 569–572.
42. Yin, H., Zhou, Y., Meng, X. *et al.* (2011) Electrochemical behavior of Sudan I at Fe_3O_4 nanoparticles modified glassy carbon electrode and its determination in food samples. *Food Chem.*, **127**, 1348–1353.

43. Romão-Sartori, E., Hissashi-Takeda, H. and Fatibello-Filho, O. (2011) Glassy carbon electrode modified with fuctionalized carbon nanotubes within a poly(allylamine hydrochloride) film for the voltammetric determination of sulfite in foods. *Electroanalysis*, **23**, 2526–2533.
44. Mehdi-Ghoreishi, S., Behpour, M. and Golestaneh, M. (2012) Simultaneous determination of Sunset yellow and Tartrazine in soft drinks using gold nanoparticles carbon paste electrode. *Food Chem.*, **132**, 637–641.
45. Isaac, A., Wain, A., Compton, R.G. *et al.* (2005) A novel electroreduction strategy for the determination of sulfite. *Analyst*, **130**, 1343–1344.
46. Ni, Y., Wang, L. and Kokot, S. (2000) Voltammetric determination of butylated hydroxyanisole, butylated hydroxytoluene, propyl gallate and *tert*-butylhydroquinone by use of chemometric approaches. *Anal. Chim. Acta*, **412**, 185–193.
47. Torabi-Kachoosangi, R., Wildgoose, G. and Compton, R. (2008) Carbon nanotube-based electrochemical sensors for quantifying the 'heat' of chilli peppers: the adsorptive stripping voltammetric determination of capsaicin. *Analyst*, **133**, 888–895.
48. Yardim, Y. (2011) Sensitive detection of capsaicin by adsorptive stripping voltammetry at a boron-doped diamond electrode in the presence of sodium dodecylsulfate. *Electroanalysis*, **23**, 2491–2497.
49. Novak, I., Šeruga, M. and Komorsky-Lovrić, Š. (2010) Characterisation of catechins in green and black teas using square-wave voltammetry and RP-HPLC-ECD. *Food Chem.*, **122**, 1283–1289.
50. Komorsky-Lovrić, S. and Novak, I. (2011) Determination of ellagic acid in strawberries, raspberries and blackberries by square-wave voltammetry. *Int. J. Electrochem. Sci.*, **6**, 4638–4647.
51. Reichart, E. and Obendorf, D. (1998) Determination of naringin in grapefruit juice by cathodic stripping differential pulse voltammetry at the hanging mercury drop electrode. *Anal. Chim. Acta*, **360**, 179–187.
52. Kilmartin, P. and Hsu, C. (2003) Characterisation of polyphenols in green, oolong, and black teas, and in coffee, using cyclic voltammetry. *Food Chem.*, **82**, 501–512.
53. Makhotkina, O. and Kilmartin, P. (2010) The use of cyclic voltammetry for wine analysis: determination of polyphenols and free sulfur dioxide. *Anal. Chim. Acta*, **668**, 155–165.
54. Gomes, J., Rodrigues, M., de Oliveira, A. *et al.* (2012) Cathodic adsorptive stripping voltammetric determination of rutin in soybean cultivars. *J. Food Compos. Anal.*, **25**, 1–8.
55. Airado, D., Galeano, T. and Durán, I. (2010) Determination of trans-resveratrol in red wine by adsorptive stripping square-wave voltammetry with medium exchange. *Food Chem.*, **122**, 1320–1326.
56. Cuartero, M., Ortuño, J., Truchado, P. *et al.* (2011) Voltammetric behaviour and square-wave voltammetric determination of the potent antioxidant and anticarcinogenic agent ellagic acid in foodstuffs. *Food Chem.*, **128**, 549–554.
57. Esposito, E., Rotilio, D., Di Matteo, V. *et al.* (2002) A review of specific dietary antioxidants and the effect on biochemical mechanisms related to neurodegenerative processes. *Neurobiol. Aging*, **23**, 719–735.

58. Wang, S., Melnyk, J., Tsao, R. and Marcone, M. (2011) How natural dietary antioxidants in fruits, vegetables and legumes promote vascular health. *Food Res. Int.*, **44**, 14–22.
59. Zamora-Ros, R., Rabassa, M., Llorach, R. *et al.* (2012) Application of dietary phenolic biomarkers in epidemiology: past, present, and future. *J. Agric. Food Chem.*, **60**, 6648–6657. doi: 10.1021/jf204742e
60. Chevion, S., Roberts, M. and Chevion, M. (2000) The use of cyclic voltammetry for the evaluation of antioxidant capacity. *Free Radic. Biol. Med.*, **28**, 860–870.
61. Roginsky, V. and Lissi, E. (2005) Review of methods to determine chain-breaking antioxidant activity in food. *Food Chem.*, **92**, 235–254.
62. Blasco, A., Rogerio, M., González, M. and Escarpa, A. (2005) "Electrochemical Index" as a screening method to determine "total polyphenolics" in foods: a proposal. *Anal. Chim. Acta*, **539**, 237–244.
63. Piljac-Žegarac, J., Valek, L., Martinez, S. and Belščak, A. (2009) Fluctuations in the phenolic content and antioxidant capacity of dark fruit juices in refrigerated storage. *Food Chem.*, **113**, 394–400.
64. Sia, J., Yee, H.-B., Santos, J. and Abdurrahman, M. (2010) Cyclic voltammetric analysis of antioxidant activity in cane sugars and palm sugars from Southeast Asia. *Food Chem.*, **118**, 840–846.
65. Keyrouz, R., Abasq, M., Le Bourvellec, C. *et al.* (2011) Total phenolic contents, radical scavenging and cyclic voltammetry of seaweeds from Brittany. *Food Chem.*, **126**, 831–836.
66. Makhotkina, O. and Kilmartin, P.A. (2009) Uncovering the influence of antioxidants on polyphenol oxidation in wines using an electrochemical method: cyclic voltammetry. *J. Electroanal. Chem.*, **633**, 165–174.
67. Šegura, M., Novak, I. and Jakobek, L. (2011) Determination of polyphenols content and antioxidants activity of some red wines by differential pulse voltammetry, HPLC and spectrophotometric methods. *Food Chem.*, **124**, 1208–1216.
68. Bordonaba, J. and Terry, L. (2012) Electrochemical behavior of polyphenol rich fruit juices using disposable screen-printed carbon electrodes; towards a rapid sensor for antioxidant capacity and individual antioxidants. *Talanta*, **90**, 38–45.
69. Gambarra-Neto, F., Marino, G., Urgulino, M. *et al.* (2009) Classification of edible vegetable oils using square wave voltammetry with multivariate data analysis. *Talanta*, **77**, 1660–1666.
70. Bhattacharyya, R., Tudu, B., Chandra, S. *et al.* (2012) Classification of black tea liquor using cyclic voltammetry. *J. Food Eng.*, **109**, 120–126.

3

Flow-Injection Analysis with Electrochemical Detection

Fabiana Silva Felix and Lúcio Angnes
Instituto de Química, Universidade de São Paulo, Av. Prof. Lineu Prestes, 748, CEP05508-000, São Paulo, SP, Brazil

3.1 Introduction

The experienced accelerated development during the last two decades has completely changed the speed at which things happen in our modern world. The advent of the personal computers, Internet, and huge explosion of services in the communication area has accelerated the commerce and shortened the distances. All types of merchandises, including foods, are actually sold worldwide. Never before foods have "traveled" as nowadays and in such large quantities. Besides this unprecedented increase in the volume of foods, the requirements of quality control are growing every day, requiring analytical methods of analysis with greater speed, higher sensitivity, and lower costs. Electroanalytical methods can provide these kinds of conditions for many applications in the food area. Until now, these techniques are still underused for analysis in the food area. In the following, we will detach the advantages of the use of these techniques.

Foods are essential to life because after being absorbed by the body, produce energy, promotes the growth, and repair of tissues and participate in multiple metabolic processes [1]. Agriculture allows us to find quicker ways to grow food, as also protect them from predators. Therefore, agricultural production still plays an important role in the world economy. The food industries, the governmental agencies, and universities work to monitor food composition and to ensure the quality and safety of the food supply. All food elements require analysis, and several methodologies are employed for this purpose and new methodologies are in progress.

Agricultural and Food Electroanalysis, First Edition.
Edited by Alberto Escarpa, María Cristina González and Miguel Ángel López.

Traditional methods such as high-performance liquid chromatography (HPLC), gas chromatography (GC), HPLC, or GC coupled to tandem mass spectrometry are widely used for the identification and quantification of nutrients (fats, carbohydrates, proteins, water, mineral elements, and vitamins), additives, or contaminants (aflatoxin, pesticides, pathogens, antimicrobial agents, growth promoters, additives, etc.), which can be found in foods [2–4].

However, these techniques require time-consuming sample preparation or expensive instrumentation [5]. Electroanalytical techniques are simple, rapid, and sensitive for many applications, including food analysis. Many active compounds (e.g., nutrients, pesticides, antimicrobial agents, hormones, etc.) found in food samples can be readily oxidized or reduced, but this property has been little explored mainly by industries. After the development of more sensitive pulse methods, the electroanalytical methodologies started to be more regularly used on the food analysis, for example, to evaluate the quality of vegetables, coconut water, and milk powder samples [6–8]. Increasing routine application of electrochemical methods can be facilitated by their use in mechanized flow analyzers or in flow systems coupled with voltammetric detectors. The association of voltammetry and flowing systems promotes the enhancement of mass transport to the working electrode, allowing significant improvement in the detection limit, compared to stationary measurements [9].

Analytical measurements in flow systems with electrochemical detection have gained increasing attention in different areas, including food and agricultural analysis. Many factors have stimulated the transition from steady state (batch systems) to flow electroanalysis among them; the analysis can be performed in an efficient and precise way with high sample throughput, low consumption of samples, better repeatability, reduction of the risk of contamination during analysis, combination of good precision with high sensitivity and relative low cost of the instrumentation, as well as good miniaturization potential of the analytical systems (e.g., apparatus and electrodes can be significantly reduced).

Besides the conventional electrodes, selectivity can be improved being associated flow injection analysis (FIA) with modified amperometric electrodes containing enzymes, polymeric, or inorganic films, porphyrin multilayer films, oxides, and so on [9, 10]. Moreover, it is possible to perform simultaneous analysis utilizing different approaches, such as differential amperometric determination with arrays of microelectrodes, employing enzyme-based techniques [11], or alternatively making electrodeposition of different noble metals onto array microelectrodes together with multivariate calibration analysis [12], or using FIA associated with multiple pulse amperometric detection [13], among other strategies.

In 1947, Muller described the first application of the flowing stream coupled to voltammetric detection using a platinum-working electrode [14]. Thereafter, many research papers have been published in scientific journals describing the injections of microliters of liquid samples into a stream of a carrier solution without the necessity of segmentation to suppress analyte dispersion allowing attaining a good sampling rate. FIA, sequential injection analysis (SIA), and batch injection analysis (BIA) are examples of the versatile techniques in flow conditions that contributed substantially to the development of automation in electroanalysis for application in various fields of science, among them, food analysis.

FIA is a well-established technique based on the injection of a sample plug into a continuously flowing carrier stream [9]. BIA was described as noncontinuous flowing injection technique due the step of sample introduction [15]. However, the detection is performed during the flow of the sample segment onto the transducer surface that suggests it be described as tubeless FIA. SIA arose as an important technique in the area of continuous flow analysis which maintained the advantages associated with FIA but reduced the inconveniences that hindered its use as a routine procedure for chemical process monitoring [16].

In this chapter, the reader will find fundamental information on voltammetric techniques coupled to flow system, mainly FIA, BIA, and SIA, with focus on food and agricultural electroanalysis. Also, fundamental principles of the three systems (FIA, BIA, and SIA) will be highlighted; examples of applications of the association in flowing conditions with voltammetric detectors and their advantages and limitations will be presented. In short, the purpose of this chapter is to give, for students and researchers involved in all areas of food analysis, an overview about electroanalysis associated with flow systems, that proportionate faster analysis, reproducible results, high sensitivity, with advantages such as the requirement of less sample (and consequent generation of less waste) and the use of simpler instrumentation.

3.2 Screening the Literature

Over the years, nations increased the requirements to acquire healthy foods. As a consequence, different analytical methods were developed to analyze from raw material to finished product. The modern agriculture introduced fertilizers and also pesticides. The explosion of the commerce of foods worldwide stimulated the blow up of the number of analysis focused on the food quality. In the past, all foods were produced and consumed in the same microregion. Nowadays, a simple meal can be composed of foods from many regions of the world. Figure 3.1 shows the evolution of the number of publications concerning the food analysis over time, based on the database of *Isi Web of Sciences* (May 2012), considering the words *food and analysis* as keywords. The impressive growth in the number of articles in the past few years can be attributed to many factors, among them the expansion of food trade (today worldwide), advance in the sanitary control of the production and marketing of food products and an increasing control of residual pesticides.

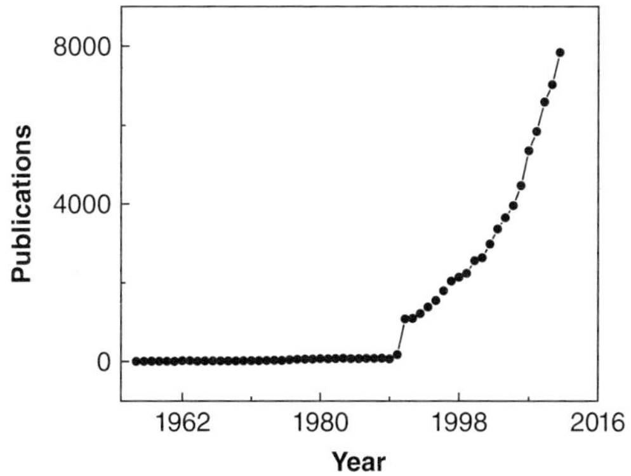

Figure 3.1 Evolution of the number of publications referring to food analysis over the years. (Source: Isi Web of Sciences, May 2012.)

Even considering that data from Figure 3.1 are not absolute (it is necessary to have in mind that the papers of many journals are not completely surveyed by the *Isi Web of Sciences* system until now) these results are still impressive. It demonstrates that in this century (period 2001–2011) the number of papers published correspond to ~70% of those in totality, indicating that in the last 11 years researchers have published much more papers than from 1887 to 2000.

3.3 Voltammetry under Flowing Stream

Articles describing the utilization of voltammetric methods are also concentrated in the last 20 years and correspond to less than 2% of all the studies. This percentage will increase in future, once the modern voltammetric methods proportionate different applications, thanks to the recent advances in instrumentation, in computerized processing of analytical data, by the use of new materials as indicator electrodes, and schemes for its electrochemical regeneration of these sensors. As a consequence, foods will be analyzed with improved sensitivity, accuracy, precision, speed, and selectivity of analysis. The miniaturization of all the system can also bring additional advantages [17]. Among several available voltammetric techniques, amperometry, differential pulse, square wave, and stripping voltammetry are the most widely used techniques in combination with flow systems in analysis of different matrixes. Examples of these associations in food and agricultural analysis are described throughout this chapter.

3.4 Flow Injection Analysis Principles

FIA arose as natural evolution of the continuous flow injection (CFA) and was originally proposed by Ruzicka and Hansen [18]. FIA is based on injection of a sample into a flowing nonsegmented continuous stream of a suitable liquid and delivered to any detector without intervention of the operator. Moreover, this technique does not require air bubbles in the stream in order to diminish the longitudinal dispersion of the sample and the interaction with other samples (carry over), as well as the measurements do not occur in the steady state, once an equilibrium condition is not necessary (since the system configuration is not changed and identical chemical and physical conditions are maintained). In an FIA system, it is possible to verify sharp peaks with fast increase and decrease of the current response, reflecting the rapid passage of the analyte onto the working electrode surface. Figure 3.2a presents a typical peak shape obtained during FIA analysis with amperometric detection. Figure 3.2b shows the excellent repeatability of the current peaks ($n = 40$) for an FIA system coupled to an amperometric detector, considering a 1.0×10^{-4} mol l^{-1} ferricyanide standard solution.

In FIA, under a laminar regime, the dispersion is defined as the amount that the chemical signal is decreased by the injection of a sample plug into a flow injection system [19], and this parameter can be affected for different factors such as the manifold components (manifold geometry and length, sample volume, and flow rate), viscosity of the fluids, and temperature between others.

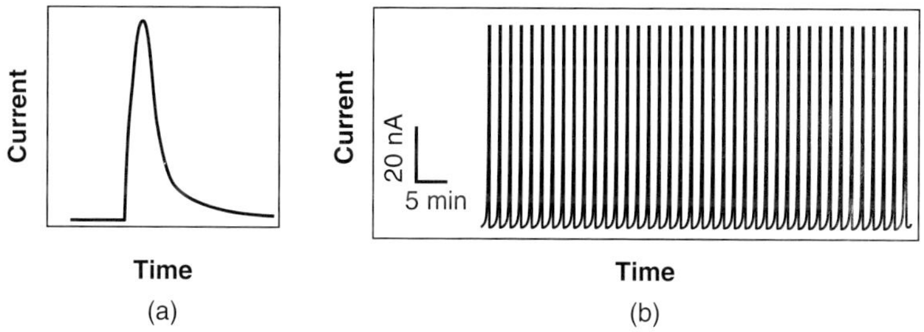

Figure 3.2 *(a) Characteristic peak signal of a FIA system with amperometric detection and (b) results obtained from repetitive injections of 1.0×10^{-4} mol l^{-1} ferricyanide standard solution during experiments in FIA-amperometry.*

Considering amperometric detection associated with flow injection system, the working potential affects not only the sensitivity and signal-to-noise ratio but also the selectivity of the current measurements. In general, a lower potential is more selective, and a higher one can suffer the interference of other species. The selection of the best working potential relies on the construction of hydrodynamic voltammograms. These voltammograms can be obtained by making repeated injections of the analyte, recording the current response at different potentials.

In FIA measurements, the amperometric signal can also alter with the volume of sample injected into the stream. Usually, volumes between 10 and 500 μl are used. The current signal increases almost linearly when sample volume increases. For higher volumes, the signal recorded has a tendency to attain a constant value. This suggests that in the center of the injected sample zone, any dispersion may occur. With increase in the sample volume, the time required for each analysis also increases. The best volume is chosen based on the compromise between reproducibility and the sampling rate. Figure 3.3a shows a sample volume study that has been developed by Daniel and Gutz [20] using 1.0×10^{-3} mol l^{-1} promethazine hydrochloride solution in 0.1 mol l^{-1} H_2SO_4. The effect of flow rate on the analytical signal can be investigated using different flow rates (typically 0.3–5.0 ml min^{-1}); the current increases with increasing flow rate until a maximum value. To elect the best condition, aspects as lower reagent consumption, improved repeatability of analytical signal and higher sample throughput must be considered.

An FIA system presents the following essential parts: the propulsion system, a sample introduction unit, the detection unit connected to the data acquisition and processing unit (normally, a computer). Figure 3.3b shows a schematic diagram of a typical single FIA system at the time when the sample was introduced in the stream. When the system works at constant flow, the most widely used means of propulsion to move the carrier solution and samples is the peristaltic pump in which it is represented by **P**, as shown in Figure 3.3b. Several types of injectors, devices used mainly to introduce the sample into the analytical path, have been reported in the literature. The most common injectors are rotary valve, which was developed by Ruzicka and Hansen [21], and proportional injector developed by researchers at the Analytical Chemistry Section of CENA/USP-Brazil [22] which are represented in Figure 3.3b by the letter **I.** Details about propulsion and sample introduction devices of the FIA system are discussed in more details in the following sections.

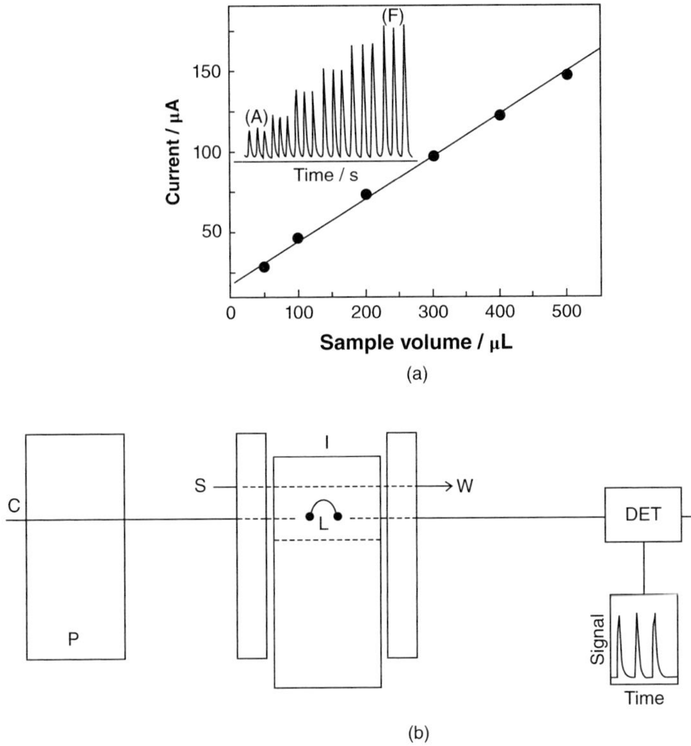

Figure 3.3 *(a) Effect of the volume of sample injected (A–F: from 50 to 500 μl) for 1.0×10^{-3} mol l^{-3} promethazine hydrochloride solution in 0.1 mol l^{-1} H_2SO_4; $E_{applied}$: +0.80 V, amperometric detector: Au electrode. (b) Schematic representation of a typical FIA system using a single line: C = carrier solution, P = peristaltic pump, I = injector, L = sample volume, S = sample, W = waste, and DET = detector.*

3.4.1 Liquid Propulsion in FIA

Fluids can be propelled in an FIA system by different mechanisms such as gravity, peristaltic pumps, micropumps, piston, or syringe pumps and gas pressure, among others [23–26]. To utilize the gravity force to propel solutions, a carrier reservoir must be positioned above the cell level (typically from 20 to 150 cm, as represented in Figure 3.4a). In this case, it is imperative to maintain the level of the solution constant; once it decreases, the flow rate also decreases. The same will occur if the solution viscosity and/or size of analytical path increase. Analytical path is defined as the way through which the carrier transports the sample to detector. This propulsion system is mainly useful for very simple assemblies such as single line manifolds (sample is merely driven to an electrochemical detector without any chemical reaction step during its trajectory inside the tubes). The major advantages are the simplicity and a very regular flow, free of variations, a fundamental condition for amperometric measurements in low concentrations.

Gas pressure-based liquid propulsion uses an inert gas (Figure 3.4b) to propel a stream with a nonpulsating flow. Due to the fact that the solubility of a given gas changes along the FIA manifold, system which is highly pressurized tends to produce disturbing bubbles.

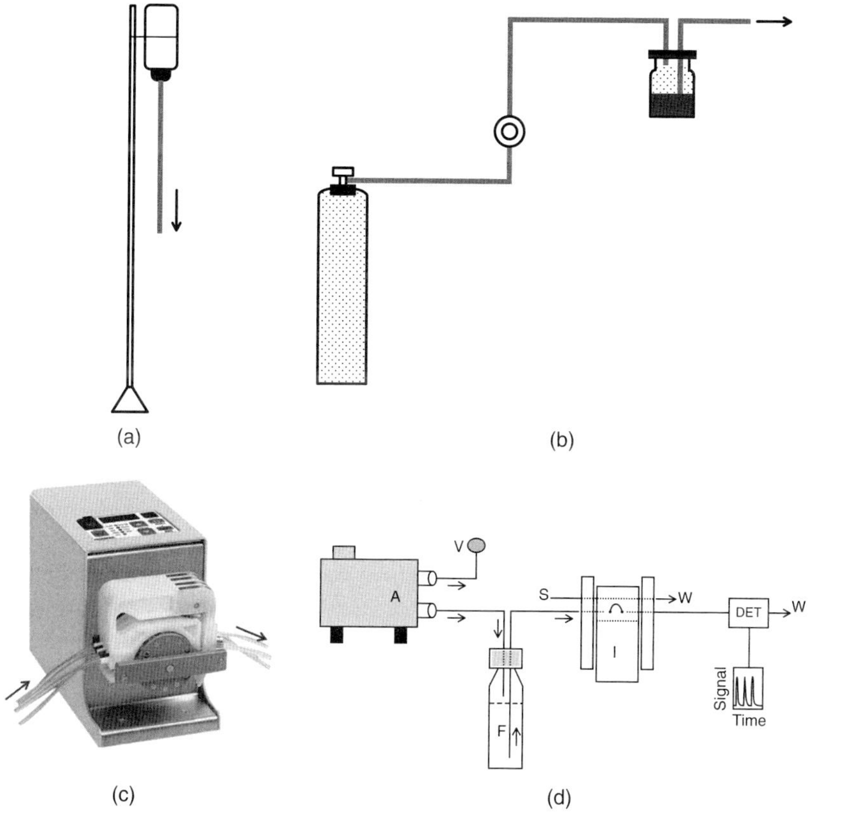

Figure 3.4 *Some propulsion systems: (a) gravity-based propulsion; (b) pressurized gas propulsion; (c) peristaltic pump; (d) compressed air, presenting the following details: (A) air compressor (aquarium pump), (F) closed flask containing the carrier, (I) injector, (S) sample, (V) outlet air control valve, and (W) waste. The arrows indicate the direction of carrier stream.*

As a consequence, oscillations in the flow rate and a decrease in the signal-to-noise ratio can appear. Other limitation is the requirement of a very precise manometer to reproduce the same pressure along different days.

Peristaltic pumps are the most widely used propulsion devices with "constant" flow rate in FIA (Figure 3.4c). In this system, the carrier and the sample are propelled by flexible tube(s) fitted in the pump. A rotor compresses the tube(s) and when it rotates the fluid is forced to move through the tube. The flexible tubes of the peristaltic pump are made with polymeric materials. The most common material is Tygon®, a transparent PVC-type material. Tygon tubes are used to propel aqueous and alcoholic solutions, where the alcohol content should not exceed more than 5% in order to ensure a long lifetime. Using peristaltic pumps the flow rates can be readily adjusted via the rotation speed of the pump or by changing the inner diameter of the flexible tube.

In reality, peristaltic pumps provide a pulsating flow, and often they are responsible for the high noise when low concentrations of analyte are analyzed amperometrically [24]. The use of a pneumatic damper can minimize the pulsation in FIA [27].

Alternatively, Figure 3.4d illustrates a simple system to propel liquids employing an aquarium air pump (A) of two channels. The air is pressurized inside a closed flask (F), forcing the solution to flow to the FIA system, passing through the injector (I), and then being carried to the detector and from there to the waste deposit (W). To the second channel is adapted a valve (V), which is used to control the air pressure inside the carrier flask, and consequently the liquid flow rate. An aquarium pump is attractive not only due to its low price, but also due to the fact that the air inside the flask is compressible and promotes a very constant flow rate of the solutions. Moreover, it can sustain a pressure around of 4.0 psi as well as is possible to reach flow rates of up to 12.5 ml min^{-1} during analysis.

Other alternative is the use of micropumps as proposed by Lapa *et al.* [23]. In this study, authors describe a strategy exploiting solenoid micropumps for each individual solution insertion (sample and carrier), controlling not only its volume but also the timing, sequence of introduction and transportation toward detector. This strategy was evaluated by using a flow network designed with two pumps.

3.4.2 Methods of Sample Introduction in an FIA System

To ensure insertion of an accurate volume of sample into the carrier stream, injectors are used during FIA experiments. Several concepts of injector have been described in the literature. Two injection devices must be detached: (a) the rotary valve [21], which two- to six-port models are the most versatile and (b) the proportional injector (manually operated) developed by researchers from CENA/USP-Brazil [22]. Their operation is based on two well-defined positions: loading and unloading positions. With the valve in the loading position, a defined volume is filled with sample. In the unloading position, the volume contained in the loop is inserted into the carrier stream.

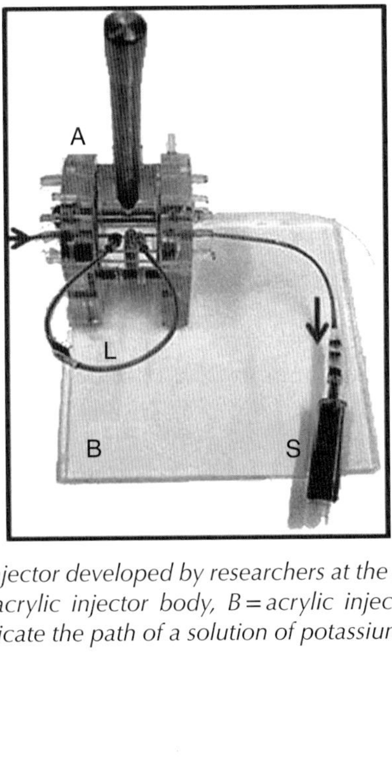

Figure 3.5 *Proportional injector developed by researchers at the analytical chemistry section of CENA/USP-Brazil: A = acrylic injector body, B = acrylic injector base, L = loop (100 µl), S = syringe. The arrows indicate the path of a solution of potassium permanganate at the aspiration step.*

These injectors are constructed with four Plexiglass® parts, one used as base, two lateral parts, and one circular mobile set. The central part of this injector can be displaced forward or backward in relation to the sides. In one position, the loop is filled with the sample and then it is moved to the other extremity to insert (the sample) into the carrier stream (Figure 3.5). The most common way to fill the sample into the injector is by aspiration since a minimum of manual intervention is required. Therefore, filling from a syringe can be acceptable if few samples are analyzed. In this system, the injection volume can be changed simply replacing the loop by other with different extension. Once the sample volume is chosen, the volume still remains perfectly constant during following experiments. Different analytes have been quantified in an FIA system with amperometric detection using proportional injector as a method of sample introduction [28–30].

3.4.3 Flow Cell Designs

There is a great variety of cell designs for electrochemical detection combined with flow analysis that can fulfill determined requirements such as high sensitivity, easy construction, operation, maintenance, minimum dead volume, speed of analysis, and small volume of sample and reagents. Designs for electrochemical cells can be quite simple – a solution

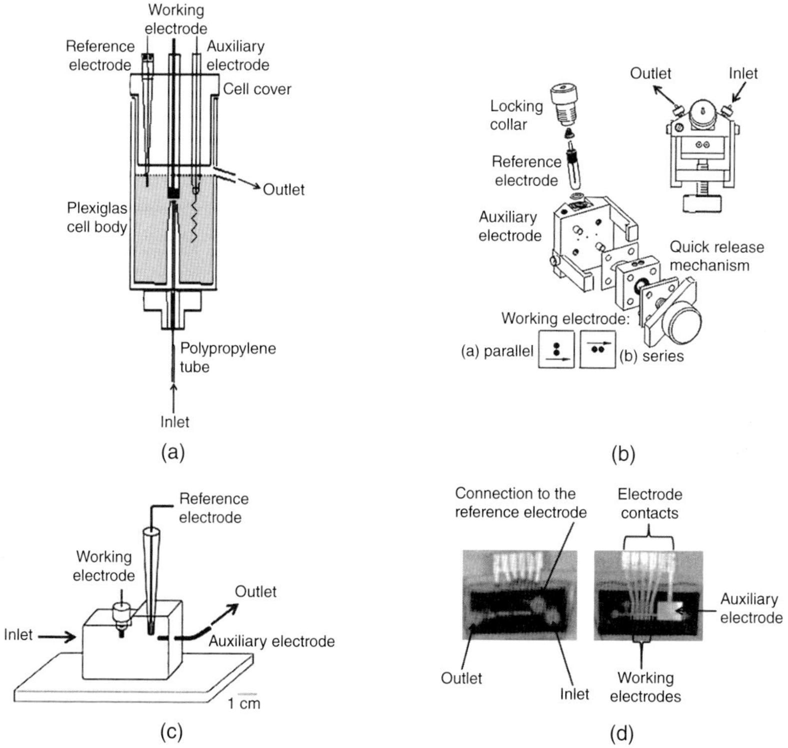

Figure 3.6 *Different FIA cell configurations: (a) Wall-jet type cell; (b) commercial thin layer flow cell; (c) Plexiglass flow cell; and (d) micro flow cell. Cells (a), (c), and (d) were built in the author's laboratory.*

inlet, a Plexiglass®, or glass vial with a cover drilled with holes for each electrode (working, reference, and auxiliary), and solution outlet – to complex, requiring significant machining skills and/or microfabrication technologies.

Most often it is possible to use different detectors, from mercury drop electrodes to microelectrodes, unmodified electrodes, or electrodes with modified surfaces, with the same electrochemical cell. Moreover, cell volumes are quite variable and range from milliliters to few nanoliters, depending upon the purpose of the analytical experiment.

There are many cell configurations employed in an FIA associated with amperometry, and some of them are depicted in Figure 3.6. Wall-jet cells (Figure 3.6a) are very simple to assemble. In addition, the solution at the surface of the detector is continuously renewed and as a consequence it is possible to perform measurements in small cells [31]. Thin layer flow cells (Figure 3.6b) are also popular; they can be acquired commercially or can also be built in the laboratory. Other arrangements (Figure 3.6c and d) with distinct working electrodes were used for the amperometric quantification of different analytes of precise and rapid way [32, 33]. Configurations have been developed for analysis with very small sample volumes with great sensitivity and reproducibility [34–36]. Moreover, numerous efforts have been directed to the development of FIA-cell for analysis of gaseous species due to the advent of microfabrication which facilitates manipulation of these samples as well as the easy access to gaseous standards and the increased demand for quantification of gaseous analytes [37].

3.5 Batch Injection Analysis Principles

BIA was introduced by Wang and Taha in 1990 [38] as an alternative to perform quick analysis in a very simplified way. In this system, the sample (typical volume: 10–100 μl) is injected exactly on the center of the electrode, positioned from the bottom of the cell. During the injection process, the transport of the analyte to the electrode surface is intense and this is the reason for the elevated currents registered in this technique. Utilizing motorized pipettes, the transport of analyte is very reproducible and the dispersion is controlled in time and space. The peak response is a transient form (similar to FIA), which grows very fast, attains in short time a steady-state signal (there the injection attained a constant velocity), and drops very rapidly to the base line (Figure 3.7a).

During the injection process in BIA, the analyte attains the electrode and disperses in the solution that fills the cell. When the injection ends, the recorded current begins to fall since no more analyte is transported to the electrode surface. BIA presents interesting features also observed in FIA, such as speed, sensitivity, and requirement of small volume of sample, with the advantages of the elimination of systems for liquid propulsion, valves, and manifold components. There is a minimum volume to attain the steady-state signal, as can be seen in Figure 3.7b. The signals obtained for 2.0×10^{-4} mol l^{-1} salbutamol injections utilizing different volumes (Figure 3.7b) showed that 40 μl was necessary to attain the maximum signal. Larger volumes of analyte generated exactly the same current. The utilization of different velocities of injection (Figure 3.7c) demonstrates that the signal grows with increase in the flow rate [39].

A simple BIA system with amperometric detection is depicted in Figure 3.8. The injection of discrete volumes of analyte onto the detector surface can be accomplished

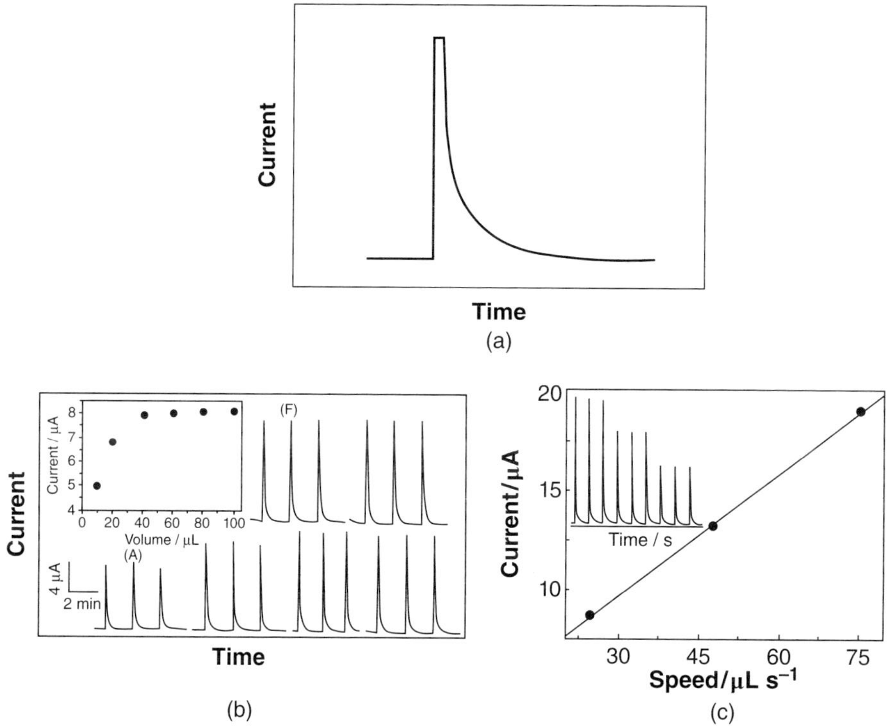

Figure 3.7 *(a) Typical peak signal of a BIA system in association with amperometric detection and the results obtained for (b) study of the influence of sample volume (A–F: 10–100 μl) on the peak performance, and (c) effect of injection velocity (from 24.5 to 75.2 μl s^{-1}) using 100 μl of 2.0×10^{-4} mol l^{-1} salbutamol.*

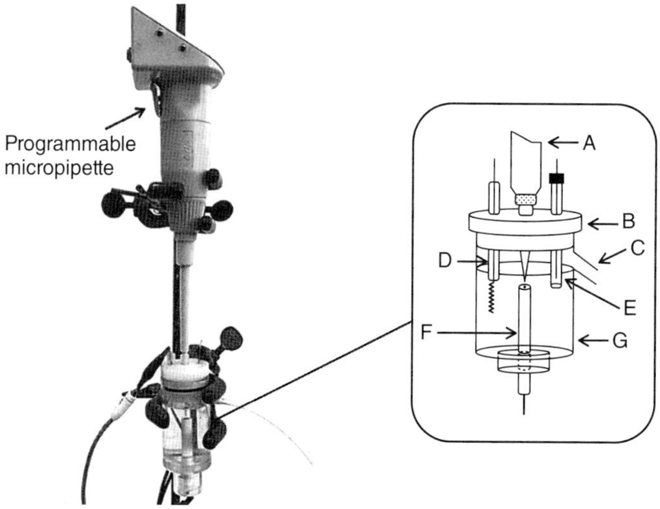

Figure 3.8 *Representation of a simple BIA system with amperometric detection. The inset shows a typical electrochemical cell used during experiments: (A) electronic micropipette; (B) cell cover; (C) outlet solution; (D) auxiliary electrode; (E) reference electrode; (F) amperometric detector; (G) Plexiglass body.*

by using of electronically propelled micropipettes (as seen in Figure 3.8). Different types of working electrode with or without modification on its surface have been used in this system [39–41]. Modification with polymeric films, porphyrins multilayer films, or incorporation of enzymes can be performed to reduce adsorption phenomena and/or improve the selectivity and sensitivity.

3.6 Sequential Injection Analysis Principles

Figure 3.9 shows the main parts of a SIA system associated with voltammetric detection used for herbicide analysis in tropical soil sample [42]. As well as in FIA, the basic principles are accurate sample injection, controlled dispersion, and reaction times with high repeatability.

In a typical SIA system both syringe and peristaltic pumps can be successfully used as propulsion device. Moreover, this system uses a multiposition rotary valve with many entry points (the one depicted in Figure 3.9 presents eight ports) to select different solutions. A computer controls the synchronization between pump and valve. After a defined volume of carrier and sample solutions are sequentially aspirated into a holding coil, the flow is reversed and then the stacked zones are pumped to the detector position through the selector valve [43, 44]. Similar to the FIA system, an increase in the current signal can be achieved by an increase in the injected sample volume. However, comparing of methodologies of the same analysis in the FIA and SIA modes shows that the sample rate of the SIA systems is about half to that of the FIA ones [45].

Among the variations of the SIA system, sequential injection chromatography (SIC) is the most recent technique and it has been proposed by Huclova *et al.* [46]. These authors describe the use of a porous monolithic column of silica as the stationary phase coupled to SIA for the separation and quantification of salicylic acid and methyl salicylate in pharmaceutical products.

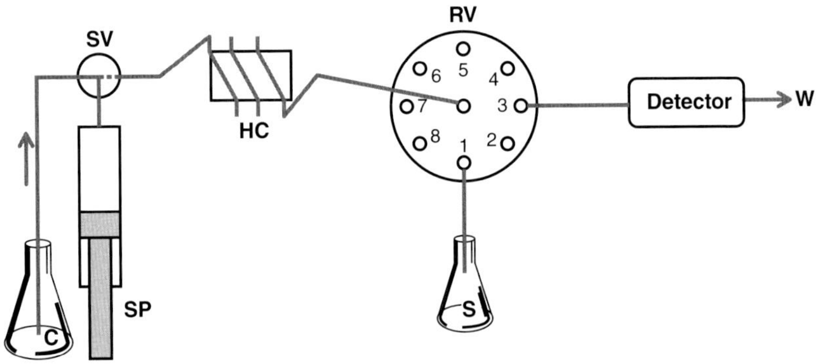

Figure 3.9 *Basic scheme of a SIA manifold: (C) carrier solution; (SV) syringe valve; (SP) syringe pump; (HC) holding coil; (RV) rotary selection valve with eight ports; (S) sample or standard solution; and (W) waste.*

3.7 Applications

Since the first application of the combination of voltammetry with flow analysis in 1947 [14], numerous publications in several areas of science have been presented. The simplicity of the treatment of samples (often requires just dilution with electrolyte) and the high sensitivity and sample throughput compared to manual procedures have caused an increasing interest for flow systems in the research field, but still very slowly in pharmaceutical, clinical, agriculture, and food laboratories. The reasons are not completely clear, but the actual state contributes the lack of knowledge about these techniques of many researchers (especially of many analysts who are today in the industries) and the pressure of the instrumentation companies, which will sell more expensive instruments. In the following sections, an overview of representative applications of the voltammetric detections coupled to flow systems for food and agricultural analysis will be presented. The main advantages of the FIA, BIA, and SIA systems such as versatility, speed, repeatability, and sensitivity will be highlighted.

3.7.1 FIA and Voltammetric Detection – A Happy Marriage

The association between FIA and amperometry offers many advantages such as wide linear working range, high sensitivity, sampling rate, low detection limit, and excellent repeatability, using a relatively inexpensive instrumentation. This perfect partnership has attracted attention for many applications, including the analysis of foodstuffs and agricultural side products.

Two procedures involving amperometric detection and FIA have been described for the determination of the antioxidants tert-butylhydroxyanisole (BHA) and *tert*-butylhydroxytoluene (BHT) in corn oil samples [47]. During oxidation of both antioxidants, a glassy carbon electrode was possible to obtain good results in the presence of a solution of perchloric acid in 5% methanol (v/v). RSD percentage values were 2.2% (BHA) and 3.6% (BHT) for the repeatability study.

A simple and quick method in FIA using a glassy carbon disk electrode modified with a molybdenum oxide film has been proposed for the quantification of iodate in commercial salt samples [48]. In this study, the film actuated as a catalytic agent in the reduction of iodate and a linear response was observed (from 1.0×10^{-6} to 1.0×10^{-3} mol l^{-1}). Other amperometric method coupled to FIA exploring the electrochemical reduction of iodate was also described [49]. Here, the authors used a more favorable working potential (+0.2 V vs. Ag/AgCl) in the presence of a solution of 1.0% w/v NaCl, 0.02% w/v KI, and 0.1 mol l^{-1} HCl as carrier.

In 2001, Buratti *et al.* [50] described an electrochemical method to evaluate the antioxidant power of lipophilic food extracts. The method was based on the association between FIA and amperometric detection and a working potential of +0.50 V (vs. Ag/AgCl) was used in the experiments. When applied to pure compounds, it was possible to observe that lycopene presented the greatest antioxidant power of all the lipophilic compounds (lycopene, β-carotene, zeaxanthin, α-carotene, β-cryptoxanthin, lutein, α-tocopherol, capsaicin, chlorophyll, astaxanthin, and canthaxanthin) used in this study. This result was compared with one of the most used methods to verify the total antioxidant capacity of foods (a radical cation decolorization assay) and a good correlation between both methods was found.

The determination of nitrite in mineral water and in cured meat samples was performed using a glassy carbon electrode modified with a film of tetraruthenated cobalt porphyrin [51], associated with an FIA configuration. With this modified electrode, the anodic process was anticipated (working potential maintained at +0.75 V rather than +1.0 V) and a 15-fold current enhancement was verified with respect to results observed with the unmodified working electrode. By incorporating a copper–cadmium column before the amperometric detector and in the FIA manifold, it was possible to succeed the quantification of nitrate in the same samples.

Masawat and Slater have presented the FIA method for determination of tetracyclines (tetracycline, chlortetracycline, and oxytetracycline) in chicken and shrimp meats and pharmaceutical samples [52]. The electrochemical oxidation of analytes was performed using a gold screen-printed electrode and good detection limits were achieved, 0.96, 0.58, and 0.35 $\mu mol\,l^{-1}$ for tetracycline, chlortetracycline, and oxytetracycline, respectively. Moreover, some advantages of the screen-printed electrodes were highlighted such as potential for miniaturization, reliability, portability, low cost, simplicity of construction, and operation.

In 2011, Toniolo *et al.* [53] described an amperometric method in FIA for the simultaneous quantification of ascorbic acid (AA) and hydrogen peroxide in orange-taste soft drinks using a thin layer dual electrode. In this system, a glassy carbon electrode and a Pt electrode coated with polyeugenol film (that prevents oxidation of ascorbate anions) were used for electrochemical oxidation of AA and H_2O_2, respectively. It was possible to obtain a wide linear range (from 5.0×10^{-6} to 1.0×10^{-3} mol l^{-1}) for both analytes with detection limits of about 1.0×10^{-6} mol l^{-1} (AA) and 5.0×10^{-7} mol l^{-1} (H_2O_2).

When the oxidation or reduction of organic compounds generates species that can adsorb or react on the detector surface decreasing significantly its response, it may be desirable to apply a series of different potentials (pulses, scans, multipulses, or other waveforms can be chosen) to overcome this problem. Pulsed amperometric detection (PAD) is a voltammetric technique that can be associated with FIA to solve detector problems. PAD is usually based on a triple potential waveform that incorporates amperometric measurements for analyte determination together with cleaning and reactivation of the detector surface. Taking advantages of PAD in association with FIA, Surareungchai *et al.* [54] have presented a method for simultaneous determination of sugars (glucose and fructose) in different fruit samples, using an Au electrode modified with Nafion® film. In this study, a quadruple-potential waveform was used: a waveform composing of two detection potentials followed by an anodic cleaning pulse (in a more positive potential) and a cathodic activation pulse. The use of PAD produced a notable improvement in terms of analysis time and sensitivity as well as the results were in agreement with those found by liquid chromatography.

Recently, Medeiros *et al.* [55] proposed a new procedure for simultaneous quantification of BHA and BHT antioxidants using multiple pulse amperometry (MPA) with FIA in food samples (commercial mayonnaise). Determination of these phenolic antioxidants was performed with a cathodically pretreated boron-doped diamond detector and a dual-potential waveform was used which first promoted the only the oxidation of BHA and second of

BHA plus BHT. To quantify BHT in the second potential, it was necessary to subtract the signal generated by BHA. This method was rapid, simple, and precise, and the detection limits were 0.030 $\mu mol\, l^{-1}$ for BHA and 0.40 $\mu mol\, l^{-1}$ for BHT. Other authors have proposed fast and simple methods involving voltammetric detection coupled to FIA for quantification of different analytes in foods [56–62] and liquids such as alcoholic beverages, wines, tea, fruit juices, and milk [63–68]. Moreover, a review about the determination of sulfite, an important preservative and antioxidant agent used in food and beverages, using FIA in association with several detectors (including amperometric sensors) has been reported [69].

A biosensor can be defined as a sensor that uses a biological material (enzymes, antibodies, antigens, plant, or animal tissues, organelles, hormones, among others) connected to a transducer that converts a biological signal in an electrical signal [70]. Moreover, the biological material is immobilized on a suitable membrane that is coupled to the detector surface, which monitors the disappearance of some reactant or the appearance of a product of the reaction between the biological material and the substrate of interest. Biosensors have been widely used in the electroanalytical field to improve selectivity and electrocatalytic properties and to increase the reproducibility of the detector response. The potential applications of biosensors in analysis of food and agriculture are numerous and reviews about these subjects have been described [70–72].

Mizutani *et al.* [73] have developed an amperometric biosensor based on immobilizing glucose oxidase on an Au electrode coated with a polyion complex layer coupled to FIA. This system was used for the quantification of glucose in beverage and biological samples by applying a working potential at 0.80 V (vs. Ag/AgCl). The biosensor revealed good stability (for more than two months) and elevated reproducibility with the sampling rate of 180 h^{-1}.

The performance of an amperometric tyrosinase biosensor based on graphite–Teflon composite electrode matrix used for the quantification of propyl gallate (PG) in olive oil samples has been reported [74]. Using an FIA system, the enzyme reaction involved the catalytic oxidation of PG to the corresponding o-quinone, and the electrochemical reduction of this o-quinone was monitored at 0.20 V (vs. Ag/AgCl) in phosphate buffer or 0.10 V (vs. Ag/AgCl) in acetonitrile–Tris buffer solution. Repeatability studies ($n = 25$) were performed using 5.0×10^{-5} $mol\, l^{-1}$ PG (in phosphate buffer) and 4.0×10^{-4} $mol\, l^{-1}$ PG (in acetonitrile–Tris buffer mixture) with RDS of 6.0 and 4.7%, respectively. Other studies have been described for quantification of various analytes in food products using biosensors in association with FIA systems [75–84].

In addition to the use of amperometric biosensors, enzymatic reactors inserted in the FIA system before the amperometric detector have been reported for the analysis of different analytes such as aspartame [85] in dietary food and AA [86] in multivitamin effervescent tablet and white wines.

The voltammetric techniques based on pulses (a sequence of pulses of different potentials are applied to the working electrode in which the Faradaic current decays more slowly than charging current) and stripping (that includes an electrolytic preconcentration step before analysis) [17] in association with FIA are mainly applied in the analysis of metals in samples of soil and river water [87–91].

Jakubowska *et al.* [91] have presented a differential pulse anodic stripping voltammetric method associated with FIA for determination of thallium in water extract of soil samples.

A concentration range of the order of pmol l^{-1} and detection limit of 0.25 pmol l^{-1} was achieved using a preconcentration potential at −0.90 V (vs. SCE) during 60 min.

The association between amperometric detection and FIA system in natural water samples has been reported for the analysis of different pesticides from river, lake, and estuary [92–97]. FIA amperometry was utilized to quantify ethiofencarb, a systemic carbamate used in agriculture to control parasites, in acetate buffer solution, at pH 5.0 [97]. Using a glassy carbon electrode potentiostated at 0.75 V (vs. Ag/AgCl), the pesticide was oxidized over a concentration range from 1.0×10^{-7} to 1.0×10^{-5} mol l^{-1} and presented a detection limit of 1.0×10^{-8} mol l^{-1}. By inhibition of the activity of selected enzymes and immobilized on surface of different detectors, biosensors can also be used in the analysis of pesticides [98–100].

Recently, Llorent-Martinez *et al.* [101] have described a review concerning applications of the FIA system coupled to electrochemical and optical detectors for analysis in different fields, including agriculture, food, and clinical.

3.7.2 BIA with Voltammetric Detection

A less flexible but simpler technique in comparison to the FIA system is BIA, which presents several advantages such as high reproducibility, elevated sampling rate, fast analysis time, and rapid response. BIA presents comparable sensitivity to FIA and does not require pumps, tubes, and commutators to introduce the samples. Moreover, motorized micropipettes provide very precise injections of small sample volumes at the constant rate due to the reproducible movement of the internal piston [40, 102, 103].

Voltammetric and amperometric measurements in BIA can be performed with the same electrochemical cells and detectors utilized in FIA, including wall-jet arrangement with injection of sample volume of a minimum 10 μl and the use of microelectrode arrays [104, 105]. Special attention has been attributed to the monitoring in agriculture due to the possibility of exploring portable instruments for *in situ* analysis. It occurs mainly through association between BIA and stripping voltammetric techniques.

Simões *et al.* [106] have presented a simple method based on BIA associated with square-wave voltammetry for detection of herbicide paraquat with good sensitivity. It was possible to quantify this herbicide at microgram per liter levels with the detection limit of 20 μg l^{-1} and using a glassy carbon electrode without modification in its surface. Paraquat analysis was carried out without any preconcentration step and with fast responses (<2 s).

BIA in association with anodic stripping voltammetry has been applied in the determination of trace metals in water samples [107]. In this study, the authors used a solid sorbent fixed in the micropipette tip for on-line preconcentration of the metals. After the optimization of experimental parameters, it was possible to obtain peak currents about 10–15% higher than without preconcentration step and linear concentration ranges at low levels such as 10–100 nmol l^{-1} for Cd analysis.

Economou and Voulgaropoulos [108] have successfully exploited the combination between stripping voltammetry and BIA system for quantification of Cd^{2+}, Pb^{2+}, Zn^{2+}, and In^{3+} (by anodic stripping voltammetry) and Ni^{2+}, Co^{2+}, U^{6+}, and Al^{3+} (by adsorptive stripping voltammetry). A bismuth-film electrode was coated with Nafion to increase the sensitivity and improve the mechanical stability to the film during the determination of Cd, Pb, and Zn. Adequate selectivity, quickness, simplicity, and low consumption of samples were also achieved in the studies involving BIA and voltammetry for analysis of trace metals in water and fertilizer samples.

The increased sensitivity in BIA through the coating on the surface of bismuth-film has also been observed when Adraoui *et al.* [109] used fibrinogen, a protein presents in blood, instead of Nafion in the quantification of Pb and Cd in tap water and tea samples. In addition, the sensitivity increased twice for both metals during anodic stripping voltammetry as well as low detection limits were obtained for Pb (0.5 nmol l^{-1}) and Cd (1.0 nmol l^{-1}) using a deposition time of 60 s and potential at −1.2 V.

The quantification of AA in tropical fruit juices has been performed in a BIA system where a mercury electrode was used as an amperometric sensor [41]. During electrochemical oxidation of AA, the capillary of the mercury drop electrode was adapted in an inverted position of the electrochemical cell in order to obtain stable and reproducible signals. Using a working potential of +0.23 V (vs. Ag/AgCl), it was possible to reach a frequency of 300 injections per hour.

3.7.3 SIA with Voltammetric Detection

SIA has been proposed as an alternative to FIA especially in monitoring of industrial processes where robustness, low reagent consumption, stability, low maintenance, versatility, and automation are essential requirements. There are various review papers describing the fundamentals, potentialities, and limitations of SIA [16, 43, 44, 110–112]. However the applications in agriculture and food analysis have still presented very few examples using electrochemical detectors, in particular voltammetric sensors.

Gutes *et al.* [113] have described an automated procedure based on voltammetric electronic tongue formed by a biosensor array for the determination of glucose in fruit juice samples. Linear sweep voltammetric signals were obtained with high selectivity and artificial neural networks were used as the modeling tool.

The SIA system in association with amperometric detection using a boron-doped diamond electrode was proposed for the quantification of sulfite in wine samples [114]. To prevent interference from sample matrices, a gas diffusion unit was used. A linear range from 0.2 to 20 mg l^{-1} and a detection limit of 0.05 mg l^{-1} were achieved during SO_3^{2-} analysis.

Exploring the excellent combination between SIA and anodic stripping voltammetry for quantification of metals in river sediment extracts and water samples, different detectors were used as well as the preconcentration step was carried out by deposition of analytes on mercury electrode and bismuth or antimony film electrode [115–117]. Reliable voltammetric signals were obtained since it was possible to reproduce the conditions of charge and mass transport with high precision by reproducible electrode area, stirring, and time control.

Injang *et al.* [118] have developed a fast method for the simultaneous determination of Pb^{2+}, Cd^{2+}, and Zn^{2+} in herb samples by association of SIA with anodic stripping voltammetry. A bismuth film on the screen-printed carbon nanotubes was prepared *in situ* and used as working electrode. Using a flow rate of 12 μl s^{-1}, low detection limits (0.2 μg l^{-1} for Pb, 0.8 μg l^{-1} for Cd and 11 μg l^{-1} for Zn) were achieved.

Masini *et al.* have proposed the use of SIA coupled to square-wave voltammetry for analysis of different pesticides (atrazine, picloram, methyl parathion, and paraquat) in water samples [119–122] and studies involving the adsorption of atrazine onto soil samples [120] in the presence of hanging mercury drop electrode. During voltammetric experiments for quantification of paraquat, 212 determinations per hour were observed since sample or standard solution was prepared off-line and the SIA was used just to inject the solution toward the electrochemical cell [122].

3.8 Advantages of Voltammetry under Flowing Stream

The ever-increasing demand on the number of samples to be analyzed, reduction of manifolds, the need for reducing the consumption of reagents and samples, and great speed of analysis have motivated the use of the voltammetric techniques under flowing stream in many branches of applications. The association between voltammetric detection and different systems of flow analysis (FIA, BIA, and SIA highlighted throughout this chapter) has presented results with high sampling throughput, wide linear concentration, precision, accuracy, and low detection limits, favoring the determination of very low concentrations. In some cases described in the extended literature, voltammetric methods in flow system can improve the sensitivity or even the selectivity compared to a nonflow system using the same methodology. The improved transport of electroactive material to the electrode surface has contributed significantly to the performance of the hydrodynamic voltammetry measurements.

The versatility of the flow system enables the analyst to perform voltammetric analysis with the same detector or even using similar apparatus. Suteerapataranon *et al.* [116] have proposed the use of components of the SIA system similar to the FIA system with exception of the syringe pump and the 10-port selection valve for the simultaneous quantification of cadmium, copper, lead, and zinc. The proposed FIA and SIA systems associated with stripping voltammetry were easily operated and required little amount of sample (30 μl provided reproducible current signals) with high degree of automation.

Economou and Voulgaropoulos [123] have described a hybrid FIA/SIA system in association with stripping voltammetry for determination of trace metals in fertilizer samples. The apparatus combined features from FIA (using a peristaltic pump for solution transport) and SIA (using the multiport selection valve as the sample introduction device without holding coil), enhancing the advantages and eliminating the shortcomings of each individual flow system. In this study, the authors utilized a bismuth-film electrode as the working electrode. Aspects as the high sensitivity, good accuracy, and elevated precision obtained by the use of the FIA/SIA methodology are highlighted. According to the authors, the utilization of the bismuth electrode, a metal with low toxicity, is also an advantage.

The voltammetric methods allow in many samples the direct analysis of electroactive compounds, eliminating procedures as extraction, filtration, the use of organic solvents or the necessity of complex and toxic reagents. This simplicity of sample preparation combined with the high speed of analysis under hydrodynamic conditions (found in FIA, BIA, or SIA) makes these procedures very attractive. In some cases, for a signal gain or improvement of selectivity different operations on line can be explored instead of using working electrodes with modifications in their surfaces.

3.9 Concluding Remarks

Voltammetric techniques associated with flow systems have been shown to be very appropriate methodologies for fast and simple analyses in different fields, including food and agriculture products. This association presents several advantages such as high sensitivity, elevated sample throughput, low detection limits, wide linear concentration, good precision, and great versatility, mainly in comparison with nonflow systems using the same voltammetric conditions. Moreover, voltammetric analysis under flowing stream requires low volumes of sample and reagents and as a consequence generates smaller amount of waste. The required instruments in hydrodynamic voltammetry measurements are relatively simple and they can be interfaced and controlled by a computer.

The high flexibility of flow system creates many possibilities to voltammetric detection, including pretreatment steps mainly for the preconcentration of the analyte or improvement of selectivity as well as the use of similar instruments for analysis in different flow systems with high sensitivity. Furthermore, chemometric strategies for optimization and calibration in analysis of food and agricultural using voltammetric methods in association with flow system can be applied to improve the resolution of signals. Advances in the automation and microfabrication technology associated with the development of new sensors and electrochemical cells might contribute for extending the application of voltammetric methods coupled to flowing for quantification of many compounds in different fields.

Acknowledgments

The authors gratefully thank the financial support from Brazilian Research Founding Agencies FAPESP – Process 2010/50303-1, CNPq – Processes 306504-2011-1 and 480311-2009-9 and CAPES).

References

1. Nielsen, S.S. (2010) *Food Analysis*, Springer, New York.
2. Khayoon, W.S., Saad, B., Lee, T.P. and Salleh, B. (2012) *Food Chem.*, **133**, 489–496.
3. Almeida, C., Fernandes, J.O. and Cunha, S.C. (2012) *Food Control*, **25**, 380–388.
4. Dong, J., Pan, Y.X., Lv, J.X. *et al.* (2011) *Chromatographia*, **74**, 109–119.
5. Gholivand, M.B., Torkashyand, M. and Malekzadeh, G. (2012) *Anal. Chim. Acta*, **713**, 36–44.
6. Sartori, E.R., Takeda, H.H. and Fatibello-Filho, O. (2011) *Electroanalysis*, **23**, 2526–2533.
7. Liao, C.W., Chen, Y.R., Chang, J.L. and Zen, J.M. (2011) *J. Agric. Food Chem.*, **59**, 9782–9787.
8. Felix, F.S. and Angnes, L. (2010) *J. Pharm. Sci.*, **99**, 4784–4804.
9. Mannino, S. and Wang, J. (1992) *Electroanalysis*, **4**, 835–840.
10. van Staden, J.F. and Matoetoe, M.C. (2000) *Anal. Chim. Acta*, **411**, 201–207.
11. Wang, J., Riviejo, A.J. and Mannino, S. (1992) *Anal. Lett.*, **25**, 1399–1409.

12. Matos, R.C., Angnes, L., Araujo, M.C.U. and Saldanha, T.C.B. (2000) *Analyst*, **125**, 2011–2015.
13. dos Santos, W.T.P., Gimenes, D.T., Richter, E.M. and Angnes, L. (2011) *Quim. Nova*, **34**, 1753–1761.
14. Muller, O.H. (1947) *J. Am. Chem. Soc.*, **69**, 2992–2997.
15. Quintino, M.S.M. and Angnes, L. (2004) *Electroanalysis*, **16**, 513–523.
16. Lenehan, C.E., Barnett, N.W. and Lewis, S.W. (2002) *Analyst*, **127**, 997–1020.
17. Uslu, B. and Ozkan, S.A. (2011) *Anal. Lett.*, **44**, 2644–2702.
18. Ruzicka, J. and Hansen, E.H. (1975) *Anal. Chim. Acta*, **78**, 145–157.
19. Trojanowicz, M. (2008) *Advances in Flow Analysis*, Wiley-VCH Verlag GmbH, Weinheim.
20. Daniel, D. and Gutz, I.G.R. (2003) *Anal. Chim. Acta*, **494**, 215–224.
21. Ruzicka, J. and Hansen, E.H. (1988) *Flow Injection Analysis*, John Wiley & Sons, Inc., New York.
22. Reis, B.F., Giné, M.F. and Kronka, E.A.M. (1989) *Quim. Nova*, **12**, 82–91.
23. Lapa, R.A.S., Lima, J.L.F.C., Reis, B.F. *et al.* (2002) *Anal. Chim. Acta*, **466**, 125–132.
24. Matos, R.C., Gutz, I.G.R., Angnes, L. *et al.* (2001) *Quim. Nova*, **24**, 795–798.
25. Dasgupta, P.K., Vithnage, R.S. and Petersen, K. (1988) *Anal. Chim. Acta*, **215**, 277–282.
26. Ferreira, L.M.C., Felix, F.S. and Angnes, L. (2012) *Electroanalysis*, **24**, 961–966.
27. Pedersen, K.M., Kummel, M. and Soeberg, H. (1990) *Anal. Chim. Acta*, **238**, 191–199.
28. Pedrosa, V.A., Lowinsohn, D. and Bertotti, M. (2006) *Electroanalysis*, **18**, 931–934.
29. Felix, F.S., Yamashita, M. and Angnes, L. (2006) *Biosens. Bioelectron.*, **21**, 2283–2289.
30. Franchini, R.A.A., Matos, M.A.C. and Matos, R.C. (2011) *Anal. Lett.*, **44**, 232–240.
31. Lima, A.W.O., Nascimento, V.B., Pedrotti, J.J. and Angnes, L. (1997) *Anal. Chim. Acta*, **354**, 325–331.
32. Felix, F.S., Brett, C.M.A. and Angnes, L. (2007) *J. Pharm. Biomed. Anal.*, **43**, 1622–1627.
33. Felix, F.S., Brett, C.M.A. and Angnes, L. (2008) *Talanta*, **76**, 128–133.
34. Daniel, D. and Gutz, I.G.R. (2006) *Anal. Chim. Acta*, **571**, 218–227.
35. Richter, E.M., Augelli, M.A., Kume, G.H. *et al.* (2000) *Fresenius J. Anal. Chem.*, **366**, 444–448.
36. Ferreira, H.E.A., Daniel, D. and Bertotti, M. (2008) *J. Braz. Chem. Soc.*, **19**, 1538–1545.
37. Pasquini, C. and da Silva, M.C.H. (1999) *J. Braz. Chem. Soc.*, **10**, 85–91.
38. Wang, J. and Taha, Z. (1991) *Anal. Chim. Acta*, **252**, 215–221.
39. Quintino, M.S.M. and Angnes, L. (2004) *Talanta*, **62**, 231–236.
40. Brett, C.M.A. (1999) *Electroanalysis*, **11**, 1013–1016.
41. de Donato, A., Pedrotti, J.J. and Gutz, I.G.R. (1999) *Electroanalysis*, **11**, 1124–1129.
42. dos Santos, L.B.O., Abate, G. and Masini, J.C. (2005) *Talanta*, **68**, 165–170.
43. dos Santos, A.C.V. and Masini, J.C. (2010) *Quim. Nova*, **33**, 1949–1956.
44. Perez-Olmos, R., Soto, J.C., Zarete, N. *et al.* (2005) *Anal. Chim. Acta*, **554**, 1–16.
45. Trojanowicz, M. (2000) *Flow Injection Analysis – Instrumentation and Applications*, World Scientific, London.

46. Huclova, J., Satinsky, D. and Karlicek, R. (2003) *Anal. Chim. Acta*, **494**, 133–140.
47. Sedeno, P.Y., Pingarron, J.M. and Diez, L.M.P. (1991) *Anal. Chim. Acta*, **252**, 153–159.
48. Kosminsky, L. and Bertotti, M. (1999) *Electroanalysis*, **11**, 623–626.
49. Jakmunee, J. and Grudpan, K. (2001) *Anal. Chim. Acta*, **438**, 299–304.
50. Buratti, S., Pellegrini, N., Brenna, O.V. and Mannino, S. (2001) *J. Agric. Food Chem.*, **49**, 5136–5141.
51. da Rocha, J.R.C., Angnes, L., Bertotti, M. *et al.* (2002) *Anal. Chim. Acta*, **452**, 23–28.
52. Masawat, P. and Slater, J.M. (2007) *Sens. Actuators B*, **124**, 127–132.
53. Toniolo, R., Dossi, N., Pizzariello, A. *et al.* (2011) *Electroanalysis*, **23**, 628–636.
54. Surareungchai, W., Deepunya, W. and Tasakorn, P. (2001) *Anal. Chim. Acta*, **448**, 215–220.
55. Medeiros, R.A., Lourenção, B.C., Rocha-Filho, R.C. and Fatibello-Filho, O. (2010) *Anal. Chem.*, **82**, 8658–8663.
56. Blasco, A.J., Rogerio, M.C., Gonzalez, M.C. and Escarpa, A. (2005) *Anal. Chim. Acta*, **539**, 237–244.
57. Possari, R., Carvalhal, R.F., Mendes, R.K. and Kubota, L.T. (2006) *Anal. Chim. Acta*, **575**, 172–179.
58. Avila, M., Crevillen, A.G., Gonzalez, M.C. *et al.* (2006) *Electroanalysis*, **18**, 1821–1826.
59. Jayasri, D. and Narayanan, S.S. (2007) *Food Chem.*, **101**, 607–614.
60. Buratti, S., Benedetti, S. and Cosio, M.S. (2007) *Talanta*, **71**, 1387–1392.
61. Prabakar, S.J.R. and Narayanan, S.S. (2010) *Food Chem.*, **118**, 449–455.
62. Akbari-adergani, B., Norouzi, P., Ganjali, M.R. and Dinarvand, R. (2010) *Food Res. Int.*, **43**, 1116–1122.
63. Paixão, T.R.L.C., Corbo, D. and Bertotti, M. (2002) *Anal. Chim. Acta*, **472**, 123–131.
64. Mannino, S., Brenna, O., Buratti, S. and Cosio, M.S. (1998) *Electroanalysis*, **10**, 908–912.
65. Gonçalves, L.M., Pacheco, J.G., Magalhães, P.J. *et al.* (2010) *Food Addit. Contam., Part A*, **27**, 175–180.
66. Volikakis, G.J. and Efstathiou, C.E. (2005) *Anal. Chim. Acta*, **551**, 124–131.
67. Casella, I.G. (1996) *Electroanalysis*, **8**, 128–134.
68. Chen, J., Gorton, L. and Akesson, B. (2002) *Anal. Chim. Acta*, **474**, 137–146.
69. Claudia, R.C. and Francisco, J.C. (2009) *Food Chem.*, **112**, 487–493.
70. Terry, L.A., White, S.F. and Tigwell, L.J. (2006) *J. Agric. Food Chem.*, **53**, 1309–1316.
71. Velasco-Garcia, M.N. and Mottram, T. (2003) *Biosyst. Eng.*, **84**, 1–12.
72. Baeumner, A.J. (2003) *Anal. Bioanal. Chem.*, **377**, 434–445.
73. Mizutani, F., Sato, Y., Hirata, Y. and Yabuki, S. (1998) *Biosens. Bioelectron.*, **13**, 809–815.
74. Morales, M.D., Gonzalez, M.C., Reviejo, A.J. and Pingarron, J.M. (2005) *Microchem. J.*, **80**, 71–78.
75. Dominguez-Sanchez, P., O'Sullivan, C.K., Miranda-Ordieres, A.J. *et al.* (1994) *Anal. Chim. Acta*, **291**, 349–356.
76. Parellada, J., Dominguez, E. and Fernandez, V.M. (1996) *Anal. Chim. Acta*, **330**, 71–77.

77. Paredes, P.A., Parellada, J., Fernandez, V.M. *et al.* (1997) *Biosens. Bioelectron.*, **12**, 1233–1243.
78. Bassi, A.S., Lee, E. and Zhu, J.X. (1998) *Food Res. Int.*, **31**, 119–127.
79. Palenzuela, B., Simonet, B.M., Garcia, R.M. *et al.* (2004) *Anal. Chim. Acta*, **524**, 167–174.
80. Nikolelis, D.P., Simantiraki, M.G., Siontorou, C.G. and Toth, K. (2005) *Anal. Chim. Acta*, **537**, 169–177.
81. Carelli, D., Centonze, D., Palermo, C. *et al.* (2007) *Biosens. Bioelectron.*, **23**, 640–647.
82. Miyanishi, N., Sato, N., Nakakita, S.I. *et al.* (2008) *Biosens. Bioelectron.*, **23**, 1347–1352.
83. Ghindilis, A.L., Gavrilova, V.P. and Yaropolov, A.I. (1992) *Biosens. Bioelectron.*, **7**, 127–131.
84. Gamella, M., Campuzano, S., Reviejo, A.J. and Pingarron, J.M. (2006) *J. Agric. Food Chem.*, **54**, 7960–7967.
85. Male, K.B., Luong, J.H.T. and Mulchandani, A. (1991) *Biosens. Bioelectron.*, **6**, 117–123.
86. Vig, A., Igloi, A., Adanyi, N. *et al.* (2010) *Bioprocess Biosyst. Eng.*, **33**, 947–952.
87. Lukaszewski, Z. and Zembrzuski, W. (1992) *Talanta*, **39**, 221–227.
88. Lukaszewski, Z., Zembrzuski, W. and Piela, A. (1996) *Anal. Chim. Acta*, **318**, 159–165.
89. Cavicchioli, A. and Gutz, I.G.R. (2001) *Anal. Chim. Acta*, **445**, 127–138.
90. Munoz, E., Palmero, S. and Garcia-Garcia, M.A. (2002) *Talanta*, **57**, 985–992.
91. Jakubowska, M., Zembrzuski, W. and Lukaszewski, Z. (2008) *Electroanalysis*, **20**, 1073–1077.
92. Zen, J.M., Kumar, A.S. and Chang, M.R. (2000) *Electrochim. Acta*, **45**, 1691–1699.
93. Zen, J.M., Chen, H.P. and Kumar, A.S. (2001) *Anal. Chim. Acta*, **449**, 95–102.
94. Cerejeira, R.P.A.G., Delerue-Matos, C. and Vaz, M.C.V.F. (2002) *Anal. Bioanal. Chem.*, **373**, 295–298.
95. Nouws, H.P.A., Delerue-Matos, C., Lima, J.L.C. *et al.* (2002) *Int. J. Environ. Anal. Chem.*, **82**, 69–76.
96. Barroso, M.F., Paiga, P., Vaz, M.C.V.F. and Delerue-Matos, C. (2005) *Anal. Bioanal. Chem.*, **383**, 880–885.
97. Barroso, M.F., Vaz, M.C.V.F., Sales, M.G.F. *et al.* (2006) *Anal. Lett.*, **39**, 2387–2403.
98. Jeanty, G., Wojciechowska, A., Marty, J.L. and Trojanowicz, M. (2002) *Anal. Bioanal. Chem.*, **373**, 691–695.
99. Shitanda, I., Takamatsu, S., Watanabe, K. and Itagaki, M. (2009) *Electrochim. Acta*, **54**, 4933–4936.
100. Trojanowicz, M. (2002) *Electroanalysis*, **14**, 1311–1328.
101. Llorent-Martinez, E.J., Ortega-Barrales, P., Fernandez-de Cordova, M.L. and Ruiz-Medina, A. (2011) *Anal. Chim. Acta*, **684**, 30–39.
102. Wang, J., Chen, L.A., Angnes, L. and Tian, B.M. (1992) *Anal. Chim. Acta*, **267**, 171–177.
103. Brett, C.M.A., Brett, A.M.O. and Mitoseriu, L.C. (1995) *Electroanalysis*, **7**, 225–229.

104. Brett, C.M.A., Brett, A.M.O. and Tugulea, L. (1996) *Anal. Chim. Acta*, **322**, 151–157.
105. Fungaro, D.A. and Brett, C.M.A. (1999) *Anal. Chim. Acta*, **385**, 257–264.
106. Simões, F.R., Vaz, C.M.P. and Brett, C.M.A. (2007) *Anal. Lett.*, **40**, 1800–1810.
107. Trojanowicz, M., Kozminski, P., Dias, H. and Brett, C.M.A. (2005) *Talanta*, **68**, 394–400.
108. Economou, A. and Voulgaropoulos, A. (2010) *Electroanalysis*, **22**, 1468–1475.
109. Adraoui, I., Rhazi, M.E. and Amine, A. (2007) *Anal. Lett.*, **40**, 349–368.
110. Masini, J.C. (2008) *Quim. Nova*, **31**, 704–708.
111. Cerda, V., Cladera, A., Oms, M.T. *et al.* (2001) *Trends Anal. Chem.*, **20**, 407–418.
112. Perez-Olmos, R., Soto, J.C., Zarate, N. *et al.* (2005) *Food Chem.*, **90**, 471–490.
113. Gutes, A., Ibanez, A.B., del Valle, M. and Cespedes, F. (2006) *Electroanalysis*, **18**, 82–88.
114. Chinvongamorn, C., Pinwattana, K., Praphairaksit, N. *et al.* (2008) *Sensors*, **8**, 1846–1857.
115. da Silva, C.L. and Masini, J.C. (2000) *Fresenius J. Anal. Chem.*, **367**, 284–290.
116. Suteerapataranon, S., Jakmunee, J., Vaneesorn, Y. and Grudpan, K. (2002) *Talanta*, **58**, 1235–1242.
117. Guzsvany, V., Nakajima, H., Soh, N. *et al.* (2011) *Electroanalysis*, **23**, 1593–1601.
118. Injang, U., Noyrod, P., Siangproh, W. *et al.* (2010) *Anal. Chim. Acta*, **668**, 54–60.
119. dos Santos, L.B.O., Silva, M.S.P. and Masini, J.C. (2005) *Anal. Chim. Acta*, **528**, 21–27.
120. dos Santos, L.B.O. and Masini, J.C. (2007) *Talanta*, **72**, 1023–1029.
121. dos Santos, L.B.O. and Masini, J.C. (2008) *Anal. Chim. Acta*, **606**, 209–216.
122. dos Santos, L.B.O., Infante, C.M.C. and Masini, J.C. (2010) *Anal. Bioanal. Chem.*, **396**, 1897–1903.
123. Economou, A. and Voulgaropoulos, A. (2007) *Talanta*, **71**, 758–765.

4

HPLC Techniques with Electrochemical Detection

Manuel Chicharro Santamaría, Mónica Moreno Barambio and Alberto Sánchez Arribas
Departamento de Química Analítica y Análisis Instrumental, Universidad Autónoma de Madrid, C/Francisco Tomás y Valiente, 7, 28049, Madrid, Spain

4.1 Introduction

The present topic of the discipline of food analysis cannot be understood without the implementation of separation techniques, both chromatographic and electrophoretic, during the twentieth century, especially the analytical ones [1]. The development of chromatography and its related instrumentation within this period has provided the broadest impact and largest application in food and agricultural analysis, including the safety, security, quality control, and basic research fields. Therefore, high-performance liquid chromatography (HPLC), GC, ion chromatography, thin-layer chromatography, flash chromatography, or capillary electrophoresis (CE) techniques are used nowadays for routine food analysis in academia, governments, or industry laboratories. For example, in 2010, the separation market accounted for about one-tenth of the total analytical instrumentation demand in the food industry [2].

In this scene, HPLC is a paramount technique in the modern age of food analysis with hundreds of applications [3], some of which include the detection of aflatoxins, amino acid analysis, vitamins separation, profiling of various food components, analysis of colorants and their residues, or the determination of the sugars content.

The detector is a fundamental component in the chromatographic system. These devices allow the visualization of the analytes once separated in the chromatographic column; thus, both the sensitivity and the efficiency that can be attained with the system are dependent

Agricultural and Food Electroanalysis, First Edition.
Edited by Alberto Escarpa, María Cristina González and Miguel Ángel López.

on detecting techniques. There are many characteristics to be considered when choosing a detector [4] and some of them are listed as follows.

- high sensitivity and reproducible, predictable response;
- respond to all solutes or have predictable specificity;
- wide linear dynamic range; response that increases linearly with the amount of solute;
- response unaffected by changes in temperature and mobile phase flow;
- response independent of the mobile phase composition;
- not contributing to extra-column band broadening;
- reliable and convenient to use;
- nondestructive of the solute;
- provide qualitative and quantitative information on the detected peak;
- fast response.

Since no detector has all these characteristics, over the time a large number of detectors have been designed, produced, and sold to answer one particular challenge or another. Ease of use, predictability, and reproducibility are all very important characteristics; however, recently there is an increased emphasis on the flow cell contribution to band broadening and faster detector responses. This emphasis is due to new, low dispersion ultrahigh-performance liquid chromatography (UHPLC) systems designed to take full advantage of sub-two micrometers particle size column packings [4].

The detection in HPLC systems devoted to food analysis is usually carried out using spectrophotometric UV–visible (UV–vis) or mass spectrometry (MS) detectors. UV–vis detector is of widespread use because it is easy-to-use, relatively cheap (a diode array detector, DAD, is significantly more expensive than a fixed wavelength one) can be tuned from universal (at 200–210 nm wavelengths) to more selective detection (at longer wavelengths), DAD extends the utility of UV–vis detection by supplying the spectra of eluting peaks (used to help in peak identification, especially during method development, and to evaluate peak homogeneity or purity) and can also serve as a multiwavelength detector. The main drawbacks of UV–vis in HPLC when compared to other detection techniques are the low sensitivity and lack of selectivity for some applications. MS detector is currently expanding its use because cost-accessible equipment has been released in the last years. These detectors are powerful tools for the identification, structural confirmation, and quantification of the different compounds found in the very complex matrixes handled in food analysis, which supported most of the progress observed in this discipline in the last decades. The more important disadvantages and limitations of MS detectors are the high costs of acquisition and maintenance of these equipment, their difficult usage, and the detrimental impact on important method parameters (limit of detection (LOD), limit of quantification, linearity, accuracy, and precision) that matrix components exert on the ionization pathways of target analytes [5].

Other important detector types are fluorescence detector (FL), light scattering detector (LS), and the refractive index detector (RI). The FL detection offers the higher sensitivity and the lower detection limits. Nevertheless, these properties are frequently counteracted by the limited number of compounds with native FL; therefore, additional derivatization reactions of analytes are usually needed, and the cost of acquisition and maintenance. In addition, when FL is used in HPLC systems, the linear dynamic range is often small for many analytes (even when the dynamic range is relatively large) and care should be taken to properly select the mobile phase components as some buffers or solvents can cause background FL or quenching (if not degassed), both affecting sensitivity. The RI detector is the original and the oldest LC detector, and it has been used successfully for the detection of many important analytes, such as sugars, triglycerides, and organic acids, since it measures a universal bulk property. Nowadays, RI detection is being replaced in many applications by LS or corona discharge detectors (as known as Charged Aerosol Detectors®) due to its limited sensitivity and gradient incompatibility. The LS detectors have found increased utility in HPLC because of the recent improvements in the ability to nebulize efficiently the column effluent, being the most popular the evaporative light scattering (ELS) one. The latter has some advantages like the independence of its response on the solvent, so that it can be used with gradients, temperature, or flow rate fluctuations. Mobile phases must be volatile, similar to those used for MS detection and linearity can be limited in some applications.

There is always a need for more sensitive and selective analytical applications, also cheaper and simpler, which can solve the day-to-day problems in food-related activities (and, in general, in chemical analysis) and this is the frame where electrochemical detection (ED) coupled to HPLC can supply important solutions for some relevant applications in food analysis. The instrumentation required for ED is quite simple and cheap when compared to other HPLC detectors and very sensitive responses can be obtained for electroactive analytes whereas selectivity can be tuned by proper election of the instrumental parameters. This chapter deals with the design and integration of electrochemical detectors, namely the amperometric, coulometric, and voltammetric ones, within the HPLC separation system and their compatibility and compromise with the chromatographic conditions necessary to achieve optimum resolution of the analytes in food and agricultural analysis. Conductometric detection, which is widely used in ion-exchange chromatography for detecting ionic species with low molar absorptivity, and potentiometric detection, with notably limited applications in HPLC, will not be covered and the reader is referred to valuable reviews in that fields for more information [6–9].

4.2 Fundamentals

4.2.1 Electrochemical Cell

When a detector is designed to be employed for measuring test substances in a flowing separation system, there are some analytical criteria and separation requirements that must be studied, optimized, and balanced. In the case of electrochemical detectors, some aspects for an appropriate electrochemical functioning should be observed in addition. The way how these properties are met or compromised, both in the detector design and in its coupling to the separation system, will characterize its performance and suitability in HPLC separations. This section describes some of these issues.

Basically, an electrochemical detector for HPLC is a flow electrochemical cell which contains all of its basic elements (working, reference and auxiliary electrodes, holder, and connections) adapted to the experimental conditions expected in the chromatographic separations. The characteristics of the materials used for the holder and the electrodes construction must be compatible with typical mobile phases, working pressures, and temperatures employed in HPLC whereas the geometric configuration must provide an easy coupling to HPLC tubing, low cell volumes and user-friendly handling, and easy replacement of its elements.

Among others, high sensitivity, low noise, low LOD, wide dynamic range, and selective responses are desired analytical properties that usually can be obtained with a proper selection of the electrode materials (especially the working electrode). In chromatographic analysis, long-term signal stability and a rapid response to concentration changes in time are mandatory; thus, some of the previously mentioned properties can be sacrificed. Another critical issue for the electrochemical detector is the minimization of the extra-column band broadening using low cell volumes.

4.2.1.1 Electrode Materials

In general, the electrode materials chosen for the construction of a working electrode are expected to provide good electrical conductivity, low potential for the required electrochemical measurement, and robustness to be mechanically formed into suitable shapes. The particular experimental conditions for HPLC separations impose some especial characteristics to working electrodes as can be the ability to be polished to a very smooth surface (especially when used in thin-layer cells) and withstanding a constant flow or jet of eluent. Additionally they must be inert toward the eluent and/or its impurities, other sample components, and any reactive species that might be formed by electrolysis of the analyte(s). In the event of electrode fouling, the working electrode assembly should be readily disassembled and the electrode easily cleaned or readily and cheaply replaced. Therefore, simple arrangements are preferred over more sophisticated electrode designs to fulfill these requirements and typically solid working electrodes made of carbon-based materials and noble metals are currently used.

The most popular material for solid electrodes is carbon because it shows the least electrode deactivation in routine use [9]. Among other allotropes of carbon, glassy carbon is widely used in HPLC because it is relatively pure, mechanically strong and inert to organic solvents, has good electrical properties, and can be readily cleaned mechanically by polishing with diamond and/or alumina pastes. Glassy carbon electrode (GCE) is usually employed for monitoring electrochemical oxidations at positive potentials but it also performs particularly well when operated at a negative potential. Other interesting carbon-based material is graphite, commonly used as part of the carbon paste electrode (CPE) which is made of graphite particles combined with a binder, such as Nujol. CPE offer excellent electrochemical performance with low background signals combined with an easy and inexpensive preparation. In addition, their electrochemical properties can be efficiently tuned and enhanced by introducing a modifier (a catalyst, an enzyme, a ligand, etc.) in the graphite-binder composite [10, 11]. Their main drawbacks are that they require considerable skill in preparation and are incompatible with many common HPLC solvents, such as acetonitrile, which dissolves the oil used to form the paste. However, under appropriate conditions and with careful use, CPE can give highly sensitive responses [12]. Boron-doped diamond (BDD) is a relatively new material, which has been incorporated to some commercial HPLC-ED equipment. This material has wide potential window in aqueous solutions and has excellent chemical inertness and enhanced stability [13]; therefore, it is being incorporated to new routine HPLC methods [14–17]. Other carbon-based nanomaterials such as carbon nanotubes [18] or graphene [19] are getting introduced as part of the working electrode (typically as a layer on top of a GCE), showing low detection overpotentials and higher resistance to passivation than the unmodified electrode [20, 21].

Noble metal electrodes satisfy many of the requirements for good electrodes, but they often show rapid deterioration in response and are thus only employed in specific analyses especially where they exhibit electro-catalytic properties. Platinum and gold electrodes, for example, are used extensively in the pulsed amperometric detection (PAD) of thiols and carbohydrates [22]. Copper electrodes have been used to measure amino acids specifically by exploiting the well-known complexation reaction between copper ions and amino acids [23]. Liquid mercury has always been a common electrode material, especially in the form of the dropping mercury electrode, used in HPLC-ED for monitoring electrochemical reduction because of its extended negative potential window. Besides the toxicity of this metal, the fact that the electrode surface is continually renewed complicates its use in flowing systems since a more robust design is required to overcome the problems of being in a pumped flowing stream of eluent.

The electrochemical characteristics of a solid electrode can be chemically, electrochemically, as well as physically modified in order to increase its stability, sensitivity, or most commonly selectivity. An important example of selectivity modification is the formation of a mercury amalgam on gold electrodes to specifically detect thiols. Other examples are the adsorption of metals, polymers, self-assembled monolayers (SAMs), or enzymes but their applications to HPLC are yet scarce.

Regarding the reference electrode, the conventional Ag/AgCl/KCl electrode is often adapted to the presence of organic solvents and the different conductivity and ionic mobility in the typical HPLC mobile phases by using a porous ceramic frit for the electric contact or by replacing the KCl internal solution by a low-maintenance gel. Solid-state reference electrodes made of Ag or Pd has also been included in commercial equipment. Finally, auxiliary electrodes are currently made of metal and they are often little more than the stainless steel outlet from the cell.

4.2.1.2 Flow-Cell Designs

Of the various flow-cell designs (Figure 4.1), the thin-layer and wall-jet configurations are commonly used in many of the HPLC-ED applications found in the literature. Their simplicity combined with their good electroanalytical performance (high sensitivity and low LODs) has led to widespread use in homemade laboratory equipment as well as they are commercially available from different manufacturers (e.g., Bioanalytical Systems Inc, BAS; CH Instruments, CHI; Antec; Eicom; Waters; Metrohm; GBC Pty Scientific Equipment). Porous flow-through designs are usually more difficult to build but they are often employed because of their commercial availability (Thermo Scientific, formerly ESA/Dionex). Tubular electrodes have not been commercially exploited and the single fiber and microdisk electrodes are chiefly confined to be used with miniaturized systems and CE. However, the trend in the development of new electrochemical cells appears to be toward including working electrodes with smaller surface areas to make the detectors compatible with microbore HPLC systems and to reduce noise, which is proportional to the surface area, making the detector less susceptible to flow-rate fluctuations.

In this section, only thin-layer, wall-jet, and porous flow-through cell designs will be described. More detailed information about cell designs and their configurations can be

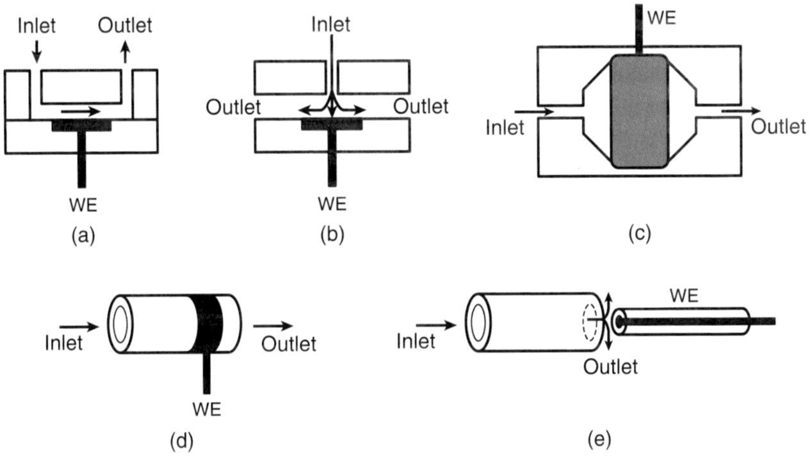

Figure 4.1 *Flow-cell schemes based on those shown in Refs. [24–26]. (a) Thin-layer, (b) wall–jet, (c) porous flow through, (d) tubular, and (e) microelectrode in end-column configuration.*

found both in the classical Štulik and Pacakova treatise [24], in the HPLC-ED book from Flanagan *et al.* [25], in Trojanowicz reviewing works [6, 9], as well as in the manufacturers' catalogs and websites.

4.2.1.2.1 Thin-Layer Cells In the thin-layer cell, which was used in the first commercial HPLC-ED detector, the eluate enters the detector cell *via* a suitable microport and then flows through a narrow channel cut into a plastic gasket over a "solid" working electrode. The flow continues past the reference and counter electrodes. The body of the cell is usually formed from two solid blocks, one of which contains a force fitted disc or short rod of electrode material. The blocks are separated using a thin gasket, typically made of Teflon® (polytetrafluoroethylene; PTFE), as a spacer around the working electrode as well as the inlet and the outlet orifices of the cell. The block containing the working electrode is usually constructed from an inert material, such as PTFE or Kel-F® (polychlorotrifluoroethylene). Since working electrodes are usually small, it is possible to incorporate more than one into this block [26]. This leads to new detection strategies, as explained below, or allows the rapid connection of a second electrode when the first becomes contaminated without having to dismantle the cell. The second cell-block may be made of stainless steel and forms the counter electrode. Connection of these electrodes to the electronics of the system is usually *via* simple gold pins and suitable electrical connectors. This simple classic design was pioneered by Kissinger [27] and has formed the basis of the majority of cells in commercial instruments, including those supplied by BAS, CHI, Waters, and Agilent (the last currently discontinued).

Typical electrode parameters are an area of 15–20 mm^2 with gaskets of 20–100 μm giving a cell volume of less than 1 μl and capable of working with flow rates of about 1 ml min^{-1}. Advantages of thin-layer cells are low cost of construction (only a means of clamping the two cell-blocks firmly together is required), ease of disassembly for maintenance (particularly polishing the working electrode), the variety of possible configurations (see below), the possible use of a variety of materials for the electrodes and cell body, the ability to be used in series with other detectors, rapid response times, and easily adjustable (low) cell volumes giving minimal extra-column dispersion. Disadvantages include small electrode area, difficulty in achieving very small distances between the electrode blocks and the inability to withstand a back pressure when used in series with other detectors.

4.2.1.2.2 Wall-Jet Cells In these types of cell, the eluate is delivered, *via* a small orifice, into the cell and directly impinges as a jet at right-angles onto the face (wall) of the working electrode before flowing radially across the surface. In these types of cells, it must be assumed that the jet does not break up before it hits the electrode. Besides, it is important not to restrict the radial flow from the zone where the jet impinges on the electrode because it is intended that the flow of electrolyzed eluate should sweep away over the electrode surface. In contrast to thin-layer designs where the jet should be confined to a thin channel, the cell can be unsealed and performance should not be degraded by trapped air bubbles. However, it is unsuitable for coupling to a second detector and should therefore be last in line if serial detection is employed.

A wall-jet cell was applied firstly to flow injection analysis (FIA) in 1973 by Yamada and Matsuda [28] and very shortly afterward applied to HPLC by Fleet and Little [29]. Currently, there are several commercial instruments that use wall-jet cells (e.g., Antec; BAS; GBC Pty; Metrohm) for HPLC. These cells are usually made of PTFE and consist of two blocks, one for the inlet and the other holds the working electrode. The space between the inlet and the working electrode, where the jet should develop, is delimited by a cavity in the inlet block or by a spacer placed between the cell blocks. Auxiliary and reference electrodes can be fitted either to one block indistinctly.

4.2.1.2.3 Porous Flow-Through Cells These types of cells are designed to maximize the area-to-volume ratio using a porous working electrode with a very large active surface. This can be achieved in various ways, for example, use of stacks of metal gauze, sintered metal tubes, reticulated vitreous carbon, or packed beds along the lines of HPLC columns composed of conducting material, such as carbon particles or silver granules. Typically, the material of choice is porous graphitic carbon, which has a very high surface area, is mechanically robust and, and more importantly it is permeable to the mobile phase. The large surface area of the porous electrode ensures that 100% of the eluted material is reacted [30]. Thus, the electrochemical reaction is no longer amperometric, but now coulometric (see Section 4.2.1.3.2).

Commercially available cells with rate constant of 500 s^{-1} and a cell volume of about 5 μl assure coulometric efficiency for typical HPLC flow rates with minimal extra-column band broadening. Each electrochemical unit has a central porous carbon electrode, on either side of which is situated a reference electrode and an auxiliary electrode. The characteristics of porous graphitic carbon electrode facilitate the construction of electrode arrays. Typical commercial systems include two units placed in series but arrays of up to 16 units are commercially available (Thermo Scientific, formerly ESA/Dionex). These cells have some degree of resistance to flow and with use can develop a significant back pressure. To minimize such back pressure changes, they need to be protected from particulate materials. Their intrinsic back pressure should also be borne in mind when connecting other types of HPLC detector cell in series.

4.2.1.3 Operation Modes

4.2.1.3.1 Amperometric In this mode the detection is carried out by applying a fixed detection potential, which allows the electrochemical oxidation or reduction of the analytes, and the recording of the corresponding current evolution versus time is the resulting chromatogram. Typical thin-layer and wall-jet cells employed in HPLC commonly work in amperometric regime. Usually it has been reported that using cells working in the amperometric detection mode provide the highest signal-to-noise ratios (S/N) and rapid response time whereas their main disadvantage is that the amperometric response can be affected by the progressive accumulation of reaction products and/or other contaminants onto the working electrode surface, therefore decreasing with time.

This mode is characterized by an incomplete electrochemical conversion of the electroactive analytes that reach the working electrode. In fact, a coulometric efficiency of only 5–10% is usually attained. Consequently, during the residence time of an analyte band flowing through the detector, there is always unreacted analyte in the vicinity of the working electrode. Therefore, the amperometric current measured under these conditions is controlled by the analyte diffusion from the bulk toward the electrode surface which is characterized by a limiting current defined as

$$i_{\mathrm{lim}} = nFADC/d$$

where A is the electrode surface area (cm^2), D is the diffusion coefficient ($\mathrm{cm}^2\,\mathrm{s}^{-1}$), d is the diffusion layer thickness (cm), and C is the concentration of analyte ($\mathrm{mol\,cm}^{-3}$).

The diffusion layer thickness depends on the flow rate and the cell geometry, which leads to a characteristic dependence of limiting current on the flow rate. Table 4.1 summarizes the limiting current equations calculated for popular electrochemical cell configurations in terms of their hydrodynamics where it can be seen that the response (current) at an amperometric electrode is proportional to the molar concentration of analyte in the eluate.

From these equations, it is clear that for maximum current (thus sensitivity) these cells should maximize the working electrode area and operate at high flow rates. In practice, increasing the surface area and the flow rates will also raise the background current, while in many cases flow rates are limited by the chromatographic system. In the case of thin-layer cells, using the thinnest possible gaskets is advantageous but the imperfections in the planarity of the working electrode blocks limit the dimensions of the gaskets that can be

Table 4.1 *Limiting current equations in a standard format for popular cell designs [31]*

Type of cell	Limiting current
Thin-layer (rectangular channel)	$1.47\ nFC\ D^{2/3}U^{1/3}(A/d)^{2/3}$
Planar (parallel flow)	$0.68\ nFC\ D^{2/3}U^{1/2}(A/d)^{1/2}\nu^{-1/6}$
Planar (perpendicular flow)	$0.903\ nFC\ D^{2/3}U^{1/2}A^{3/4}\nu^{-1/6}$
Wall-jet	$0.898\ nFC\ D^{2/3}U^{3/4}A^{3/8}\nu^{-5/12}a^{-1/2}$
Tubular	$1.61\ nFC\ D^{2/3}U^{1/3}(A/r)^{2/3}$

A = electrode area (cm^2), U = mean volume flow-rate ($\mathrm{cm}^3\,\mathrm{s}^{-1}$), C = analyte concentration ($\mathrm{mol\,l}^{-1}$), D = diffusion coefficient ($\mathrm{cm}^2\,\mathrm{s}^{-1}$), F = Faraday constant, n = number of electrons, a = inlet diameter (cm), d = channel depth (or gasket thickness) (cm), r = radius of channel (cm), and ν = kinematic viscosity ($\mathrm{cm}^2\,\mathrm{s}^{-1}$).

employed. It should be noted that cells with thin-layer geometric configurations behave as a theoretical planar cell with parallel flow when the channel geometry and/or dimensions do not allow a fully development of a laminar flow inside them. Similarly wall-jet type cells show a pure behavior when the working electrode diameter is at least 10-fold greater than the jet diameter and at high flow velocities. Otherwise, their response obeys the relationship observed for theoretical planar cell with perpendicular flow [31].

4.2.1.3.2 Coulometric The electrochemical response under the coulometric detection mode is obtained in the same way as shown previously for the amperometric mode (current measured under a fixed detection potential) with a fundamental difference: the complete electrochemical conversion of the electroactive analytes that reach the working electrode. The key to obtain a full electrolysis of analytes is the use of cells with high working electrode area-to-volume ratio and high reaction rate constant or cell constant. Therefore, this mode is the characteristic of porous flow-through cells which include porous working electrodes of the large surface area. Under these conditions, the steady-state current i is dependent on the rate at which analyte is delivered to the electrode and is given by

$$i = nFCU$$

Thus, increasing the flow-rate U will increase the peak height but the peak area remains constant over a wide range of flow rates. Coulometric detection is therefore an absolute method, such that peak area can provide a means of quantification by relating area directly to sample mass using Faraday's law, when the number of electrons being transferred is known.

When using the coulometric mode, the analytical signals are usually higher than those obtained with the amperometric one. Nevertheless, it does not mean directly an improvement of LOD of such detection, as it is often accompanied by an increase of the background current and the instrumental noise amplitude, which is associated with the increase of the surface of the working electrode [9]. It is also expected that high coulometric efficiency may lead to rapid electrode poisoning and signal decreasing. A careful design of the cell and selection of the operating conditions can minimize these effects and it has been estimated that up to 96% of the electrode surface can be fouled whilst only reducing the efficiency of electrolysis by 0.5% in commercially available cells [25].

4.2.1.3.3 Pulsed Detection Pulsing the applied potential at the working electrode using simple waveforms has long been employed in the electrochemical analysis and it has proved to be very useful for several types of analytes when using metallic electrodes [22]. This scheme was also proposed in some of the earliest patents for electrochemical detectors for liquid chromatography and finally became commercially available in the mid-1980s with the introduction of electronically controlled PAD by Dionex, which was developed for the measurement of carbohydrates at gold electrodes in very alkaline eluents. Modern electronics are capable of reproducibly generating a variety of waveforms [32].

The main idea of application of variety of potential-time waveforms for polarization of working electrodes is employing a decrease of the activation barrier for oxidation of determined analytes. The electrocatalytic electrode process is frequently based on stabilization of reaction intermediates *via* the adsorption at clean noble electrode surface. Typical pulse sequences are shown in Figure 4.2.

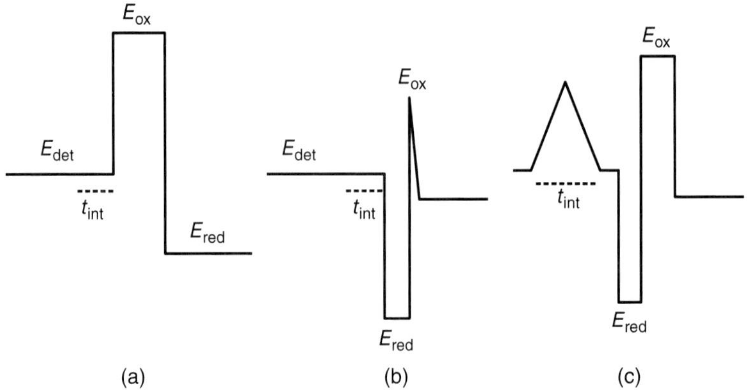

Figure 4.2 *Usual potential waveforms employed in PAD [22, 32]. (a) Triple-pulse PAD, (b) quadruple pulse PAD, and (c) triangular scan integrated pulsed amperometric detection (IPAD). E_{det}, detection potential; E_{ox}, oxidation potential for electrode conditioning; E_{red}, reduction potential for electrode conditioning; t_{int}, time for signal acquisition.*

In conventional triple-pulse PAD, the initial detection potential is held for a few hundred milliseconds, which is followed by a short period in which the response is sampled to give the measured output. This delay in the response recording minimizes the contribution of the capacitive component. During this detection stage, the noble metal electrode is fouled when being used to detect aliphatic compounds, such as sugars. The next step of the waveform is a short large oxidation pulse which generates an oxide layer on the electrode plus complete oxidation of the analyte residues to water and CO_2. The next step reduces the electrode surface back to the clean noble metal and finally the detection potential is re-established. Alternatively, quadruple-pulse PAD restores the fouled metal surface by applying a large reduction potential followed by a short oxidative pulse that activates the surface. Then a potential lower than that applied for detection is set to promote de adsorption of analytes before detection potential is re-established.

The most important advantage of PAD compared to the direct measurement of the amperometric current is that pulsed potential cleaning eliminates the necessity of frequently mechanical cleaning of the working electrodes and its response will be reproducible so allowing quantitative performance. Subsequently pulsing profiles have been employed to maintain electrode performance for some other difficult analytes, such as thiols, and inorganic ions, such as sulfite in foods [22].

An improvement of LOD in PAD can be achieved by the integration of the current at appropriate period of time [22, 32], and such integrated pulsed amperometric detection (IPAD) has found numerous applications. IPAD is similar to PAD, the difference is that in IPAD the potential during current sampling is not held constant but is scanned between a high and low value. The advantage of IPAD compared with PAD relates to maximization of the analyte signal and minimization of baseline magnitude and drift for oxide-catalyzed detections. The IPAD waveform has been highly successful for the determination of amine- and sulfur-containing compounds whereas for carbohydrates no significant difference in S/N was shown when compared to PAD.

4.2.1.3.4 Multiple Detection Systems

4.2.1.3.4.1 Dual Electrode In recent years, thin-layer cells with built-in dual electrodes (typically made of glassy carbon) arranged either in parallel or in series have been developed and used for many applications [26]. In this instance, several modes of operation become possible [33] for coulometric operation as discussed below. In the first approach, the eluate simultaneously passes over two electrodes, which can be held at different working potentials. The benefits of this approach are that the ratio of currents at each electrode can provide peak purity and identification information. Moreover, oxidation and/or reduction reactions can be carried out simultaneously and can be monitored at different sensitivities. Eventually a difference signal can be used to remove a common interference. In the dual series configuration, the eluate first passes over one electrode held at one potential, and then over a second electrode possibly held at a different potential. In this configuration, the selectivity can be enhanced by partial oxidation or reduction at the first electrode and the unreacted substances or the products generated can be detected in the second electrode. Other benefits similar to that obtained using the parallel configuration are the peak purity and identification information that can be acquired from the ratio of currents at each electrode and the possibility of subtracting the signal of one of the electrodes to remove a common interference or correct the baseline.

Similarly, two coulometric cell units can be placed in series and this arrangement has been employed widely. All recent applications have been carried out using the Coulochem® cell models supplied by ESA/Dionex (currently within the Thermo Scientific group) which include dual porous graphitic carbon electrode either both of them working under coulometric regime or the first coulometric and the last amperometric. Coulochem® cells can be operated in three modes, often referred to as "screen," "difference," and "redox." Their operation is similar to that described earlier for thin-layer cells but since full electrochemical conversion occurs for analytes detected in the first cell unit, some improvements in the final readout can be obtained [25]. In the *screen mode* the potential of the first electrode (E1) is set, typically, 0.2–0.3 V below than that of the second electrode (E2). The aim is to reduce the concentration of any unwanted readily oxidizable/reducible components reaching the second electrode. This in turn should reduce the standing current and thus improve the S/N obtained from the second electrode. With serial dual electrode in the amperometric detection, only a small proportion of the compounds would have been electrolized at the first electrode.

A further way of reducing interference is to use *difference mode*. The Coulochem® has three outputs: those for the two electrodes, El and E2, and the difference (El – E2). If the electrodes are set to similar potentials, such that the analyte of interest is oxidized at the first electrode with little or no signal at the second, then the difference will contain a signal due to the analyte. If impurities produce similar responses at each electrode then their contribution to the signal in the difference channel will be small. The difference mode is especially useful for quantifying peaks on the solvent front or on a sloping base line.

The third mode of operation, *redox mode,* can be used if a product of oxidation or reduction is stable and it is possible to reduce or oxidize the reaction product at the second electrode. Depending on how other species respond at El and E2, either output may be used for quantification. As the reduction signal is negative with respect to the oxidation signal, the difference, El–E2, in effect sums the two to produce a larger response. It is usually possible to use one or all three channels for quantification. In general, experience suggests that the relative standard deviations (RSDs) for replicate analyses are lowest when the oxidation signal is used. However, the additional information provided by the reduction channel gives some indication of peak purity. Occasionally the oxidation signal may be lost beneath an impurity, but the reduction channel is unaffected and can thus be used in the analysis.

4.2.1.3.4.2 Array Detectors This arrangement consists of *n* working electrodes (amperometric or coulometric) disposed in series into one housing. Each electrode is held at incremental working voltages and a series of chromatograms were obtained simultaneously and displayed by a computer in a 3D format together with time and working potential. The overall concept of the electrode array system is similar to that of the DAD, in order to acquire more detailed information for peak purity assignment and identification. The eluting analytes flow through the array until each one reaches the electrode that has the required potential to permit its oxidation or reduction. The response will depend on coulometric efficiency of the working electrodes. In amperometric mode only a small part of the solute is reacted and so the remainder can proceed to the next electrode system and be detected again and again. In this way the electrodes set at low overpotentials will detect easily oxidizable or reducible analytes (more selective) whereas the electrodes at high overpotentials will detect all the analytes. The evolution of the electrochemical signal with the electrode potential for each peak has a sigmoidal shape characteristic for each analyte and the ratio of peak currents at different potentials can be used for its identification [34]. In this case, it is usually more difficult to assign the identity and to resolve the contribution of co-eluting substances in each peak. At difference, in coulometric mode each analyte is completely reacted and detected in that electrode that has the required potential and thus it will not be sensed by other units [30, 35]. Chromatographic peaks therefore will be produced only at those electrodes with the appropriate potentials for each analyte whereas the evolution of the electrochemical signal with the electrode potential produces a peak at a potential that is characteristic of the analyte being detected. This leads to an apparent improvement in the chromatographic resolution in a similar way to that of the DAD. Two peaks that have not been chromatographically resolved and are eluted together can still be shown as two peaks that are resolved electrochemically and can be quantitatively estimated. Another advantage is that high oxidation potentials can be used without the high background currents and noise that usually accompany such operating conditions. The electrodes that are operating at high voltages are "buffered" by the previous electrodes operating at lower voltages, which results in reduced background currents and noise. Similarly to the amperometric arrays, the ratio of chromatographic peak currents at different potentials produces a characteristic pattern for each analyte and can be used to confirm its identity. The coulometric electrode array design is commercially available from ESA/Dionex (currently within the Thermo Scientific group). The so called CoulArray® system contains 8–16 electrodes in series, each independently under full computer control. This package has proved very effective controlling the baseline drifting in HPLC-ED under gradient elution [36].

4.2.1.3.4.3 Rapid Scanning Detectors Rapid scanning voltammetry forms the basis of a type of electrochemical detector in which the working potential at a single amperometric electrode is varied with time and the resulting current is measured as a function of analysis time and potential [37, 38]. Effectively it is a system for continuously obtaining voltammograms. Such systems were originally developed for dropping mercury electrodes in static polarography systems. To gain useful information from such a system, when it is being used to monitor a chromatographic eluate requires that the potential range to be scanned is approximately 1 V with a scan rate of about 1 V s^{-1}. A number of waveforms can be used, that is, square-wave, normal pulse, and staircase. The staircase form is considered the most suitable for use with solid electrodes because the equal voltage steps lead to more consistent background currents, which can then be subtracted from the signal. The resulting chromato-voltammogram is the ED equivalent of the rapid scanning optical detector. The most obvious advantage of rapid scanning detection is that a voltammogram for each chromatographic peak is obtained which is used to confirming peak purity and helping with identification. Unfortunately, this mode has had little application in food analysis.

4.2.1.3.4.4 Series with Other Detectors The most common dual arrangement with ED is in series with a UV detector. The order in which the detectors are placed will be dictated by the requirements of the assay and by the design of the detectors. Theoretically, the detector with the smallest cell volume, that is, lowest dispersion, should be placed first. The working volume of electrochemical detectors is usually very small, but the dead volume of many electrochemical flow-cells after the working electrode is often high because of the presence of large, poorly swept volumes around the reference electrode. Secondly, unlike many electrochemical cells, most UV flow-cells can withstand modest back pressures, that is, up to 400–500 psi, and similar back pressures are often inadvertently created when coupling flow cells. Even so, as in all coupled systems care must be taken to keep pressures low to avoid leaks or cracked flow-cell windows. This is even more important if a FL detector is employed since their flow cells can often only withstand a few bar before leaking.

Special care should be taken when coupling detectors with porous flow-through cells (or electrochemical guard cells). Contamination of analytical porous graphite electrodes can lead to rapid increases in resistance and back pressure. Consequently, it is recommended that coulometric cells are first in any series of cells. Besides, the ED in the coulometric mode produces new chemical species that may have a marked effect on the spectral characteristics of the parent compound(s) and thus of the response. This may be exploited to produce species with improved spectral characteristics, or it may lead to a loss of signal at subsequent detectors. The stoichiometric conversion of analytes at coulometric cells under appropriate conditions can also be exploited for nonelectrochemical detection of the electrolyzed analytes, especially when the standing current background noise is too high to permit direct ED.

4.2.2 Development of HPLC-ED Methods

Optimization of an HPLC system for a specific analysis can be a complex task and many additional factors must be considered when using ED [25]. These can be divided into system dependent variables and analyte dependent variables. System-dependent variables include the nature of the working, reference and auxiliary electrodes, and associated electronics, and the applied potential. In addition, there may be influences from the eluent (solvent composition, pH, ionic strength), temperature, and flow rate. Analyte-dependent variables are specifically the presence of one or more electroactive moieties or functional groups on the compound(s) of interest. With compounds which are not inherently electroactive it may be possible to derivatize them, or to chemically modify them in some other way, to impart electrochemical properties.

4.2.2.1 *Getting Started*

As with all HPLC method development, it is important to start work with a new column and fresh eluent at an optimum flow rate. Frequently, the HPLC separation is initially optimized using a UV–vis or other detector that relies on a physical property of the analyte which is usually independent of the performance of the ED. Transferring the analysis to an electrochemical detector and then to real samples often brings problems. Firstly, the operational sensitivity may be completely different and it may be necessary to compare separations of standards performed using UV–vis with those done with ED. Although the selectivity of ED will probably be better than with UV–vis when analyzing real samples, many times the HPLC conditions have to be adjusted to assure that the electrochemical signals are not disturbed by system peaks (especially for those compounds eluting close to the void sample solvent front) and/or the presence of other electroactive compounds in the sample.

4.2.2.2 *Hydrodynamic Voltammograms*

One of the first tasks is to ascertain whether the compound(s) of interest are electroactive and to define the optimum working potentials. Some guidance may be found in the literature, but it is important to evaluate the effect of the working potential upon the analyte(s) signal using the HPLC system chosen for the analysis. The plot of the electrochemical response versus the applied potential in a flowing system is referred to as a hydrodynamic voltammogram (HDV) and it is highly system dependent. The shape and voltage relationships shown in these HDVs may vary with the eluent pH and composition, the electrode material, electrode contamination, the nature and condition of the reference electrode, the physical dimensions and positioning of electrodes, and the electrochemical process itself. Therefore, to avoid misleading results, working with clean electrodes, new reference electrodes (where relevant) and fresh eluent is highly recommended.

4.2.2.2.1 *Using the HPLC-ED* A HDV is readily obtained after successive injections of analyte(s) at a suitable concentration at different working potentials. The amount of analyte being injected should be such that the detector sensitivity is set in the mid-range. It is important that the HPLC–ED system is stable and reproducible before the HDV recording. When high negative or positive overpotentials are fixed, that is, $> \pm 0.8$ V, it takes long time to establish a stable electrode response. Therefore, it is recommended to allow the electrode to stabilize at the highest voltage useable with the appropriate eluent flowing through the system, possibly overnight. Once stabilized, repeated injections of a standard analyte solution are carried out and the peak responses (height or area) are measured using successive 0.1 V reductions in the applied overpotential each time. After each change it is necessary to allow time for the detector output to re-stabilize, but this stabilization time will be shorter for reduced voltages than if the voltage was successively increased. For trace analysis, it is also important to measure the baseline noise when the detector is operating at high sensitivity. At each voltage, the peak-to-peak baseline noise should be measured with the detector at a suitably high sensitivity level.

4.2.2.2.2 *Using Flow Injection Analysis* HDVs can be constructed faster using a FIA approach on the HPLC system, just by replacing the analytical column with a length of narrow bore stainless steel tubing, that is, about 2 m of 0.1 mm i.d. tubing. This creates sufficient back pressure to allow most HPLC pumps to continue to pump reproducibly whilst also broadening the injection bolus so that a reasonable response is obtained in about 30 s. It will be necessary to use a relatively low eluent flow rate (about 0.3 ml min^{-1}) to minimize baseline noise and the analytes should be dissolved in the mobile phase to avoid artifacts due to differences in the solvent conductivity that can give false data. This approach is better employed to rapidly explore conditions for use as a basis on which to build more detailed studies with the column in place as described earlier.

4.2.2.2.3 Operating Potentials Using either of the aforementioned procedures the optimum working potential for maximum sensitivity can be established from the HDV. It is usual to work on the first part of the plateau if maximum stability is required at moderate sensitivity with some selectivity. If maximum sensitivity is required the best working potential will be that giving the maximum S/N and this is often about 0.1 V below the plateau. If complex extracts are to be injected, it is possible that the number of electroactive species found in a sample increases with applied voltage, thus, increases the possibility of signal overlapping. In that case, obtaining HDV of the sample can help to observe the number and size of peaks at each working potential and to select a detection potential that provides an acceptable selectivity and sensitivity. Equally important is that higher working potentials can lead to the appearance of late eluting peaks from endogenous compounds or impurities under some conditions, which will increase analysis times and hence decrease throughput. Another important issue is that the risk of electrode deactivation, as a result of accumulation of reaction products and possibly other contaminants on the electrode surface, increases at high detection potentials. Therefore, the use of an applied potential as low as possible is recommended to avoid these problems as well as the background current and noise are lower and the stabilization time is shorter.

4.2.2.3 Mobile Phase Composition

ED imposes certain restrictions on the type of chromatography that is employed and the mobile phase that is used. The detecting system requires a conducting mobile phase and thus must contain water. Therefore, the majority of "normal phase" systems are not usable. Reversed-phase high performance liquid chromatography (RP-HPLC), however, is ideally suited to ED. Nevertheless, certain precautions must be taken for its effective use.

4.2.2.3.1 Purity Solvents and salt buffers developed for HPLC analysis are normally purified by the manufacturers for their use with UV–vis. These products are not necessarily ideal for working with an electrochemical detector because of their low purity or the presence of additives or stabilizers; the experience suggests that HPLC grade reagents may have high UV–vis purity, but not ED purity. Therefore, it is highly recommended to prepare mobile phases with organic modifiers, such as methanol, and other additives, such as ion-pairing agents and buffer salts, of the highest quality available and use distilled and/or de-ionized water of the highest possible quality as well. To test whether a high background signal arises from the eluent, the standing current at the working potential should be monitored with the flow switched off. If the current drops sharply, this indicates the presence of electroactive compounds. Periodic monitoring of the standing current provides a simple means of monitoring eluent contamination and/or electrode deactivation during normal use.

Taking into account this advice, it is necessary to observe some tips in the election of the components of the mobile phase and its afterward preparation

The presence of antioxidants in eluents and extraction solvents: Antioxidants can be readily oxidized electrochemically and generate high background currents or interfering broad peaks. Thus, eluents and extraction solvents containing such compounds should be either avoided or purified before use. For example, ethers, such as diethyl ether, di-isopropyl ether, and tetrahydrofuran are likely to contain up to 0.1% (w/v) pyrogallol or quinol (hydroquinone) as stabilizer. If the stabilizer is removed, peroxides will form and their concentration will increase with time unless the solvent is stored under nitrogen. Not only do peroxides present a hazard from explosion, but they may also oxidize susceptible analytes. Methyl *t*-butyl ether (MTBE), on the other hand, is stable to oxidation.

The nature of salts might alter the electrochemical response: It is important to remember that perchlorates are strong oxidizing agents and appropriate caution must be exercised in their use. Ammonium salts provide buffering capacity at higher pH values, but electrochemical oxidation of ammonia may become a limiting factor. Amine modifiers, such as dimethyloctylamine or triethylamine, added in relatively high concentration to improve the peak shapes of basic compounds on conventional reversed-phase HPLC columns, are relatively easily electrochemically oxidized and their use may preclude the utility of ED.

The effect of ionic strength: Eluents containing salt concentrations of 10 mmol l^{-1} or so are generally suitable for use with ED. At higher ionic strengths an increase in the standing current may reduce the linear range of coulometric detectors. For some applications, higher ionic strengths may be necessary with certain amperometric detectors to maintain stability and so the effect of ionic strength of the eluent should be considered when adapting an amperometric method for coulometric use.

Presence of metals: ED is sometimes particularly susceptible to increases in background noise caused by metal ions present in the eluent, the samples, or introduced from components of the HPLC system. Some analysts add EDTA to their eluents, typically at concentrations of 30–150 mg l^{-1} (0.1–0.5 mmol l^{-1}), with the aim of chelating iron and possibly other metal ions. EDTA may also enhance the stability of analytes, such as catecholamines, during sample storage and analysis.

Oxygen removal: For work at high negative potentials, eluents and samples should be oxygen free to avoid the interference of its electrochemical reduction. This can be achieved by sparging with helium or oxygen-free nitrogen, although there is the risk that too much sparging will alter the eluent solvent composition. Regardless of the deoxygenating process, the eluent must then be pumped *via* stainless steel tubing to the column and hence to the detector. This is because Teflon tubing is porous to air and the eluent can be reoxygenated during pumping through the HPLC system.

Recycling the eluent is often effective in removing reactive components: Once stable isocratic chromatography has been achieved, the eluent can be recycled from a vessel covered for minimizing solvent evaporation and dust collection. It is usual to recycle the eluent for some hours up to a few weeks and, provided that the solvent reservoir is sealed, the eluent composition only changes slowly with time. This enables the detector to remain switched on continuously in order to achieve maximal stability since the electrolyzed mobile phase contains lower amounts of dissolved air and electroactive substances. The column performance will also be more stable. Recycling is best for eluents containing 10% or more organic modifier since this inhibits bacterial growth. The background current and column performance should be monitored to decide when fresh eluent should be prepared.

4.2.2.3.2 Compatibility with the HPLC System One aspect of eluent compatibility with ED is that there should be no effect on the components of the detector. Detector cell bodies are now routinely constructed of PTFE, other fluoroplastics, glass, or stainless steel, and seem stable to most eluents. Nevertheless, electrodes are vulnerable to chemical attack. Problems with the longer term use of some eluents at potentials around +1 V vs. Ag/AgCl have been experienced. For example, ammonium acetate buffers have caused flaking of the surface of GCEs held at as little as +0.1 V for one batch of electrodes [25]. Noble metal electrodes are easily contaminated by a number of eluents unless the electrode is cleaned on-line by using PAD as in carbohydrate analysis. Conventional reference electrodes, like the Ag/AgCl one, contain electrolytes that can either diffuse into, or be contaminated by ions diffusing from, the mobile phase. Greater the difference between the concentration of salts in the eluate and that in the reference cell, faster will be the contamination of the reference electrode, thus, changing the reference potential and/or performing incorrectly. It is important that the cell design permits the reference electrolyte to be renewed regularly or, if that is not possible, to soak overnight the glass parts of the electrode in the reference electrolyte (typically saturated potassium chloride). For some eluents, such as those containing lithium salts (as used in some vitamin measurements), it is necessary that the reference electrolyte be changed to the lithium form prior to use. The use of perchlorates can lead to deposition of potassium perchlorate on contact with potassium chloride.

4.2.2.3.3 pH In HPLC-ED eluent pH often affects not only the analyte retention but also its electrochemical process; thus, its electrochemical signal. For example, aliphatic amines cannot be detected under strongly acidic conditions since only the fully protonated (nonoxidizable) species is present. In contrast, phenols can be oxidized at most pH values. Therefore, it is necessary to ascertain the effect of pH on the electrochemical response and it can be useful to observe it independently of the chromatographic system by employing the FIA approach shown previously.

It is recommended to fix the eluent pH using buffer salts at or near their pK_a for maximum system stability. The pH of the eluent should be measured and defined (in the method) for the aqueous component in the absence of the organic modifier, preferably without even ion-pairing agents, because the presence of organic solvents will change not only the measured pH, but also the apparent pH of the reference electrode in the pH meter. When, as is often the case, a "final eluent" pH is quoted, it should be remembered that it is not only an "apparent pH," but also it is pH meter dependent.

4.2.2.4 Temperature

Ideally all HPLC-ED components, that is, injector, eluent, column, and detector cell should be held at a constant temperature. In many systems the column, injector and cell are held in a temperature control unit. At a very minimum, the column and cell should be shielded from draughts and placed in the same compartment. There is no optimum temperature for ED and the best working temperature should be found by plotting changes in S/N for the analyte(s) of interest against varying temperatures.

4.2.2.5 *Flow Rate*

The response of an electrochemical detector, as was shown in Section 4.2.1.3, is very sensitive to pressure/flow fluctuations; therefore, it is very important that the HPLC pump provide a controllable, reproducible, and stable flow rate. Besides, the piston motion of the HPLC pump induces a periodic pulsation in the flow which results in another contribution to noise that depends on both the magnitude of the flow rate fluctuation and the background current. This flow rate fluctuation varies with the design of the HPLC pump but there have been large system improvements to reduce this pulsation and the recent pumps create much less pulse compared to the older ones. However, pulse damping systems are often necessary when using ED, particularly in high sensitivity work, and they can be introduced independently in the chromatographic system. Most pumps with low-speed pistons, whether single or dual headed, will require additional pulse damping when used with an electrochemical detector. For moderate sensitivity, flow fluctuations of dual headed pumps can often be reduced by introducing long coils of narrow bore stainless steel tubing between the pump and injector. More efficient pulse damping is achieved with mechanical pulse dampeners, such as the liquid membrane dampeners incorporated into some HPLC pumps and in the equipment manufactured by Antec.

4.2.2.6 *Electrode Treatment*

It has already been stressed that the electrode surface should be clean and very smooth for efficient ED to occur. Even brand new electrodes can sometimes function less than optimally due to the conditions of storage once manufactured. Manufacturers quote various "optimal" methods for cleaning electrodes. A new HDV should be constructed after any electrode treatment or replacement since the cell behavior may not be exactly comparable to that shown previously.

4.2.2.6.1 Cleaning Thin-Layer Electrodes In the case of solid electrodes that can be dismantled, the electrodes can be cleaned by polishing, using jewelry grade alumina dust and suitable gem polishing pads. A small amount of alumina is placed on the pad and then wetted with distilled or deionized water and the electrode cleaned manually using a rotary motion for a few minutes. The electrode should be briefly washed with pure water and inspected under a low powered magnifying glass until it looks smooth and clean. It can then be briefly sonicated in deionized or distilled water to remove the final traces of alumina. The whole cell can then be re-assembled having checked for any damage to the spacer. Damaged spacers should be replaced. Solid metal electrodes can, in addition to the polishing process, also be cleaned by a brief exposure to a strong acid such as nitric acid.

4.2.2.6.2 Maintaining and Cleaning Porous Graphite Electrodes Successful use of porous graphite electrodes is dependent on preventing blockage and keeping the electrode clean. Before storage, cells should be washed with water and then methanol to prevent bacterial growth. Eluents should be filtered through a 0.22 μm filter before use. Appropriate in-line carbon filters are supplied for use with the Coulochem® instrument and any increase in back pressure can usually be traced to blockage of a filter. If the working electrode(s) becomes contaminated, as indicated by a reduced response, then applying a potential of +1 V (*vs.* Pd) for 30 min (or −0.4 V for 10 min) may restore the response [25]. Eluent should not be recirculated during such treatment.

Washing with a series of solvents of decreasing polarity (e.g., water, methanol, acetonitrile, and tetrahydrofuran) and then reversing the procedure may serve to remove lipophilic contaminants. If contamination by silica is suspected, then the cell should be washed with aqueous sodium hydroxide ($2\,mol\,l^{-1}$) or aqueous ammonium hydroxide ($1\,mol\,l^{-1}$). If the cell appears to be partially blocked, then flushing with phosphoric acid (60% w/v) is recommended. Flushing with nitric acid ($6\,mol\,l^{-1}$), which was once recommended by the manufacturers, should only be used as a last resort [25]. Of course, these procedures may not work and the use of appropriate sample purification procedures, high quality solvents, appropriate filters, low electrode potentials, and so on, should be beneficial in the long term.

4.2.2.7 *Gradient Elution*

The instability of ED signal toward changes in eluent ionic strength implies that isocratic elution is to be preferred, especially for applications requiring high sensitivity. Gradient elution can, however, be used in conjunction with HPLC-ED if the stability of the baseline and of the response is acceptable at low to moderate sensitivities. The use of dual or array electrode systems, especially the coulometric ones, has been demonstrated useful to reduce and correct the slope of the baseline in the resulting chromatogram by subtracting the baseline obtained at a selected electrode unit (see Section 4.2.1.3.4) and/or employing an appropriate software in signal processing.

4.2.2.8 *Maintenance of HPLC-ED Systems*

There is no doubt that electrochemical detectors are more prone to operational difficulties than most other HPLC detectors, and it is also more difficult to take the corrective action. The close inter-relationship between the chromatograph and the detector makes fault diagnosis and correction even more difficult. A good understanding of the basic principles underlying ED is essential to a logical and hopefully successful approach to system maintenance. Nevertheless some problems associated with ED can be difficult to resolve even for the most experienced workers.

Assuming that the chromatographic system is working satisfactorily, the main problems associated to ED in HPLC are the increase in the baseline noise and drifting and the loss of the sensitivity. Many sources of these adverse effects have been described within this section. Initially, improper selection of system and detection settings usually leads to high background noise. Too high flow-rate, detection potential, and/or sensitivity range, also low time constant for signal filtering are typical examples of that. When those issues are solved, high background signals and noise are typically related to the presence of electroactive substances in the mobile phase and it was advised about the purity and replenishment of solvents and reagents used, the presence of dissolved oxygen, and the release of metals from the components of the system. Judicious recycling of the mobile phase, which also reduces the presence of air bubbles (the use of a degasser system is highly recommended anyway), minimizes these effects by electrolyzing impurities. Alternatively, the use of an electrochemical guard cell with a large working electrode has proven to be very effective to remove potential interferences from eluents and is quite commonly used with coulometric detector with successful background diminution and subsequent improvement of LOD. Other source of signal instability can come from deterioration of the electrode surface and connections; therefore, their condition should be checked periodically. Eventually pulsation from pump or interactions with other equipment in the lab as well as electrostatic effects can induce to increase or vary the noise. The last can be solved by proper electrical shielding and using a common earthing (grounding) of all parts of the chromatographic system. Baseline stability is essential for the day-to-day work; therefore, it is a good practice to allow signal stabilization overnight and, if possible, the thermostatization of the detector. Finally, the sensitivity loss is frequently the symptom of electrode fouling. Among other reasons, it is important to carry out frequent analysis of appropriate quality assurance analyte solutions to confirm the electrode fouling and to ensure the reliability of quantitative data. Although the response of a deactivated electrode may sometimes be restored temporarily by using a slightly higher applied potential, the most reliable way of restoring the response is to disassemble the cell and to clean the surface of the electrode (Section 4.2.2.6).

4.3 Analytical Designs and Performance

Food chemistry has found valuable solutions to several analytical problems in the exploitation of the versatility of HPLC as separation technique together with the special sensitivity and selectivity characteristics conferred by the ED. The HPLC-ED has been shown to be selective and sensitive in the determination of many constituents found in foods, of great importance either in the investigation, food technology, and processing as well as quality control fields. In this section, a list of selected applications developed in the late years regarding ED in HPLC for the analysis of food products is reported and commented. It has been structured in analysis of the natural constituents or the contaminants (non anthropogenic and anthropogenic) found in foods.

4.3.1 Natural Constituents

4.3.1.1 Carbohydrates

While the main impact of carbohydrates occurrence in a food preparation is linked to the sweetener effect related to the presence of mono- and disaccharides, the occurrence of oligo- and polysaccharides gives to the food different properties depending on the type of linkage between the monomeric units that may be attacked by the human enzymes or not [39]. The analysis of carbohydrates is a milestone of HPLC-ED and it is presently being applied to a variety of routine and research applications. In particular, high-performance anion-exchange chromatography (HPAEC) coupled with PAD has proven to be a very useful method for this issue, including those carbohydrates of interest found in foods. Applications regarding HPAEC-PAD of carbohydrates determination in several food products have been extensively reviewed by Cataldi *et al.* [40], El Rassi [41], and more recently by Corradini *et al.* [39]. Many of the references listed in these works are reflected in Table 4.2.

Carbohydrates are weak acids with pK_a values in the range 12–14 and, consequently, at high pH can be transformed into oxyanions, and can be readily separated using highly efficient anion-exchange columns. ED in the PAD mode resolves sensitivity problems in carbohydrate analysis derived from the lack of chromophore or fluorophore groups in those molecules and the disadvantages found using RI or ELS when combined with

Table 4.2 *HPLC-ED analysis of carbohydrates*

Analyte	Analyzed sample	HPLC mode	Electrochemical detection	References
FOSs	Agave	HPAEC	PAD (Au)	[42]
Fructans	Agave americana	HPAEC	PAD (Au)	[43]
Water-soluble carbohydrates	Agave tequiliana	HPAEC	PAD (Au)	[44]
IOSs	Artichoke	HPAEC	PAD (Au)	[45]
FOSs	Banana	HPAEC	PAD (Au)	[46]
Oligosaccharides	Barley	HPAEC	PAD (Au)	[47]
Oligosaccharides	Brewer's spent grain	HPAEC	PAD (Au)	[48]
Maltose	Dietary supplements	HPAEC	PAD (Au)	[49]
Oligosaccharides	Fermented beverage of plant	HPAEC	PAD (Au)	[50]
Fructopyranose oligosaccharides	Fermented beverage of plant	HPAEC	PAD (Au)	[51]
Oligosaccharides	Fermented beverage of plant	HPAEC	PAD (Au)	[52]
Galactooligo-saccharides (GOSs)	Fermented milk	HPAEC	PAD (Au)	[53]
FOSs and IOSs	Fermented milk	HPAEC	PAD (Au)	[54]
FOSs and inulins	Fermented milk, cooked ham, food ingredients	HPAEC	PAD (Au)	[55]
FOSs and IOSs	Fermented milk (synbiotic)	HPAEC	PAD (Au)	[56]
Sucralose and FOSs	Fruit juice	HPAEC	PAD (Au)	[57]
Simple sugars and FOSs	Grape	HPAEC	PAD (Au)	[58]
Oligosaccharides	Honey	HPAEC	PAD (Au)	[59]
Oligosaccharides	Honey	HPAEC	PAD (Au)	[60]
Monosaccharides and oligosaccharides	Honey	HPAEC	PAD (Au)	[61]
Polysaccharides	Honey	HPAEC	PAD (Au)	[62]
Glucose and sylose	Instant coffee	HPAEC	PAD (Au)	[63]
GOSs	Prebiotic mixture	HPAEC	PAD (Au)	[64]
Maltodextrins	Rice husk	HPAEC	PAD (Au)	[65]
XOSs	Rice husk	HPAEC	PAD (Au)	[66]
XOSs	Rice husk	HPAEC	PAD (Au)	[67]
Oligosaccharide	Rice bran (defatted)	HPAEC	PAD (Au)	[68]
Glucose, fructose, sucrose, raffinose, stachyose	Soybean	HPAEC	PAD (Au)	[69]
Monosaccharides	Soybean	HPAEC	PAD (Au)	[70]
AOSs	Sugar beet	HPAEC	PAD (Au)	[71]
Monosaccharides, AOSs, and oligosaccharides	Sugar beet	HPAEC	PAD (Au)	[72]
Oligosaccharides	Sugar beet	HPAEC	PAD (Au)	[73]
Arabinose and AOSs	Sugar beet	HPAEC	PAD (Au)	[74]
AOSs, XOSs, and FOSs	Wheat bran, chicory root	HPAEC	PAD (Au)	[75]

(continued overleaf)

Table 4.2 *(continued)*

Analyte	Analyzed sample	HPLC mode	Electrochemical detection	References
Arabinoxylans	Wheat flour	HPAEC	PAD (Au)	[76]
Sugars and alditols	Plant tissues	HPAEC	PAD (Au)	[77]
Monosaccharide	Sweet potato fiber, starch residues of sweet potato, cassava, potato	HPAEC	PAD (Au)	[78]
Sorbitol, flucose, fructose, and sucrose	Fruit-based beverages	HPAEC	PAD (Au)	[79]
Free sugars and myo-inositol	Citrus juice	HPAEC	PAD (Au)	[80]
Glucose (and ethanol)	Sweet wine (sherry)	RP-HPLC	Amperometric (trienzyme biosensor)	[81]
Sugars (glucose, fructose, and sucrose) and myo-inositol	Citrus fruits	HPAEC	Amperometric (Ni–Ti alloy)	[82]
Glucose, fructose, and sucrose	Honey	HPAEC	Amperometric (Ni–Cr alloy)	[83]

FOSs, Fructooligosaccharides; AOSs, arabinooligosaccharides; IOSs, Inulooligosaccharides; XOSs, Xylooligosaccharides; *HPAEC*, high-performance anion exchange chromatography; *RP-HPLC*, reversed-phase HPLC; *PAD*, pulsed amperometric detection; Au, gold working electrode.

gradient elution. PAD of carbohydrates is favorably performed by taking advantage of their electrocatalytic oxidation mechanism at a gold working electrode in alkaline media. The traditional triple-pulse potential waveform used in PAD has been replaced by a quadruple-potential waveform, which improves significantly the long-term reproducibility and stability of the gold electrode [22, 32]. The compatibility of PAD with gradient elution, coupled with the high selectivity of the anion-exchange stationary phases, allows mixtures of simple sugars, oligo-, and polysaccharides to be separated with high resolution in a single run.

4.3.1.2 Amino Acids

Precise knowledge of amino acid composition, types, and abundance is essential for an unambiguous identification of a protein or peptide as well as for the elucidation of their structure and function. For those industries that produce edible products, the detection and determination of amino acids may also be an important quality control feature ensuring product safety and manufacturing efficiency. Chromatographic methods for the determination of amino acids have been available for many years [84] and UV–vis, FL, and MS have been utilized for the detection of these compounds. Similarly to carbohydrates, HPLC-ED represents a very attractive detection possibility for underivatized amino acids (Table 4.3) due to the lack of natural chromophore or fluorophore groups in these compounds. The underivatized amino acids can undergo electrocatalytic oxidation in alkaline solution through simultaneous adsorption onto the oxides electrochemically formed on several

Table 4.3 *HPLC-ED analysis of amino acids*

Analyte	Analyzed sample	HPLC mode	Electrochemical detection	References
Secondary amino acids	Gelatine	HPAEC	IPAD (Au)	[85]
Lys, Ala, Thr, Gly, Val, Ser, Proline, Ile, Met, His, Phe, and carbohydrates	Liquid condiment	HPAEC	IPAD (Au)	[86]
Free amino acids	Soy sauce	HPAEC	IPAD (Au)	[87]
–	Sugar beet	HPAEC	IPAD (Au)	[88]
Ala, Arg, Asn, Glu, Gly, Lys, Met, Ser, Thr, and Val	Cheese	HPAEC	Amperometric (NiOx-Au)	[93]
Met, Leu, Phe, Asp, Lys, His, Arg Gly, Cys, and Tyr	Meat	RP-HPLC	Amperometric (CuNP)	[94]
Lys, carnosine, anserine, and glutamine	Ostrich meat	RP-HPLC	Amperometric detection (CuNP)	[95]
D-Ala	Fruit concentrate or purees	HPCEC	ER-amperometric detection (Pt)	[96]
–	Foods	HPCEC	ER-amperometric detection (Pt)	[97]
L- and D-amino acids	Beer, wine, sherry, and fruit juice	HPCEC	ER-amperometric detection (Pt)	[98]

HPAEC, High-performance anion exchange chromatography; *RP-HPLC,* Reversed-phase HPLC; *HPCEC,* High-performance cation exchange chromatography; *ER,* Enzymatic post-column reactor; *IPAD,* Integrated pulse amperometric detection; Au, gold working electrode; NiOx-Au, nickel(III) oxyhydroxide electrodeposited onto the gold working electrode; CuNP, copper nanoparticle plated electrode; *Pt,* Platinum working electrode.

transition metal electrodes, such as copper, nickel, and gold. For this purpose, IPAD has shown the best results regarding to sensitivity and stable long-term operation, typically using a gold working electrode. This detection setup in connection to HPAEC has been employed in the analysis of secondary amino acids in a gelatin hydrolysate solution to distinguish gelatin sources [85], for simultaneous determination of 17 amino acids and 9 carbohydrates in a liquid condiment [86], in the analysis of 18 free amino acids in soy sauce [87], and for the screening of the amino acid component in beet brei samples of different origin [88]. Alternatively, several transition metal-based electrodes and modified traditional electrodes have been proposed as amperometric sensors for the determination of amino acids at constant applied potential [89–92]. A gold electrode substrate modified by anodic deposition of a thin film of nickel has been used as working electrode in the HPAEC determination of several free amino acids in cheese from two different brands through their electrocatalytic oxidation by electrogenerated nickel(III) oxyhydroxide [93]. Chou *et al.* have proposed a copper nanoparticle plated electrode for the fast differentiation of meats from 15 animal species [94] and the differentiation of ostrich meat from chicken, pork, and beef [95] at low detection potential with reverse phase HPLC at neutral pH. In these cases, the detection is based on the electrochemical formation of Cu(II) species in the presence of free amino acids. Postcolumn enzymatic reactors have also been coupled to HPLC equipment, leading to a promising extension in the field of biosensors [96–98].

These works are based on the principle that the separation efficiency of HPLC is coupled with the specificity of amino acid oxidase enzymes and the sensitivity of ED of the hydrogen peroxide enzymatically generated.

4.3.1.3 Vitamins

Vitamins are a group of food constituents with very different chemical structure and properties that are found in a wide range of concentration levels. They are classified as water-soluble and fat-soluble vitamins and are usually found bounded to carbohydrates, lipids, or proteins. Vitamins are most often determined in free form which involves hydrolysis of the phosphorylated forms and/or those bound to proteins (and optionally glycosylated) during the extraction step performed prior to the chromatographic separation. Many of these molecules have oxidizable functional groups, thus, can be readily detected electrochemically. Because of this, many methods for the determination of vitamins in different food matrices by reversed-phase HPLC-ED have been developed to date, typically using carbon-based working electrodes, as can be seen in Table 4.4.

Ascorbic acid (vitamin C) possesses a high antioxidant activity and can be readily detected at low potentials at GCEs. Iwase [99, 114] has studied its occurrence in several foods and beverages by isocratic elution using mobile phases containing amino acids and nucleic acids instead of inorganic salts to increase the stability and reproducibility of such assays. Similar examples using conventional mobile phases have been reported [100–102].

Vitamin E group have similar antioxidant properties as ascorbic acid and can be detected by HPLC-ED at low detection potentials after sample treatment and extraction. Delgado-Zamarreño *et al.* have exhaustively studied the occurrence of tocopherols isomers and tocotrienol isomers in different foods using both the amperometric mode at single [103, 104] and dual GCE [105] and the coulometric mode using the Coulochem® cells [106–108]. This group has integrated hydrolysis and extraction processes coupled on-line with the HPLC-ED [106–108]. Separations were carried out using isocratic elution in the reversed-phase mode and took advantage of using a an electrochemical guard cell at high oxidation potentials in order to eliminate interferences and decrease the background current whereas the detection was carried out using a redox mode with the first cell set at negative potential and the second working at mild oxidation potential for the detection.

Vitamin B group includes a heterogeneous number of compounds, some of them electrochemically oxidizable, which have been analyzed in several foods by isocratic reversed-phase HPLC-ED after acid or alkaline digestion and subsequent enzymatic extraction. The coulometric mode using Coulochem® cells have been employed in these cases using an electrochemical guard cell at an oxidation potential closer to the final detection potential [96, 97]. In all cases, the potential of the coulometric cell units were set to perform in the screening mode. Similar experimental scheme has been employed elsewhere for vitamin D3 and provitamin D3 analysis [98].

Vitamin K (phylloquinone) structure contains quinone groups that can be electrochemically reduced to hydroquinone counterparts. Therefore, Koivu *et al.* [110–113] employed isocratic reversed-phase HPLC-ED schemes for vitamin K determination in digested extracts of different foods. They used Coulochem® cells in the redox mode with the first cell unit at a large negative reduction potential and the second unit detecting the vitamin K reduction products at very low potential.

Table 4.4 HPLC-ED analysis of vitamins

Vitamin	Analyzed sample	HPLC mode	Electrochemical detection	References
Folic acid (B9)	Fruit juices and cereal products	RP-HPLC	Coulometric	[96]
B1 (thiamine)	Seafood (B6 and B12)	RP-HPLC	Coulometric	[97]
B6 (pyridoxamine, pyridoxal, and pyridoxine)	Food products of plant origin (B1, B6, and B12)			
B12				
D	Fish	RP-HPLC	Coulometric	[98]
Ascorbic acid (C)	Athlete food, Japanese tea, orange juice, grated radish, grated carrot, vinegar, honey, soybean sauce	RP-HPLC	Amperometric (GCE)	[99]
C	Fruit juices	RP-HPLC	Amperometric (GCE)	[100]
C	Meat and processed food	RP-HPLC	Amperometric (GCE)	[101]
C	Fruit and vegetables	RP-HPLC	Coulometric	[102]
A, D3, E	Milk	RP-HPLC	Amperometric (GCE)	[103]
A, D3, and E	Yogurt	RP-HPLC	Amperometric (GCE)	[104]
A, E, K1	Milk	RP-HPLC	Dual amperometric detection (GCE)	[105]
E	Butter	RP-HPLC	Coulometric	[106]
E	Seeds and nuts	RP-HPLC	Coulometric	[107, 108]
E	Barley and oat grains	RP-HPLC	Coulometric	[109]
K1	Oils and margarines	RP-HPLC	Coulometric	[110, 111]
K1	Vegetables, fruits, and berries	RP-HPLC	Coulometric	[112]
K1	Cereal products	RP-HPLC	Coulometric	[113]

RP-HPLC, Reversed-phase HPLC and *GCE*, Glassy carbon working electrode.

4.3.1.4 Natural Phenolic Compounds

Natural phenolic compounds are a complex group of substances present in food products and beverages from plant origin. There is a considerable current interest on them because they were found to exhibit antisclerotic [115] and anticarcinogenic effects [116] since they neutralize the effects of free radicals, acting as antioxidants, and, thereby, protect the body from diseases and human aging. All of them possess a common structural: an aromatic ring with one or more hydroxyl substituents. These groups can be electrochemically oxidized at relatively low potentials using carbon-based working electrodes. Therefore, ED is especially well suited for the analysis of such phenolic compounds with high sensitivity and selectivity, thus, reducing the sample treatment, if any, before the chromatographic analysis. The main problem found in the analysis of these phenolics when using HPLC-ED is the need of gradient elution in the reversed-phase mode to enable the separation in a single run of compounds with strongly differing polarities [117]. Coulochem® and CoulArray® electrochemical detectors and their corresponding controlling software have found especial importance for the compensation of baseline drift during gradient run while maintaining good sensitivity and signal stability [118]. A wide number of applications are listed in Table 4.5. The ED using the amperometric mode has been exhaustively used with interesting results. The use of square arrays with four channel-detection [129, 130, 131] and dual electrodes [127, 128] has been proven very efficient for the selective identification, detection, and screening of phenolics with low oxidation potentials. HPLCE-ED performed with single GCEs has shown good results but it has been usually limited to the detection of a small number of compounds with isocratic elution and somewhat reduced signal stability [133].

HPLC with coulometric-array detection has been successfully used for the analysis of phenolic compounds in meads [119], natural beverages and plant extracts [117, 120], juice beverages [121], beers [118, 122, 123], and wines [124].

Amperometric detection was used in isocratic HPLC of phenolic acids and flavonoids in beer and other beverages [127, 129], honey [131, 132], and jujube [134, 136]. Oliveira-Brett *et al.* [135] developed a RP-HPLC method for the analysis of 11 different flavonoids in red grape skins which employed an elution program combining isocratic and gradient conditions whereas ED in the amperometric mode with a GCE working electrode was carried out. Higher sensitivity and LOD when compared to UV–vis combined with stable background signals were reported.

4.3.2 Nonanthropogenic Contaminants

4.3.2.1 Biogenic Amines

Biogenic amines (BAs) (histamine, tyramine, putrescine, cadaverine, spermine, spermidine, etc.) are formed because of microbiological spoilage in fish, cheese, and other foods, and they are often responsible for food poisonings. Therefore, it is important to monitor BA levels for human health and food safety and their content in various foods and beverages has been widely studied [137]. The two highest drawbacks in the analysis of BAs in food are: (i) the complexity of the sample matrix and (ii) the low concentration levels at which the compounds are present in the sample. The more successful contributions of HPLC-ED schemes to BA analysis have been provided by high-performance cation-exchange liquid chromatography (HPCEC) coupled with IPAD at gold working electrodes (Table 4.6), where electrocatalytic oxidation of BA occurs once metal oxides are electrogenerated. Favaro *et al.* and Pastore *et al.* have reported the determination of BAs in fresh and processed meat [138] and in chocolate [139] using HPCEC with integrated square-wave detection (ISWD), a waveform variation of IPAD which minimizes background contribution and provides long-term stability by improved stabilization of the electrode surface after oxide generation for these analytes [22, 32]. Separation was carried out by gradient elution with a mobile phase containing increased amounts of methanesulfonic acid and a postcolumn

Table 4.5 *HPLC-ED analysis of natural phenolics*

Analyte	Analyzed sample	HPLC mode	Electrochemical detection	References
25 Phenolic compounds	Mead (honey wines)	RP-HPLC	Coulometric-array	[119]
Flavonoids and simple and complex phenols	Beer, red and white wines, lemon juice, and soya, forsythia and tobacco extracts	RP-HPLC	Coulometric-array	[120]
Phenolic acids and flavonoid	Beer, wine, tea, and plant extracts	RP-HPLC	Coulometric-array	[117]
Phenolic acids and flavonoid	Juice beverages	RP-HPLC	Coulometric-array	[121]
Free phenolic acid	Beer	RP-HPLC	Coulometric-array	[122]
Free and bound phenolic acid	Beer	RP-HPLC	Coulometric-array	[123]
Lignans	Wine	RP-HPLC	Coulometric-array	[124]
27 Phenolic and flavonoid	Beer	RP-HPLC	Coulometric-array	[118]
Carnosinic acid	Meat extracts	RP-HPLC	Coulometric	[125]
Rutin, catechin, epicatechin, and epicatechin gallate	Different parts of buckwheat	μHPLC	Amperometric (GCE)	[126]
Phenolic acids and flavonoids	Wine and grape juice	RP-HPLC	Amperometric with dual electrode (GCE)	[127]
Phenolic acids and flavonoids	Wine	RP-HPLC	Amperometric with dual electrode (GCE)	[128]
Resveratrol	Wine, grape juice, and grape seed capsules	RP-HPLC	Four-channel ED (GCE)	[129]
Phenolic acids	Honey	RP-HPLC	Four-channel ED (GCE)	[130, 131]
Caffeic acid, p-coumaric, ferulic acid, and hesperetin	Chinese citrus honey	RP-HPLC	Amperometric (GCE)	[132]
Hydroxycinnamic acids and volatile phenols	Wort and beer	RP-HPLC	Amperometric (GCE)	[133]
Phenolic acids	Jujube	RP-HPLC	Amperometric (GCE)	[134]
Flavonoids	Grape	RP-HPLC	Amperometric (GCE)	[135]
Gallic acid, protocatechuic acid, caffeic acid, *p*-coumaric acid, rutin, and quercetin	Jujube	RP-HPLC	Amperometric (GCE)	[136]

RP-HPLC, Reversed-phase HPLC and *GCE*, Glassy carbon working electrode.

Table 4.6 HPLC-ED analysis of biogenic amines

Analyte	Analyzed sample	HPLC mode	Electrochemical detection	References
Tyramine (Ty), putrescine (Put), cadaverine (Cad), histamine (His), agmatine (Agm), spermidine (Spd), spermine (Spm), natural polyamines	Fresh and processed meat	HPCEC	IPAD (Au)	[138]
Dopamine (Dop), serotonin (Se), Ty, His, phenylethylamine (Phe)	Chocolate	HPCEC	IPAD (Au)	[139]
Dop, Ty, Put, Cad, His, Agm, Spm, Spd, Phe	Alcoholic beverages	HPCEC	IPAD (Au)	[140]
Cad, Put, Agm, Spd, Spm	Soybean seeds	HPCEC	ISWD (Au)	[141]
Put, Cad, His, Ty, Spm	Kiwi, cheese, salami, pilchard	HPCEC	IPAD (Au)	[142]
His, Put, Cad, Spd, natural polyamine	Fish samples	HPCEC	Amperometric biosensor	[143]

HPCEC, High-performance cation exchange chromatography; IPAD, integrated pulse amperometric detection; ISWD, integrated square wave detection; and Au, gold working electrode.

addition of a strong base was necessary to obtain the alkaline pH required for ED. Similar schemes have been reported elsewhere for BA analysis in alcoholic beverages [140], soybean seed [141] and kiwi, salami, pilchard and cheese [142] showing that the advantage of these methods is the good sensitivity and selectivity without a prior step of derivatization as required in UV and FL detections. An alternative scheme for the ED was proposed by Muresan *et al.* who developed a sensitive amperometric biosensor as working electrode for BAs detection in HPCEC [143]. A graphite electrode was modified with amine oxidase, peroxidase, and an osmium containing-polymer to detect the hydrogen peroxide generated enzymatically at low potential whereas a postcolumn addition of phosphate buffer was necessary to obtain the neutral pH required for optimum enzymatic activity. Authors claimed important advantages such as: simple and reproducible biosensor preparation, low applied potential, highly selective detection, and extended linear range.

4.3.2.2 *Mycotoxins*

Mycotoxins are toxic chemical metabolic products produced by microscopic mold fungi that colonize cereal crops, fruits, and vegetables and some of them exhibit carcinogenic and mutagenic properties. The allowed maximum levels in foodstuffs are very low and exhaustive extraction, preconcentration, and clean-up processes are required to attain the desired sensitivity. The alternative of ED (amperometric detection) combined with HPLC can be useful to simplify the detection of such low concentration levels. However, it has been used rarely for the analysis of these compounds in food because many of these compounds can be electrochemically oxidized or reduced at extreme potentials. Visconti *et al.* [144], analyzed

Table 4.7 HPLC-ED analysis of mycotoxins

Analyte	Analyzed sample	HPLC mode	Electrochemical detection	References
Altertoxin-I and altertoxin-II	Maize, rice, and tomatoes	RP-HPLC	Coulometric	[144]
Altenuene and isoaltenuene	Artificially contaminated samples	RP-HPLC	Coulometric (ID-PCBr)	[145]
Aflatoxins B1, B2, G1, and hdG2	Peanut butter	RP-HPLC	RS-DPV (DME)	[146]
Aflatoxins B1	Several food	RP-HPLC	RS-DPV (DME)	[147]
Zeranol and zearalenone metabolites	Rice, corn flakes, and soybean	RP-HPLC	Dual amperometric-UV (SPE)	[148]
Trenbolone and zeranol	Cattle	RP-HPLC	Dual amperometric-UV (GCE)	[149]
Zeranol and zearalenone and their isomers and metabolites	Edible animal tissue	RP-HPLC	Amperometric (GCE)	[151]
Roquefortine	Blue cheese and blue cheese dressing	HPLC	Dual amperometric-UV (GCE)	[150]

RP-HPLC, reversed-phase HPLC; *ID-PCBr*, Indirect detection of post-column electrogenerated bromine; *RS-DPV*, Rapid scanning differential pulse voltammetry; *GCE*, glassy carbon working electrode; *SPE*, Screen printed electrode; *DME*, Dropping mercury electrode.

subparts per million levels of altertoxins in sample maize, rice, and tomatoes infected by Alternaria alternate by reversed-phase HPLC with gradient elution with dual in-series electrodes operating in the coulometric "redox" mode (generator electrode at +1.0 V for oxidizing the analytes and indicator electrode at −0.1 V to detect quinone derivates). The same authors have proposed a method for determining of electroinactive alternaria mycotoxins by RP-HPLC-ED, with indirect detection using on-line generated bromine [145]. Duhart *et al.* [146] and Holak *et al.* [147] proposed the HPLC-ED of aflatoxins by rapid scanning voltammetry at mercury electrode following the analytes reduction signal.

Zeranol and zearalenone and their metabolites have been recently analyzed by Hsieh *et al.* by dual UV-vis and ED in rice, corn flakes, and soybean [148]. The method provided quick and reliable semiconfirmative and quantitative information on the occurrence of these analytes and supplement to ELISA screening method for total mycotoxins. Other HPLC-ED schemes for these mycotoxins have been applied to cattle [149], cheese [150], and edible animal tissue [151]. In the last work, these mycotoxins were enzymatically hydrolyzed, in order to be easily oxidized on GCE (Table 4.7).

4.3.3 Anthropogenic Contaminants

4.3.3.1 Antibiotics

Detection of antibiotic residuals in food is becoming an important concern since they can cause serious health consequences for consumers that range from allergic reactions to

Table 4.8 HPLC-ED analysis of antibiotics

Analyte	Analyzed sample	HPLC mode	Electrochemical detection	References
Sulfur-containing	Chicken feed	RP-HPLC	IPAD (Au)	[152]
Cephapirin and ampicillin	Raw milk	RP-HPLC	IPAD (Au)	[153]
Macrolide antibiotics	Animal feeding stuff	RP-HPLC	Coulometric-array	[154]
Tetracyclines	Ovine milk	RP-HPLC	Coulometric-array	[155]
Tetracyclines	Milk	RP-HPLC	Amperometric (Au)	[156]
lincomycin	Pig feed, honey, and milk	RP-HPLC	Amperometric (pSPE)	[157]
Frenolicin B	Poultry feed	RP-HPLC	Coulometric	[158]

RP-HPLC, reversed-phase HPLC; *IPAD*, integrated pulse amperometric detection; Au, gold working electrode; *pSPE*, pre-anodized screen printed electrode.

the evolution of resistant strains of bacteria, which may eventually lead to the need for stronger antibiotics. Antibiotics are classified in several groups with different structure and properties. Some HPLC-ED methods have been developed for the detection of particular groups of antibiotics (Table 4.8). For example, Dasenbrock and LaCourse have demonstrated the utility of IPAD for the detection of sulfur-containing antibiotics separated by reversed-phase HPLC [152, 153, 159]. For these particular analytes, IPAD can be carried out in acid conditions; thus, it was unnecessary any postcolumn modification of mobile phase composition. On the other hand, González de la Huebra *et al.* [154] studied the occurrence of macrolide antibiotics in different foods by HPLC-ED exploiting their oxidation signal at single GCE and using a multichannel CoulArray® coulometric cell with special attention to the sample treatment and preconcentration strategies. Similar procedures were also carried out for the analysis of tetracyclines in ovine milk [155] whereas Casella and Picerno [156] proposed an alternative use of a single gold working electrode in the amperometric mode at fixed potential for the same purpose. More recently, a preanodized screen-printed carbon electrode has been developed for the determination of lincomycin by HPLC with amperometric detection in pig feed, milk, and honey after solid phase extraction [157]. Finally, Frenolizin B and related compounds, all of them containing quinone groups in their structure, have been determined in Poultry Feed using the redox mode approach in coulometric cells [158] by initial electrochemical reduction of analytes and subsequent detection at low oxidation potential.

4.3.3.2 Pesticides (Herbicides, Insecticides, and Fungicides)

Due to the indiscriminate use of pesticides, serious problems in the environment are emerging and they are an important risk to human health. Therefore, the determination of pesticides in environmental and food-related samples is one example of the general and more frequently occurring problem of determining traces of a single component or a group of special components in different matrices. Hence, there is an increase in the number of analytical methods that are being developed to detect the presence of those compounds and their degradation products, some of which are more toxic that the original pesticide.

Table 4.9 *HPLC-ED analysis of pesticides*

Analyte	Analyzed sample	HPLC mode	Electrochemical detection	References
Glyphosate (herbicide)	Mineral water	HPAEC	Amperometric (Cu)	[160]
Glyphosate (herbicide)	Tomato juice	HPAEC	Amperometric (Cu)	[161]
Chlorophenoxy acid (herbicides)	Drinking water	RP-HPLC	Coulometric	[162]
Phenylurea (herbicides)	Mineral water	RP-HPLC	Coulometric	[163]
Organophosphorus (insecticides)	Water samples	RP-HPLC	Amperometric (GCE)	[164]
Carbamates (insecticides)	Apples, cabbage, grapes, and tomatoes crops	RP-HPLC	Coulometric	[165]
Carbamates (fungicides)	Apples	RP-HPLC	Amperometric (G-PTFE)	[166]

HPAEC, high-performance anion exchange chromatography; *RP-HPLC*, reversed-phase HPLC; Cu, copper working electrode; *GCE*, glassy carbon working electrode; *G-PTFE*, Graphite-polytetrafluoroethylene electrode.

Several applications in HPLC-ED in the coulometric mode were reported for pesticide analysis (Table 4.9). A highly selective and sensitive method using HPLC in conjunction with coulometric detection, with dual electrode analytical cell and a single electrode guard cell, was developed for the analysis for chlorophenoxy acid herbicides in ground and drinking water [162]. In this method, after photoconversion by UV irradiation, detection was carried out at potentials 0.3 V lower than without photolysis. To obtain efficient preconcentration with good reproducibility and accuracy a C18 solid-phase extraction system was applied. Phenylurea herbicides have been determined in mineral and tap water after extraction form the sample on solid-phase cartridges by RP-HPLC using an array of four cell modules, each containing four coulometric electrochemical cells [163].

Direct current amperometric detector in a thin-layer configuration equipped with a single GCE was used for the determination of fenitrothion and its main metabolites in environmental water samples [164]. In this work, the combination of SPME treatment with HPLC with on-line DAD and ED allowed to carry out environmental studies which monitored the impact of fenitrothion after aerial or spray applications.

Another interesting example is the HPAEC determination of herbicide glyphosate and its metabolite aminomethyl phosphonic acid which is based on the electrocatalytic oxidation on metallic cooper electrode. The detection is based on the enhancement in the anodic current of Cu electrode experienced in the presence of a Cu(II)-complexing agent like these analytes. These methods were successfully applied in analysis of tomato juice [161] and mineral water samples [160] without any derivatization, clean-up, or preconcentration step.

Carbamate pesticides have been determined by HPLC-ED usually following two strategies for the ED. One is the direct detection of the pesticides through their electrochemical oxidation at high positive potentials [166–168] whereas the second method is based on the alkaline hydrolysis of these pesticides to produce phenolic derivatives which were detected at relatively lower oxidation potential [165, 168, 169]. These methods have been typically applied to different water samples whereas Fernández *et al.* [166] also examined apples samples and Krause [165] extended the utility for the determination of six carbamates in apples, cabbage, grapes, and tomatoes crops.

4.4 Concluding Remarks

HPLC is nowadays an invaluable analytical tool in food analysis as a result of the versatility and efficiency of the various available separation modes currently employed. Many detection methods have been coupled to HPLC looking for a universal and sensitive response as that provided by flame ionization detection to GC. However, despite none of them supplies such almost ideal performance, far from being a detrimental, the development and evolution of these detection systems have expanded the utility of HPLC to practically all groups of known chemical compounds and food-related fields are not an exception. Among these detection systems, the largest potential in terms of analytical information is provided by MS detection, which is especially valuable in research studies; however, for routine applications in various areas, most probably for many years, it may not be commonly employed due to high cost of equipment and required high qualifications of personnel. Routine work involving HPLC in food analysis is typically carried out using UV–vis (or RI or LS in less extent and for specific applications) due to the commercial availability at affordable costs of these equipment. Thus, what place is left for ED in food analysis? Definitely, the answer is within those areas of applications involving the detection of electroactive analytes that show weak absorption of UV–vis radiation without derivatization. The HPLC-ED methods developed in that cases are very often more selective than absorptive spectrophotometric detections, which is favorable for samples with complex matrices. Besides, sample treatment can be significantly simplified whereas analysis time and cost can be reduced as well. For these reasons HPLC-ED is well accepted in the analysis of carbohydrate and natural polyphenols in beverages and food commodities. In these cases, the availability of HPLC-ED equipment and technical applications for these purposes from Dionex and ESA (now part of Thermo Scientific group) is very important. However, HPLC-ED did not succeed equally for other analytes with similar detection possibilities, which is reflected in the observed decay in the number of related publications in the last years. In general it can be said that HPLC-ED possesses limited acceptance in the field of food analysis, especially when compared to that shown in biomedical or pharmacological fields, and its analytical potential has not been properly exploited yet.

Why this is the case? It is generally assumed that although ED can give high sensitivity and selectivity and can be relatively inexpensive in operation, it is not an easy technique to use, the ever-present concerns being electrode deactivation or other more subtle factors that may act to influence response. These issues generate a lack of confidence in the technique to the nonexpertise users. As has been stressed along this chapter, undoubtedly a good knowledge of electrochemistry and a previous evaluation of the compound behavior are fundamental aspects for the successful application of HPLC-ED. However, this idea should apply when using another type of detector too but commonly an unskilled new HPLC user is not afraid to starting a new spectrophotometric methodology (or even dealing with MS) whereas refuses in the use of ED. Therefore, one of the challenges for the widespread usage of HPLC-ED is to overcome these prejudges and to implement a facile launching to the technique for the nonspecialists.

The future prospects of HPLC-ED in the food analysis field (and in general) are associated to a further development of most widely used amperometric and coulometric detections, focused on preserving the long-term activity of working electrodes, which is indispensable for obtaining reliable analytical results. The present search for novel materials (such as nanomaterials, nanotubes, multi-oxides, new forms of carbon, conductive polymers, and so on) or ways of modification of working electrodes (polymeric films, self-assembling, gels, entrapment or chemical bonding of bio-species, and much more), as well as the improvement of chemical and electrochemical methods will be the key to attain the required stability combined with appropriate sensitivity and selectivity of the analyses. Finally, another important issue for food analysis is the full exploitation of possibilities offered by ED, either those well known and those currently under development. One example is the coupling of electrochemical detectors to UHPLC systems, which is currently offered by Antec. A further example is the coupling of photochemical reaction systems to HPLC-ED, which has proven to be adaptable for routine analysis, including use in automated systems. Although photochemical reactors offer the ability to be coupled with numerous modes of detections (i.e., UV, FL, and CL) in HPLC, ED has benefited the most by the addition of a postcolumn photochemical reactor on-line, which further expands the range of analytes able to take advantage of ED inherent sensitivity and selectivity.

References

1. McGorrin, R.J. (2009) One hundred years of progress in food analysis. *J. Agric. Food Chem.*, **57**, 8076–8088.
2. Cudiamat, G. (2011) Separations in the food industry, *The Column*, **7**, 17, http://digital.findanalytichem.com/nxtbooks/advanstaruk/thecolumn100711/#/18 (accessed 12 December 2014).
3. Nollet, L.M.L. (2000) *Food Analysis by HPLC*, 2nd edn, Marcel Dekker, Inc., New York.
4. Swartz, M. (2010) HPLC detectors: a brief review. *J. Liq. Chromatogr. Related Technol.*, **33**, 1130–1150.
5. Trufelli, H., Palma, P., Famiglini, G. and Cappiello, A. (2011) An overview of matrix effects in liquid chromatography–mass spectrometry. *Mass Spectrom. Rev.*, **30**, 491–509.
6. Trojanowicz, M., Szewczyńska, M. and Wcisło, M. (2003) Electroanalytical flow measurements – recent advances. *Electroanalysis*, **15**, 347–365.
7. Haddad, P.R., Jackson, P.E. and Shaw, M.J. (2003) Developments in suppressor technology for inorganic ion analysis by ion chromatography using conductivity detection. *J. Chromatogr. A*, **1000**, 725–742.
8. Kubáň, P. and Hauser, P.C. (2008) A review of the recent achievements in capacitively coupled contactless conductivity detection. *Anal. Chim. Acta*, **607**, 15–29.
9. Trojanowicz, M. (2011) Recent developments in electrochemical flow detections – a review: part II. Liquid chromatography. *Anal. Chim. Acta*, **688**, 8–35.
10. Švancara, I., Vytřas, K., Kalcher, K. *et al.* (2009) Carbon paste electrodes in facts, numbers, and notes: a review on the occasion of the 50-years jubilee of carbon paste in electrochemistry and electroanalysis. *Electroanalysis*, **21**, 7–28.

11. Zima, J., Švancara, I., Barek, J. and Vytřas, K. (2009) Recent advances in electroanalysis of organic compounds at carbon paste electrodes. *Crit. Rev. Anal. Chem.*, **39**, 204–227.
12. Zima, J., Dejmkova, H. and Barek, J. (2007) HPLC determination of naphthalene amino derivatives using electrochemical detection at carbon paste electrodes. *Electroanalysis*, **19**, 185–190.
13. Angus, J.C. (2011) Electrochemistry on diamond: history and current status, in *Synthetic Diamond Films: Preparation, Electrochemistry, Characterization, and Applications* (eds E. Brillas and C.A. Martínez-Huitle), John Wiley & Sons, Inc., Hoboken, NJ, pp. 3–19.
14. Pecková, K., Musilová, J. and Barek, J. (2009) Boron-doped diamond film electrodes – new tool for voltammetric determination of organic substances. *Crit. Rev. Anal. Chem.*, **39**, 148–172.
15. Terashima, C., Rao, T.N., Sarada, B.V. *et al.* (2002) Electrochemical oxidation of chlorophenols at a boron-doped diamond electrode and their determination by high-performance liquid chromatography with amperometric detection. *Anal. Chem.*, **74**, 895–902.
16. Yosypchuk, O., Barek, J. and Vyskočil, V. (2012) Determination of 1-hydroxypyrene in human urine by HPLC with electrochemical detection at a boron-doped diamond film electrode. *Anal. Bioanal. Chem.*, **404**, 693–699.
17. Maixnerová, L., Barek, J. and Pecková, K. (2012) Thin-layer and wall-jet arrangement of amperometric detector with boron-doped diamond electrode: comparison of amperometric determination of aminobiphenyls in HPLC-ED. *Electroanalysis*, **24**, 649–658.
18. Yáñez-Sedeño, P., Pingarrón, J.M., Riu, J. and Rius, F.X. (2010) Electrochemical sensing based on carbon nanotubes. *Trends Anal. Chem.*, **29**, 939–953.
19. Pumera, M., Ambrosi, A., Bonanni, A. *et al.* (2010) Graphene for electrochemical sensing and biosensing. *Trends Anal. Chem.*, **29**, 954–965.
20. Vega, D., Agüí, L., González-Cortés, A. *et al.* (2007) Voltammetry and amperometric detection of tetracyclines at multi-wall carbon nanotube modified electrodes. *Anal. Bioanal. Chem.*, **389**, 951–958.
21. Chailapakul, O., Wonsawat, W., Siangproh, W. *et al.* (2008) Analysis of sudan I, sudan II, sudan III, and sudan IV in food by HPLC with electrochemical detection: comparison of glassy carbon electrode with carbon nanotube-ionic liquid gel modified electrode. *Food Chem.*, **109**, 876–882.
22. LaCourse, W.R. and Modi, S.J. (2005) Microelectrode applications of pulsed electrochemical detection. *Electroanalysis*, **17**, 1141–1152.
23. Zen, J.-M., Hsu, C.-T., Senthil Kumar, A. *et al.* (2004) Amino acid analysis using disposable copper nanoparticle plated electrodes. *Analyst*, **129**, 841–845.
24. Štulik, K. and Pacakova, V. (1987) *Electroanalytical Measurements in Flowing Liquids*, Ellis Horwood, Chichester.
25. Flanagan, R.J., Perrett, D. and Whelpton, R. (2005) *Electrochemical Detection in HPLC. Analysis of Drugs and Poisons*, Royal Society of Chemistry, London.
26. Lunte, S.M., Lunte, C.E. and Kissinger, P.T. (1996) Electrochemical detection in liquid chromatography and capillary electrophoresis, in *Laboratory Techniques in*

Electroanalytical Chemistry, P. T. Kissinger and W. R. Heineman eds, 2nd edn, Marcel Dekker, Inc., New York, 1996, pp. 813–853.
27. Kissinger, P.T. (1977) Amperometric and coulometric detectors for high-performance liquid chromatography. *Anal. Chem.*, **49**, 447A–456A.
28. Yamada, J. and Matsuda, H. (1973) Limiting diffusion currents in hydrodynamic voltammetry. *J. Electroanal. Chem.*, **44**, 189–198.
29. Fleet, B. and Little, C.J. (1974) Design and evaluation of electrochemical detectors for HPLC. *J. Chromatogr. Sci.*, **12**, 747–752.
30. Cullison, J.K. and Gamache, P.H. (1997) Qualitative, utility of coulometric array detection: theoretical and applied aspects, in *Coulometric Electrode Array Detectors in HPLC*, Progress in HPLC-HPCE, vol. **6** (eds I.N. Acworth, M. Naoi, H. Parvez and S. Parvez), VSP, Zeist.
31. Elbicki, J.M., Morgan, D.M. and Weber, S.G. (1984) Theoretical and practical limitations on the optimization of amperometric detectors. *Anal. Chem.*, **56**, 978–985.
32. LaCourse, W.R. (2011) Pulsed electrochemical detection: waveform evolution, *LC-GC North Am.*, **29**, 584–593. http://www.chromatographyonline.com/pulsed-electrochemical-detection-waveform-evolution (accessed 27 April 2015)
33. Roston, D.A. and Kissinger, P.T. (1982) Series dual-electrode detector for liquid chromatography/electrochemistry. *Anal. Chem.*, **54**, 429–434.
34. Chao, M.-H. and Huang, H.-J. (1997) Application of a 32-microband electrode array detection system for liquid chromatography analysis. *Anal. Chem.*, **69**, 463–470.
35. Sontag, G., Bernwieser, I. and Krach, C. (1997) HPLC with electrode array detection and its application in analytical food chemistry, in *Coulometric Electrode Array Detectors in HPLC*, Progress in HPLC-HPCE, vol. **6** (eds I.N. Acworth, M. Naoi, H. Parvez and S. Parvez), VSP, Zeist.
36. P. Jandera, Methods for the HPLC analysis of phenolic compounds and flavonoids in beer, in *Beer in Health and Disease Prevention*, V. R. Preedy Eds, Academic Press, Elsevier, Boston, MA, 2009 , pp. 1003–1014
37. Lunte, C.E. (1989) Voltammetric detection for liquid-chromatography. *LC–GC*, **7**, 492–498.
38. Long, J.T. and Weber, S.G. (1992) Voltammetry in static and flowing solutions with a large-amplitude sine wave potential. *Electroanalysis*, **4**, 429–437.
39. Corradini, C., Cavazza, A. and Bignardi, C. (2012) High-performance anion-exchange chromatography coupled with pulsed electrochemical detection as a powerful tool to evaluate carbohydrates of food interest: principles and applications, *Int. J. Carbohydr. Chem.*, doi: 10.1155/2012/487564.
40. Cataldi, T.R.I., Campa, C. and De Benedetto, G.E. (2000) Carbohydrate analysis by high-performance anion-exchange chromatography with pulsed amperometric detection: the potential is still growing. *Fresenius. J. Anal. Chem.*, **368**, 739–758.
41. El Rassi, Z. (2002) in *Carbohydrate Analysis by Modern Chromatography and Electrophoresi*, Journal of Chromatography Library, vol. **66** (ed Z. El Rassi), Elsevier Science, Amsterdam.
42. Ávila-Fernández, A., Galicia-Lagunas, N., Rodríguez-Alegría, M.E. *et al.* (2011) Production of functional oligosaccharides through limited acid hydrolysis of agave fructans. *Food Chem.*, **129**, 380–386.

43. Ravenscroft, N., Cescutti, P., Hearshaw, M.A. *et al.* (2009) Structural analysis of fructans from agave americana grown in South Africa for spirit production. *J. Agric. Food Chem.*, **57**, 3995–4003.
44. Arrizon, J., Morel, S., Gschaedler, A. and Monsan, P. (2010) Comparison of the water-soluble carbohydrate composition and fructan structures of Agave tequilana plants of different ages. *Food Chem.*, **122**, 123–130.
45. Ronkart, S.N., Blecker, C.S., Fourmanoir, H. *et al.* (2007) Herck and Michel Paquot, isolation and identification of inulooligosaccharides resulting from inulin hydrolysis. *Anal. Chim. Acta*, **604**, 81–87.
46. Der Agopian, R.G., Soares, C.A., Purgatto, E. *et al.* (2008) Identification of fructooligosaccharides in different banana cultivars. *J. Agric. Food Chem.*, **56**, 3305–3310.
47. Yoo, D.H., Lee, B.H., Chang, P.S. *et al.* (2007) Improved quantitative analysis of oligosaccharides from lichenase-hydrolyzed water-soluble barley β-glucans by high-performance anion-exchange chromatography. *J. Agric. Food Chem.*, **55**, 1656–1662.
48. Treimo, J., Westereng, B., Horn, S.J. *et al.* (2009) Enzymatic solubilization of brewers' spent grain by combined action of carbohydrases and peptidases. *J. Agric. Food Chem.*, **57**, 3316–3324.
49. Mosca, M., Boniglia, C., Carratù, B. *et al.* (2008) Determination of α-amylase inhibitor activity of phaseolamin from kidney bean (*Phaseolus vulgaris*) in dietary supplements by HPAEC-PAD. *Anal. Chim. Acta*, **617**, 192–195.
50. Kawazoe, N., Okada, H., Fukushi, E. *et al.* (2008) Two novel oligosaccharides isolated from a beverage produced by fermentation of a plant extract. *Carbohydr. Res.*, **343**, 549–554.
51. Okada, H., Fukushi, E., Yamamori, A. *et al.* (2010) Novel fructopyranose oligosaccharides isolated from fermented beverage of plant extract. *Carbohydr. Res.*, **345**, 414–418.
52. Okada, H., Fukushi, E., Yamamori, A. *et al.* (2011) Isolation and structural confirmation of the oligosaccharides containing α-d-fructofuranoside linkages isolated from fermented beverage of plant extracts. *Carbohydr. Res.*, **346**, 2633–2637.
53. Martínez-Villaluenga, C., Cardelle-Cobas, A., Corzo, N. and Olano, A. (2008) Study of galactooligosaccharide composition in commercial fermented milks. *J. Food Compos. Anal.*, **21**, 540–544.
54. Borromei, C., Cavazza, A., Merusi, C. and Corradini, C. (2009) Characterization and quantitation of short-chain fructooligosaccharides and inulooligosaccharides in fermented milks by high-performance anion-exchange chromatography with pulsed amperometric detection. *J. Sep. Sci.*, **32**, 3635–3642.
55. Borromei, C., Careri, M., Cavazza, A., Corradini, C., Elviri, L., Mangia, A. and Merusi, C. (2009) Evaluation of fructooligosaccharides and inulins as potentially health benefiting food ingredients by HPAEC-PED and MALDI-TOF MS, *Int. J. Anal. Chem.*, doi:10.1155/2009/530639.
56. Borromei, C., Cavazza, A., Corradini, C. *et al.* (2010) Validated HPAEC-PAD method for prebiotics determination in synbiotic fermented milks during shelf life. *Anal. Bioanal. Chem.*, **397**, 127–135.

57. Feinberg, M., San-Redon, J. and Assié, A. (2009) Determination of complex polysaccharides by HPAE-PAD in foods: validation using accuracy profile. *J. Chromatogr. B*, **877**, 2388–2395.
58. Blanch, M., Sanchez-Ballesta, M.T., Escribano, M.I. and Merodio, C. (2011) Fructo-oligosaccharides in table grapes and response to storage. *Food Chem.*, **129**, 724–730.
59. Morales, V., Sanz, M.L., Olano, A. and Corzo, N. (2006) Rapid separation on activated charcoal of high oligosaccharides in honey. *Chromatographia*, **64**, 233–238.
60. Morales, V., Corzo, N. and Sanz, M.L. (2008) HPAEC-PAD oligosaccharide analysis to detect adulterations of honey with sugar syrups. *Food Chem.*, **107**, 922–928.
61. Ouchemoukh, S., Schweitzer, P., Bachir Bey, M. *et al.* (2010) HPLC sugar profiles of Algerian honeys. *Food Chem.*, **121**, 561–568.
62. Megherbi, M., Herbreteau, B., Faure, R. and Salvador, A. (2009) Polysaccharides as a marker for detection of corn sugar syrup addition in honey. *J. Agric. Food Chem.*, **57**, 2105–2111.
63. Girard, P., Stöber, P., Blanc, M. and Prodolliet, J. (2006) Carbohydrate specification limits for the authenticity assessment of soluble (Instant) coffee: statistical approach. *J. Assoc. Off. Agric. Chem.*, **89**, 999–1003.
64. Coulier, L., Timmermans, J., Bas, R., Van Den Dool, R., Haaksman, I., Klarenbeek, B., Slaghek, T. and Van Dogen, W. (2009) In-depth characterization of prebiotic galactooligosaccharides by combination of analytical techniques, *J. Agric. Food Chem.*, **57**, 8488–8495.
65. Patindol, J., Wang, Y.J. and Jane, J.L. (2005) Structure-functionality changes in starch following rough rice storage. *Starch/Staerke*, **57**, 197–207.
66. Gullón, P., Moura, P., Esteves, M.P. *et al.* (2008) Assessment on the fermentability of xylooligosaccharides from rice husks by probiotic bacteria. *J. Agric. Food Chem.*, **56**, 7482–7487.
67. Gullón, P., González-Muñoz, M.J., van Gool, M.P. *et al.* (2010) Production, refining, structural characterization and fermentability of rice husk xylooligosaccharides. *J. Agric. Food Chem.*, **58**, 3632–3641.
68. Patindol, J., Wang, L. and Wang, Y.J. (2007) Cellulase-assisted extraction of oligosaccharides from defatted rice bran. *J. Food Sci.*, **72**, C516–C521.
69. Giannoccaro, E., Wang, Y.J. and Chen, P. (2008) Comparison of two HPLC systems and an enzymatic method for quantification of soybean sugars. *Food Chem.*, **106**, 324–330.
70. Bainy, E.M., Tosh, S.M., Corredig, M. *et al.* (2008) Varietal differences of carbohydrates in defatted soybean flour and soy protein isolate by-products. *Carbohydr. Polym.*, **72**, 664–672.
71. Zaidel, D.N.A., Arnous, A., Holck, J. and Meyer, A.S. (2011) Kinetics of enzyme-catalyzed cross-linking of feruloylated arabinan from sugar beet. *J. Agric. Food Chem.*, **59**, 11598–11607.
72. Holck, J., Lorentzen, A., Vigsnæs, L.K. *et al.* (2011) Feruloylated and nonferuloylated arabino-oligosaccharides from sugar beet pectin selectively stimulate the growth of bifidobacterium spp. In human fecal in vitro fermentations. *J. Agric. Food Chem.*, **59**, 6511–6519.

73. Martínez, M., Gullón, B., Yáñez, R. *et al.* (2009) Direct enzymatic production of oligosaccharide mixtures from sugar beet pulp: experimental evaluation and mathematical modeling. *J. Agric. Food Chem.*, **57**, 5510–5517.
74. Westphal, Y., Kühnel, S., de Waard, P. *et al.* (2010) Branched arabino-oligosaccharides isolated from sugar beet arabinan. *Carbohydr. Res.*, **345**, 1180–1189.
75. Courtin, C.M., Swennen, K., Verjans, P. and Delcour, J.A. (2009) Heat and pH stability of prebiotic arabinoxylooligosaccharides, xylooligosaccharides and fructooligosaccharides. *Food Chem.*, **112**, 831–837.
76. Virkki, L., Maina, H.N., Johansson, L. and Tenkanen, M. (2008) New enzyme-based method for analysis of water-soluble wheat arabinoxylans. *Carbohydr. Res.*, **343**, 521–529.
77. Cataldi, T.R.I., Margiotta, G., Iasi, L. *et al.* (2000) Determination of sugar compounds in olive plant extracts by anion-exchange chromatography with pulsed amperometric detection. *Anal. Chem.*, **72**, 3902–3907.
78. Salvador, L.D., Suganuma, T., Kitahara, K. *et al.* (2000) Monosaccharide composition of sweetpotato fiber and cell wall polysaccharides from sweetpotato, cassava, and potato analyzed by the high-performance anion exchange chromatography with pulsed amperometric detection method. *J. Agric. Food Chem.*, **48**, 3448–3454.
79. Corradini, C., Cristalli, A. and Corradini, D. (1994) Determination of carbohydrates in fruit-based beverages by high-performance anion-exchange chromatography with pulsed amperometric detection (HPAEC-PAD). *Ital. J. Food Sci.*, **6**, 103–111.
80. Lee, H.S. and Coates, G.A. (2000) Quantitative study of free sugars and myo-inositol in citrus juices by HPLC and a literature compilation. *J. Liq. Chromatogr. Related Technol.*, **23**, 2123–2141.
81. Guzmán-Vázquez de Prada, A., Peña, N., Parrado, C. *et al.* (2004) Amperometric multidetection with composite enzyme electrodes. *Talanta*, **62**, 896–903.
82. Masuda, T., Kawano, A., Kitahara, K. *et al.* (2003) Quantitative determination of sugars and myo-inositol in citrus fruits grown in Japan using high-performance anion-exchange chromatography. *J. Nutr. Sci. Vitaminol.*, **49**, 64–68.
83. Mora, M.I. and Marioli, J.M. (2001) Honey carbohydrate analysis by HPLC, with electrochemical detection, using a Ni–Cr alloy electrode. *J. Liq. Chromatogr. Related Technol.*, **24**, 711–720.
84. Molnár-Perl, I. (2000) Role of chromatography in the analysis of sugars, carboxylic acids and amino acids in food. *J. Chromatogr. A*, **891**, 1–32.
85. Russell, J.D., Dolphin, J.M. and Koppang, M.D. (2007) Selective analysis of secondary amino acids in gelatin using pulsed electrochemical detection. *Anal. Chem.*, **79**, 6615–6621.
86. Yu, H., Ding, Y.S., Mou, S.F. *et al.* (2002) Simultaneous determination of amino acids and carbohydrates by anion-exchange chromatography with integrated pulsed amperometric detection. *J. Chromatogr. A*, **966**, 89–97.
87. Shu-Min, M., Li-Na, L., Ya-Qui, C. *et al.* (2006) Analysis of amino acids in soy sauce with high performance anion-exchange chromatography and electrochemical detector. *Fenxi Shiyanshi*, **25**, 36–39.
88. De Bruijn, J.M. and Bout, M. (2000) HPAEC analysis of amino acids in sugar beet samples: method development and application. *Zuckerindustrie*, **125**, 604–609.

89. Kok, W.T., Hanekamp, H.B., Bos, P. and Frei, R.W. (1982) Amperometric detection of amino acids with a passivated copper electrode. *Anal. Chim. Acta*, **142**, 31–45.
90. Stulik, K., Pacakova, V., Weingart, M. and Podolak, M. (1986) Operation parameters of voltammetric high-performance liquid chromatographic detectors with copper electrodes and application to a determination of some fodder biofactors. *J. Chromatogr.*, **367**, 311–321.
91. Luo, J.P., Zhang, F. and Baldwin, R.P. (1991) Constant-potential amperometric detection of underivatized amino acids and peptides at a copper electrode. *Anal. Chem.*, **63**, 1702–1707.
92. Xie, Y. and Huber, C.O. (1991) Electrocatalysis and amperometric detection using an electrode made of copper oxide and carbon paste. *Anal. Chem.*, **63**, 1714–1719.
93. Casella, I.G., Gatta, M. and Cataldi, T.R.I. (2000) Amperometric determination of underivatized amino acids at a nickel-modified gold electrode by anion-exchange chromatography. *J. Chromatogr. A*, **878**, 57–67.
94. Chou, C.-C., Lin, S.-P., Lee, K.-M. *et al.* (2007) Fast differentiation of meats from fifteen animal species by liquid chromatography with electrochemical detection using copper nanoparticle plated electrodes. *J. Chromatogr. B*, **846**, 230–239.
95. Hung, C.-J., Ho, H.-P., Chang, C.-C. *et al.* (2011) Electrochemical profiling using copper nanoparticle-plated electrode for identification of ostrich meat and evaluation of meat grades. *Food Chem.*, **126**, 1417–1423.
96. Voss, K. and Galensa, R. (2000) Determination of L- and D-amino acids in foodstuffs by coupling of high-performance liquid chromatography with enzyme reactors. *Amino Acids*, **18**, 339–352.
97. Mögele, R., Pabel, B. and Galensa, R. (1992) Determination of organic acids, amino acids and saccharides by high-performance liquid chromatography and a postcolumn enzyme reactor with amperometric detection. *J. Chromatogr.*, **591**, 165–173.
98. Schultheiss, J. and Galensa, R. (1999) Determination of L- and D-amino acids by using HPLC-biosensor coupling. *Lebensmittelchemie*, **53**, 8.
99. Lebiedzińska, A., Dąbrowska, M., Szefer, P. and Marszałł, M. (2008) High-performance liquid chromatography method for the determination of folic acid in fortified food products. *Toxicol. Mech. Method.*, **18**, 463–467.
100. Lebiedzińska, A., Leszek Marszałł, M., Kuta, J. and Szefer, P. (2007) Reversed-phase high-performance liquid chromatography method with coulometric electrochemical and ultraviolet detection for the quantification of vitamins B1 (thiamine), B6 (pyridoxamine, pyridoxal and pyridoxine) and B12 in animal and plant foods. *J. Chromatogr. A*, **1173**, 71–80.
101. Ostermeyer, U. and Schmidt, T. (2006) Vitamin D and provitamin D in fish. Determination by HPLC with electrochemical detection. *Eur. Food Res. Technol.*, **222**, 403–413.
102. Iwase, H. (2003) Routine high-performance liquid chromatographic determination of ascorbic acid in foods using-methionine for the pre-analysis sample stabilization. *Talanta*, **60**, 1011–1021.
103. Berchmands, S., Gomathi, H. and Rao, G.P. (1989) Estimation of vitamin C in fruit juices by high-performance liquid chromatography with electrochemical detection (HPLC-EC). *Bull. Electrochem.*, **5**, 630–632.

104. Hidiroglou, N., Madere, R. and Behrens, W. (1998) Electrochemical determination of ascorbic acid and isoascorbic acid in ground meat and in processed foods by high pressure liquid chromatography. *J. Food Compos. Anal.*, **11**, 89–96.
105. Franke, A.A., Custer, L.J., Arakaki, C. and Murphy, S.P. (2004) Vitamin C and flavonoid levels of fruits and vegetables consumed in Hawaii. *J. Food Compos. Anal.*, **17**, 1–35.
106. Delgado Zamarreño, M.M., Sánchez Pérez, A., Gómez Pérez, C. and Hernández Méndez, J. (1992) High-performance liquid chromatography with electrochemical detection for the simultaneous determination of vitamin A, D3 and E in milk. *J. Chromatogr.*, **623**, 69–74.
107. Delgado Zamarreño, M.M., Sánchez Pérez, A., Sánchez Rodríguez, M. *et al.* (1996) Determination of fat-soluble vitamins in yogurt by HPLC with electrochemical detection. *Talanta*, **43**, 1555–1563.
108. Delgado Zamarreño, M.M., Sánchez Pérez, A., Gómez Pérez, M.C. *et al.* (1995) Determination of vitamins A, E and K1 in milk by high-performance liquid chromatography with dual amperometric detection. *Analyst*, **120**, 2489–2492.
109. Delgado Zamarreño, M.M., Sánchez Pérez, A., Bustamante Rangel, M. and Hernández Méndez, J. (1999) Automated analysis for vitamin E in butter by coupling sample treatment – continuous membrane extraction – liquid chromatography with electrochemical detection. *Anal. Chim. Acta*, **389**, 99–106.
110. Delgado-Zamarreño, M.M., Bustamante-Rangel, M., Sánchez-Pérez, A. and Hernández-Méndez, J. (2001) Analysis of vitamin E isomers in seeds and nuts with and without coupled hydrolysis by liquid chromatography and coulometric detection. *J. Chromatogr. A*, **935**, 77–86.
111. Delgado-Zamarreño, M.M., Bustamante-Rangel, M., Sánchez-Pérez, A. and Carabias-Martínez, R. (2004) Pressurized liquid extraction prior to liquid chromatography with electrochemical detection for the analysis of vitamin E isomers in seeds and nuts. *J. Chromatogr. A*, **935**, 249–252.
112. Colombo, M.L., Marangon, K. and Bugatti, C. (2009) CoulArray® electrochemical evaluation of tocopherol and tocotrienol isomers in barley, oat and spelt grains. *Nat. Prod. Commun.*, **4**, 251–254.
113. Piironen, V., Koivu, T., Tammisalo, O. and Mattila, P. (1997) Determination of phylloquinone in oils, margarines and butter by high-performance liquid chromatography with electrochemical detection. *Food Chem.*, **59**, 473–480.
114. Koivu, T., Piironen, V., Lampi, A.-M. and Mattila, P. (1999) Dihydrovitamin K1 in oils and margarines. *Food Chem.*, **64**, 411–414.
115. Koivu, T.J., Piironen, V.I., Henttonen, S.K. and Mattila, P.H. (1997) Determination of phylloquinone in vegetables, fruits, and berries by high-performance liquid chromatography with electrochemical detection. *J. Agric. Food Chem.*, **45**, 4644–4649.
116. Koivu, T., Piironen, V. and Mattila, P. (1998) Phylloquinone (vitamin K1) in cereal products. *Cereal Chem.*, **75**, 113–116.
117. Iwase, H. (2000) Use of nucleic acids in the mobile phase for the determination of ascorbic acid in foods by high-performance liquid chromatography with electrochemical detection. *J. Chromatogr. A*, **881**, 327–330.
118. Yang, C.S. (1997) Inhibition of carcinogenesis by tea. *Nature*, **389**, 134–135.

119. Shi, S.T., Wong, Z.Y., Smith, T.J. *et al.* (1994) Effects of green tea and black tea on 4-(methylnitrosamino)-1-(3-pyridyl)-1-butanone bioactivation, DNA methylation, and lung tumorigenesis in A/J mice. *Cancer Res.*, **54**, 4641–4647.
120. Jandera, P., Škeříková, V., Řehová, L. *et al.* (2005) RP-HPLC analysis of phenolic compounds and flavonoids in beverages and plant extracts using a CoulArray detector. *J. Sep. Sci.*, **28**, 1005–1022.
121. Rehová, L., Skerídová, V. and Jandera, P. (2004) Optimisation of gradient HPLC analysis of phenolic compounds and flavonoids in beer using a CoulArray detector. *J. Sep. Sci.*, **27**, 1345–1359.
122. Kahoum, D., Rezkova, S., Veskrnova, K. *et al.* (2008) Determination of phenolic compounds and hydroxymethylfurfural in meads using high performance liquid chromatography with coulometric-array and UV detection. *J. Chromatogr. A*, **1202**, 19–33.
123. Achilli, G., Cellerino, G.P., Gamache, P.H. and Melzi d'Eril, G.V. (1993) Identification and determination of phenolic constituents in natural beverages and plant extracts by means of a coulometric electrode array system. *J. Chromatogr.*, **632**, 111–117.
124. Gamache, P., Ryan, E. and Acworth, I.N. (1993) Analysis of phenolic and flavonoid compounds in juice beverages using high-performance liquid chromatography with coulometric array detection. *J. Chromatogr.*, **635**, 143–150.
125. Montanari, L., Perretti, G., Natella, F. *et al.* (1999) Organic and phenolic acids in beer. *Lebensm. Wiss. Technol.*, **32**, 535–539.
126. Nardini, M. and Ghiselli, A. (2004) Determination of free and bound phenolic acids in beer. *Food Chem.*, **84**, 137–143.
127. Nurmi, T., Heinonen, S., Mazur, W. *et al.* (2003) Lignans in selected wines. *Food Chem.*, **83**, 303–309.
128. Škrinjar, M., Hadolin Kolar, M., Jelšek, N. *et al.* (2007) Application of HPLC with electrochemical detection for the determination of low levels of antioxidants. *J. Food Compos. Anal.*, **20**, 539–545.
129. Danila, A.-M., Kotani, A., Hakamata, H. and Kusu, F. (2007) Determination of rutin, catechin, epicatechin, and epicatechin gallate in buckwheat fagopyrum esculentum moench by micro-high-performance liquid chromatography with electrochemical detection. *J. Agric. Food Chem.*, **55**, 1139–1143.
130. Lunte, S.M. (1987) Structural classification of flavonoids in beverages by liquid chromatography with ultraviolet-visible and electrochemical detection. *J. Chromatogr.*, **384**, 371–382.
131. Brenna, O., Buratti, S., Cosio, M.S. and Mannino, S. (1998) A new HPLC method for the determination of polyphenols in wines based on the use of less aggressive eluents and a coupled revelation system. *Electroanalysis*, **10**, 1204–1207.
132. Zhu, Y., Coury, L.A., Long, H. *et al.* (2000) Liquid chromatography with multichannel electrochemical detection for the determination of resveratrol in wine, grape juice, and grape seed capsules with automated solid phase extraction. *J. Liq. Chromatogr. Related Technol.*, **23**, 1555–1564.
133. Solomon, B.P., Long, H., Zhu, Y. *et al.* (2000) Four channel liquid chromatography/electrochemistry. *Curr. Sep.*, **18**, 113–116.
134. Long, H., Zhu, Y., Coury, L.A. *et al.* (2001) Simultaneous multiple electrode liquid chromatography/electrochemistry of phenolic acids in honey. *LC-GC Eur.*, **14**, 323–328.

135. Liang, Y., Cao, W., Chen, W.-J. *et al.* (2009) Simultaneous determination of four phenolic components in citrus honey by high performance liquid chromatography using electrochemical detection. *Food Chem.*, **114**, 1537–1541.
136. Vanbeneden, N., Delvaux, F. and Delvaux, F.R. (2006) Determination of hydroxycinnamic acids and volatile phenols in wort and beer by isocratic high-performance liquid chromatography using electrochemical detection. *J. Chromatogr. A*, **1136**, 237–242.
137. Wang, B.-N., Liu, H.F., Zheng, J.B. *et al.* (2011) Distribution of phenolic acids in different tissues of jujube and their antioxidant activity. *J. Agric. Food Chem.*, **59**, 1288–1292.
138. Novak, I., Janeiro, P., Seruga, M. and Oliveira-Brett, A.M. (2008) Ultrasound extracted flavonoids from four varieties of Portuguese red grape skins determined by reverse-phase high-performance liquid chromatography with electrochemical detection. *Anal. Chim. Acta*, **630**, 107–115.
139. Wang, B.-N., Cao, W., Gao, H. *et al.* (2010) Simultaneous determination of six phenolic compounds in jujube by LC-ECD. *Chromatographia*, **71**, 703–707.
140. Önal, A. (2007) A review: current analytical methods for the determination of biogenic amines in foods. *Food Chem.*, **103**, 1475–1486.
141. Favaro, G., Pastore, P., Saccani, G. and Cavalli, S. (2007) Determination of biogenic amines in fresh and processed meat by ion chromatography and integrated pulsed amperometric detection on Au electrode. *Food Chem.*, **105**, 1652–1658.
142. Pastore, P., Favaro, G., Badocco, D. *et al.* (2005) Determination of biogenic amines in chocolate by ion chromatographic separation and pulsed integrated amperometric detection with implemented wave-form at Au disposable electrode. *J. Chromatogr. A*, **1098**, 111–115.
143. De Borba, B.M. and Rohrer, J.S. (2007) Determination of biogenic amines in alcoholic beverages by ion chromatography with suppressed conductivity detection and integrated pulsed amperometric detection. *J. Chromatogr. A*, **1155**, 22–30.
144. Pineda, R., Knapp, A.D., Hoekstra, J.C. and Johnson, D.C. (2001) Integrated square-wave electrochemical detection of biogenic amines in soybean seeds following their separations by liquid chromatography. *Anal. Chim. Acta*, **449**, 111–117.
145. Draisci, R., Giannetti, L., Boria, P. *et al.* (1998) Improved ion chromatography–integrated pulsed amperometric detection method for the evaluation of biogenic amines in food of vegetable or animal origin and in fermented foods. *J. Chromatogr. A*, **798**, 109–116.
146. Muresan, L., Ronda Varela, R., Frébort, I. *et al.* (2008) Amine oxidase amperometric biosensor coupled to liquid chromatography for biogenic amines determination. *Microchim. Acta*, **163**, 219–225.
147. Visconti, A., Sibilia, A. and Palmisano, F. (1991) Selective determination of altertoxins by high-performance liquid chromatography with electrochemical detection with dual "in-series" electrodes. *J. Chromatogr.*, **540**, 376–382.
148. Palmisano, F., Sibilia, A. and Visconti, A. (1990) Determination of altenuene and isoaltenuene by liquid chromatography-electrochemical detection with on-line generated bromine. *Chromatographia*, **29**, 333–337.
149. Duhart, B.T., Shaw, S., Wooley, M. *et al.* (1988) Determination of aflatoxins B1, B2, G1 and G2 by high-performance liquid chromatography with electrochemical detection. *Anal. Chim. Acta*, **208**, 343–346.

150. Holak, W., DiProssimo, V. and Malek, E.G. (1997) Reductive voltammetric hplc detection of aflatoxins: determination of aflatoxin B1 in foods. *J. Liq. Chromatogr. Related Technol.*, **20**, 1057–1065.
151. Hsieh, H.-Y., Shyu, C.-L., Liao, C.-W. *et al.* (2012) Liquid chromatography incorporating ultraviolet and electrochemical analyses for dual detection of zeranol and zearalenone metabolites in mouldy grains. *J. Sci. Food Agric.*, **92**, 1230–1237.
152. Taguchi, S., Yoshida, S., Tanaka, Y. and Hori, S. (2001) Simple and rapid analysis of trenbolone and zeranol residues in cattle muscle and liver by stack-cartridge solid-phase extraction and HPLC using on-line clean-up with EC and UV detection. *J. Food Hyg. Soc. Jpn.*, **42**, 226–230.
153. Ware, G.M., Thorpe, C.W. and Pohland, A.E. (1980) Determination of roquefortine in blue cheese and blue cheese dressing by high pressure liquid chromatography with ultraviolet and electrochemical detectors. *J. Assoc. Off. Anal. Chem.*, **63**, 637–641.
154. Roybal, J.E., Munns, R.K., Morris, W.J. *et al.* (1988) Determination of zeranol/zearalenone and their metabolites in edible animal tissue by liquid chromatography with electrochemical detection and confirmation by gas chromatography/mass spectrometry. *J. Assoc. Off. Anal. Chem.*, **71**, 263–271.
155. LaCourse, W.R. and Dasenbrock, C.O. (1999) Pulsed electrochemical detection of sulfur-containing antibiotics following high performance liquid chromatography. *J. Pharm. Biomed. Anal.*, **19**, 239–252.
156. Dasenbrock, C.O. and LaCourse, W.R. (1998) Assay for cephapirin and ampicillin in raw milk by high-performance liquid chromatography-integrated pulsed amperometric detection. *Anal. Chem.*, **70**, 2415–2420.
157. González de la Huebra, M.J., Vincent, U. and von Holst, C. (2007) Sample preparation strategy for the simultaneous determination of macrolide antibiotics in animal feeding stuffs by liquid chromatography with electrochemical detection (HPLC-ECD). *J. Pharm. Biomed. Anal.*, **43**, 1628–1637.
158. Zhao, F., Zhang, X. and Gan, Y. (2004) Determination of tetracyclines in ovine milk by high-performance liquid chromatography with a coulometric electrode array system. *J. Chromatogr. A*, **1055**, 109–114.
159. Casella, I.G. and Picerno, F. (2009) Determination of tetracycline residues by liquid chromatography coupled with electrochemical detection and solid phase extraction. *J. Agric. Food Chem.*, **57**, 8735–8741.
160. Chiu, M.-H., Yang, H.-H., Liu, C.-H. and Zen, J.-M. (2009) Determination of lincomycin in urine and some foodstuffs by flow injection analysis coupled with liquid chromatography and electrochemical detection with a preanodized screen-printed carbon electrode. *J. Chromatogr. B*, **877**, 991–994.
161. Mazur, B. and Haroon, Y. (1995) Redox mode electrochemical detection approach with chemically modified electrodes for the measurement of frenolicin B (antibiotic) and related compounds in poultry feed. *J. Agric. Food Chem.*, **43**, 3042–3045.
162. La Course, W.R. and Dasenbrock, C.O. (1998) High-performance liquid chromatography-pulsed electrochemical detection for the analysis of antibiotics. *Adv. Chromatogr.*, **38**, 189–232.
163. Coutinho, C.F.B., Coutinho, L.F.M., Mazo, L.H. *et al.* (2008) Rapid and direct determination of glyphosate and aminomethylphosphoric acid in water using anion-exchange chromatography with coulometric detection. *J. Chromatogr. A*, **1208**, 246–249.

164. Coutinho, C.F.B., Coutinho, L.F.M., Mazo, L.H. and Lancas, F.M. (2007) Copper microelectrode as liquid chromatography detector for herbicide glyphosate. *Electroanalysis*, **19**, 1223–1226.
165. Wintersteing, R., Goger, B. and Krautgartner, H. (1999) Quantitation of chlorophenoxy acid herbicides by high-performance liquid chromatography with coulometric detection. *J. Chromatogr. A*, **846**, 349–357.
166. Achilli, G., Cellerino, G.P., d'Eril, G.M. and Bird, S. (1995) Simultaneous determination of 27 phenols and herbicides in water by high-performance liquid chromatography with multi-electrode electrochemical detection. *J. Chromatogr. A*, **697**, 357–362.
167. Sánchez-Ortega, A., Sampedro, M.C., Unceta, N. *et al.* (2005) Solid-phase microextraction coupled with high performance liquid chromatography using on-line diode-array and electrochemical detection for the determination of fenitrothion and its main metabolites in environmental water samples. *J. Chromatogr. A*, **1094**, 70–76.
168. Krause, R.T. (1988) High-performance liquid chromatographic determination of aryl *N*-methylcarbamate residues using post-column hydrolysis electrochemical detection. *J. Chromatogr.*, **442**, 333–343.
169. Fernández, C., Reviejo, A.J., Polo, L.M. and Pingarrón, J.M. (1996) HPLC-electrochemical detection with graphite-poly(tetrafluoroethilene) electrode. Determination of the fungicides thiram and disulfiram. *Talanta*, **43**, 1341–1348.
170. Anderson, J.L., Whiten, K.K., Brewster, J.D. *et al.* (1985) Microarray electrochemical flow detectors at high applied potentials and liquid chromatography with electrochemical detection of carbamate pesticides in river water. *Anal. Chem.*, **57**, 1366–1373.
171. Rao, T.N., Loo, B.H., Sarada, B.V. *et al.* (2002) Electrochemical detection of carbamate pesticides at conductive diamond electrodes. *Anal. Chem.*, **74**, 1578–1583.
172. Galeano Diaz, T., Guiberteau, A., Salinas, F. and Ortiz, J.M. (1996) Rapid and sensitive determinations of carbaryl, carbofuran and fenobucarb by liquid chromatography with electrochemical detection. *J. Liq. Chromatogr. Related Technol.*, **19**, 2681–2690.

5

Capillary Electrophoresis with Electrochemical Detection

Gang Chen
Department of Pharmaceutical Analysis, School of Pharmacy, Fudan University, 826 Zhanheng Road, Shanghai, 201203, China

5.1 Introduction

The quality and safety of foods and agricultural products is crucially important for food industry, agricultural and rural economic development, the health of human, and social stability. The analysis of the constituents in foods and agricultural products is an effective approach to safeguard their quality in terms of safety, ingredient quality, and nutritional composition. Agricultural and food analysis involves the determination of a variety of components that can be classified into nutritions, phenolic compounds, purines, food additives, residues, and so on. Foods and agricultural products are complex mixtures, usually containing hundreds of different constituents. The trace target analytes are usually mixed with a large amount of coexistent substances, such as proteins, sugars, or tannins, which make the analysis extremely difficult [1]. Therefore, it is of high interest to develop highly selective and sensitive analytical methods for agricultural and food analysis.

Liquid chromatography (LC) and gas chromatography (GC) are two commonly used techniques for the analysis of foods and agricultural products due to their powerful separation efficiency and sensitive detection [2, 3]. Since capillary electrophoresis (CE) in its modern form was first described by Joegenson and Luckas in 1981 [4], its application for the separation and determination of a variety of samples has become increasingly widespread because of its low cost, minimal sample volume requirement, short analysis time, and high separation efficiency. As an alternative method, CE has been investigated for many years to analyze various constituents in foods and agricultural products mainly

Agricultural and Food Electroanalysis, First Edition.
Edited by Alberto Escarpa, María Cristina González and Miguel Ángel López.

in capillary zone electrophoresis (CZE) and micellar electrokinetic capillary chromatography (MECC) modes. It seems extremely suitable for the measurement of the constituents in foods and agricultural products based on the fact that the separation capillary is easy to be regenerated by flushing with a suitable solution after it was contaminated by the coexistent interferences in the sample solutions. The interferences may deteriorate or even damage LC column completely.

After CE separation, ultraviolet (UV) detector has been commonly employed for the detection of the constituents extracted from foods and agricultural products [1, 2]. Because the absorbance path length of the capillary (the inner diameter, 25–100 μm i.d.) is very short, the low sensitivity of the UV detector used results in a low detection limit (typically 10 μM). Usually, the content of some constituents in the foods and agricultural products is very low. High-sensitive detection methods are highly demanded. To meet the requirement, CE-mass spectrometry (MS) has been applied for the determination of constituents in foods and agricultural products [3]. Although it can provide high sensitivity and structure information, a cost problem arises compared with other modes of detection. Recently, electrochemical detection (ECD) has been coupled with CE for the sensitive detection of foods and agricultural products [5]. ECD offers great promise for such CE systems, with features that include remarkable sensitivity, inherent miniaturization of both the detector and the control instrumentation, independence of sample turbidity or optical path length, low cost, minimal power demands, and high compatibility with the CE separation system. CE–ECD offers effective alternatives to analytical methodologies currently employed for characterizing many constituents. ECD can also provide higher selectivity as only electroactive substances can be detected so that the electropherograms are simplified, which is important for the analysis of foods and agricultural products because the constituents in them are usually complex.

This chapter focuses on recent advances in the application of CE–ECD for analyzing bioactive constituents in foods and agricultural products. This field of study has grown considerably in the past decade. The following sections will cover the commonly used separation modes of CE in agricultural and food analysis, CE–ECD system, the applications of CE–ECD in the determination of nutritions, phenolic compounds, purines, and food additives in foods and agricultural products as well as future prospects.

5.2 Separation Techniques in Agricultural and Food Analysis

CE encompasses a family of related separation techniques that use narrow-bore fused-silica capillaries to separate a variety of large and small molecules. High electric field strengths are used to separate molecules based on their differences in charge, size, and hydrophobicity. Sample introduction is accomplished by immersing the inlet of the capillary into a sample vial and applying pressure, vacuum, or voltage. Depending on the types of capillaries and electrolytes used, the technology of CE can be classified into several separation techniques, such as CZE, MECC or MEKC, capillary gel electrophoresis (CGE), capillary isoelectric focusing (CIEF), microemulsion electrokinetic chromatography (MEEKC), capillary electrochromatography (CEC), and so on. Among them, CZE and MECC are the two commonly used methods for the separation of the various constituents in foods and agricultural products [2, 3].

CZE, also known as free-solution CE, is the simplest form of CE. The separation mechanism is based on the differences in the charge-to-mass ratio of the analytes. The separation relies principally on the pH-controlled dissociation of acidic groups in the analyte or on the protonation of basic functions in the analyte. CZE is the most widely employed method for the separation of the constituents in foods and agricultural products presented in ions partially in the running buffer. If several analytes in the samples are neutral, MECC has to be employed for their separation. MECC is a mode of electrokinetic chromatography in which surfactants are added to the buffer solution at concentrations that form micelles. It was first developed by Terabe in 1984 [6]. The commonly used surfactant for the separation of the neutral constituents in foods and agricultural products is sodium dodecylsulfate (SDS) [6]. The separation of nonionic analytes is based on their different distribution coefficients between the aqueous phase and the micellar phase [7]. MECC has great utility in separating mixtures that contain both ionic and neutral species.

5.3 ECD in the CE Analysis of Foods and Agricultural Products

As with other analysis systems for foods and agricultural products, sensitive and selective detection techniques are also highly required for CE. Currently, several detection techniques, such as UV-visible absorption, laser-induced fluorescence (LIF), MS, and ECD have been used in combination with CE. However, only UV, MS, and ECD have been employed for the detection of the constituents of foods and agricultural products. The UV detector is the most widely used method for the detection of the constituents in foods and agricultural products after CE separation [5]. However, its application in CE was restricted by its lower sensitivity due to the extremely short optical path-lengths of the capillary. In addition, the method cannot be employed for the detection of some constituents such as carbohydrates containing no chromophoric groups. LIF and MS are other two important detection methods for CE with the detection limit ranges of 1–100 pM and 1–100 nM, respectively. However, both LIF and MS need sophisticated and expensive instrumentation. LIF typically requires precapillary or postcapillary derivatization of the sample with a fluorophore and is limited to fluorescent analytes and analyte derivatives [5]. Furthermore, only a select number of wavelengths can be used for excitation. An MS detector provides not only satisfactory sensitivity but also structure information. However, it is expensive and complicated.

Recently, ECD has attracted considerable interest for the detection of the bioactive constituents in foods and agricultural products because it can be coupled with CE to provide sensitive and selective detection. ECD can be classified into three general modes: conductimetry, potentiometry, and amperometry.

5.3.1 Amperometric Detection

Amperometry is the most widely reported ECD method for CE systems. It is accomplished by applying a constant potential to the working electrode and measuring the resulting current that is proportional to the concentration of analytes oxidized or reduced at the electrode surface. It is a sensitive detection approach with the detection limits (evaluated on the basis of a single-to-noise ratio of 3, i.e., $S/N = 3$) ranging from 1 to 10 nM and is extremely valuable for the detection of low-content electroactive constituents.

In the conventional three-electrode system, a reference and an auxiliary electrode are also present. Sometimes, only a working and counter electrode may be employed in a two-electrode configuration. In CE, the amperometric detection (AD) is usually performed in the off-column and end-column modes. In the case of off-column AD, a decoupling junction between the separation and detection capillaries is employed to minimize the effect of the separation voltage from the AD system. In the case of the end-column AD, a microdisk detection electrode is placed directly opposite to the outlet of the separation capillary without a conductive joint. However, a gap between the capillary outlet and the detection capillary has been left to reduce the interference of the high separation voltage. AD is selective in that it detects only electroactive constituents in foods and agricultural products that undergo oxidation (or reduction) at the applied potential [5]. It depends on the electrode material whether a constituent is electroactive. Many different electrode materials have been employed for CE–ECD. Carbon and copper disk electrodes have been commonly employed for the detection of the constituents in foods and agricultural products. The diameter of the disk electrode was ~300 μm [5]. This is primarily due to the size compatibility of the electrodes and the orifice of the separation capillary (25–50 μm i.d., 360 o.d.). The disk electrode was made of a piece of a graphite rod of 300 μm diameter or a copper wire coated with epoxy.

By now, the ADs of the constituents in foods and agricultural products after CE separation were usually performed in oxidative mode due to the interference from the dissolved oxygen in the reductive mode. Flavonoids, vitamins, alkaloids, colors, phenolic acids, glycosides of phenols, phenols and polyphenols, and aromatic amines in foods and agricultural products can be detected by a carbon electrode in the oxidation mode. The detection potential (E) was in the range of +0.9 to +1.1 V (vs. SCE (saturated calomel electrode)). Usually, phenolic compounds, aromatic amines, and thiols are electroactive when they are detected in the oxidation mode by the carbon electrode [5]. In the case of amino acid, peptides, carbohydrates, and constituents that own nearby hydroxyl groups, the copper and nickel electrodes have to be employed for their detection in a diluted NaOH aqueous solution because they are not electroactive at the carbon electrode. The electroactive substances at the copper and nickel electrodes usually contain a nearby hydroxyl group or a primary amino group. The detection potential ranges from +0.6 to +0.65 V (vs. SCE). Cyclic voltammetry was usually employed to know if a constituent was electroactive preliminarily [5]. The potential applied to the detection electrode directly affects the sensitivity and the detection limits of CE–ECD, and it is necessary to determine the hydrodynamic voltammograms for the target constituents to obtain the optimum detection potential. The sensitivity usually increases with raising the detection potential while both the baseline noise and the background current of the electropherograms increase simultaneously. An optimum detection potential can be obtained after evaluating the sensitivity, the background current, and baseline noise; offering the most favorable signal-to-noise characteristics. High background current at a higher potential may lead to an unstable baseline, which is the disadvantage for the sensitive and stable detection.

5.3.2 Conductivity Detection and Potentiometric Detection

For the detection of small inorganic and organic ions in foods and agricultural products by CE, conductivity detection (CD) is particularly promising. Conductimetry measures the conductance of a solution in the separation capillary, and the response is proportional to the concentration of the analyte ions. CD can be classified into contact and contactless modes. Contact CD is carried out by a galvanic contact between the measurement electrodes and the solution, while contactless CD is carried out through the use of external electrodes that are coupled capacitively to the electrolyte. Comparatively, contact CD is more sensitive. However, the galvanic contact with the solution is a potential source of problems for microchip CE. The bubble generation caused by electrolysis and electrode fouling may disturb the CE analyses. Furthermore, a special protection of the detector circuit and electronics is required to prevent damage from the high electric field. Contactless CD of CE avoids the problems associated with the direct solution contact by insulating the measurement electrodes electrically from the electrolyte.

Potentiometric detection for CE systems is based on a detection electrode bearing sensing membranes, allowing the buildup of a measurable potential, which follows the Nernst equation. Such an electrode is known as the ion-selective electrode (ISE). In order to be useful for detection in electro-separation methods, the selectivity should extend to a range of ions to form universal potential sensors. However, potentiometric detection was seldom employed in the CE analysis of foods and agricultural products.

5.4 Instrumentations of CE-ECD

The schematic of the commonly used CE–AD system is discussed in this section shown in Figure 5.1. The system consists of a ±30 kV high-voltage dc power supply (a), a piece of a fused-silica capillary coated with a layer of polyimide (n), an amperometric detector (h), a data acquisition system (i), a three-dimensional manipulator, an anode of the high voltage (m), and a detection reservoir (g) [8]. The high-voltage dc power supply (a) provides an injection or a separation voltage (typically 9–20 kV) between both the ends of the capillary. The inlet of the capillary is held at a positive potential and the outlet of the capillary is maintained at the ground potential. The separations are usually carried out in a 40–80 cm length of a 25 μm i.d. and a 360-μm o.d. fused-silica capillary (Figure 5.1n).

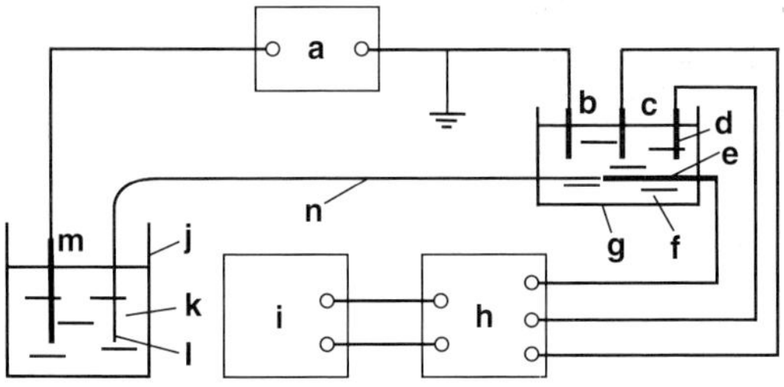

Figure 5.1 *Schematic of capillary electrophoresis with electrochemical detection system. (a) High voltage supply; (b) grounded platinum electrode; (c) reference electrode; (d) auxiliary platinum electrode; (e) detection electrode; (f) BGE; (g) electrochemical detection reservoir; (h) amperometric detector; (i) recorder; (j) anode reservoir; (k) sample solution or BGE; (l) the inlet of capillary; (m) platinum electrode for the high voltage; and (n) fused silica capillary. Reprinted from Ref. [5] with permission*

The detection reservoir contains a platinum grounded electrode (Figure 5.1b) and a three-electrode system consisting of a 300-μm diameter carbon or copper disk detection electrode (Figure 5.1e), a platinum auxiliary electrode (Figure 5.1d), and an SCE as the reference electrode (Figure 5.1c). Note that all electrodes are immersed in the background electrolyte (BGE, Figure 5.1f) for the CE separation. The three-electrode system is connected to the amperometric detector (Figure 5.1h) to monitor the analytes eluted out of the outlet of the capillary. The commonly used device for the AD of CE is also reviewed in this section. Before use, the disk detection electrode is successively polished with emery paper and alumina powder, and sonicated in doubly distilled water. It is then positioned carefully opposite to the outlet of the capillary with the aid of a three-dimensional micromanipulator, and arranged in a wall–jet configuration. The interface between the capillary outlet and the surface of the detection electrode is observed. The gap between the outlet and the detection electrode is usually adjusted to be 25–50 μm using a micrometer caliper of the micromanipulator [8]. The high voltage is applied between the anode for the high voltage in the anode reservoir and the grounded platinum electrode in the detection reservoir. In order to prevent the operator from the high voltage and ensure the safety of the CE–ECD system, the entire capillary, the buffer reservoirs, all electrodes, and the three-dimensional manipulator are enclosed in a Plexiglas box with a safety switch wired to turn off the power supply whenever the box was opened.

A CE–ECD system mainly consists of a ±30 kV high-voltage dc power supply, a piece of a fused-silica capillary, a conductivity detector, a data acquisition system, and a pair of electrodes for contact or contactless detection.

5.5 Determination of Nutritions by CE-ECD

Since dietary contains important nutritions including vitamins, minerals, proteins, fats, and carbohydrates, a variety of analytical approaches have been developed for their analysis. CE–ECD has also been applied in the rapid of these nutritions in foods and agricultural products.

5.5.1 Amino Acids and Peptides

Amino acids, peptides, and proteins are important nutritions in foods and agricultural products. Amino acids and peptides in foods and agricultural products have been determined by CE coupled with a carbon and a copper detection electrodes.

In 2003, Wang *et al.* determined tryptophan, tyrosine, tyramine, and tryptamine in rice spirit by MECC-AD [9]. The detection electrode was a carbon disk electrode that worked at a detection potential of 0.8 V (vs. SCE). The four analytes were well separated within 14 min in a piece of a 70-cm-long fused-silica capillary at a separation voltage of 20 kV in a 20 mM borate buffer (pH 10.4) containing 30 mM SDS.

In 2002, Jin *et al.* investigated the precapillary derivatization of 20 amino acids with naphthalene-2,3-dicarboxaldehyde (NDA) and CN^- [10]. All these derivatized amino acids could be oxidized on the carbon fiber microdisk bundle electrode except proline at a detection potential of 1.15 V (vs. SCE). CZE with ECD was employed for the analysis of 19 amino acids (including arginine, lysine, ornithine tryptophan, histidine isoleucine, leucine phenylalanine, methionine, glutamine, tyrosine, valine, threonine, serine, alanine, glycine,

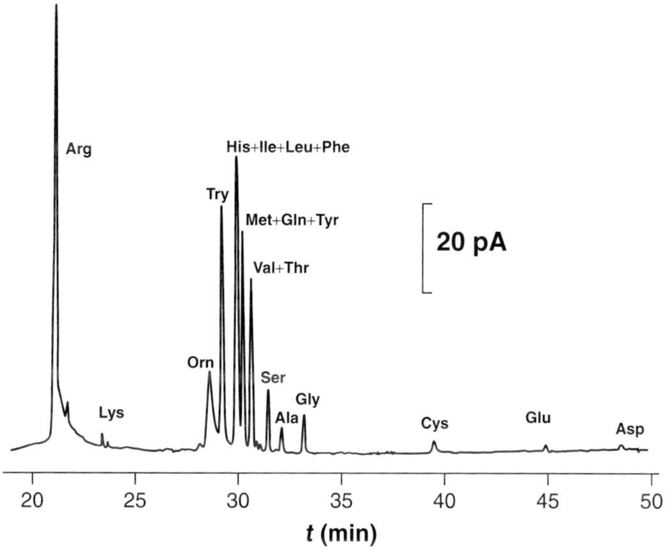

Figure 5.2 *Electropherogram of 19 amino acids derivatives. 12.5 mM borax to 3.13 mm NaOH; 20 μm amino acids. 1.2 mM NDA; 1.2 mM capillary; 120 cm in length, 20 μm I.D.; detection potential, 1.15 V (vs. SCE). Reproduced with permission from [10]. © 2002 Wiley-VCH Verlag GmbH, Weinheim, Germany*

cysteine, glutamic acid, and aspartic acid) in a piece of a 120-cm-long fused-silica capillary at a separation voltage of 18 kV in a 50 mM borate buffer (pH 9.48). The electropherogram is illustrated in Figure 5.2. Limits of detection of concentration for individual amino acids were between 0.17 and 1.8 μmol/l. The relative standard deviations were between 0.80% and 2.3% for the migration times and 1.4% and 6.4% for the electrophoretic peak currents, respectively. The method has been successfully employed in the determination of amino acids in beer by the standard addition method. The recovery for the amino acids in beer was 91%–109%.

As a biogenic amine, histamine is a small molecule derived from the decarboxylation of the amino acid histidine. It involves local immune responses as well as regulating physiological function in the gut and acts as a neurotransmitter. Histamine-rich foods can cause food-intolerance symptoms. Histamine-rich foods include red wine and beer, pizza, soya sauce, bacon, and many other preferred foods. It is of high interest to monitoring the content of histamine in foods and agricultural products. In 2002, Zhang *et al.* employed CZE for the determination of histamine using end-column AD with a carbon fiber micro-electrode [11]. Histamine was well separated and detected in a 10 mM phosphate buffer (pH 5.6) at a separation voltage of 15 kV and a detection potential of 1.35 V (vs. SCE), respectively. The proposed method was successfully applied to the direct determination of histamine in the beer samples without any sample clean-up procedures.

Besides a carbon electrode, a copper electrode has also been employed in the direct detection of amino acids and peptides in combination of CZE. In 1994, Baldwin *et al.* developed a method based on a copper disk electrode for the direct detection of amino acids and peptides after they were separated by CZE [12]. The separation medium was 50 or 100 mM

NaOH aqueous solution. The use of copper electrodes permitted the direct detection of amino acid species in CE at constant applied potentials and without derivatization. The detection limits they obtained varied from 1 to 10 fmol for most amino acids. The feasibility and the application of the approaches they developed have been demonstrated included the determination of amino acids in human urine, of the dipeptide aspartame in diet soft drinks, and of pentapeptide products of a solid-phase synthesis procedure.

5.5.2 Carbohydrates

As the primary metabolites, carbohydrates such as sucrose, glucose, and fructose are found widely presented in foods and agricultural product. Carbohydrates are neutral compounds within the physiological pH range. Strongly alkaline separation medium (50–100 mM NaOH) was used to keep all the analytes in anionic form. The separation of these analytes by CE is based on their degrees of dissociation. Carbohydrates and their derivatives are not normally electroactive at carbon electrodes, the most commonly used detection electrode in ECD. Thus, a variety of metal electrodes made of copper, nickel, gold, and platinum have been employed for the ECD of carbohydrates and their derivatives [5]. Among them, copper and nickel are most widely utilized. Carbohydrates and their derivatives can be detected by copper electrodes at a constant applied potential in strongly alkaline media based on electrocatalytic oxidation. It has been revealed that the minimum structural requirement for facile oxidation is the presence of at least two nearby hydroxyl groups in the compound to be oxidized. All of these oxidations are found to consist of many electron processes with C–C bond cleavage involved. Usually, a copper disk electrode was employed for the ECD of sugars and compounds containing nearby hydroxyl groups at constant applied potentials (+0.60 to +0.65 V (vs. SCE)) in a 50 or 75 mM NaOH aqueous solution.

Baldwin *et al.* determined carbohydrates, sugar acids, and alditols by CE and ECD at a copper electrode [13]. CE–ECD at copper electrodes have been demonstrated to provide a simple and sensitive method for the direct analysis of samples containing a wide range of carbohydrate compounds including simple sugars, sugar acids, and alditols. Both the separation and the detection required the use of a strongly alkaline medium. The hydroxide content could be varied to optimize conveniently the migration times and resolution obtained. Detection consisted of a direct oxidation that required no derivatization and yielded detection limits at or below the femtomole level for most of the carbohydrate species. The developed method has been applied in the analysis of the sugars in drinks and foods.

Recently, a novel graphene–copper nanoparticle composite was prepared by the *in situ* chemical reduction of a mixture containing graphene oxides and copper(II) ions using potassium borohydride as a reductant. It was mixed with paraffin oil and packed into one end of the fused capillary to fabricate microdisk electrodes for sensing carbohydrates [14]. The results indicated that copper nanoparticles with an average diameter of 20.8 nm were successfully deposited on the graphenes to form a well-interconnected hybrid network.

The suitability of the graphene–copper nanoparticle composite paste electrode for measuring real samples was demonstrated by detecting carbohydrates in milk and fruits after CE separation. Figure 5.3A illustrates the electropherogram of a mixture containing 0.25 mM mannitol, 0.5 mM sucrose, 0.5 mM lactose, 0.5 mM glucose, and 0.5 mM fructose at the graphene–copper nanoparticle composite paste electrode. The five analytes

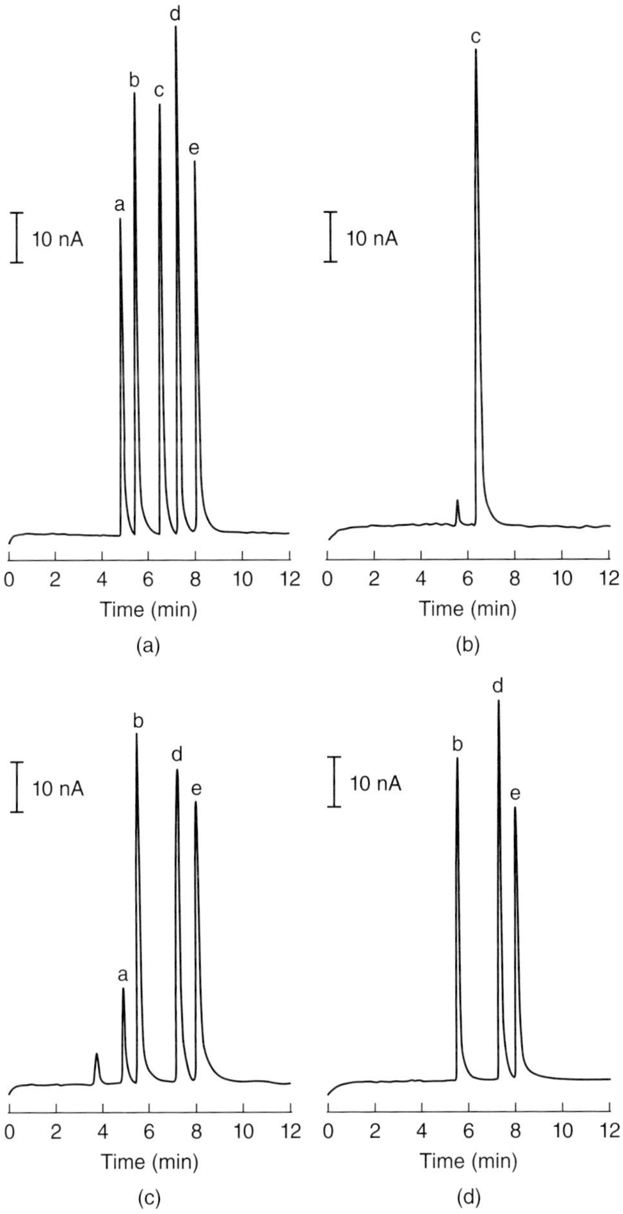

Figure 5.3 *(A) Electropherogram of a mixture containing 0.25 mM mannitol (a), 0.5 mM sucrose (b), 0.5 mM lactose (c), 0.5 mM glucose (d), 0.5 mM fructose, (e) the typical electropherograms of (B) the diluted bovine milk and the diluted extracts from (C) honey peach and (D) banana. Conditions: fused-silica capillary, 25 μm i.d. × 40 cm long; detection electrode, 320 μm diameter grapheme–copper nanoparticle composite paste electrode; running buffer, 75 mM NaOH; separation and injection voltage, 12 kV; injection time, 6 s; and detection potential, +0.65 V (vs. Ag/AgCl). Reproduced with permission from Ref. [14]. Copyright © 2012, American Chemical Society*

could be separated resulting in well-defined and resolved peaks within 9 min under the selected conditions. Figure 5.3B and C illustrates the typical electropherograms of the diluted bovine milk and the diluted extracts from honey peach and banana. The content of lactose in the bovine milk was determined to be 4.735 g l^{-1} (relative standard deviation (RSD) 3.2%, $n = 3$) that was in well agreement with the label amount (4.7 g l^{-1}).

In 2001, Ye *et al.* determined three sugar-free milk powder samples by CE-AD using a copper disk electrode as the working electrode in a 0.1 M NaOH medium [15]. There was acceptable linearity (0.9969) between the peak current and concentration of lactose in the range from 0.005 to 5 mg ml^{-1} with the detection limit of 0.1 μg ml^{-1}. In 2001, Hu *et al.* developed a simple and fast method based on CE-AD, for the separation and determination of six monosaccharides, including glucose, galactose, arabinose, fructose, xylose, and ribose, in soluble coffees [16]. A copper disk electrode was used as the working electrode. The optimum conditions for separation and detection were 50 mM NaOH, separation voltage of 5 kV, and the detection potential of 0.65 V (vs. Ag/AgCl). The proposed method was applied directly to the separation and determination of six sugars without prior derivatization, and the assay results were satisfactory. Besides AD, contactless CD has also been coupled with CE for the determination of glucose, fructose, galactose, mannose, ribose, sucrose, and lactose in drinks and foodstuffs [17].

5.5.3 Vitamins

Vitamins are a class of organic compounds required by an organism as a vital nutrient in limited amounts. They cannot be synthesized in sufficient quantities by an organism and must be obtained from the diet. Vitamins are widely present in foods and agricultural products. Their contents are mainly determined by LC. In the past two decades, CE–ECD has been employed in the analysis of a variety of vitamins in foods and agricultural products.

In 1997, Li *et al.* separated ascorbic acid, thiamine, and pyridoxine, three water-soluble vitamins by CZE and MECC with AD [18]. The detection electrode was made of carbon fiber. A constant detection potential of 510–540 mV (vs. SCE) was applied to the detection electrode for AD. The developed analytical methods have been successfully applied in the determination of the three vitamins in health-caring foods.

In 2000, Chen *et al.* [8] developed a method based on CE–ECD for the determination of melatonin and pyridoxine in pharmaceutical preparations for health-caring purposes. CE was performed in a 25 mM phosphate buffer (pH 8.3). The two compounds were separated and detected within 12 min. The working electrode used was a carbon electrode operated in a wall–jet configuration. Excellent linearity was observed between peak current and concentration of analytes with the detection limits ranging from 1.3 to 2.7 μM. The proposed method was successfully applied to analyze the actual samples with satisfactory assay results.

Hu *et al.* determined nicotinamide, pyridoxine, and ascorbic acid in beverages by CE–AD [19]. The detection electrode was a carbon disk electrode that worked at a detection potential of 1.2 V (vs. SCE). The three water-soluble vitamins were determined with satisfactory results.

In 2005, contactless conductivity was coupled with CE for the determination of ascorbic acid and two preservatives (benzoate and sorbate) in soft drinks and vitamin C tablets [20]. The three analytes in real samples were well separated and detected within 400 s.

5.5.4 Ions

Ions are essential nutritions for human health. They are important constituents in foods and agricultural products. Usually, CD was employed in the CE analysis of ions.

Klampfl *et al.* have developed a method based on CE coupled with CD for the simultaneous determination of chloride, sulfate, oxalate, formate, malate, citrate, succinate, pyruvate, acetate, lactate, phosphate, and pyroglutamate in beer [21]. The anionic species were well separated and detected using a piece of a 60-cm-long fused-silica capillary at a separation voltage of −30 kV in a 7.5 mM 4-aminobenzoic acid containing 0.12 mM tetradecyltrimethylammonium bromide, pH adjusted to 5.75 with histidine.

Kaniansky *et al.* have investigated the CZE separations of inorganic anions (chloride, bromide, iodide, sulfate, nitrite, nitrate, fluoride, phosphate) based on their host–guest complex equilibria with α-cyclodextrin (α-CD) [22]. α-CD employed as a host was found to influence selectively the effective mobilities of the anions. The anions were detected by conductivity after CE separation. The developed method has been successfully applied in the anions in milk samples with satisfactory results.

5.6 Determination of Phenolic Compounds by CE-ECD

Phenolic compounds are widely present in foods and agricultural products as essential constituents. They usually serve as antioxidants and flavor additive in foods, bioactive constituents in foods and plants. Although LC has found a wide range of application in their analysis, CE-AD is an efficiently for the rapid and sensitive determination of phenolic species in foods and agricultural products.

5.6.1 Phenols in Tea

In 2009, Fang *et al.* determined kaempferol, apigenin, rutin, quercetin, luteolin, and ferulic acid herbal tea using CZE-AD [23]. β-Cyclodextrin (β-CD) was added to the phosphate running buffer to enhance the separation. The method was applied to analyze tea samples with recoveries in the range of 90.0%–107.0%.

5.6.2 Phenols in Coffee

Ye *et al.* have determined catechin, rutin, ferulic acid, *o*-dihydroxybenzene, chlorogenic acid, caffeic acid, gallic acid, and protocatechuic acid in coffee by CE-AD [24]. The working electrode was a carbon disk electrode at an applied potential of +0.95 V (vs. SCE). Under the optimum conditions, the analytes can be well separated within 26 min in a 75-cm-long fused-silica capillary. The current response was linear over three orders of magnitude with detection limits ranging from 60 to 360 ng/ml for all analytes.

5.6.3 Phenols in Wines

In 2002, Chicharrol *et al.* developed a method for the simultaneous detection of five polyphenols (caffeic, chlorogenic, ferulic acid, gallic acid, and catechin) by CZE with ECD using a 100 mM borate buffer (pH 9.2) [25]. The analytes were detected using a carbon nanotube (CNT)-modified glassy carbon electrode. The excellent electrochemical properties of this electrode allowed the detection of the selected polyphenols at 1200 mV with improved efficiency and resolution. The developed method has been successfully applied for the detection (ferulic, caffeic and gallic acids, and catechin) and the quantification (gallic acid and catechin) of polyphenols in two white wines without any preconcentration step. In 2004, Peng *et al.* developed a method based on CE-AD for the simultaneous determination of trans-resveratrol, (−)-epicatechin, and (+)-catechin in red wine [26]. Under the optimum conditions, the analytes could be separated in a 100 mM borate buffer (pH 9.2) within 20 min.

5.6.4 Phenols in Herbal Drugs

Phenolic compounds are an important group of secondary metabolites, which are synthesized by plants as a result of plant adaptation to biotic and abiotic stress conditions (infection, wounding, water stress, cold stress, high visible light, etc.) and are the main bioactive constituents in some herbal drugs. It has been proved by modern medicine that small phenols in herbal drugs have a broad range of physiological activities, such as antioxidation, antibacteria, anti-inflammation, antiallergy, alleviating pain, and enhancing the immune system [5]. They have found a wide range of applications in medicine, incense, and chemistry. For example, paeonol has been used clinically as a therapeutic medicine for myalgia, rheumatic pain, neuralgia, coetaneous pluribus, and so on. Its content is an important parameter for evaluating the quality of Moutan Cortex. We have established a method based on CZE-ECD for the identification and determination of paeonol, benzoyloxypaeoniflorin, and oxypaeoniflorin in Moutan Cortex [27].

In 2006, a novel carbon nanotube/poly(methylmethacrylate) (CNT/PMMA) composite electrode was developed as a sensitive amperometric detector of CE [28]. The composite electrode was fabricated on the basis of the *in situ* polymerization of a mixture of CNT and prepolymerized methylmethacrylate in the microchannel of a piece of a fused-silica capillary under heat. The performance of this unique system has been demonstrated by separating and detecting honokiol and magnolol in traditional Chinese medicine, Cortex Magnoliae Officinalis. As illustrated in Figure 5.4, Honokiol and magnolol have been well separated within 7 min in a 40-cm-long capillary at a separation voltage of 15 kV using a 50 mM borate buffer (pH 9.2). The new CNT-based CE detector offered significantly lower operating potentials, yielded substantially enhanced signal-to-noise characteristics, and exhibited resistance to surface fouling, and, hence, enhanced stability. It demonstrated long-term stability and reproducibility with relative standard deviations of less than 5% for the peak current ($n = 9$). In addition, we have developed a CNT/poly(ethylene-*co*-vinyl acetate) composite electrode for the capillary electrophoretic determination of esculin and esculetin in Cortex Fraxini in 2009 [29].

5.6.5 Flavones in Herbal Drugs

Flavonoids have been found widely presented in herbal drugs and have a broad range of physiological applications, such as anti-inflammatory, antitumor, anti-oxidant,

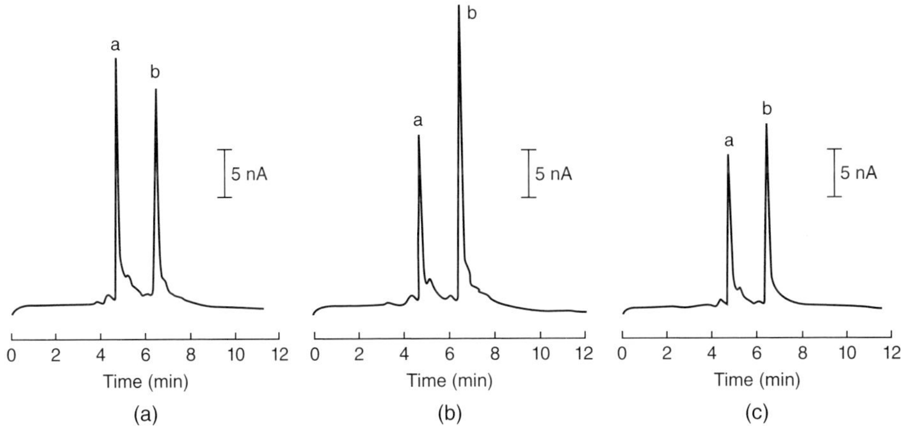

Figure 5.4 *Typical electropherograms for the diluted extracts from three samples of Cortex Magnoliae Officinalis ((A) Sample 1, (B) Sample 2, and (C) Sample 3). Detection electrode, CNT/PMMA composite electrode; Fused-silica capillary, 25 μm i.d. × 40 cm length; detection electrode, 320 μm diameter composite disc electrode; running buffer, 50 mM borate buffer (pH 9.2); separation and injection voltage, 15 kV; injection time, 6 s; detection potential, +0.8 V (vs. Ag/AgCl wire). Reproduced with permission from Ref. [28]. Copyright © 2006 Wiley-VCH Verlag GmbH & Co. KGaA, Weinheim*

antibacteria, and so on. According to the molecular structure, they can be classified into flavones, flavonols, flavanones, flavanonols, isoflavones, isoflavonol, chalcones, biflavones, and so on. Flavonoids usually contain several phenolic hydroxyl groups and are easy to be oxidized on carbon electrodes at moderate detection potentials ranged from +0.9 to +1.0 V (vs. SCE). The BGE for the CE separation of flavonoids was usually an alkaline borate buffer because the nearby hydroxyl groups in some flavonoids could react with borate to form anionic complex so that the separation might be improved.

Chinese traditional medicine, Scutellariae Radix, is the dried root of *Scutellaria baicalensis* Georgi. Baicalin and baicalein are two active flavones in Radix Scutellariae. In China, the Public Health Department requires that the content of baicalin in Scutellariae Radix should not be less than 8.0%. A method based on CZE-ECD has been developed for the determination of baicalein and baicalin in Scutellariae Radix and its pharmaceutical preparations [30]. The contents of baicalin and baicalein in the investigated samples of Radix Scutellariae were determined to be 44.4–85.5 and 10.7–20.6 mg g^{-1}, respectively.

In 2012, we have developed a carbon nanotube/poly(ethyl 2-cyanoacrylate) (CNT/PECA) composite electrode for enhanced AD [31]. The composite electrode was fabricated on the basis of the water vapor-initiated polymerization of a mixture of CNTs and ethyl 2-cyanoacrylate in the bore of a piece of fused-silica capillary. The analytical performance of this unique CNT-based detector has been demonstrated by separating and detecting six flavones in combination with CE (Figure 5.5). The advantages of the CNT/PECA composite detector include the lower operating potential, higher sensitivity, low expense of fabrication, satisfactory resistance to surface fouling, and enhanced stability, indicating great promise for a wide range of applications.

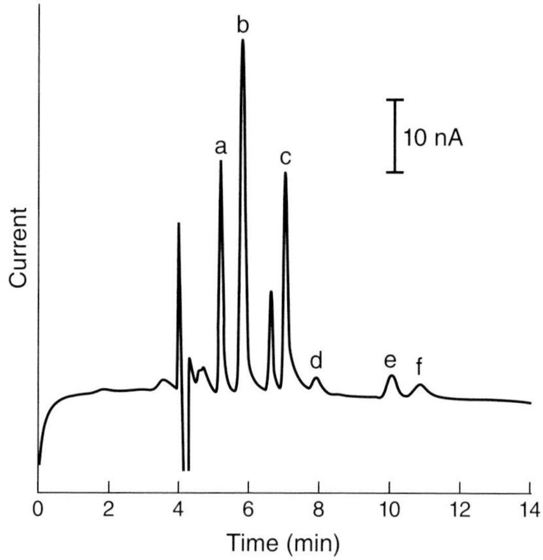

Figure 5.5 *Typical electropherogram for the diluted extract from a sample of Fructus Sophorae. Conditions: fused-silica capillary, 25 μm i.d. × 40 cm length; detection electrode, 320 μm diameter disc electrode; running buffer, 50 mM borate buffer (pH 9.8); separation and injection voltage, 12 kV; injection time, 6 s; and detection potential, 0.8 V (vs. Ag/AgCl wire electrode). Peak identification, (a) genistin, (b) sophoricoside, (c) rutin, (d) genistein, (e) kaempferol, and (f) quercetin. Reproduced with permission from Ref. [31]. Copyright © 2011 Wiley-VCH Verlag GmbH & Co. KGaA, Weinheim*

5.7 Determination of Purines by CE-ECD

As an important purine compound, caffeine in cola drink has been determined by CE coupled with a carbon disk electrode [32] at a high detection potential of 1.45 V (vs. SCE). In addition, we have determined theophylline in green tea by CE using a carbon electrode. Theophylline was well separated from other purines including adenine, guanine, thypoxanthine, xanthine, and uric acid within 14 min using a 40-cmlong fused-silica capillary at a separation voltage of 10 kV in a 100 mM borate buffer (pH 10.0) [33].

5.8 Determination of Food Additives by CE-ECD

Food additives are substances added to food to preserve its flavor or enhance its taste and appearance. However, their contents in foods should be strictly controlled. CE–ECD has found applications in the determination of food additives.

5.8.1 Preservatives

In 2010, Ye *et al.* employed CE-AD for the simultaneous determination of our electroactive preservatives (methylparaben, ethylparaben, propylparaben, and butylparaben) and two nonelectroactive preservatives (potassium sorbate and sodium lactate) in soft drinks [34]. An electroactive additive, 3,4-dihydroxybenzylamine, was added to the running buffer, so that both electroactive and nonelectroactive food preservatives were simultaneously detected.

5.8.2 Antioxidants

A new and efficient method for the determination of synthetic phenolic antioxidants has been developed by Ye *et al.* using MECC-AD [35]. Under the optimum conditions, all analytes were successfully separated within 13 min at the separation voltage of 18 kV in a 20 mM borate running buffer (pH 7.4) containing 25 mM SDS. The excellent linearity was obtained in the concentration range from 2 to 500 µM, and the detection limits of propyl gallate, tert-butylhydroquinone butylated hydroxyanisole, and butylated hydroxytoluene ranged from 0.29 to 2.7 µM. This method has been proved to be effective and successfully applied for the determination of these antioxidants in vegetable oil, providing a promising and convenient entry to monitor the superscale use of phenolic antioxidants.

5.8.3 Colors

A new analytical method, pressurized capillary electrochromatography (pCEC) with AD using 1.5 mm reversed phase nonporous silica packed columns has been developed for the rapid separation and determination of four Sudan dyes in hot chili [36]. The influence of several experimental parameters on the retention behavior has been investigated. The electrochemical oxidation of Sudans I–IV separated by pCEC can be reliably monitored with a carbon electrode at 0.95 V (vs. Ag/AgCl). Fast and efficient separation of the analytes was achieved within 7 min by pCEC under the optimum conditions. To evaluate the feasibility and reliability of this method, the proposed pCEC-AD method was further demonstrated with hot chili samples spiked with Sudan dyes.

5.8.4 Artificial Sweeteners

CE with capacitively coupled contactless C^4D was used for the simple, rapid, and simultaneous determination of aspartame, cyclamate, saccharin, and acesulfame-K in commercial samples of soft drinks and tabletop sweetener formulations [37]. A buffer solution containing 100 mM tris(hydroxymethyl)aminomethane and 10 mM histidine was used as BGE. A complete separation of the analytes could be attained in less than 6 min. The detection limit was considered to be better than those usually obtained by CE with photometric detection. Recoveries ranging from 94% to 108% were obtained for samples spiked with standard solutions of the sweeteners.

5.9 Summary

In conclusion, CE–ECD has become a powerful tool for agricultural and food analysis. Electrochemical detection leads to the improvements in the sensitivity of CE analysis, which will undoubtedly pave the way for more routine applications of the technique, particularly for the analysis of trace components. It is certain that analytical methods that are simple, rapid, and environmentally friendly will play an important role in the effort of providing high-quality foods and agricultural products to consumers worldwide. CE–ECD is an alternative approach to meet this requirement because of its high separation efficiency and detection sensitivity. Another important advantage of CE as an analytical tool for food and agricultural analysis is that the capillary is much easier to wash to acquire high reproducibility. The ongoing maturation of CE and further developments in ECD offer a promising technique for the separation, quantitation, and identification of the constituents. It can be concluded that CE–ECD is simple, efficient, and sensitive, providing a way for evaluating the quality of foods and agricultural products in marketplace. Advances in the field of microfluidic chip-based separations will allow many of the developed CE–ECD methods to be miniaturized to a smaller scale, requiring less sample, reagents, and time.

Abbreviations

CE	capillary electrophoresis
CNT	carbon nanotube
ECD	electrochemical detection
LC	liquid chromatography
GC	gas chromatography
CZE	capillary zone electrophoresis
MECC	micellar electrokinetic capillary chromatography
UV	ultraviolet
MS	mass spectrometry
CGE	capillary gel electrophoresis
CIEF	capillary isoelectric focusing
MEEKC	microemulsion electrokinetic chromatography
CEC	capillary electrochromatography
SDS	sodium dodecylsulfate
i.d.	inner diameter
o.d.	outer diameter
SCE	saturated calomel electrode
BGE	background electrolytes
LIF	laser-induced fluorescence

Acknowledgments

This work was financially supported by NSFC (21075020 and 21375023), Shanghai Science Committee (12441902900), State Oceanic Administration (201105007), and the Education Ministry of China (NCET-08-0134).

References

1. Escarpa, A. (2012) Food electroanalysis: sense and simplicity. *Chem. Rec.*, **12**, 72–91.
2. Frazier, R.A., Ames, J.M. and Nursten, H.E. (1999) The development and application of capillary electrophoresis methods for food analysis. *Electrophoresis*, **20**, 3156–3180.
3. Frazier, R.A. (2001) Recent advances in capillary electrophoresis methods for food analysis. *Electrophoresis*, **22**, 4197–4206.
4. Jorgenson, J.W. and Lukacs, K.D. (1981) Zone electrophoresis in open-tubular glass capillaries. *Anal. Chem.*, **53**, 1298–1302.
5. Chen, G., Zhu, Y., Wang, Y. *et al.* (2006) Determination of bioactive constituents in traditional Chinese medicines by CE with electrochemical detection. *Curr. Med. Chem.*, **13**, 2467–2485.
6. Terabe, S., Otsuka, K., Ichikawa, K. *et al.* (1984) Electrokinetic separations with micellar solutions and open-tubular capillaries. *Anal. Chem.*, **56**, 111–113.
7. Terabe, S., Otsuka, K. and Ando, T. (1985) Electrokinetic chromatography with micellar solution and open-tubular capillary. *Anal. Chem.*, **57**, 834–841.
8. Chen, G., Ding, X.G., Cao, Z.G. and Ye, J.N. (2000) Determination of melatonin and pyridoxine in pharmaceutical preparations for health-caring purposes by capillary electrophoresis with electrochemical detection. *Anal. Chim. Acta*, **408**, 249–256.
9. Wang, Q.J., Yu, H., Ding, F. *et al.* (2003) Simultaneous determination of food-related biogenic amines and precursor amino acids by micellar electrokinetic capillary chromatography with electrochemical detection. *Food Chem.*, **83**, 311–317.
10. Dong, Q., Jin, W.R. and Shao, J.H. (2002) Analysis of amino acids by capillary zone electrophoresis with electrochemical detection. *Electrophoresis*, **23**, 559–564.
11. Zhang, L.Y., Huang, W.H., Wang, Z.L. and Cheng, J. (2002) Determination of histamine by capillary zone electrophoresis with amperometric detection. *Anal. Sci.*, **18**, 117–1120.
12. Ye, J.N. and Baldwin, R. (1994) Determination of amino acids and peptides by capillary electrophoresis and electrochemical detection at a copper electrode. *Anal. Chem.*, **66**, 2669–2674.
13. Ye, J.N. and Baldwin, R. (1994) Determination of carbohydrates, sugar acids and alditols by capillary electrophoresis and electrochemical detection at a copper electrode. *J. Chromatogr. A*, **687**, 141–148.
14. Chen, Q.W., Zhang, L.Y. and Chen, G. (2012) Facile preparation of graphene–copper nanoparticle composite by in situ chemical reduction for electrochemical sensing of carbohydrates. *Anal. Chem.*, **84**, 171–178.
15. Zhang, X., Cao, Y.H. and Ye, J.N. (2001) Determination of lactose in sugar-free milk powder by capillary electrophoresis with electrochemical detection. *Food Chem.*, **72**, 385–388.
16. Hu, Q., Zhou, T.S., Zhang, L. and Fang, Y.Z. (2001) Study of the separation and determination of monosaccharides in soluble coffee by capillary zone electrophoresis with electrochemical detection. *Analyst*, **126**, 298–301.
17. Tuma, P., Malkova, K., Samcova, E. and Stulik, K. (2011) Rapid monitoring of mono- and disaccharides in drinks, foodstuffs and foodstuff additives by capillary electrophoresis with contactless conductivity detection. *Anal. Chim. Acta*, **698**, 1–5.

18. Li, G., Fan, C. and Xing, C. (1997) Preliminary study on separation of three vitamins by capillary zone electrophoresis and micellar electrokinetic chromatography with amperometric electrochemical detection. *Se Pu*, **15**, 65–66.
19. Hu, Q., Zhou, T.S., Zhang, L. *et al.* (2001) Separation and determination of three water-soluble vitamins in pharmaceutical preparations and food by micellar electrokinetic chromatography with amperometric electrochemical detection. *Anal. Chim. Acta*, **437**, 123–129.
20. Law, W.S., Kuban, P., Zhao, J.H. *et al.* (2005) Determination of vitamin C and preservatives in beverages by conventional capillary electrophoresis and microchip electrophoresis with capacitively coupled contactless conductivity detection. *Electrophoresis*, **26**, 4648–4655.
21. Klampfl, C.W. and Katzmayr, M.U. (1998) Determination of low-molecular-mass anionic compounds in beverage samples using capillary zone electrophoresis with simultaneous indirect ultraviolet and conductivity detection. *J. Chromatogr. A*, **822**, 117–123.
22. Masar, M., Bodor, R. and Kaniansky, D. (1999) Separations of inorganic anions based on their complexations with α-cyclodextrin by capillary zone electrophoresis with contactless conductivity detection. *J. Chromatogr. A*, **834**, 179–188.
23. Chi, L.Z., Li, Z.C., Dong, S.Q.g. *et al.* (2009) Simultaneous determination of flavonoids and phenolic acids in Chinese herbal tea by beta-cyclodextrin based capillary zone electrophoresis. *Microchim. Acta*, **167**, 179–185.
24. Chu, Q.C., Lin, M., Yu, X.Q. and Ye, J.N. (2008) Study on extraction efficiency of natural antioxidant in coffee by capillary electrophoresis with amperometric detection. *Eur. Food Res. Technol.*, **226**, 1373–1378.
25. Moreno, M., Arribas, A.S., Bermejo, E. *et al.* (2011) Analysis of polyphenols in white wine by CZE with amperometric detection using carbon nanotube-modified electrodes. *Electrophoresis*, **32**, 877–883.
26. Peng, Y.Y., Chu, Q.C., Liu, F.H. and Ye, J.N. (2004) Determination of phenolic constituents of biological interest in red wine by capillary electrophoresis with electrochemical detection. *J. Agric. Food Chem.*, **52**, 153–156.
27. Chen, G., Zhang, L.Y. and Yang, P.Y. (2005) Determination of three bioactive constituents in moutan cortex by capillary electrophoresis with electrochemical detection. *Anal. Sci.*, **21**, 1161–1165.
28. Yao, X., Xu, X.J., Yang, P.Y. and Chen, G. (2006) Carbon nanotube/poly(methyl methacrylate) composite electrode for capillary electrophoretic measurement of honokiol and magnolol in Cortex Magnoliae Officinalis. *Electrophoresis*, **27**, 3233–3242.
29. Chen, Z., Zhang, L.Y.N. and Chen, G. (2009) Carbon nanotube/poly(ethylene-*co*-vinyl acetate) composite electrode for capillary electrophoretic determination of esculin and esculetin in Cortex Fraxini. *Electrophoresis*, **30**, 3419–3426.
30. Chen, G., Zhang, H.W. and Ye, J.N. (2000) Determination of baicalein, baicalin and quercetin in Scutellariae Radix and its preparations by capillary electrophoresis with electrochemical detection. *Talanta*, **53**, 471–479.
31. Chen, Q.W., Gan, Z.B., Wang, J. and Chen, G. (2011) Facile preparation of carbon nanotube/poly(ethyl 2-cyanoacrylate) composite electrode by water vapor-initiated polymerization for enhanced amperometric detection. *Chem. Eur. J.*, **17**, 12458–12464.

32. Jin, W.R., Yu, D.Q., Dong, Q. and Ye, X.Y. (2000) Quantitation of caffeine by capillary zone electrophoresis with end-column amperometric detection at a carbon microdisk array electrode. *J. Chromatogr. Sci.*, **38**, 11–15.
33. Chen, G., Chu, Q.C., Zhang, L.Y. and Ye, J.N. (2002) Separation of six purine bases by capillary electrophoresis with electrochemical detection. *Anal. Chim. Acta*, **457**, 225–233.
34. Wang, W.Y., Wang, Y.P., Zhang, J.B. *et al.* (2010) Simultaneous determination of electroactive and non-electroactive food preservatives by novel capillary electrophoresis with amperometric detection. *Anal. Chim. Acta*, **678**, 39–43.
35. Guan, Y.Q., Chu, Q.C., Fu, L. *et al.* (2006) Determination of phenolic antioxidants by micellar electrokinetic capillary chromatography with electrochemical detection. *Food Chem.*, **94**, 157–162.
36. Liu, S.F., Zhang, X., Lin, X.C. *et al.* (2007) Development of a new method for analysis of Sudan dyes by pressurized CEC with amperometric detection. *Electrophoresis*, **28**, 1696–1703.
37. Bergamoa, A.B., Silvaa, J.A.F. and Jesusa, D.P. (2011) Simultaneous determination of aspartame, cyclamate, saccharin and acesulfame-K in soft drinks and tabletop sweetener formulations by capillary electrophoresis with capacitively coupled contactless conductivity detection. *Food Chem.*, **124**, 1714–1717.

Part II

Electrochemical Sensing in Food Analysis

6

Microelectrode Designs

Jonathan P. Metters and Craig E. Banks
School of Chemistry and the Environment, Division of Chemistry and Environmental Science, Manchester Metropolitan University, Chester Street, Manchester, M1 5GD, UK

6.1 Introduction

Microelectrodes are a useful tool in the electrochemist's arsenal and have at least one dimension of the order of microns. Typically, these microelectrodes are fabricated by taking the desired microwire and encasing this within glass which is heated to seal the microelectrode. The microelectrode is then revealed by grinding down to a planar surface and is consequently polished to a mirror finish with suitable polishing agents. Problems can arise from ineffective sealing and handling of the microwire, but these can be overcome with due care and diligence. Microelectrodes originally arose from the requirement to measure oxygen concentrations in biological organisms with the first microelectrode configuration of tungsten fabricated by Husel, a Nobel laureate [1–3].

The small geometric size of the microelectrode naturally gives rise to small currents and as a result decreases the ohmic drop and also has the benefit that the counter electrode is not necessary. In decreasing the electrode dimension from macro to that of micro, the rate of mass transport is quantitatively changed, being effectively enhanced allowing fast electrode kinetics to be studied [1, 4–6]. In terms of the usefulness in measuring analytes, the above benefits of using a microelectrode give rise to enhanced response times and allowing greater sensitivities with lower limits of detection to be achieved than would be possible utilising a macroelectrode.

Typically, at a macroelectrode, a peak-shaped response is observed which is governed by the well-known Randles–Ševćik equation given by

$$I_p = 0.4463nFAC\left[D\upsilon\left(\frac{nF}{RT}\right)\right]^{1/2}, \tag{6.1}$$

Agricultural and Food Electroanalysis, First Edition.
Edited by Alberto Escarpa, María Cristina González and Miguel Ángel López.

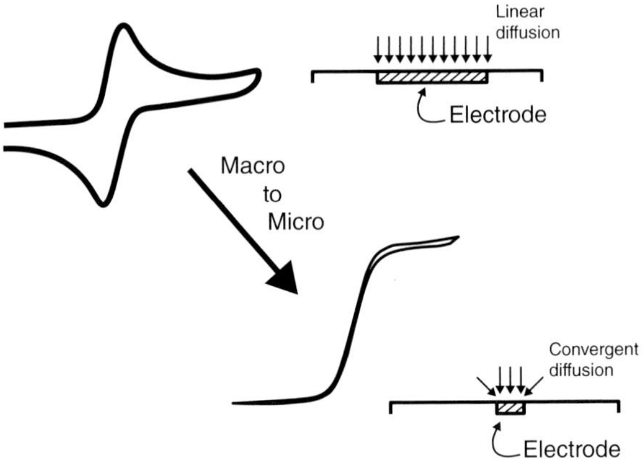

Figure 6.1 *Typical cyclic voltammetric responses obtained at both macro and micro electrodes originating from their mode of diffusion*

where n is the number of electrons transferred, A is the geometric area of the electrode, υ is the voltammetric scan rate, C is the bulk concentration of the analyte/redox probe and F, R and T possess their standard values. Due to the size of the macroelectrode, the diffusion layer has zero curvature and is consequently termed 'linear' or 'planar' diffusion. In changing the electrode size, such as reducing this to a microelectrode, the diffusion layer curves and as such the diffusion occurring is termed 'convergent', and in this case, the peak-shaped response is replaced with a steady-state-type response; Figure 6.1 shows such typical voltammetric responses and the associated mass transport regimes.

For the case of a microelectrode (a microdisc), a steady-state (limiting) current (I_L) is observed (see Figure 6.1) which is quantitatively described by

$$I_L^{\text{disc}} = 4nFDCr, \tag{6.2}$$

where r is the electrode radius. Related to microdisc electrodes are microband electrodes which do not show true steady-state behaviour and exhibit a steady-state at long times as the current changes very slowly with time, which is given by

$$I_L^{\text{band}} = \left(\frac{2\pi nFADC}{wln\left(\frac{64D\tau}{w^2}\right)} \right), \tag{6.3}$$

where w is the width of the band. In such electrode configurations, the dimensions of the microband are usually micron in width giving quasi-steady-state responses where the current is multiplied by the length of the band, thus giving rise to larger current outputs comparable to macroelectrodes [5, 6]. Such electrodes thus utilise the benefits of microelectrodes whilst overcoming the small output currents.

Another approach to produce voltammetric signals on the same current scale as macroelectrodes is through the use of microelectrode arrays, which give rise to significantly improved signal-to-noise ratios. In the case of a microelectrode array, the current is described by

$$I_L^{\text{disc}} = 4nFDCrN, \tag{6.4}$$

where N is the total number of electrodes comprising the array [7]. Note that the requirements for this equation to be valid are that no diffusional interaction/overlap between microelectrodes comprising the array occurs; diffusion must be independent at each of the applied voltammetric scan rates. In order to determine diffusional interaction, the diffusion layer can be estimated from

$$\delta \sim \sqrt{6D\tau}, \tag{6.5}$$

where D is the diffusion coefficient and τ is the timescale of the experimental analysis (the time to perform the voltammetric scan from a non-Faradaic process to that of a Faradaic process), such that

$$\delta = \sqrt{6D\left(\frac{\Delta E}{\upsilon}\right)}, \tag{6.6}$$

where ΔE is the potential width of the voltammograms as measured from the point corresponding to a non-Faradaic process and a Faradaic process. In order for Equation 6.4 to be valid, there should be no diffusion layer interaction between neighbouring microelectrodes such that the diffusion layer, δ, must be less than f_{greater} as given by

$$f_{\text{greater}} = \left(\frac{d_{\text{centre}}}{2}\right) - \left(\frac{d_{\text{electrode}}}{2}\right), \tag{6.7}$$

where d_{centre} is the centre-to-centre separation between the electrodes and $d_{\text{electrode}}$ is the diameter of the electrode. Thus, for an array to exhibit its full potential, the diffusional interaction should be minimal [8]. Indeed, it has been shown that a larger number of microelectrodes comprising an array do not necessarily provide improved analytical performance and an array with a smaller overall number of electrodes can outperform an array comprising a larger number if it exhibits diffusional independence [9].

In other electrode configurations, the electrodes, either a microdisc or an array of electrodes, can be recessed, that is, the electrode is not planar to the insulating material and consequently the diffusional profiles will quantitatively change [10]; recessed electrodes are typically produced unintentionally when photolithography is used. In the case of an array of recessed microdiscs,

$$I_L^{\text{Recess}} = 4\pi nFDC\left(\frac{r^2}{4L + \pi r^2}\right), \tag{6.8}$$

where L is the recess depth. Due to the nature of recessed electrodes, at slower scan rates, the transition from diffusional independence (sigmoidal response) to diffusional dependence (peak-shaped response) is extended over that possible at planar microelectrode arrays due to the time required for the diffusion layer to reach the edge of the recess [11]. Thus, it should be easier to achieve a sigmoidal response at a recessed microelectrode array over a planar microelectrode array which has the same disc size, geometry, distance between neighbouring electrodes and the total number of electrodes comprising the array. The advantages are a greater sensitivity, an increased current density and an improved signal-to-noise ratio [11] allowing an enhanced electroanalytical sensing [12, 13].

The next sections consider the various approaches by researchers to fabricate microelectrodes and arrays.

6.2 Microfabrication Techniques

6.2.1 Lithography and Related Processes (Deposition, Sputtering, Other Relevant Technologies)

In this section, we overview the various approaches by researchers utilised in order to fabricate microelectrodes and arrays of microelectrodes which inherently have their individual advantages and disadvantages.

A useful and innovative approach reported by Fletcher and Horne [14] involves sealing ~3200 randomly dispersed conducting carbon microdiscs (3.5 μm radius) embedded into epoxy resin. This electrode configuration is known as a RAM^{TM} electrode (random assembly of microelectrodes); however, only a small fraction of these are actually working [15]. The term assembly is used to indicate no regular spacing between the microelectrodes. Other microdiscs can, of course, be readily employed allowing noble metal RAM^{TM} electrodes to be readily derived.

A more common approach that also allows one to control a critical parameter, the separation between neighbouring microelectrodes, is through the use of photolithography which is based upon photoresists (light sensitive chemicals) and exposure tools such as UV sources [1, 16]. The fabrication process of gold disc microelectrode arrays is highlighted within Figure 6.2 [17].

The gold disc microelectrode arrays are fabricated using 4-in. silicon wafers and standard microfabrication techniques in a class 100 clean room. The first step in the fabrication is the growth of a thermal oxide layer across the wafer (Figure 6.2.2). This thermal oxide is grown in a furnace at 1100 °C within a water atmosphere, and the resulting oxide layer

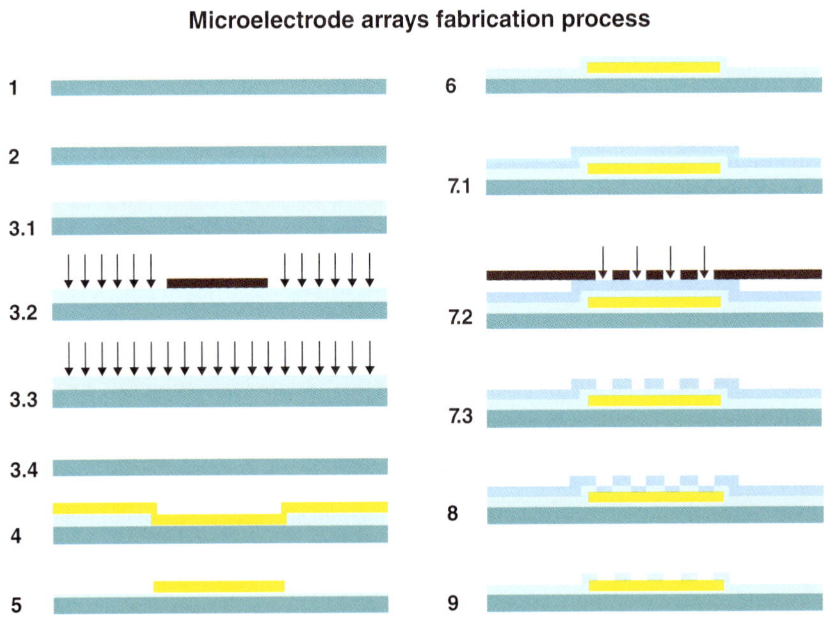

Figure 6.2 Typical fabrication process of gold disc microelectrode arrays. Reproduced from Ref. [17] with permission from Elsevier

is 1 mm in thickness and insulates the microelectrodes from the semiconducting silicon substrate to avoid leak currents. After this, the metal parts of the device are defined in a process known as lift-off. This process combines a photolithographic step and metallisation. Image reversal photoresist is then used to define the metal parts using a bright field mask (Figures 6.2.3.1–6.2.3.3). After developing the resist using the AZ-726 developer, the wafer is then rinsed using deionised water and dried in a nitrogen stream (Figure 6.2.3.4). Following this lithographic step, the wafers are metallised by DC-sputtering using a tri-layer composed of Ti (10 nm), Ni (10 nm) and Au (150 nm), as shown in Figure 6.2.4. Following metallisation, the wafers are immersed in a sonicated acetone bath to remove the resist, leaving on the wafer only the electrodes, contact pads and connection lines (Figure 6.2.5). After stripping the resist, the wafers are rinsed again in deionised water and dried in nitrogen. Next, a passivation layer consisting of 525-nm-thick silicon nitride over 500 nm of silicon oxide is deposited over the wafer surface in a plasma enhanced chemical vapour deposition (PECVD) reactor (Figure 6.2.6). Next, the passivation needs to be open to expose the contacts and define the final geometry of the microelectrodes. This is done with a new photolithographic step (Figures 6.2.7.1–6.2.7.3) followed by a sequence of etching steps. The areas to be etched are defined with a positive resist isolated through a dark-field photomask (Figure 6.2.7.2). After development, the resist is hard baked for a few minutes to make it more resistant to the conditions inside the reactive ion etch, RIE, reactor. The silicon nitride layer is completely etched by RIE, while the silicon oxide underneath it is only partially etched, leaving layer of a few nanometres over the gold surface. This thin layer is subsequently etched in a wet step by the immersion of the wafers in the Sioetch (BASF) solution for less than 3 min (Figure 6.2.8). It is important that the time in contact with the Sioetch solution, which is a buffered HF solution, is strictly controlled in order not to alter the geometry of the resulting microelectrodes in the array. Because HF etching is isotropic, a slight undercut may appear in passivation openings. This undercut is generally kept well below the micron, so it does not affect the electrochemical response of the arrays [17].

After the microelectrodes and the contacts are exposed, the excess resist is removed again in acetone in a sonicated ultrasonic bath, followed by rinsing with deionised water and drying under nitrogen (Figure 6.2.9). Next, the wafers are diced into 3×3 mm^2 individual microchips that are then attached to suitable print circuit boards (PCBs). The electrodes are connected to these PCBs by wire bonding and encapsulated using Epotek H-77 (Epoxy Technology, USA), a thermocurable resin [18]. Following this, the fabrication of the gold disc microelectrode arrays is completed producing a cubic arrangement of the microelectrodes comprising 135 individual microelectrodes, which are 20 μm in diameter and are separated from their nearest neighbour by 100 μm [17].

Despite the long fabrication process, high-quality microelectrode arrays are produced, where the density of microelectrodes and inter-electrode spacing from the nearest neighbour can potentially be controlled. Bai used such fabricated microelectrodes and modified them to produce a sensitive aptamer sensor [19]. Clearly, while this approach will produce arrays, single microelectrodes can be realised. The photolithographic technique is useful and can be extended to produce unique structures such as gold micropillar arrays, [20] as shown in Figure 6.3 and ring-recessed disc electrodes [21]. Problems encountered with arrays fabricated *via* the lithography are that the arrays comprise defective electrodes and

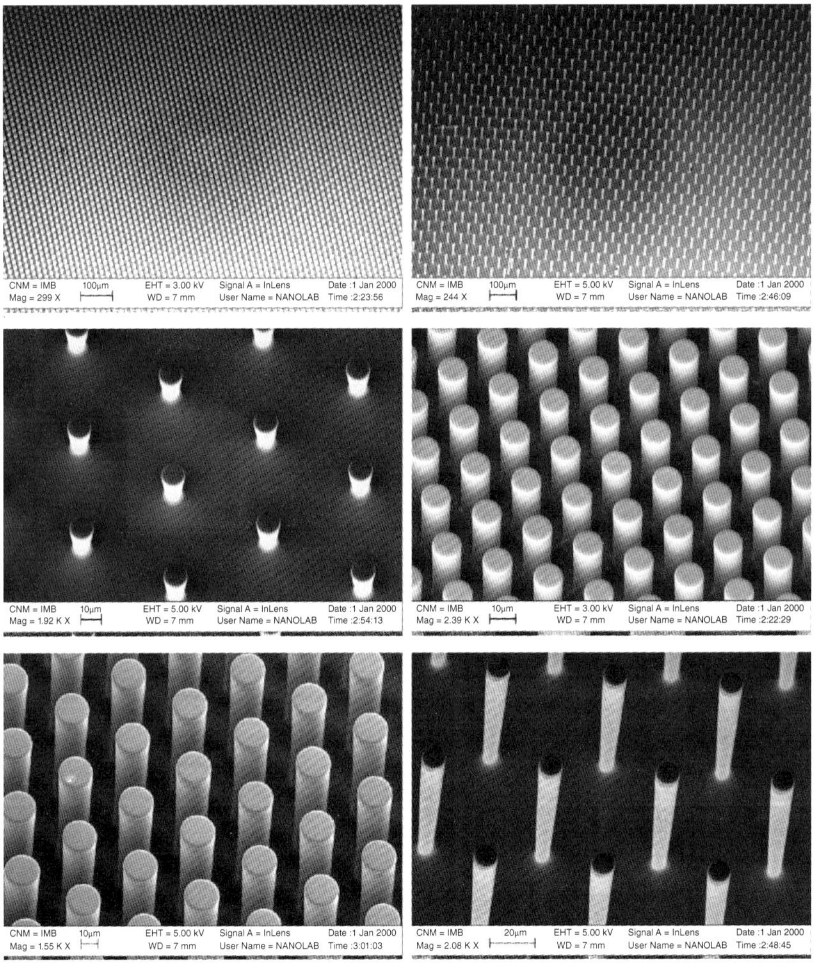

Figure 6.3 *SEM images of several fabricated gold micropillar array electrodes showing different radius, centre-to-centre distances and heights. Reproduced from Ref. [20]. Copyright (2012) The American Chemical Society*

it is not easy to work out how many these equate to without experiments and fitting of the obtained voltammetric data with numerical simulations [22, 23].

The exact costs and scalability of such microelectrodes produced through photolithography are still unknown, particularly when prerequisites such as clean-room facilities remain; nevertheless, unique electrode structures can be designed and fabricated with great care in regards electrode spacing and density. Inevitably, each fabrication approach using photolithography has subtle differences and will likely give rise to differing electrochemical performances.

Lançon *et al.* demonstrate the electrochemical differences in two arrays fabricated using two different methods: electroplating versus electron beam metallisation [24]. Electron-beam evaporation consists of exposing the metal to a beam of high-energy electrons in a high vacuum so that the metal evaporates and then condenses on the substrate. Electrochemical metallisation is achieved by applying an external electric potential between the substrate and a counter electrode while they are immersed in an electrolyte solution [24]. When the electrical potential is applied, the metal ions of the solution are reduced to their solid state at the surface of the electrode, to form a layer of material. Deposition is isotropic. In particular, for e-beam metallisation, the clean insulated substrate was coated with chromium (10 nm) and then with gold (200 nm), using an e-beam evaporator (Lesker six source). The electrodes were patterned using a standard photolithography process, which is described in detail in many standard sources found from the literature. For electroplating metallisation, the clean substrate was firstly coated with chromium (10 nm) and then with gold (50 nm). Positive photoresist S1813 was used to pattern the part of the substrate to be metallised with electroplated gold. The photoresist and unwanted seed layers were removed at the end of the electroplating process. SEM analysis showed that the thickness of the gold layer varied between 1 and 3 μm. This variation was not a problem for the electrode characterisation as the roughness of the gold, measured by atomic force microscopy (AFM), showed good consistency. The microdisc electrodes were shaped by the SU8-2 photoresist (Chestech), employed as an insulator. The insulating layer was then baked a final time at 120 °C for 10 min to settle its solidity. The surface of the electronic device was cleaned from residual material by oxygen plasma etching (Oxford Plasma Technology RIE80) [24]. Interestingly, it was found that the surface roughness significantly affected the output of the electrodes produced with the authors highlighting for the first time that this is a factor that needs consideration when making and implementing such electrodes.

Other uses of photolithography include making boron-doped diamond arrays [25–28]. Boron-doped diamond electrodes are attracting attention for use in electroanalysis due to their large potential windows (~3 V in KCl), low background currents, ruggedness, low susceptibility to fouling and long-term stability [29–37]. The observed low background currents and wide potential window allow low detection limits and quantification of species that would normally be voltametrically invisible, that is, outside the accessible potential window [38]. Additionally, the resulting flat electrode surface allows AFM to study nucleation phenomena [39]. Thus, it can be readily envisaged that a microelectrode array consisting of boron-doped diamond microelectrode discs is a versatile electrode device which harnesses the advantages of both microelectrode arrays and boron-doped diamond as an electrode material.

Boron-doped diamond microelectrode arrays have been reported [25–28, 40–43]. Fujishima and co-workers [40] fabricated a boron-doped diamond array by creating a photoresist pattern on silicon onto which the boron-doped diamond film is deposited. After this, the surface is spin coated with a polyimide film which is then mechanically etched producing a protruding boron-doped diamond array consisting of 200 boron-doped diamond discs with sizes between 25 and 30 μm diameter separated by a distance of 250 μm. Another approach is where first a boron-doped diamond film is formed onto which a Si_3N_4 layer (5 μm thick) is patterned via standard photolithography techniques [41]. This effectively produces recessed boron-doped diamond arrays. A hexagonal array of 106 boron-doped diamond discs, which are 5 μm in diameter separated by a distance of 150 μm, has been reported. However, the Si_3N_4 layer may not be stable under harsh conditions such as elevated temperatures and pressures where boron-doped diamond is utilised due to its ruggedness [41]. Swain and co-workers [42] have reported diamond ultramicroelectrode arrays which are based on forming a pattern via photolithography onto a silicon wafer with the CVD of diamond grown into the mould. These boron-doped diamond electrodes again are effectively recessed squares which can be produced to be separated by 20 or 100 μm. In other work, Madore *et al.* have reported on boron-doped diamond microdisc electrodes fabricated using CVD deposition and photolithographic techniques producing 200 microdisc electrodes which are 5 μm in diameter and separated by a 100 μm distance [44]. Similarly, Pickles and co-workers have reported on an all-diamond electrode which consists of individual boron-doped diamond discs which can be controlled between 10 and 25 μm in diameter, separated from the nearest neighbour by 10 times as much as the discs diameter. The array is in a hexagonal pattern which has been micro-machined using a combination of state-of-the-art microwave-induced plasma growth and laser ablation shaping techniques to prepare and coat a patterned boron-doped diamond substrate with an intrinsic diamond insulating layer which has the advantage of the array being co-planar to the dielectric surroundings [45]. Thus, in this configuration, the array is completely made from diamond; the insulation is diamond, while the discs comprise boron-doped diamond. Fabricating such a device is advantageous since the resulting electrode has no seals, recesses or elevations in contrast to those already reported. Additionally, the arrays can form the basis of an electrode array substrate where metals can be electroplated and re-used due to the ruggedness of boron-doped diamond [25].

A boron-doped diamond array of notability is shown in Figure 6.4, which depicts the schematic plots of the fabrication of integrated all-diamond ultramicroelectrode arrays (UMEAs). As shown in Figure 6.4a, an insulating diamond (iD) film with a thickness of a

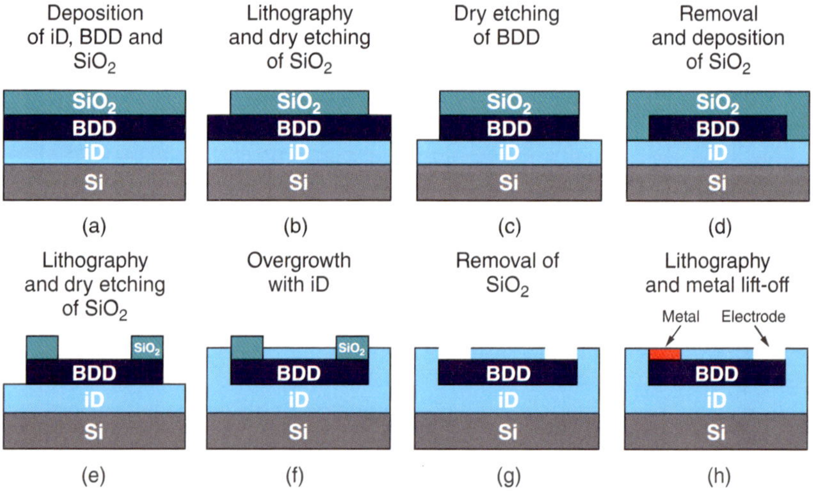

Figure 6.4 *Schematic plots of the fabrication of integrated all-diamond ultramicroelectrode arrays. The short names of iD represent insulating diamond and of BDD boron-doped polycrystalline diamond. (a) Deposition of iD, BDD and SiO_2, (b) lithography and dry etching of SiO_2, (c) dry etching of BDD, (d) removal and deposition of SiO_2, (e) lithography and dry etching of SiO_2, (f) overgrowth with iD, (g) removal of SiO_2 and (h) lithography and metal lift-off. Reproduced from Ref. [46]. Copyright (2011) The American Chemical Society*

few micrometres (8–10 μm) is first grown on a Si wafer of 2 in. in diameter to electrically disconnect the Si substrate from the electrochemical sensor. The iD film is then polished, followed by the deposition of 200–500 nm thick boron-doped polycrystalline diamond. The boron-doped diamond film was cleaned wet-chemically in the mixture of concentrated sulfuric acid (98%) and concentrated nitric acid (65%; v/v = 3 : 1) at 200 °C for 1.5 h, resulting in an oxidised surface. To generate SiO_2 patterns on the boron-doped diamond film, an evaporated SiO_2 layer with a thickness of 350 nm is structured by means of optical photolithography and dry etching using SF_6 gas (Figure 6.4b). These SiO_2 patterns are then used as a mask to structure the boron-doped diamond film during a dry-etching process, which was conducted using reactive ion etching in a gas mixture of oxygen and hydrogen (Figure 6.4c). In Figure 6.4d, a subsequent deposition of SiO_2 on the whole wafer was performed. The second optical photolithography step and dry etching of SiO_2 using SF_6 gas are then applied (Figure 6.4e). The protected parts by SiO_2 patterns are the electrodes and the contacts of the chip. The areas that were not protected by SiO_2 patterns are then overgrown with the iD film as shown in Figure 6.4f. The overgrowth of the iD film on the boron-doped diamond film formed boron-doped diamond channels which were used for the electric connection between electrodes and the contacts. After the wet-chemical removal of the SiO_2 (Figure 6.4g), the third optical photolithography step was applied and then Ti/Pt/Au (20/60/200 nm) metal layers were deposited on the contacts by the lift-off technique (Figure 6.4h).

Figure 6.5a shows the picture of an integrated all-diamond fabricated UMEA. The realised chip has a size of 5 × 5 mm^2 and has a reference electrode, a counter electrode

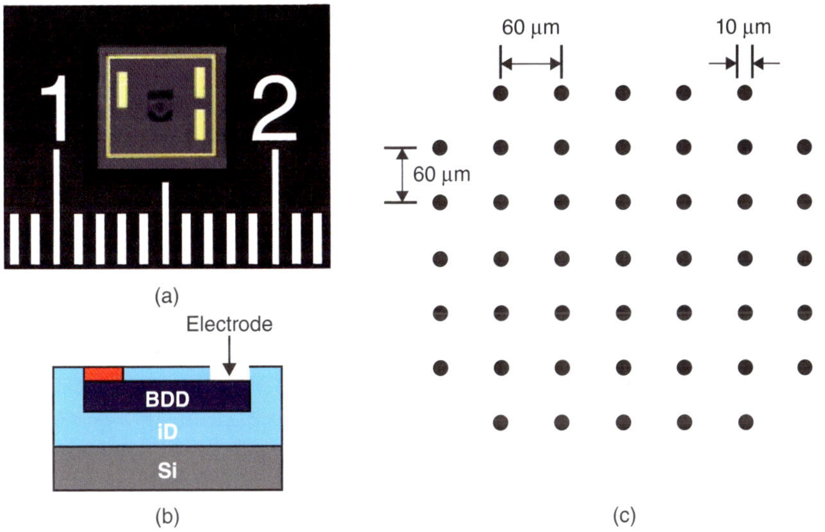

Figure 6.5 *(a) Photography of an integrated all-diamond ultramicroelectrode array chip. The yellow parts are metal contacts. The dark part with a semicircle is the counter electrode, the dark rectangle is the quasi reference electrode, and the centre electrode is the working electrode. (b) Schematic plot of the structure of the electrode. (c) Arrangement of the array which is composed of 45 ultramicroelectrodes (10 μm in diameter) in a 500 μm (diameter) circle. Reproduced from Ref. [46]. Copyright (2011) The American Chemical Society*

and an array of ultramicroelectrodes (working electrode). Such electrodes are connected with metal contacts via boron-doped diamond channels which are hermetically covered by insulating diamond (Figure 6.5b) [46]. The arrangement of the boron-doped diamond ultramicroelectrodes is shown in Figure 6.5c. The diameters of the resulting boron-doped diamond ultramicroelectrodes are 10 μm (diameter), with the vertical and horizontal centre-to-centre spacing between electrodes being 60 μm (width). The total number of ultramicroelectrodes on the chip was reported to be 45. For one 2 in. wafer, more than 40 integrated diamond UMEA chips can be obtained. Note that this fabrication approach produces recessed boron-doped diamond microelectrodes.

Clearly, boron-doped diamond arrays are viable, though each possesses its own pros and cons. Again, the cost of these arrays is yet to be confirmed and time will tell whether mass production is viable; both factors will dictate their market implementation within diagnostic devices, for example.

Other approaches to fabricate arrays include electrolytic deposition onto electrode substrates resulting in the surface decoration of micro/nanoparticles [47], or the use of polymers to cover a macroelectrode that only reveals the underlying electrodes where the exposed electrode is of microdomain size [48]. As mentioned earlier, in microelectrode arrays, microelectrodes are at a fixed distance from their nearest neighbour and an alternative is to have a random array of microelectrodes which are termed as ensembles due to no regular spacing between neighbouring microelectrodes [14, 9]. It has been shown by simulations that a random array can produce the same (but never greater) current–potential response as that of a regular array of equal macroscopic coverage [49] and, due to their nature, random arrays are generally easier to fabricate. Such an approach is cost-effective only if screen-printed electrodes are utilised and consequently can be disposed of the following use, rather than having to clean and pre-treat the electrode surface before a fresh micro/nanoparticle-decorated electrodes can be obtained, as is the case of a solid electrode such as glassy carbon or boron-doped diamond.

Following this, it is evident that a deciding factor for the implementation of the electrodes is, of course, cost, that is, the cost per sensor to the end user. As such, a really cost-effective approach is the sputtering of noble metals onto surfaces for use as electrodes. Saito first reported the sputtering of platinum films for the polarographic sensing of oxygen where in comparison to a platinum wire electrode, a beneficial response was observed at the sputter platinum film [50]. Brett *et al.* have reported a thin-film gold electrode fabricated via radio-frequency sputtering onto PVC sheets which was used in batch injection analysis (BIA). They found that this approach produced polycrystalline-like platinum and exhibited beneficial responses for the sensing of hydrogen peroxide which is of significant interest in biosensor applications. Clearly, such a configuration is highly cost-effective; yet batch-to-batch variation/control needs to be diligently considered [51].

Other notable work from Mie [52] detailing gold sputtered electrodes which provided direct electrochemistry with human cytochrome P450 mimicking a gold nanoparticle electrode. Given the cost and easy of fabrication, this electrode configuration is highly promising. Configurations into arrays are yet to be realised and are challenging to produce; hence, lithography is still favoured for this approach. Additionally, sputtered SnO_2 and platinum have been used in fuel cell applications [53, 54].

In the above section, we have seen some fascinating approaches to produce electrodes and electrode arrays. It can be observed that lithography and related techniques allow arrays to be produced with controllable electrode separations which are key for their beneficial use. However, cost is still a factor yet to be overcome for such configurations. Alternatively, sputtered electrodes are highly cheap, mimicking polycrystalline electrodes, but are yet to be successfully implemented for the production of arrays, and their mass production needs to be considered. One known approach that meets the criteria of being cheap, disposable and practical on a large scale, whilst maintaining reproducibility, is that of screen-printing. In the following sections, we consider these versatile electrodes.

6.3 Screen-Printing for Producing Electrochemical Sensors

Screen-printing technology is a well-established technique dating back to AD 960–1279 as a form of stencilling that first appeared in a recognisable manner in China during the Song Dynasty [55]. It was then adapted by other Asian countries like Japan and was furthered by creating newer methods [55]. More recently, it has seen a plethora of applications when utilised for the fabrication of both chemical and biosensors [56–59]. One of the most notable applications of screen-printed sensors involves the commercialisation of screen-printed electrodes for the monitoring of blood glucose levels in diabetics; a market which is a billion dollar per annum global market [60–63]. As self-monitoring sensing devices continue to hold a vital role within our ever expanding society, it would be naive to ignore the inevitable growth and demands for sensing devices related to clinical and industrial applications.

De-centralised sensing is ever more necessary and, thus, traditional techniques utilising highly expensive, largely immobile analytical equipment such as gas chromatography-mass spectrometers are not feasible for sensing outside the realms of standard laboratories nor in the evermore concerning economic climate [64]. Scenarios in which the implementation of portable, economical and sensitive sensors would be highly sought after include utilisation in hospitals where there is a suspected drug overdose, the personal monitoring of diseases such as diabetes [60] which utilises a simplistic glucose biosensing system (as mentioned earlier), the detection of potential pollutants or toxins within environmental samples [65] such as river water, the screening of drinking water at different sources [66] and also the rapid determination of naturally occurring biomolecules [57].

In addition to offering a truly cost-effective device, screen-printed electrodes also satisfy the previously much sought after criterion of highly reproducible and sensitive methods of detection towards target analytes, whilst maintaining the low cost production even through scales of economy on a commercial basis. The adaptability of screen-printed electrodes is also of great benefit in areas of research; the ability to modify the electrodes with ease, through differing inks commercially available for the reference, counter and working electrodes, allows for highly specific and finely calibrated electrodes to be produced for specific target analytes [56]. Similarly, the fluidity and ease of stencil design also allow for the tailoring of sensors to enable improvements in analytical performance, for example through the improvement of mass transport [67].

Screen-printing involves the movement of a thixotropic fluid through a mesh screen, which defines the shape and size of the desired electrode, on to the desired substrate material beneath. The thixotropic fluid contains a variety of substances typically including graphite, carbon black, solvents, and polymeric binder with the exact constituents and ratios

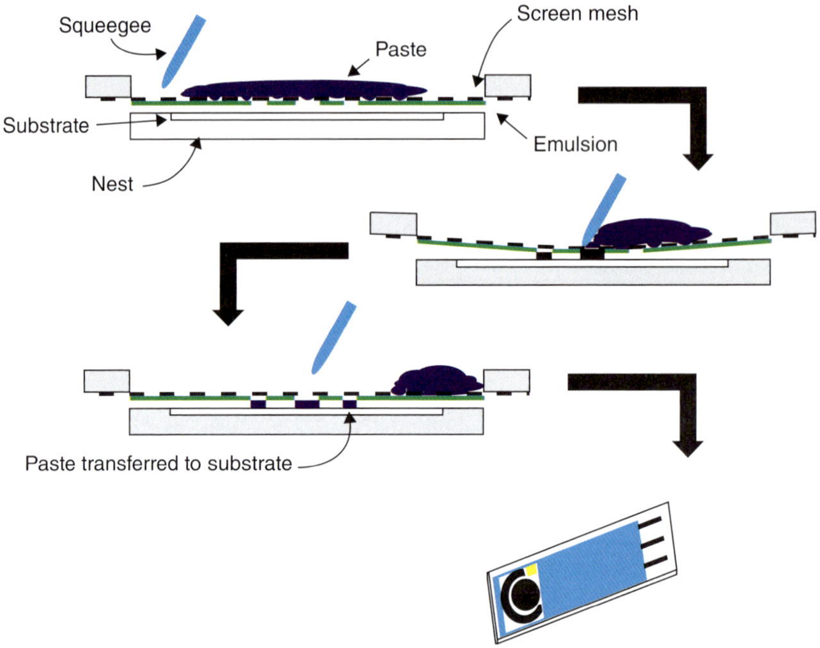

Figure 6.6 *Schematic representation of the process of screen printed manufacturing of electrodes (cross-sectional side view) outlining the basic processes involved. Adapted from Ref. [72] with permission from The Royal Society of Chemistry*

being withheld as propriety information [59, 68, 69], although it is important to note that the electrode material of choice can be easily substituted, for example, using a noble metal in place of graphite [70, 71]. Figure 6.6 depicts a schematic representation of the process of screen printing, where multiple layers and complex designs can be implemented. The inks utilised have a high viscosity (3–10 Pa s at a sheer rate of 230 s^{-1}) but when forced through the screen mesh by the squeegee blade, the ink undergoes sheer thinning which in turn allows it to penetrate through the screen mesh which defines the final shape/design. Upon contact with the substrate, typically a ceramic or plastic material, the ink returns to its viscous state forming the intended shape/design, with the definition achieve reliant upon the viscosity of the ink utilised. Such final shapes/designs have thicknesses in the range of 20–100 μm and as such are thicker than those obtained by other printing methodologies; thus, they are consequently termed 'thick film technology'. Note that the thickness can be readily controlled by the thickness of the stencil design and the mesh [73, 74]. As is explained, such designs can be printed onto ceramic substrates [75] or plastic substrates [75–77] depending on the intended application, though paper-based [78] substrate materials have also been reported. In the penultimate case, the cost is generally lower and the carbon (typically) print is better adhered to the substrate than in the former case.

Conventionally, these electrodes are utilised in aqueous solutions, but as there are many applications in non-aqueous systems, it is useful for screen-printed electrodes to operate in this type of media, but such designs are yet to be fully realised [79].

The effect of mechanical stress on screen-printed sensors fabricated upon differing electrode substrates has been studied by Wang *et al.* [80]. The different substrates utilised within the study included Mylar, polyethylene naphthalate (PEN) and Kapton. In this instance, the bending of the electrodes to extremely small radii of curvature was determined to have minimal effect upon the electrochemical behaviour, resulting in only a small increase in the electrical resistance of the sensors. Additionally, the study determined that the electrochemical signal was not adversely impacted from cyclic bending; this illustrates the robust nature of screen-printed electrodes upon flexible substrates [80].

Developing on their initial study, Wang and colleagues studied the influence of mechanical bending of carbon screen-printed electrodes [81] modified with Nafion and glucose oxidase has been explored by Wang and co-workers, whereby screen-printed sensors were placed under extreme stress, ~180° pinch, and were found to display an enhancement in the sensitivity which is attributed to a change in the bioactivity, viz. the glucose oxidase/Nafion layer rather than that of the carbon electrode surface [81]. This work has implications for on-body wearable sensors where the screen-printed electrode will undergo great strain and duress and is utilised for monitoring the wearer's health and the surrounding environment [81–83], and has additionally been shown to be useful for 'biosensors in briefs' where screen-printed fabricated amperometric sensors have been incorporated into clothing; directly into the elastic waist of the underwear [84]. Note that there are many research efforts into developing screen-printed biosensors and consequently there is a substantial wealth of literature and as such there are authoritative reviews on utilising screen-printed enzymatic electrode biosensors, see, for example, [58, 72, 85–88].

6.3.1 Improving Mass Transport

As has been highlighted earlier, the improvement of mass transport is critical for the continued development and enhancement of electrochemical sensors desired for analytical applications. Evidently, the fabrication of sensors which allow for this, for example microelectrodes, is of key interest and an area (see Section 6.1). This area holds a great potential through the incorporation of screen printing which allows for the production of highly specific and tailored electrode designs and shapes.

Recently, the fabrication of microelectrode arrays entirely via screen printing has been explored where a large carbon working is first printed. Onto this carbon layer, a dielectric is printed which has predefined holes, as defined by the mesh screen, exposing the underlying carbon electrode in the form of carbon microelectrode arrays. It was found that the optimum printing allowed microdiscs with radii of 116 (± 6) μm; attempts at smaller microelectrodes result in a large range of sizes and ill-defined geometries [12]. Due to the fabrication methodology, recessed microelectrodes are produced, and in our recent case, this is ~4 μm, equating to a shallow recessed microelectrode. If the number of microelectrodes on the array is known (N), the average disc radius, $\bar{r}$, could be estimated from a single measurement of the limiting current as demonstrated earlier in Equation 6.8. This should be the case due to the fabrication process via screen-printing which defines the geometry and number of electrodes. However, it was found that this was overestimated with the error arising due to the diffusional overlap between neighbouring microelectrodes. The electroanalytical utility of the screen-printed electrode array was explored in relation to the sensing of Mn(II) allowing a limit of detection of ~81 nM. Note that this was achieved using an array comprising 16 electrodes separated by 1250 μm from its nearest neighbour. A six microelectrode array was also constructed which has a separation distance of 2500 μm from its nearest neighbour and exhibited a detection limit of ~64 nM. Both responses have significant analytical benefits over the current literature in terms of both the electrochemical and general analytical methodologies. In comparison of the two shallow recessed microelectrode arrays, the six microelectrode array exhibits an improved sensitivity and a wider linear range over the 16 microelectrode array. This is due to the distance between neighbouring electrodes being larger on the six microelectrode array than the 16 microelectrode array, where the radial diffusion of the former is greater than that on the latter. Based on these results, the screen-printed shallow microelectrode arrays clearly have beneficial analytical utility in comparison to conventional carbon electrodes [12]. Note that this has been demonstrated for the sensing of lead in river samples at the EC Dangerous Substance Directive (76/464/EEC) [89] and for the detection of nitrite in river water samples where Khairy *et al.* [90] demonstrated a protocol which was deemed feasible for the sensing of levels indicated by the World Health Organization. The disposable nature and low cost of the sensor offer an economical and portable screening tool for nitrite.

Other work by Kadara *et al.* [91] reported the fabrication of disposable and flexible screen-printed microelectrodes characterised by microscopy and cyclic voltammetry. The advantages presented by the fabricated electrodes included reduced expenditure and cleaning processes, as each of the microelectrodes is designed to be disposable; the removal of the requirement of cleaning stages, but also pre-treatment between analyses, allows for a much more efficient and rapid analysis of samples; the work also boasted exceptional detection limits with the screen-printed electrodes providing comparable detection limits to that obtained in the literature at insonated boron-doped diamond electrodes [91].

Elegant work by Ball *et al.* [92] highlighted the concept of coupling screen printing and laser micromachining technology in order to fabricate a nanovial with a 'built-in' working and reference electrodes. A nanolitre cell was created through the incorporation of both mechanisms which successfully simplifies manipulations necessary for small volume experiments. Additionally, the ease of manipulation of both screen printing and laser micromachining techniques allows for a wide array of applications in electrochemistry in ultrasmall environments in the future.

Other notable reports of using screen printing for adventurous designs have been reported by Sch*ü*ler *et al.* [93] who explored the fabrication of DNA chips. Figure 6.7 shows one such chip fabricated using a gold paste screen printed onto glass. This methodology offers a cost-effective alternative for the production of chips with electrode structures as needed for an electrical chip-based DNA detection [93].

Electrodes of dimensions lower than the millimetre size with diffusion-limited processes have attracted considerable interest to electrochemists (see the earlier section). Their unique electrochemical properties have been used in many applications preferable to electrodes of conventional size [94]. Zen and his group [94–96] have developed edge band ultramicroelectrodes (SPUME), building upon the initial work by Craston and co-workers [97] where the electrode is built up as shown in Figure 6.8. Through slicing the edge of the working electrode, a band-type ultramicroelectrode is formed and can easily be 'renewed' after use by cutting the edge off. The length of the band-type ultramicroelectrode can be varied in the range of 0.18–1.35 mm with a width of ~20 μm [94]. This electrode configuration has been extended and explored in flow injection analysis for the sensing of nitrite in lake and

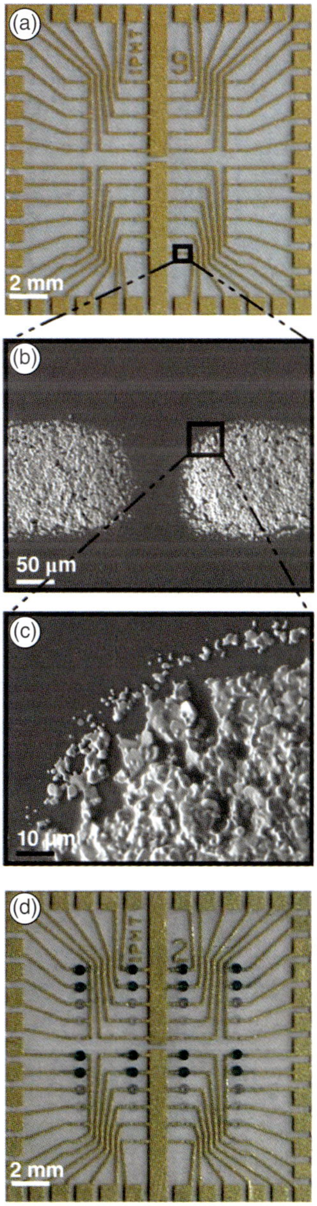

Figure 6.7 *Screen printed chips with gold (a) electrode structures. The black spots (d) represent the signals observed on chips after enzyme induced silver deposition. SEM images show the ultrastructure of gold (b,c) electrodes. Reproduced from Ref. [93] with permission from Elsevier*

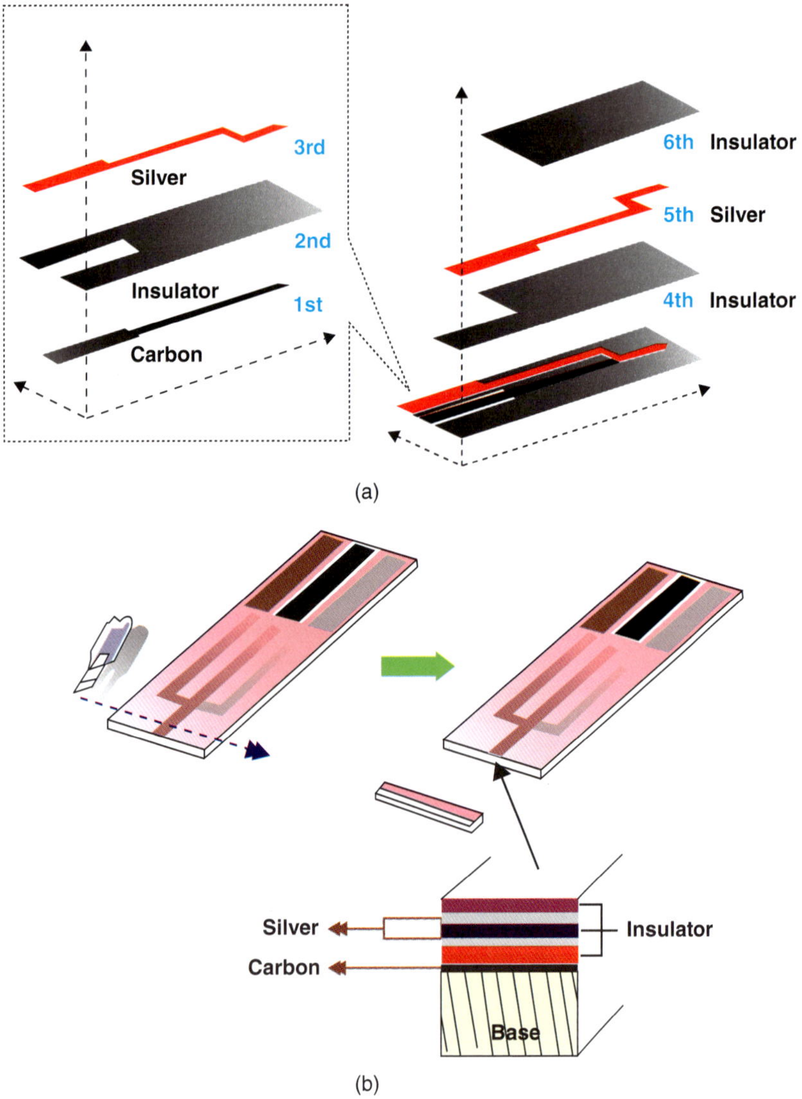

Figure 6.8 *(a) An alternating layer-by-layer pattern structure of the SPUME (a disposable screen printed edge band carbon ultramicroelectrode) assembly. (b) Cross-sectional diagram of a typical SPUME with a built-in three-electrode configuration. Reproduced from Ref. [94] with permission from Elsevier*

group waters [98, 99], and has recently been modified with platinum nanoparticles for the amperometric sensing of carbon monoxide [100] and formaldehyde [101].

This novel design has been extended by Hart *et al.* using water-based inks containing cobalt phthalocyanine for the sensing of lactate [102], toxins [103] and glucose [104]. Other advantageous designs involve the fabrication of wall–jet screen-printed ring disc electrodes which, compared to traditional methods of fabrication, are inexpensive and due to the disposable nature do not suffer from extensive pre-treatment in the form of electrode polishing [95, 96]. Additionally, such electrodes could be modified with nano- and micro-sized electro-catalytic particles or produced using metallic inks.

In a similar fashion, Metters *et al.* designed and fabricated a carbon-based microband produced entirely through screen printing [67]. Unlike that described by Zen [94] and Hart [105], that proposed by Metters consists of a screen-printed graphite microbands printed upon a flexible plastic substrate which is 10 mm in length and 100 μm in width (see Figure 6.9). The graphite microbands have a critical microscopic dimension imparting microelectrode behavior, whilst the length yields an increase in the current output. Electro-analytical characterisation with the electrochemical detection of NADH and nitrite reveal that experimental limits of detection (3σ) of 0.48 and 0.05 μM, respectively, are possible. The screen-printed graphite band electrodes are benchmarked against screen-printed graphite macroelectrodes as well as graphite ultramicrobands, where in terms of current density, the band electrodes exhibit a greater response over the former, but not the latter. Critically, however, the study is extended to explore the potential pitfalls of the fabrication of sensors utilising the method described by Zen [94]. As is demonstrated in Figure 6.10, during the preparatory steps required for the fabrication of ultramicroband electrodes, that is, the mechanical shearing of the electrode to reveal the ultramicroband

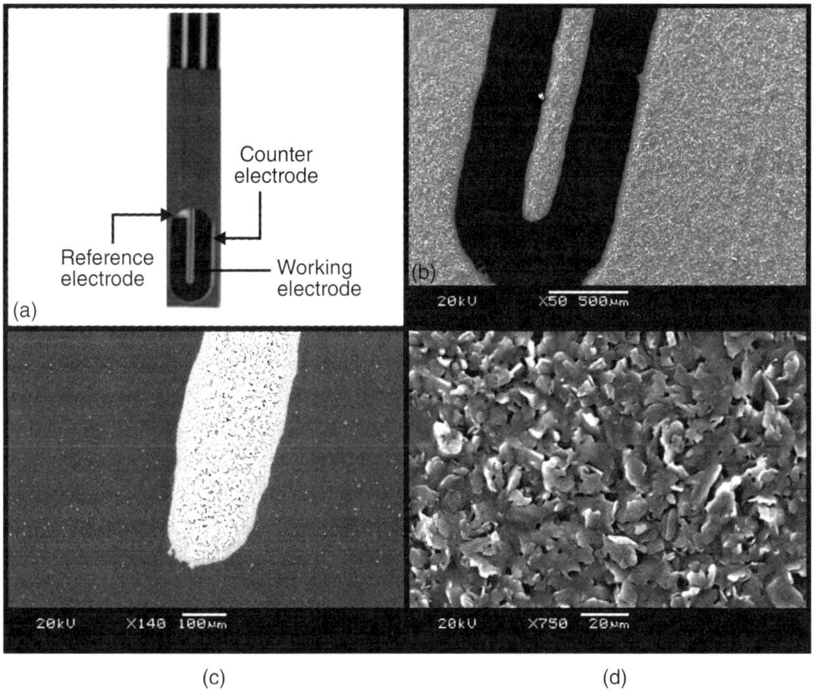

Figure 6.9 *Optical and scanning electron microscopy images of the screen-printed microband electrode (bSPE). (a) An optical microscopic image of the screen printed band electrode, (b) SEM 50× magnification of the bSPE working area tip, (c) SEM 140× magnification of the bSPE working area tip and (d) SEM 750× magnification of the bSPE working area tip. Reproduced from Ref. [67] with permission from Elsevier*

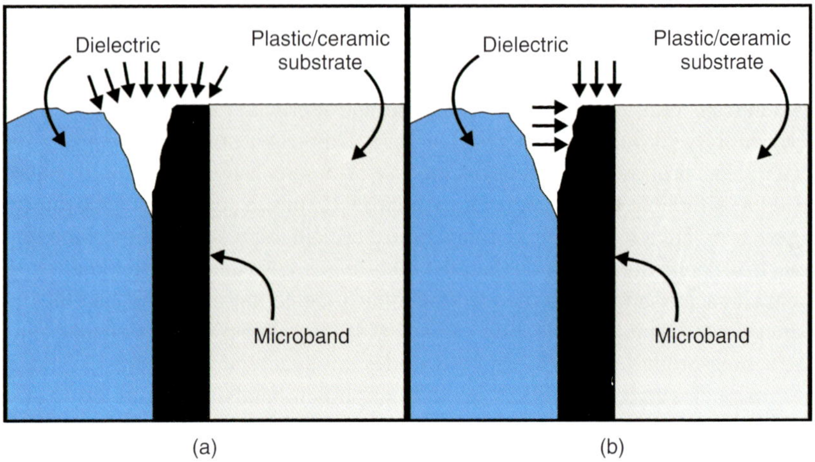

Figure 6.10 A schematic representation of convergent diffusion (a) and linear diffusion (b) at μbSPEs which arise from cutting with scissors. Reproduced from Ref. [67] with permission from Elsevier

(see Figure 6.8b), the cutting pulls the dielectric away from the graphite microband, such that at slow scan rates, convergent diffusion dominates as soon as the volume of solution contained within the expanded/raised area between the dielectric and graphite is electrochemically exhausted. At faster scan rates, linear diffusion dominates, and it is the summation of horizontal and vertical mass transport. Such a response has been obtained for thin metals trapped within glass which are not sealed properly [106]. Such changes in the mass transport regime, which are not controllable and dependent upon cutting, could clearly contribute to the observed (variable) electrochemical response [67].

As such, the band reported by Metters *et al.* aims to alleviate such problems whilst also reducing any preparatory steps required as upon the completion of the screen-printing process, the sensor requires no further treatment or modification. The mass production of such disposable microband electrodes holds great promise in electrochemistry, particularly as a base transducer for utilisation in electroanalytical sensing where the benefits of the microscopic domain of the band electrode imparting improvements in the analytical performance can be realised.

Very recently, the Banks group expanded the horizons of screen-printed microband electrodes through the fabrication of a graphite microband of 50 μm in width [107]. The reduction in microband width was made possible through the development of a new 'V-mesh', an alternative to the more traditionally utilised stainless steel and polyester meshes for the production of screens [108]. The V-mesh is produced by using filaments of Vecry, a hybrid fibre with both unique construction and characteristics. Vecry is a sheathed filament surrounding a liquid crystal-based polymer core, and can be woven into a high 'Open Area' construction having very fine filaments [108]. This new type of screen for use within the electrode fabrication steps provided the advantage of very high tensile strength – greater than stainless steels; critically, this higher tension enables screen peel to occur with less snap-off, enhancing paste release without reducing gasketing and the risk to both edge definition and 'overfilling' of the print volume [108]. Further attributes

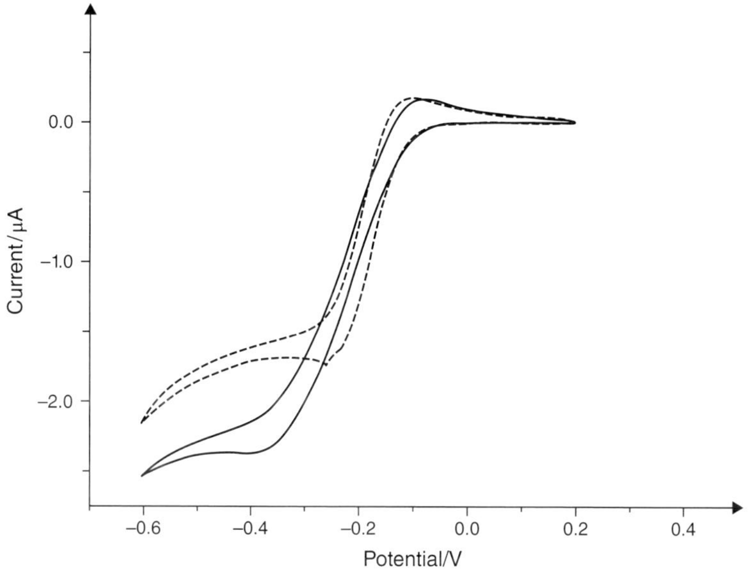

Figure 6.11 *Typical cyclic voltammetric traces obtained in 1 mM hexaammine-ruthenium(III) chloride/0.1 M KCl using the a graphite-bSPE of 50 μm (solid line) and 100 μm (dashed line) in length. Scan rate: 5 mVs^{-1}. Reproduced from Ref. [107] with permission from the Royal Society of Chemistry*

such as very low elongation with a high elastic recovery, low sensitivity to changes in temperature and humidity, outstanding abrasion resistance and to the fibrillation typical in other 'super fibres' and excellent response to the UV wavelengths used in pattern exposure are also offered through the utilisation of a V-mesh within the screen-printing process [108]. The fabricated band was compared to that discussed earlier of 100 μm in width through a comparison between the observed voltammetric responses obtained in hexa-ammine-ruthenium(III) chloride. Figure 6.11 shows an overlay of the cyclic voltammetric responses at each of the sensors upon the implementation of a scan rate of 5 mV s^{-1}. Interestingly, a greater voltammetric peak current is readily observed at the 50 μm graphite microband rather than at that of 100 μm which possesses a greater working electrode area (twice as much as the 50 μm microband). Such a response would suggest that the reduction in the microband width to that of 50 μm is sufficient enough to result in a '*true*' microelectrode response, whereby a reduction in the working electrode area results in an increase in the observed voltammetric current. Further to this, the group also benchmark the microband towards other sensors available within the literature finding the screen-printed microband to offer highly competitive limits of detection for NADH, but also against their own 100 μm microband electrode where they determining an improvement offered by the 50 μm microband with a limit of detection (3σ) of 0.48 μM for the sensing of NADH.

Further notable designs from Karousos *et al.* [109] have been shown where the carbon fibre matting (and foils) are trapped between alternating layers of the insulating polymer, where a hole is created through the centre of the laminate exposing the conductive carbon – each layer effectively forming a microtube-type electrode within the wall of the hole which serves as a three-dimensional detection well with the volume of the sample controlled through the manipulation of the well diameter. The bottom carbon layer is left intact and acts as a counter electrode. One major advantage of this type of methodology is the step-wise assembly of the various layers such that pre-patterned layers can be created and then laminated as is necessary when adding the base (counter) electrode layer [109]. This work has been shown to be useful where the base fibre mat is modified with copper oxide, where the sample permeates and selectively removes ascorbate. Whilst the methodology does not strictly involve screen printing (but rather pad printing), this approach can be readily adapted for mass production using screen-printing technology with the detection well volume readily tailored to specific applications.

Exciting new sensors described by Tan *et al.* explore the potential utilisation of screen-printing technology for the production of recessed electrodes [110]. The fabrication of novel carbon-based screen-printed disc-shaped recessed electrodes of 250 μm radius is electrochemically characterised and contrasted to other screen-printed sensors previously reported upon within the literature. The carbon-based recessed electrode is explained by Tan and colleagues to exhibit a greater current density over their coplanar counterparts. The potential electroanalytical applications of these carbon-based disc-shaped shallow recessed electrodes are explored through the sensing of NADH and nitrite exhibiting analytically relevant limits of detection (3σ) of 5.2 and 7.28 mM, respectively. The application of the screen-printed sensors towards the sensing of nitrite in an analytically relevant sample like canal water is also described demonstrating the robust nature of the sensors [110].

Further improvement upon the disc-shaped recessed screen-printed sensor is sought by Tan and co-workers through the fabrication of a highly novel pentagon-shaped carbon-based shallow recessed electrodes, which are compared and contrasted with the shallow recessed disc electrodes for the electroanalytical sensing of manganese(II) [110]. In comparison of the observed current density, the disc-shaped shallow recessed electrode offers greater sensitivity over a coplanar screen-printed electrode, whilst in addition to this, a pentagon-shaped recessed electrode offers improved sensitivity over even that of the disc-shaped shallow recessed screen-printed electrode. The ultra-low nM sensing of manganese(II) is shown to be possible at both the disc and pentagon shallow recessed electrodes though the pentagon shallow recessed sensor is found to exhibit an improved limit of detection (3σ) found to correspond to 36 nM. The fabrication methodology utilised within the study of the shallow recessed electrodes is shown to be generic in nature such that the underlying carbon layer, which defines the composition of the shallow recessed working electrode, can be replaced with electrocatalytic surfaces [110].

One approach towards screen-printed microelectrode arrays has reported the coating of screen-printed electrodes with polymeric films. These are then subjected to sonochemical ablation producing a random ensemble of microelectrodes, albeit with a large size distribution [12], which are also irregularly shaped and distorted which is likely to hamper the reproducibility between sensors from the same batch [111, 112]. Another approach is the screen printing of a carbon layer onto which an inert material is placed, which has been laser ablated to produce micron-sized holes which expose the underlying screen-printed carbon surface [113]. However, problems may be encountered between sealing the patterned material to the carbon layer leading to leakages. Other approaches involve patterning inert material *in situ* but can change the microstructure of the underlying carbon surface, ultimately affecting the electrochemical performance of the sensor. A novel new kind of microelectrode array based on femtosecond laser ablation and the screen-printing process was reported by Cugnet *et al.* [114] which demonstrated an enhanced diffusion behaviour and a greater mass transport in comparison with macroelectrodes. These approaches, while noteworthy, are unlikely to be used in the mass production of microelectrode arrays [115]. An individually addressable array comprising eight electrodes of 1 mm diameter has been reported by Dock *et al.* [116] made entirely by screen printing. It is evident that pushing the boundaries of screen printing is the only way for new and diverse electrochemical platforms to be produced via this technology [115].

6.3.2 Metal Oxide Electrodes, Metal and Nano and Micro Modified Screen-Printed Sensors

Modification of the electrode surface with electro-catalytic metals is a common approach, such as decoration or through the use of metallic screen-printed inks producing film-modified screen-printed electrodes [117]. Wu *et al.* [118] have utilised ruthenium oxide which is commonly used in resistive pastes for screen printing and demonstrated this towards the sensing of ascorbic acid which provided an analytical output of 0–4 mM with little interference from uric acid and hydrogen peroxide [118]. This work has been extended, for example, for the determination of hydrogen peroxide and with enzymatic modification for sensing hypoxanthine and glucose, whilst also being applied to food analysis [119]. Along these lines, the Banks group has reported the bulk modification of screen-printed electrodes with copper oxide for carbohydrate sensing [120], nickel oxide for hydroxide detection [121], manganese oxide for nitrite, oxygen and ascorbic acid [122], and finally bismuth oxide [115].

This last application of using bismuth oxide is particularly interesting as modifying an electrode substrate with bismuth is well documented to improve the electroanalytical performance and can act akin to a mercury-modified electrode yet having negligible toxicity [123–126]. To generate a truly disposable mass produced sensor, bismuth oxide has been reported [115] rather than modifying the surface *in situ* via the addition of a bismuth salt or *ex situ*, where it is electrochemically deposited in an external solution, bismuth oxide is incorporated into the bulk of the screen-printed electrode such that when the potential is held sufficiently negative, bismuth oxide at the electrode surface is electrochemically reduced forming bismuth metal. The electrochemical deposition of the target species preferentially occurs on the bismuth metal surface, in comparison to the underlying graphite surface, allowing for an enhanced sensing in the case of target metals; alloy formation beneficially assists. On the anodic sweep, the *in situ* formed $Bi_{(s)}$ should be maintained until at sufficiently high potentials when the transition from Bi^0 to Bi^{3+} will occur. Such an approach alleviates the need for *ex situ* bismuth film plating or *in situ* bismuth film formation greatly simplifying the analytical protocol [115, 126]. Bismuth domains act as a preferential nucleation site which due to the nature of electrode fabrication are randomly distributed across the electrode surface such that each bismuth microdomain has its own diffusional zone and due to the reduced spacing of the microdomains and partial diffusional overlap of diffusion zones, the mass transport is akin to that observed at a macroelectrode made entirely from bismuth yet it has a very low coverage [8]. The advantages of mimicking a bismuth film macroelectrode are that similar analytical performances can be obtained with a cheap, easily mass produced, disposable electrode which does not require *ex situ* or *in situ* plating of bismuth [126].

Similarly, Metters and co-workers have reported upon the fabrication of screen-printed sensors comprising gold [70] and platinum [71] macro screen-printed sensors. These screen-printed electrodes were produced through the modification of an existing graphite screen-printed electrode with a working area of 3 mm in diameter, whereby a further screen-printed layer comprising the noble metal ink of choice is printed upon the existing working electrode as is outlined in Figure 6.12. Although it would be possible to print the initial layer consisting of the working electrode and underlying tracks in the metal ink, rather than modifying an existing graphitic surface, such a decision was made to ensure the economical utilisation of the highly expensive noble metal inks. Both the gold- and platinum-based screen-printed sensors were determined to exhibit electrochemical characteristics such as electron transfer rate kinetics of a similar order to those recognised at traditional polycrystalline macro electrodes comprising the same noble metals. Additionally, the screen-printed sensors were determined to be analytically useful with the gold screen-printed electrode being benchmarked towards the sensing of the highly toxic and notable analyte chromium(VI) [70], whilst the platinum screen-printed sensor is demonstrated to be analytically useful for the key analytes of interest hydrazine and hydrogen peroxide [71].

A further implementation of such screen-printing processes and the ever expanding possibilities of screen-printing technologies, which continue to broaden due to developments within ink and screen technologies, is the fabrication of screen-printed single-walled carbon nanotube electrodes as reported by Metters *et al.* [127] The single-walled carbon nanotube sensors were fabricated as detailed above (see Figure 6.12) and after both microscopic and electrochemical characterisation were compared and contrasted with commercially available counterparts comprising both single- and multi-walled carbon nanotubes. The sensors were benchmarked towards the sensing of capsaicin; capsaicin belongs to the family of capsaicinoids, the main reason for the sharp flavour of hot peppers [128]. The capsaicinoid family consists of around 11 components such as capsaicin and dihydrocapsaicin which are responsible for about 90% of the pungency flavour in hot

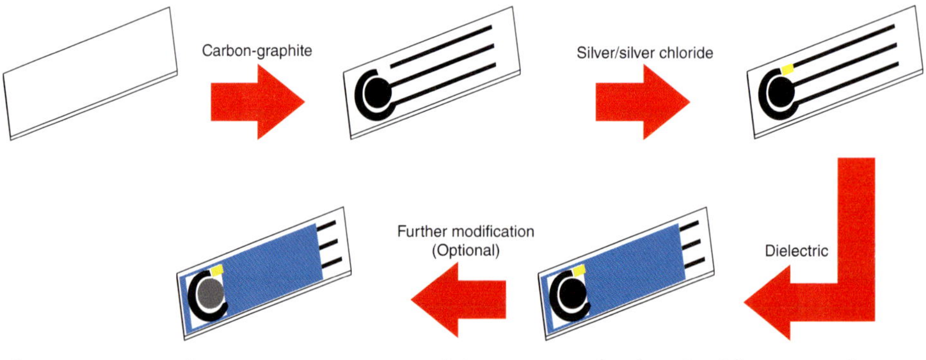

Figure 6.12 *A schematic representation of the steps involved in the fabrication of screen printed sensors and their further modification. Adapted from Ref. [127] with permission from Elsevier*

peppers [129]. Capsaicinoids also possess some biological properties that are thought to be beneficial for human health [130, 131]. Interestingly, the single-walled screen-printed sensors fabricated by Banks and co-workers outperformed their commercial counterparts towards the sensing of capsaicin achieving a low limit of detection (3σ) of 1.9 μM. The beneficial response obtained at the screen-printed sensors was attributed to the single-walled carbon nanotubes within the sensors which are hypothesised to allow the adsorption of the capsaicin on the nanotube walls [127].

6.4 Conclusions and Perspectives

We have seen in this chapter that microelectrodes and microelectrode arrays have unique benefits that can be harnessed by researchers, as, for example, in the field of food and agricultural electroanalysis. For sensors to be realised and implemented into this area, this chapter has shown that while there are many useful techniques, the main driving force; cost, means that only a few approaches have merit. These are sputtered electrodes and screen-printed sensors. Given the versatility of the latter, provided through the wide range of designs that have been produced, screen printing holds significant promise since mass produced, reproducible yet economical sensors can be realised.

References

1. Said, N.A.M., Twomey, K., Ogurtsov, V.I. *et al.* (2011) Fabrication and electrochemical characterization of micro- and nanoelectrode arrays for sensor applications. *J. Phys. Conf. Ser.*, **307**, 012052.
2. Bond, A.M., Oldham, K.B. and Zoski, C.G. (1989) Steady-state voltammetry. *Anal. Chim. Acta*, **216**, 177.
3. Hubel, D.H. (1971) Specifity of responses of cells in the visual cortex. *J. Psychiatr. Res.*, **8**, 301.
4. Henstridge, M.C. and Compton, R.G. (2012) Mass transport to micro- and nanoelectrodes and their arrays: a review. *Chem. Rec.*, **12**, 63.
5. Huang, X.-J., O'Mahony, A.M. and Compton, R.G. (2009) Microelectrode arrays for electrochemistry: approaches to fabrication. *Small*, **5**, 776.
6. Guo, J. and Linder, E. (2009) Cyclic voltammograms at coplanar and shallow recessed microdisc electrode arrays: guidelines for design and experiment. *Anal. Chem.*, **81**, 130.
7. Compton, R.G. and Banks, C.E. (2010) *Understanding Voltammetry*, 2nd edn, Imperial College, London.
8. Compton, R.G. and Banks, C.E. (2007) *Understanding Voltammetry*, World Scientific, Singapore.
9. Hoood, S.J., Kampouris, D.K., Kadara, R.O. *et al.* (2009) Why 'the bigger the better' is not always the case when utilising microelectrode arrays: high density vs. low density arrays for the electroanalytical sensing of chromium(vi). *Analyst*, **134**, 2301.
10. Guo, J. and Linder, E. (2009) Cyclic voltammetry at shallow recessed microdisc electrode: theoretical and experimental study. *J. Electroanal. Chem.*, **629**, 180.

11. Berduque, A., Lanyon, Y.H., Beni, V. *et al.* (2007) Voltammetric characterisation of silicon-based microelectrode arrays and their application to mercury-free stripping voltammetry of copper ions. *Talanta*, **71**, 1022.
12. Kadara, R.O., Jenkinson, N. and Banks, C.E. (2009) Screen printed recessed microelectrode arrays. *Sens. Actuators, B*, **142**, 342.
13. Bartlett, P.N. and Taylor, S.L. (1998) An accurate microdisc simulation model for recessed microdisc electrodes. *Electroanal. Chem.*, **453**, 49.
14. Fletcher, S. and Horne, M.D. (1999) Random assemblies of microelectrodes (RAM™ electrodes) for electrochemical studies. *Electrochem. Commun.*, **1**, 502.
15. Ordeig, O., Banks, C.E., Davies, T.J. *et al.* (2006) The linear sweep voltammetry of random arrays of microdisc electrodes: fitting of experimental data. *Electroanal. Chem.*, **592**, 126.
16. Ordeig, O., del Campo, J., Munoz, F.X. *et al.* (2007) Electroanalysis utilizing amperometric microdisk electrode arrays. *Electroanalysis*, **19**, 1973.
17. Bai, H.-Y., del Campo, F.J. and Tsai, Y.-C. (2013) Sensitive electrochemical thrombin aptasensor based on gold disk microelectrode arrays. *Biosens. Bioelectron.*, **42**, 17.
18. Huang, X.-J., Aldous, L., O'Mahony, A.M. *et al.* (2010) Toward membrane-free amperometric gas sensors: a microelectrode array approach. *Anal. Chem.*, **82**, 5238.
19. Jiang, Y., Fang, X. and Bai, C. (2004) Signaling aptamer/protein binding by a molecular light switch complex. *Anal. Chem.*, **76**, 5230.
20. Sanchez-Molas, D., Esquivel, J.P., Sabate, N. *et al.* (2012) High aspect-ratio, fully conducting gold micropillar array electrodes: silicon micromachining and electrochemical characterization. *J. Phys. Chem. C*, **116**, 18831.
21. Menshykau, D., O'Mahony, A.M., del Campo, F.J. *et al.* (2009) Microarrays of ring-recessed disk electrodes in transient generator-collector mode: theory and experiment. *Anal. Chem.*, **81**, 9372.
22. Ordeig, O., Banks, C.E., Davies, T.J. *et al.* (2006) Regular arrays of microdisc electrodes: simulation quantifies the fraction of 'dead' electrodes. *Analyst*, **131**, 440.
23. Davies, T.J., Ward-Jones, S., Banks, C.E. *et al.* (2005) The cyclic and linear sweep voltammetry of regular arrays of microdisc electrodes: fitting of experimental data. *J. Electroanal. Chem.*, **585**, 51.
24. Lancon, A.D., Katay, R., Wood, D. and Gallant, A.J. (2011) Microelectrode arrays for electroanalytical sensing: comparison of electroplating and electron-beam metallisation. *Electochem. Commun.*, **13**, 414.
25. Simm, A.O., Banks, C.E., Ward-Jones, S. *et al.* (2005) Boron-doped diamond microdisc arrays: electrochemical characterisation and their use as a substrate for the production of microelectrode arrays of diverse metals (Ag, Au, Cu) via electrodeposition. *Analyst*, **130**, 1303.
26. Lawrence, N.S., Pagels, M., Meredith, A. *et al.* (2006) Electroanalytical applications of boron-doped diamond microelectrode arrays. *Talanta*, **69**, 829.
27. Batchelor-McAuley, C., Banks, C.E., Simm, A.O. *et al.* (2006) The electroanalytical detection of hydrazine: a comparison of the use of palladium nanoparticles supported on boron-doped diamond and palladium plated BDD microdisc array. *Analyst*, **131**, 106.

28. Ward-Jones, S., Banks, C.E., Simm, A.O. *et al.* (2005) An in situ copper plated boron-doped diamond microelectrode array for the sensitive electrochemical detection of nitrate. *Electroanalysis*, **17**, 1806.
29. Chailapakul, O., Popa, E., Tai, H. *et al.* (2000) The electrooxidation of organic acids at boron-doped diamond electrodes. *Electochem. Commun.*, **2**, 422.
30. Haymond, S., Babcock, G.T. and Swain, G.M. (2003) Electron transfer kinetics of ferrocene at microcrystalline boron-doped diamond electrodes: effect of solvent and electrolyte. *Electroanalysis*, **15**, 249.
31. Wang, J., Chen, G., Chatrathi, M.P. *et al.* (2003) Microchip capillary electrophoresis coupled with a boron-doped diamond electrode-based electrochemical detector. *Anal. Chem.*, **75**, 935.
32. Granger, M.C., Witek, M., Xu, J. *et al.* (2000) Standard electrochemical behavior of high-quality, boron-doped polycrystalline diamond thin-film electrodes. *Anal. Chem.*, **72**, 3793.
33. Nekrassova, O., Lawrence, N.S. and Compton, R.G. (2004) Direct oxidation of haemoglobin at bare boron-doped diamond electrodes. *Analyst*, **129**, 804.
34. Herlambang, O., Sarada, B.V., Rao, T.N. and Fujishima, A. (2002) Electroanalytical applications of highly boron-doped diamond microfiber electrode. *Chem. Sens.*, **18**, 160.
35. Pedrosa, V.A., Codognoto, L. and Avaca, L.A. (2003) Electroanalytical determination of 4-nitrophenol by square wave voltammetry on diamond electrodes. *J. Braz. Chem. Soc.*, **14**, 530.
36. Haymond, S., Babcock, G.T. and Swain, G.M. (2002) Direct electrochemistry of cytochrome c at nanocrystalline boron-doped diamond. *J. Am. Chem. Soc.*, **124**, 10634.
37. Spataru, N., Sarada, B.V., Popa, E. *et al.* (2001) Voltammetric determination of l-cysteine at conductive diamond electrodes. *Anal. Chem.*, **73**, 514.
38. Ivandini, T.A., Sarada, B.V., Rao, T.N. and Fujishima, A. (2003) Ectrochemical oxidation of underivatized-nucleic acids at highly boron-doped diamond electrodes. *Analyst*, **128**, 924.
39. Hyde, M.E., Jacobs, R. and Compton, R.G. (2002) In situ AFM studies of metal deposition. *J. Phys. Chem. B*, **106**, 11075.
40. Tsunozaki, K., Einaga, Y., Rao, T.N. and Fujishima, A. (2002) Fabrication and electrochemical characterization of boron-doped diamond microdisc array electrodes. *Chem. Lett.*, **5**, 502.
41. Provent, C., Haenni, W., Santoli, E. and Rychen, P. (2004) Boron-doped diamond electrodes and microelectrode-arrays for the measurement of sulfate and peroxodisulfate. *Electrochim. Acta*, **49**, 3737.
42. Soh, K.L., Kang, W.P., Davidson, J.L. *et al.* (2004) Diamond-derived microelectrodes array for electrochemical analysis. *Diamond Relat. Mater.*, **13**, 2009.
43. Cvacka, J., Quaiserova, V., Park, J.W. *et al.* (2003) Boron-doped diamond microelectrodes for use in capillary electrophoresis with electrochemical detection. *Anal. Chem.*, **75**, 2678.
44. Madore, C., Duret, A., Haenni, W. and Perret, A. (2000) Detection of trace silver and copper at an array of boron-doped diamond microdisk electrodes. *Proc.-Electrochem. Soc.*, **19**, 159. (Micorfabricated Systems and MEMS V).

45. Pagels, M., Hall, C.E., Lawrence, N.S. *et al.* (2005) All-diamond microelectrode array device. *Anal. Chem.*, **77**, 3705.
46. Smirnov, W., Yang, N., Hoffman, R. *et al.* (2011) Integrated all-diamond ultramicro-electrode arrays: optimization of Faradaic and capacitive currents. *Anal. Chem.*, **83**, 7438.
47. Toghill, K.E. and Compton, R.G. (2010) Metal nanoparticle modified boron doped diamond electrodes for use in electroanalysis. *Electroanalysis*, **22**, 1947.
48. Choudhry, N.A., Kadara, R.O. and Banks, C.E. (2010) "Cosmetic electrochemistry": the facile production of graphite microelectrode ensembles. *Phys. Chem. Chem. Phys.*, **12**, 2285.
49. Davies, T.J. and Compton, R.G. (2005) The cyclic and linear sweep voltammetry of regular arrays of microdisc electrodes: fitting of experimental data. *J. Electroanal. Chem.*, **585**, 63.
50. Saito, Y. (1967) A sputtered Pt film electrode for polarographic O2 measurement. *J. Appl. Physiol.*, **23**, 979.
51. Brett, A.M.O., Matysik, F.-M. and Vieira, M.T. (1997) Thin-film gold electrodes produced by magnetron sputtering. Voltammetric characteristics and application in batch injection analysis with amperometric detection. *Electroanalysis*, **9**, 209.
52. Mie, Y., Ikegami, M. and Komatsu, Y. (2010) Gold sputtered electrode surfaces enhance direct electron transfer reactions of human cytochrome P450s. *Electochem. Commun.*, **12**, 680.
53. Kurisawa, I., Shiomi, M., Ohsumi, S. *et al.* (2001) Development of positive electrodes with an SnO_2 coating by applying a sputtering technique for lead-acid batteries. *J. Power Sources*, **95**, 125.
54. Gasda, M.D., Teki, R., Lu, T.-M. *et al.* (2009) Sputter-deposited Pt PEM fuel cell electrodes: particles vs layers. *J. Electrochem. Soc.*, **156**, B614.
55. Sheng, A. (1999) Why ancient silk is still gold: issues in Chinese textile history. *Ars. Orientalis*, **29**, 147.
56. Hart, J.P. and Wring, S.A. (1997) Recent developments in the design and application of screen-printed electrochemical sensors for biomedical, environmental and industrial analyses. *TrAC, Trend Anal. Chem.*, **16**, 89.
57. Hart, J.P. and Wring, S.A. (1994) Screen-printed voltammetric and amperometric electrochemical sensors for decentralized testing. *Electroanalysis*, **6**, 617.
58. Albareda-Sirvent, M., Merkoci, A. and Alegret, S. (2000) Configurations used in the design of screen-printed enzymatic biosensors. A review. *Sens. Actuators, B*, **69**, 153.
59. Morrin, A., Killard, A.J. and Smyth, M.R. (2003) Electrochemical characterization of commercial and home-made screen-printed carbon electrodes. *Anal. Lett.*, **36**, 2021.
60. Honeychurch, K.C. and Hart, J.P. (2003) Screen-printed electrochemical sensors for monitoring metal pollutants. *TrAC, Trend Anal. Chem.*, **22**, 456.
61. Heller, A. and Feldman, B. (2008) Electrochemical glucose sensors and their applications in diabetes management. *Chem. Rev.*, **108**, 2482.
62. Newman, J.D. and Turner, A.P.F. (2005) Home blood glucose biosensors: a commercial perspective. *Biosens. Bioelectron.*, **20**, 2435.
63. Wilson, R. and Turner, A.P.F. (1992) Glucose oxidase: an ideal enzyme. *Biosens. Bioelectron.*, **7**, 165.

64. Brainina, K.Z. (2001) Electroanalysis: from laboratory to field versions. *Anal. Chem.*, **56**, 303.
65. Gurban, A.M., Rotariu, L., Tudorache, M. *et al.* (2009) Development of biological sensors based on screen-printed electrodes for environmental pollution monitoring. *Sens. Environ. Health Secur.*, **2**, 401.
66. Simm, A.O., Banks, C.E. and Compton, R.G. (2005) Sonoelectroanalytical detection of ultra-trace arsenic. *Electroanalysis*, **17**, 335.
67. Metters, J.P., Kadara, R.O. and Banks, C.E. (2012) Electroanalytical properties of screen printed graphite microband electrodes. *Sens. Actuators, B*, **169**, 136.
68. Kadara, R.O., Jenkinson, N. and Banks, C.E. (2009) Characterisation of commercially available electrochemical sensing platforms. *Sens. Actuators, B*, **138**, 556.
69. Wang, J., Tian, B., Nascimento, V.B. and Anges, L. (1998) Performance of screen-printed carbon electrodes fabricated from different carbon inks. *Electrochim. Acta*, **43**, 3459.
70. Metters, J.P., Kadara, R.O. and Banks, C.E. (2012) Electroanalytical sensing of chromium(III) and (VI) utilising gold screen printed macro electrodes. *Analyst*, **137**, 896.
71. Metters, J.P., Tan, F., Kadara, R.O. and Banks, C.E. (2012) Platinum screen printed electrodes for the electroanalytical sensing of hydrazine and hydrogen peroxide. *Anal. Methods*, **4**, 1272.
72. Metters, J.P., Kadara, R.O. and Banks, C.E. (2011) New directions in screen printed electroanalytical sensors: an overview of recent developments. *Analyst*, **136**, 1067.
73. Kipphan, H. (2001) *Handbook of Print Media*, Springer, Heidelberg.
74. Barlow, F.D. and Elshabini, A. (2007) *Ceramic Interconnect Technology Handbook*, CRC Press, Orlando, FL.
75. Dropsens www.dropsens.com (accessed January 2013).
76. Kanichi www.kanichi-research.com (accessed January 2013).
77. Zensor Research and Developement www.zensor.com.tw (accessed January 2013).
78. Metters, J.P., Houssein, S.M., Kampouris, D.K. and Banks, C.E. (2013) Paper-based electroanalytical sensing platforms. *Anal. Methods*, **5**, 103.
79. Kroger, S. and Turner, A.P.F. (1997) Solvent-resistant carbon electrodes screen printed onto plastic for use in biosensors. *Anal. Chim. Acta*, **347**, 9.
80. Cai, J., Cizek, K., Long, B. *et al.* (2009) Flexible thick-film electrochemical sensors: impact of mechanical bending and stress on the electrochemical behavior. *Sens. Actuators, B*, **137**, 379.
81. Chuang, M.C., Yang, Y.L., Tseng, T.F. *et al.* (2009) Flexible thick-film glucose biosensor: influence of mechanical bending on the performance. *Talanta*, **81**, 15.
82. Coyle, S., Wu, Y., Lau, K.T. *et al.* (2007) Smart nanotextiles: a review of materials and applications. *MRS Bull.*, **32**, 434.
83. Diamond, D., Coyle, S., Scarmagnani, S. and Hayes, J. (2008) Wireless sensor networks and chemo-biosensing. *Chem. Rev.*, **108**, 652.
84. Yang, Y.L., Chuang, M.C., Lou, S.L. and Wang, J. (2010) Thick-film textile-based amperometric sensors and biosensors. *Analyst*, **135**, 1230.
85. Tudorache, M. and Bala, C. (2007) Biosensors based on screen-printing technology, and their applications in environmental and food analysis. *Anal. Bioanal. Chem.*, **388**, 565.

86. Renedo, O.D., Alonso-Lomillo, M.A. and Martinez, M.J.A. (2007) Recent developments in the field of screen-printed electrodes and their related applications. *Talanta*, **73**, 202.
87. Hart, J.P., Crew, A., Crouch, E. *et al.* (2004) Some recent designs and developments of screen-printed carbon electrochemical sensors/biosensors for biomedical, environmental, and industrial analyses. *Anal. Lett.*, **37**, 789.
88. Newman, J.D. and Turner, A.P.F. (1992) Biosensors: principles and practice, essays in biochemistry. *Essays Biochem.*, **27**, 147.
89. Hood, S.J., Kadara, R.O., Kampouris, D.K. and Banks, C.E. (2010) High throughput screening of lead utilising disposable screen printed shallow recessed microelectrode arrays. *Analyst*, **135**, 76.
90. Khairy, M., Kadara, R.O. and Banks, C.E. (2010) Electroanalytical sensing of nitrite at shallow recessed screen printed microelectrode arrays. *Anal. Methods*, **2**, 851.
91. Kadara, R.O., Jenkinson, N. and Banks, C.E. (2009) Characterization and fabrication of disposable screen printed microelectrodes. *Electochem. Commun.*, **11**, 1377.
92. Ball, J.C., Scott, D.L., Lumpp, J.K. *et al.* (2000) Electrochemistry in nanovials fabricated by combining screen printing and laser micromachining. *Anal. Chem.*, **72**, 497.
93. Schuler, T., Asmus, T., Fritzsche, W. and Moller, R. (2009) Screen printing as cost-efficient fabrication method for DNA-chips with electrical readout for detection of viral DNA. *Biosens. Bioelectron.*, **24**, 2077.
94. Chang, J.L. and Zen, J.M. (2006) Fabrication of disposable ultramicroelectrodes: characterization and applications. *Electochem. Commun.*, **8**, 571.
95. Yan, T.H., Liao, C.Y., Chang, J.L. *et al.* (2009) A simple electrochemical approach based on inexpensive wall-jet screen-printed ring disk electrode to evaluate oxygen reduction catalysts. *Electroanalysis*, **21**, 2390.
96. Liao, C.Y., Chang, C.C., Ay, C. and Zen, J.M. (2007) Flow injection analysis of chloramphenicol by using a disposable wall-jet ring disk carbon electrode. *Electroanalysis*, **19**, 65.
97. Craston, D.H., Jones, C.P., Williams, D.E. and Murr, N.E. (1991) Microband electrodes fabricated by screen printing processes: applications in electroanalysis. *Talanta*, **38**, 17.
98. Chang, J.L. and Zen, J.M. (2007) A poly(dimethylsiloxane)-based electrochemical cell coupled with disposable screen printed edge band ultramicroelectrodes for use in flow injection analysis. *Electochem. Commun.*, **9**, 2744.
99. Chang, J.L. and Zen, J.M. (2006) Disposable screen-printed edge band ultramicroelectrodes for the determination of trace amounts of nitrite ion. *Electroanalysis*, **18**, 941.
100. Chou, C.H., Chang, J.L. and Zen, J.M. (2009) Homogeneous platinum-deposited screen-printed edge band ultramicroelectrodes for amperometric sensing of carbon monoxide. *Electroanalysis*, **21**, 206.
101. Chou, C.H., Chang, J.L. and Zen, J.M. (2010) Effective analysis of gaseous formaldehyde based on a platinum-deposited screen-printed edge band ultramicroelectrode coated with Nafion as solid polymer electrolyte. *Sens. Actuators, B*, **147**, 669.
102. Rawson, F.J., Purcell, W.M., Xu, J. *et al.* (2009) A microband lactate biosensor fabricated using a water-based screen-printed carbon ink. *Talanta*, **77**, 1149.

103. Pemberton, R.M., Rawson, F.J., Xu, J. *et al.* (2010) Application of screen-printed microband biosensors incorporated with cells to monitor metabolic effects of potential environmental toxins. *Microchim. Acta*, **170**, 321.
104. Pemberton, R.M., Xu, J., Pittson, R. *et al.* (2009) Application of screen-printed microband biosensors to end-point measurements of glucose and cell numbers in HepG2 cell culture. *Anal. Biochem.*, **385**, 334.
105. Honeychurch, K.C., Al-Berezanchi, S. and Hart, J.P. (2011) The voltammetric behaviour of lead at a microband screen-printed carbon electrode and its determination in acetate leachates from glazed ceramic plates. *Talanta*, **84**, 717.
106. Porat, Z., Crooker, J.C., Zhang, Y. *et al.* (1997) IR_{UNC} advantages and real geometrical dimensions of microband electrodes. *Anal. Chem.*, **69**, 5073.
107. Metters, J.P., Kadara, R.O. and Banks, C.E. (2013) Fabrication of co-planar screen printed microband electrodes. *Analyst*, **138**, 2516.
108. MCI Precision Screens Ltd www.mciprecisionscreens.com (accessed Janurary 2013).
109. Karousos, N.G., Carrington, R., Wilkins, S.J. *et al.* (2005) Molecular sieving of antioxidants at a copper–carbon laminate assembly. *Electochem. Commun.*, **7**, 500.
110. Metters, J.P., Tan, F., Kadara, R.O. and Banks, C.E. (2012) Electroanalytical properties of screen printed shallow recessed electrodes. *Anal. Methods*, **4**, 3140.
111. Gornall, D.D., Collyer, S.D. and Higson, S.P.J. (2010) Evaluation of poly (*o*-phenylenediamine) films for application as insulating layers upon carbon substrates for use within sonochemically fabricated microelectrode arrays. *Electroanalysis*, **22**, 384.
112. Barton, A.C., Collyer, S.D., Davies, F. *et al.* (2004) Sonochemically fabricated microelectrode arrays for biosensors offering widespread applicability: part I. *Biosens. Bioelectron.*, **20**, 328.
113. Wilke, S., Osborne, M.D. and Girault, H.H. (1997) Electrochemical characterisation of liquid–liquid microinterface arrays. *Electroanal. Chem.*, **436**, 53.
114. Cugnet, C., Zaouak, O., Rene, A. *et al.* (2009) A novel microelectrode array combining screen-printing and femtosecond laser ablation technologies: development, characterization and application to cadmium detection. *Sens. Actuators, B*, **143**, 158.
115. Kadara, R.O. and Tothill, I.E. (2008) Development of disposable bulk-modified screen-printed electrode based on bismuth oxide for stripping chronopotentiometric analysis of lead(II) and cadmium(II) in soil and water samples. *Anal. Chim. Acta*, **623**, 76.
116. Dock, E., Christenson, A., Sapelnikova, S. *et al.* (2005) A steady-state and flow-through cell for screen-printed eight-electrode arrays. *Anal. Chim. Acta*, **531**, 165.
117. Stozhko, N.Y., Malakova, N.A., Fyodorov, M.V. and Brainina, K.Z. (2008) Modified carbon-containing electrodes in stripping voltammetry of metals: part I. Glassy carbon and carbon paste electrodes. *J. Solid State Electrochem.*, **12**, 1219.
118. Wu, J., Suls, J. and Sansen, W. (2000) Amperometric determination of ascorbic acid on screen-printing ruthenium dioxide electrode. *Electochem. Commun.*, **2**, 90.
119. Kotzian, P., Brazdilova, P., Kalcher, K. and Vytras, K. (2005) Determination of hydrogen peroxide, glucose and hypoxanthine using (bio)sensors based on ruthenium dioxide-modified screen-printed electrodes. *Anal. Lett.*, **38**, 1099.

120. Choudhry, N.A., Kampouris, D.K., Kadara, R.O. *et al.* (2009) Next generation screen printed electrochemical platforms: non-enzymatic sensing of carbohydrates using copper(II) oxide screen printed electrodes. *Anal. Methods*, **1**, 183.
121. Hallam, P.M., Kampouris, D.K., Kadara, R.O. *et al.* (2010) Nickel oxide screen printed electrodes for the sensing of hydroxide ions in aqueous solutions. *Anal. Methods*, **2**, 1152.
122. Beyene, N.W., Kotzian, P., Schachl, K. *et al.* (2004) (Bio)sensors based on manganese dioxide-modified carbon substrates: retrospections, further improvements and applications. *Talanta*, **64**, 1151.
123. Kokkinos, C. and Economou, A. (2008) Stripping analysis at bismuth-based electrodes: a review. *Curr. Anal. Chem.*, **4**, 183.
124. Wang, J. (2005) Stripping analysis at bismuth electrodes: a review. *Electroanalysis*, **17**, 1341.
125. Enconomou, A. (2005) Bismuth-film electrodes: recent developments and potentialities for electroanalysis. *TrAC, Trend Anal. Chem.*, **24**, 334.
126. Khairy, M., Kadara, R.O., Kampouris, D.K. and Banks, C.E. (2010) Disposable bismuth oxide screen printed electrodes for the sensing of zinc in seawater. *Electroanalysis*, **22**, 1455.
127. Metters, J.P., Gomez-Mingot, M., Iniesta, J. *et al.* (2013) The fabrication of novel screen printed single-walled carbon nanotube electrodes: electroanalytical applications. *Sens. Actuators, B*, **177**, 1043.
128. Barbero, G.F., Liazid, A., Palma, M. and Barroso, C.G. (2008) Fast determination of capsaicinoids from peppers by high-performance liquid chromatography using a reversed phase monolithic column. *Food Chem.*, **107**, 1276.
129. Laskaridou-Monnerville, A. (1999) Determination of capsaicin and dihydrocapsaicin by micellar electrokinetic capillary chromatography and its application to various species of Capsicum, Solanaceae. *J. Chromatogr. A*, **838**, 293.
130. Robbins, W. (2000) Clinical applications of capsaicinoids. *Clin. J. Pain*, **16**, S86.
131. Chanda, S., Erexson, G., Frost, D. *et al.* (2007) 26-Week dermal oncogenicity study evaluating pure trans-capsaicin in Tg.AC hemizygous mice (FBV/N). *Int. J. Toxicol.*, **26**, 12.

7

Potentiometric Sensors

Geza Nagy[1] *and Lívia Nagy*[2]
[1]*General and Physical Chemistry, Faculty of Science, University of Pécs, 7624, Pécs, Ifjúság útja 6, Hungary*
[2]*Szentagothai Research Centre, University of Pécs, 7624, Pécs, Ifjúság útja 20, Hungary*

7.1 Introduction

Potentiometry is one of the most often applied methods of instrumental analysis. Thinking about pH adjustment, we can say that the potentiometric measuring apparatus is almost as an important tool of chemical laboratories as the analytical balance. Due to the following reasons it makes direct potentiometry a popular method in analytical laboratories:

- It does not require expensive instruments.
- No special skill or experience of the laboratory personals are needed for making the potentiometric analysis.
- The dynamic concentration range of the potentiometric sensors is broad. It can reach over several orders of magnitude.
- Direct potentiometry does not destroy the sample.
- It does not require special sample preparation steps.
- The potentiometric signal is not affected by color, viscosity, or turbidity of the sample.
- The potentiometric signals are not affected by the size of the sensor or by the convection in the sample.
- Selective potentiometric sensors can be made with different shapes or sizes for adapting to special requirements of the analysis.

7.2 The Types of Potentiometry

In potentiometry, two electrodes are employed in a measuring cell, and the electric potential difference between them is measured, and used as an analytical signal for evaluation.

Agricultural and Food Electroanalysis, First Edition.
Edited by Alberto Escarpa, María Cristina González and Miguel Ángel López.

The potentiometric measurements most often are carried out at no or negligible current flowing through the cell, $i = 0$. In this case, the measuring cell is actually a galvanic cell, composed of two half-cells. The potential difference is called the electromotive force (emf).

Constant current potentiometry is employed sometimes, mostly for the location of end point in redox titrations. In those cases, small constant current (i = const) is forced through the cell and the potential difference is followed between the electrodes. Often two identical metal electrodes, such as platinum needles are used. If both forms of electrochemically reversible couples are present in the titration vessel, then the voltage difference between the electrodes is small. However, if one is missing, then higher voltage shows up. This kind of potentiometry is often called bipotentiometry. Its application is advantageous in iodometric titrations where sharp voltage change indicates the disappearance of the oxidized form of the I^-/I^-_3 couples.

7.2.1 Potentiometric Stripping Analysis

In 1976, a new kind of analytical method has been introduced by Jagner and Graneli [1], it is called the potentiometric stripping analysis. In the early times, this method not only became very popular in environmental analysis, but it has been also successfully applied in measuring the trace-level concentration of different metals in agricultural and food samples. Usually, in stripping analysis, three electrodes are employed: a working electrode, a reference, and an auxiliary electrode. The first step of the procedure is a constant potential reduction of the analytes. During this, the potentiostat is connected to the electrodes and intensive stirring is employed. The measuring surface of the working electrode is made of mercury or coated with a mercury film. The constant potential step can last from a few seconds to a few minutes. In this period, the metals to be measured are deposited, accumulated on the working electrode surface.

$$\mathrm{Me}^{n+} + n\mathrm{e}^- \rightarrow \mathrm{M}^0(\mathrm{Hg})$$

At the end of the first step, the potentiostat is disconnected and a millivolts measuring instrument is connected for measuring the potential difference between the reference and the working electrode. The electrode potential–time transient is recorded and used for evaluation. (This is why this is a potentiometric method.) In the measuring cell, an oxidizing reagent is present, or a weak current is employed. Therefore, as the potentiostat is disconnected, the electrode potential starts changing to the anodic direction. When the oxidizing potential of a deposited species is achieved, the potential change slows down. The higher the amount is deposited, the longer time it takes to oxidize the metal:

$$\mathrm{M}^0(\mathrm{Hg}) + \text{oxidizing agent}(\text{or} - n\mathrm{e}^-) \rightarrow \mathrm{Me}^{n+}$$

After the total amount of the analyte 1 is oxidized, the potential changes become fast again as long as the oxidizing potential of the next analyte is achieved. Therefore, a transient slightly resembling a multiple redox titration curve is obtained. The transition times needed for the oxidation of the deposited metals are used for the evaluation. The peak shaped dt/dE versus E plot makes the evaluation more convenient.

By changing the deposition time, the oxidant concentration, or the electrode size, the method can be adjusted to the different analytical tasks. Most often, the trace levels of Cu, Pb, Bi, Sn, and Cd ion concentrations are measured with potentiometric stripping analysis.

7.2.2 Zero Current Potentiometry

In zero current potentiometry, two different kinds of galvanic measurement cells are used. One of them is a small sole compartment cell, while the other contains two half-cells joined together with a current bridge. The cell made of one compartment is called cell without transference. The sample solution and the measuring part of two electrodes are inside this cell during measurements. The other cell is called cell with transference. In this case, the two electrodes are placed in different half-cells. In the reference half-cell, the electrolyte of the appropriate solution of constant concentration is kept, while into the lumen of the other half-cell, the measuring electrode and the sample solution are introduced. The signal is the galvanic potential difference between the sample solution and the electrode surface; however, its exact measurement would be difficult. Therefore, in practice, the potential difference between the measuring electrode E_m and a reference electrode E_r ($\Delta E = E_m - E_r$) is taken as signal.

As is well known, the standard hydrogen electrode has been selected as reference for potential measurements. However, because of technical difficulties and safety hazard associated with its application, the hydrogen electrode is not often employed in every day measurements. Instead of hydrogen electrodes, electrodes of the second kind like calomel electrodes or silver/silver-chloride electrodes are used in potentiometric analysis. The two halves of the measuring cell are connected through a current bridge that allows charged ions passing through, but it avoids mixing of the two solutions: porous glass beads, U-shaped tubes filled up with salt containing agar gel, sol–gel layer, filter paper stripes, and so on, can serve for salt bridge preparation. The two electrodes are connected to the inputs of a high impedance millivolts meter, and the potential difference is measured. It is given as the voltage difference, ΔE. In reporting the ΔE value, the type of the reference electrode is usually indicated. If a reference saturated calomel electrode (SCE) is used, then after the ΔE value, the term "vs. SCE" follows.

If the input impedance of the measuring instrument is high ($>10^{12}\ \Omega$), then the negligible current passes through the cell, and, therefore, the measured ΔE value is close to the emf.

There can be several solid/liquid, liquid/liquid, or solid/solid interfaces in the potentiometric measuring circuit. Charge separation takes place producing electrode potential gradients at every interface. In a well-devised potential measuring circuit, all these gradients are constant, except the one at the measuring electrode/analyte interface. Therefore, the change of the measured ΔE shows the change of this boundary potential.

Unfortunately, one type of interfacial potential can bring uncertainty into the potentiometric analysis. It is the diffusion potential that is often called liquid junction potential. At interfaces between two electrolytes, like at the sample solution/current bridge, the differences of chemical potential drive diffusion. As the electrolytes diffuse, ion mobility differences result in charge separation. Let us consider the diffusion of hydrochloric acid. If the concentration in solution 1 is higher than that in solution 2, then both H^+ and Cl^- ions will diffuse from solution 1 to 2. However, the mobility of hydrogen ions is far higher than that of the chlorides, and, therefore, hydrogen ions move faster. The positive ions will be in excess at the boundary in solution 2 and the junction potential arises. The junction potential counteracts with the mobility differences. In potentiometric analysis, often the major source of error is the uncertainty of this diffusion potential. The higher the difference between the mobility of the anions and cations taking part in diffusion transport, the higher the diffusion potential. Therefore, in current bridges, electrolytes with ions of similar mobility, for example, KCl, are employed for decreasing the E_{diff} value.

If the diffusion potential is resulted by the concentration difference of one salt, then the value of E_{diff} can be given by the following equation:

$$E_{\mathrm{diff}} = \frac{u_{\mathrm{A}} - u_{\mathrm{C}}}{u_{\mathrm{A}} + u_{\mathrm{C}}} \frac{RT}{zF} \ln \frac{c_2}{c_1}$$

where u_{A} and u_{C} are the mobilities of anions and cations, respectively, and c_1 and c_2 are the ion concentrations of side 1 and side 2, respectively, as indicated by the lower indexes.

The chemical potential differences of all ions between the two solutions as well as the mobility differences take part in generating the E_{diff}. For the multicomponent electrolyte, the Henderson form describes its value:

$$E_{\mathrm{diff}} = -\frac{RT}{zF} \cdot \frac{\sum (u_i/z_i)(c_{i2} - c_{i1})}{\sum u_i(c_{i2} - c_{i1})} \ln \frac{\sum c_{i2} u_i}{\sum c_{i1} u_i}$$

Based on this equation, we can see that, if a salt with $u_{\mathrm{A}} \approx u_{\mathrm{C}}$ is added in a high and equal concentration into both sides of the liquid junction, then only small E_{diff} will show up.

The signal used in potentiometric analysis is generated at the boundary between the electrode and the sample solution. When no current passes through the cell then, the charge separation process is in dynamic equilibrium. For equilibrium, we can write

$$\Delta G_{\mathrm{r}} = \left(\frac{\partial G}{\partial \xi} \right)_{p,T} = 0$$

where G is the Gibbs energy of the charge separation process and ξ is the reaction coordinate. Since

$$\Delta G_{\mathrm{r}} = \sum_j \nu_j \mu_j = \sum_j \nu \mu_j^0 + RT \sum_j \nu \ln a_j = \Delta G_{\mathrm{r}}^0 + RT \sum_j \ln a_j^{\nu_j}$$

$$\Delta G_{\mathrm{r}} = \Delta G_{\mathrm{r}}^0 + RT \ln Q$$

where Q is the reaction quotient $Q = \sum_j \ln a_j^{\nu_j}$. In equilibrium $Q = K$,

$$\Delta G_{\mathrm{r}} = 0 = \Delta G_{\mathrm{r}}^0 + RT \ln K$$

where ΔG_{r} is the reaction Gibbs energy, ΔG^0 is the standard Gibbs energy change, and K is the thermodynamic equilibrium coefficient. From this

$$\Delta G_{\mathrm{r}}^0 = -RT \ln K$$

However, we know that between the electrode potential and the change of the Gibbs energy, the following relation exists: $-\nu FE = \Delta G_{\mathrm{r}}$.

Therefore, we can get to the well-known Nernst equation:

$$E = -\frac{\Delta G_{\mathrm{r}}}{\nu F} = -\frac{\Delta G_{\mathrm{r}}^0}{\nu F} - \frac{RT}{\nu F} \ln Q = E^0 - \frac{RT}{\nu F} \ln Q$$

where v is the stochiometric coefficient of the reaction and E^0 is the standard electrode potential of the reaction.

In potentiometric measurements, the cell voltage $\Delta E = E_m - E_r$ is detected; therefore, when the Nernst equation is used for analytical evaluation, then E_r contains the reference electrode potential, and all other boundary potential differences of the measuring circuit. They must be constant, while

$$E_m = E^0 - \frac{RT}{vF}\ln Q \text{ or } E_m = E^0 - \frac{2.3026 \cdot RT}{vF}\log Q$$

$$\text{for } 25\ {}^{\circ}\text{C} \quad \text{it will be} \quad E_m = E^0 - \frac{0.0591}{v}\log Q \text{ in volts.}$$

The value holds the analytical information.

Depending on the nature of the electrode reaction, different forms of the Nernst equation are used. Very often, the measuring electrode itself is an inert metal serving as an electron sink or a source for the redox reaction taking place at the metal/solution interface. An example for this is a platinum metal electrode dipped into a solution containing the Fe(III) and Fe(II) ions. The electrode process will be as follows:

$$Fe^{3+} + e^- \rightarrow Fe^{2+} \text{ with } E^0 \text{of } 0.771\,\text{V vs. SCE}$$

In this case, the electrode potential can be given in volts by the $E_m = E^0 + \frac{a_{ox}}{a_{red}} = 0.771 + 0.059\log\frac{a_{Fe^{3+}}}{a_{Fe^{2+}}}$ form, where a_{ox} and a_{red} stand for the activities of the oxidized (Fe^{3+}) and reduced (Fe^{2+}) ions, respectively.

In the case of electrodes of the first kind, the metal electrode is in contact with the solution of its ions:

$$Me^{n+} + ne^- \rightarrow Me^0$$

Then the activity of the reduced species (Me^0) is 1; therefore, the Nernst equation can be simplified accordingly:

$$E_m = E^0 + \frac{0.059}{n}\log a_{Me^{n+}}$$

The potential of the hydrogen electrode is also determined by a redox equilibrium. The oxidized species is the hydrogen ion, while the reduced one is the hydrogen gas. The E^0 value of this electrode has been accepted as zero for all temperatures. All other potential values are assigned on that basis.

The symbol of the half-cell is the following:

$$Pt(s)/H_2(g)/H^+(aq)//$$

In this case, $E_m = 0.059\log\frac{a_{H^+}}{\sqrt{p_{H_2}}}$; for obtaining $E_m = E^0$, the hydrogen ion activity must be adjusted to 1 and the pressure (correctly the fugacity) of the hydrogen gas 1 must be 1 bar.

Not only redox reactions can produce well-defined interfacial charge separation but also the equilibrium of selective ion adsorption on different crystal surfaces, or differences in the extraction character of cations and anions, as well as the interaction of surface bound host molecules with the guest ionic analyte can well result in charge separation. The measuring functions of different selective potentiomeric electrodes are based on these equilibriums. Most of the ion-selective electrodes (ISEs) use these mechanisms in which redox equilibrium is not involved. The activity of the sample ion affects the distribution of the measured ion between the electrode surface and the sample solution. The dependence of the electrode potential of these electrodes on the activity of the detected ion can also be given by the Nernst equation.

The ISEs are often use a special thin membrane for signal generation. The membrane separates the tubular lumen of the electrode from the sample solution. There is an internal filling solution inside containing the a_1 activity of the detected ion. If the activity of the detected ion in the sample is a_2, then the Nernst equation for the membrane potential E is

$$E = -\frac{RT}{\nu F} \ln \frac{a_2}{a_1}$$

For this, we need to say that the E^0 values are equal at both sides of the membrane and the membrane selective for the species is detected. In potentiometric analysis with ion-selective membrane electrodes, the following measuring cell is used:

$$Hg(l)/Hg_2Cl_2(s)/KCl\ (aq)//C^{\nu+}(aq)(a_2)/membrane/C^{\nu+}(aq)(a_1)$$
$$KCl(aq)/Hg_2Cl_2(s)/Hg(l)$$

As can be seen, two calomel electrodes are on the two sides and the potential difference (ΔE) between them is measured.

It is important to decide how the electrode potential and the cell voltage should be expressed and which sign for the measured cell voltage should be given. In 1953, in Stockholm, the International Union of Pure and Applied Chemistry (IUPAC) convention has been accepted. Accordingly, the electrode potential is a potential of the half-reaction written as reduction:

$$Li^+ + e^- \rightarrow Li_{(s)} E^0 = -3.05\,V\ (25^\circ C)$$

For obtaining the sign of the electrode potential, the electrode is coupled with a standard hydrogen electrode in the galvanic cell and, thus, the sign of the measured potential will be the actual sign detected.

According to the convention, we write the symbol of the reference electrode on the left side, while that of the measuring electrode or its internal reference on the right side. We take the difference between them as cell voltage, $\Delta E = E_{right} - E_{left}$, that is, emf. As mentioned earlier, the activities of ions inside the filling solutions of the outer and inner references are kept constant. Therefore, the change of the ΔE depends on the change of the boundary potential at the sample solution/membrane interface.

For expressing the cell voltage, we need to write the reduction process on the right side of the cell symbol and the oxidation on the left side. Thus, the cell voltage will be

$$\Delta E = E_{right} - E_{left}$$

In potentiometric analysis, the indicator and the reference electrodes are in the measurement cell and the ΔE cell voltage is measured between them. In most of the cases, the reference electrode with a constant potential is selected. The potential of the indicating electrode depends on the concentration of the analyte. In a favorable case, the applied indicating electrode is selective, which means that other ionic species being present in the sample do not influence its potential. In such a case, the simple direct potentiometric analysis can be carried out. If an appropriate ISE is not available, then the titrimetric method with potentiometric end point detection is usually employed.

7.2.3 Direct Potentiometry

In direct potentiometry, the electrode is calibrated measuring the ΔE cell voltages upon introducing calibrating standards of the known concentration. Plotting ΔE values against the negative logarithm of the analyte concentration, a straight line should be obtained. Measuring ΔE in the sample solution, the concentration can be obtained from the calibration data. There are, however, some properties of the potentiometric signal that are worth considering, and special care must be taken in direct potentiometric analysis due to the following challenges:

- One problem is the existence of the diffusion potential (E_{diff}). Depending on the chemical potential differences between the solution in the reference half-cell and the calibrating or sample solution, the unknown value of the diffusion potential contributes to ΔE. It gives uncertainty in direct potentiometry. In order to decrease this, a high concentration of KCl is used as an electrolyte in the current bridges, if it is possible (see the Henderson form).
- The other difficulty we need to discuss is originated from the differences between the value of activity and that of the concentration. As is well known, the Nernst equation gives the dependence between the electrode potential and the negative logarithm of the ionic activity. Most of the analysts, however, are interested in the determination of the concentrations of the ionic analyte species. In order to overcome this contradiction, a high concentration of ionic strength adjusting background electrolyte is added to the calibrating standards, as well as to the sample solutions. The high and close to constant ionic strength provides almost constant main ionic activity coefficients. Therefore, using the ΔE versus log concentration plot, obtained with ionic strength adjusted standard solutions, if we adjust the same ionic strength in the sample solutions, the sample concentration can be obtained.
  The dependence between the main ionic activity coefficient ($\gamma_{\pm}$) and the ionic strength (I) is usually given by one of the Debye–Hückel forms:

$$\log \gamma_{\pm} = -\frac{A|z_{+}z_{-}|\sqrt{I}}{1 + Ba\sqrt{I}} + CI$$

  where A, B, and C are temperature- and permittivity-dependent coefficients, and a is a main ionic diameter-dependent term.
- The selective potentiometric electrodes measure the activity of the free ions. In the presence of complexing agents, only a part of the ions are in the free form. If we want to know the total concentration of the ionic species, then we need to adjust the same degree of dissociation in the calibrating standards and in the samples. The analytical method often recommends the addition of special agents to the calibrating and a sample solution for providing the same degree of dissociation.

- The accuracy of direct potentiometric analytical methods depends on the selectivity of the indicating electrode. No other ions than the analyte can be present in concentration affecting the electrode potential. The addition of the certain complexing agents or pH buffer can help to decrease the interfering ion activity below the tolerable level.
- In the case of certain samples and calibrating solutions, redox reactions can change the concentration of the ionic species. The addition of certain reagents can avoid this.

Usually, the composition of the background electrolyte is given in well-developed analytical recipes. A good example for this is the TISAB (total ionic strength adjustment buffer) used in the fluoride analysis of different samples. Generally, the TISAB solution is given by the 1:1 volume ratio to samples and calibrating standards. It controls the pH, provides constant ionic strength, and complexes interfering ferric (Fe^{3+}) and aluminum (Al^{3+}) ions. Sulfide anti-oxidizing buffer (SAOB) is used in sulfide analysis. This glycine-ascorbic acid-based solution prevents the oxidation of sulfide in ISE assays.

For certain samples, the ΔE values measured in samples cannot be evaluated correctly with calibration data taken in aqueous standards. For example, if the ionic strength of the sample is high and unknown, then getting the identical activity coefficient ($\gamma_{\pm}$) in samples and calibrating standards would be difficult. If the slope of the response of the ISE is not identical in the sample and in the aqueous calibrating solutions, then the standard addition technique is the right choice. So the cell voltage ΔE_1 is measured in the sample, knowing the volume of the sample (V_s). After a known, small volume (V_c) of known concentration (C_c) standard is added to V_s, and the ΔE_2 cell voltage is measured. If we assume that the small volume of the aqueous standard changes the ionic strength and other ion concentration only by a superfluous amount, and, furthermore, the cell voltage–log C dependence in the actual range is linear with the known slope (S), then the sample concentration can be obtained using the mentioned data:

$$C_s = \frac{V_c C_c}{10^{\frac{\Delta E}{S}} (V_S + V_C) - V_S}$$

where

$$\Delta E = E_2 - E_1$$

The reliability of the standard addition method can be increased if several V_C standard values are applied and several E_2 voltages are taken. Of course, care must be taken avoiding too much change in the slope or in ionic strength.

There is an interesting standard addition method variety that can be used in cases when the slope of the ISE is not known. So according to the method called "standard addition, slope by dilution" three cell voltage values are considered. E_1 is measured in the sample of the known small volume, V_s. E_2 is obtained after the addition of a small known volume (V_c) of standard solution of known (C_c) concentration. Furthermore, for obtaining the slope (S) valid in the actual concentration range of the analysis, we add the $V_s + V_c$ volume of distilled water and take the (E_3) cell voltage value. The second addition causes 1:2 dilution after taking E_2 into consideration. So we can write

$$\Delta E_2 = E_2 - E_3 = E_0 + S \ln \frac{C_S V_S + C_C V_C}{V_S + V_c} - \left(E_0 + S \ln \frac{C_S V_S + C_C V_C}{2 \times (V_S + V_c)} \right)$$

$$\Delta E_2 = S \ln \frac{\dfrac{C_S V_S + C_C V_C}{V_S + V_c}}{\dfrac{C_S V_S + C_C V_C}{2 \times (V_S + V_c)}} = S \ln 2 = S \times 0.6931$$

$$S = \frac{\Delta E_2}{0.6931}$$

Knowing S, from $\Delta E = E_2 - E_1$, we can get C_S using the above given equation.

Two important properties of direct potentiometry need to be mentioned here: the dynamic concentration range and the accuracy. Generally speaking, the optimal concentration range of direct potentiometry is between 0.01 and 200 mM. The exception is the pH measurement with glass or metal oxide electrodes. The potential response of most ion-selective membrane electrodes ceases around 1 μM concentration. It is so if the function is tested in unbuffered solutions. Checking the electrode function with an ion-buffered solution, the measuring range could be extended toward the smaller concentrations. For example, Blum and Fog [2] obtained the calibration curve of the ion-selective copper electrode, p_{Cu}, in the range of 4–15 using complex equilibrium-based ionic buffer solutions. In analytical practice, however, the determination of free-ion concentration in ion-buffered samples is seldom needed.

Investigations started in the end of the last century [3, 4] proved that trans-membrane ionic currents are responsible for this ceasing off the ISE response in the low ranges of concentration. Chemical potential differences drive these currents. Decreasing trans-membrane transports by carefully selecting the concentration of the internal filling solution, immobilizing the species involved, or using small intensity current against ion transport can extend the lower limit of detection by orders of magnitudes. Bakker and Pretsch [5] published a detailed report about ways and chances of the application of ISEs in trace analysis. However, these needs very carefully selected electrode conditioning, pretreatment, and very carefully trimmed experimental conditions. Therefore, trace-level analytes were not analyzed routinely with direct potentiometric methods in food or agriculture samples in earlier days.

When thinking about the accuracy and precision of direct potentiometry, it must be kept in mind that 0.5 mV difference in cell voltage translates to almost 2% difference in sample concentration for univalent ions. For divalent ions, it is close to 4%, while for a trivalent ion, in which case $S \approx 20\text{mV/decade}$, it is close to 6%. Therefore, for high accuracy, very well controlled measuring parameters such as cell temperature, junction potential, electrode selectivity, and so on, are needed.

7.2.4 Titrimetric Methods

Titrimetric methods with potentiometric end point location can be applied when an electrode with the needed selectivity is not available. The precision and accuracy of potentiometric titrations are superior comparing it with the properties of direct potentiometry. However, the concentration range where potentiometric titration can be used effectively is narrower. A solution with analyte concentration below 1 mM seldom is determined by potentiometric titrations. Potentiometric end point location is most often employed in the case of acid-, base-, precipitate-, redox-, or complexometric titrations.

7.3 The Selectivity of Ion-selective Electrodes and Its Determination

The selectivity of ISEs is an important feature. It indicates that the potential response of the electrode depends only on the activity of one ionic species in the presence of others. We call the ions that the electrode measures the primary or own ions, while all the others are called interfering ions. If the activity of the primary ion is constant, then the potential of an ideally selective electrode would not change upon changing the concentration of any other ionic species. Unfortunately, no ISE has ever been made with total selectivity. The electrode potential of an ISE depends only on the activity of the primary ion if the activity of interfering ions is small; however, increasing the interfering ion activity over a certain level, interfering ions start affecting the electrode potential. It is obvious that if the unknown activity of the interfering ions exceeds the mentioned "certain" level, then direct potentiometry of the primary ion cannot be determined in a simple and reliable way.

Two things can be concluded from these. On the one hand, the selectivity of an ISE must depend on the activity ratio of the primary and interfering ions. On the other hand, the selectivity of an electrode is a feature that relates to one primary and one interfering ion. The selectivity of the electrode shows how far the electrode prefers a primary ion over a certain interfering ion.

7.3.1 The Selectivity Coefficient

The selectivity is usually expressed as a number showing the preference like $K^{\mathrm{pot}}_{A,B}$ or its negative logarithm $pK_{A,B}$, where A means the primary ion, while B is the interfering ion. In an ideal case, $K^{\mathrm{pot}}_{A,B} \ll 1$. A list of $K^{\mathrm{pot}}_{A,B}$ values relating to different B species can be used for giving the information about the selectivity of an ISE.

The selectivity coefficient of an ISE is an important property. One can guess knowing the $K^{\mathrm{pot}}_{A,B}$ values of an electrode that whether the electrode is applicable for solving the planned analytical problem or not. For this only, the ion activity range of the analytes in the sample and that of the interfering ions in the matrix have to be known or guessed. However, the selectivity coefficient is not a physical constant. Its value depends very much on the experimental condition used in its determination.

7.3.2 Separate Solution and Mixed Solution Methods

Several methods have been proposed for the experimental determination of ion-selectivity coefficient. They can be listed in the two groups: *separate solution methods and mixed solution methods*. Applying both methods, the cell voltages (emf) are measured in a cell when the reference electrodes and the ISEs are dipped in solutions of known composition. When the separate solution method is used, pure primary ion solutions or pure solutions of the tested interfering ion are introduced into the measuring cell, while in mixed solution methods, both the primary and the selected interfering ions are introduced.

The Nicolsky–Eisenmann equation or its different varieties [6, 7] give a theoretical background for the determinations. The equation was proposed originally for explaining the alkaline ion interferences (the alkaline error) observed in the case of pH-selective glass electrodes:

$$E = E_0 + \frac{RT}{Z_A F} \ln\left(a_A + \sum_{B \neq A}^{B} K_{A,B}^{\text{pot}} (a_B)^{\frac{Z_A}{Z_B}} \right)$$

where E is the measured emf value, E_0 is a constant containing the standard potential of the ISE, the potential of the reference, the junction potential, Z means the charges of the primary (A) and the interfering ions (B), and R, T, and F have the usual meaning.

The Nicolsky–Eisenmann equation supposes that the response of the primary and the interfering ions is similar. It is generally not true. More complex equations, for example [8], can give better fitting to the experimental observations.

For an easy understanding, let us suppose that $Z_A = Z_B = 1$. Then a simplified equation is obtained:

$$E = E_0 + \frac{RT}{F} \ln(a_A + K_{A,B}^{\text{pot}} a_B)$$

7.3.2.1 Separation Solution Methods

In separation solution methods, a calibration curve is prepared for the primary ion and the other one for the selected interfering one. The simplified equation will give the E – log dependence. Since for the calibration curve of the primary ion $a_{\text{B}} = 0$ and for the interfering ion $a_{\text{A}} = 0$, we can get two equations as follows:

$$E_1 = E_0 + \frac{RT}{F} \ln(a_{\text{A}})$$

$$E_2 = E_0 + \frac{RT}{F} \ln(K_{A,B}^{\text{pot}} a_{\text{B}})$$

There are two ways for obtaining the $K_{A,B}^{\text{pot}}$ value from the calibration curves:

1. In the *equal potential method*, we can pick in the calibration curves an appropriate E value (horizontal line) with $E_1 = E_2$; then reading the corresponding activity values since

   $$a_A = K_{A,B}^{\text{pot}} a_B$$

   the selectivity coefficient is obtained:

   $$K_{A,B}^{\text{pot}} = \frac{a_A}{a_B}$$

2. In the *equal activity (concentration) method*, a vertical line is drawn in the dynamic range of the electrode response of the calibration graph. We read the E_1 and E_2 values and get the $\Delta E = E_1 - E_2$ difference. From the simplified Nicolsky–Eisenmann equation, we obtain

   $$\frac{\Delta E}{RT/F} = \ln\left(\frac{a_A}{K_{A,B}^{\text{pot}} a_B}\right)$$

   $$e^{\frac{\Delta E}{RT/F}} = \frac{a_A}{K_{A,B}^{\text{pot}} a_B}$$

Usually, the *RT/F* is substituted by the slope (*S*) of the experimentally obtained calibration curve.

Therefore, the selectivity coefficient is calculated as $K_{A,B}^{\text{pot}} = \frac{a_A}{a_B} e^{-\Delta E/S}$; since $a_A = a_B$, we obtain that $K_{A,B}^{\text{pot}} = e^{-\Delta E/S}$.

We can simply give the common logarithm that

$$pK_{A,B}^{\text{pot}} = \Delta E_s / S_s$$

Indexes *s* just indicate that the ΔE is measured in mV and S in mV/decade units. In reality, however, the two calibration curves are not parallel. The Nicolsky–Eisenmann equation is not rigorously followed. Therefore, depending on where the horizontal or vertical lines are drawn, different selectivity coefficient values can be obtained with the separate solution methods. Therefore, it is recommended that the selectivity coefficient be measured in the range of the ionic concentration of the samples using *mixed solution methods*.

7.3.2.2 Mixed Solution Methods

The so-called *matched potential method* [9] is the simplest from *the mixed solution methods*. It is completely independent of the Nicolsky–Eisenmann equation. The method is based on a practical definition; it says that the selectivity coefficient is the ratio of the activity of the primary and interfering ions that gives the same potential changes in the same solution. Applying it, a solution is selected in the lower concentration range of the analyte ion expected in the sample. The electrode potential measured E_1. After Δa_A a known incremental change of the primary ion activity is made in the solution and the E_2 electrode potential value is measured. Finally, the known volume of the originally selected solution is introduced into the cell. Again, the E_1 potential value is observed. Following the changes of the electrode potential, slowly, the concentrated solution of the interfering ion is added to the solution as long as the E_2 potential value is detected again with the same electrode system. Δa_B ion activity change that results the same $\Delta E = E_2 - E_1$ potential change can be calculated knowing the solution volumes and the concentration of the interfering ion. In this case, the selectivity coefficient can be given by the following expression:

$$K_{A,B}^{\text{pot}} = \frac{\Delta a_A}{\Delta a_B}$$

The method does not deal with differences between the charges of the primary and interfering ions. It does not require close to the Nernstian response of the electrode to the different ionic species. The selectivity coefficient value obtained with this method has only practical significance. The $K_{A,B}^{\text{pot}}$ values can be used for the estimation of the reliability of direct potentiometry in the case of the electrode available and samples of approximately known concentration.

The other version of the mixed solution method is the *fixed interference method*. It is based on the

$$E = E_0 + S \lg(a_A + K_{A,B}^{\text{pot}} a_B)$$

version of the Nicolsky–Eisenmann equation. For this method, interfering ion activity is selected in the range of the ionic composition of the perspective samples. The potential of the ISE is measured in that solution. After keeping the activity of the interfering ion constant, the primary ion concentration is increased in the solution. As long as the $K^{\text{pot}}_{A,B} a_B \gg a_A$, the electrode potential does not change. Increasing a_A, the primary ion starts affecting the voltage. Furthermore, $K^{\text{pot}}_{A,B} a_B \ll a_A$. Plotting the cell voltage against the $-\log a_A$, a horizontal line and a straight line with S slope will be obtained with a curved section in between them. Extrapolating the two straight lines, the intersection point refers to the equal E values for the pure primary ion solution (a_A) and for the selected interfering ion activity (a_B). Then the actual selectivity

$$K^{\text{pot}}_{A,B} = \frac{a_A}{a_B}$$

Similarly, the primary ion activity can be fixed and the same procedure is repeated. This so-called *fixed primary ion activity* method most often is applied for the investigation of the pH range in which the tested ISE can be applied.

7.4 Measuring Electrodes Used in Potentiometric Analysis

Electrodes made of different metals have been used for long times for the location of end point in potentiometric titrations. In redox titrations often platinum electrodes, in argentometric titration silver electrodes are used. The measuring surface of these electrodes is usually made of the needle, disk, plate, wire, and so on, shaped platinum or silver metal. They are attached into the nonconductive electrode body.

A selective sensor is needed when the direct analytical method is employed. The history of selective potentiometric sensors started more than 100 years ago with the pioneering work of Cremer [10] resulting the development of the pH-sensitive glass electrodes. As is well known, the selective pH measuring function of this electrode is produced at the outer swollen layer of the special thin glass layer separating the tubular lumen of the electrode body and the sample solution. The glass electrodes concerning their analytical values such as selectivity, stability, dynamic concentration range, simplicity of measuring procedure, and so on, are fare among the best chemical sensors ever made. We dare saying that the pH-sensitive glass electrodes are the most often used tools of chemical analysis.

For pH measurements, different metal–metal oxide electrodes such as oxide film cowered antimony, stainless steel, iridium [11] have also been used.

Intensive research work has been carried out for trying to develop glass membrane electrodes for potentiometric measurements of the concentration of other cations. First, Eisenmann [6] reported about working out of a well-functioning glass-membrane-based ion-selective sodium electrode. Among many others, Lengyel [12], Csákvári and Boksai [13] should be mentioned as successful contributors of the glass electrode research.

An important landmark in the history of ion-selective potentiometric sensors has been the paper of Pungor and Hollós-Rokosinyi [14] published in1961. It reported the development of a silver iodide precipitate-based iodide-selective membrane electrode. The membrane of the electrode was made of especially prepared silver iodide precipitate particles dispersed in the thin paraffin matrix. Similar membrane electrodes were made for measuring different anionic components using other silver halides of or metal sulfide precipitates. Later, the Pungor-type electrodes employed the mechanically more stable silicon rubber matrix for measuring membrane preparation. Interestingly, Pungor-type electrodes manufactured by the Radelkis company were the first anion-selective electrodes ever commercialized. In earlier days, the measuring surface of the inorganic precipitate-based electrodes was made of homogeneous pellets prepared by high-pressure molding of the crystals of the sensing material.

Frant and Ross [15] invented an ion-elective fluoride-type electrode. Its measuring membrane was made of europium-doped homogeneous lanthanum fluoride crystal membrane. Measuring fluoride ion concentration is not an easy task; on the other hand, the contradiction concerning the beneficial and the toxic physiological effects of fluorine ion intake created intensive debates. Therefore, the electrode that made the simple potential measuring technique applicable for fluoride measurements became very popular in practice.

Measuring the function of a large group of ISEs is based on the supramolecular host–guest interaction between organic host molecules and different analyte ions. The host species is incorporated in a membrane that is most often attached onto a tubular electrode body. The membrane separates the internal reference and the sample. Selective interaction between the host ionophore and the guest–ion generates charge separation at the interface, that is, potentiometric response. Early versions of these organic ionophore-based electrodes were prepared with liquid state measuring membrane. A porous plastic film, most often Millipore filter membrane, served as support for the liquid membrane. Measuring membrane forming liquid, that is, called ion-selective cocktail, was made of a nonvolatile organic solvent in which the ionophore and other necessary ingredients were dissolved. The filter membrane was soaked with the cocktail. The electrode had a special tubular body with a small volume circular container for holding the measuring cocktail. The other part of the disk-shaped porous membrane reached into the cocktail container, so the pores were filled with the organic liquid. The electrode could be taken apart and the cocktail, the membrane, and the internal reference solution could be, and periodically had to be, renewed. By changing the cocktail, the electrode selectivity could be changed. Electrodes for measuring different ions could be obtained just changing the cocktail and putting a new filter membrane in.

Later, the liquid state measuring membrane was changed by an elastic plastic disk. This made the electrode form mechanically more stable. In the early times, well working potassium [16] and ammonium [17] ISEs were made with a silicon rubber matrix. Later, however, plasticized polyvinyl chloride (PVC) matrices became the most popular in ISE fabrication. Their preparation procedure was well worked out and clearly described by Shatkay [18] and Thomas [19]. The high molecular weight PVC used in membrane preparation can be dissolved in tetrahydrofurane. In this solution, the plasticizer, the selective ionophore, and other ingredients like special lipophilic salt are dissolved. The solution is transferred into a ring-shaped flat bottomed mold. Care is taken to let the tetrahydro furane (THF) slowly evaporate. In a few days, a circular, flexible membrane is obtained. A disk with appropriate diameter is cut from the membrane and pasted on the tubular electrode body.

The PVC matrix provides the mechanical stability. It is chemically inert and does not dissolve in water. The ionophores are lipophilic organic molecules that selectively interact with the analyte ions. The plasticizer and the lipophilic salt also play an important role in the ion-selective function. They decrease the electric resistance, increase selectivity, and influence the detection limit. As a plasticizer, bis-(2-ethylhexyl)-sebacate (DOS) or 2-nitrophenyl-octyl-ether (*o*NPOE) is used, while for lipophilic salt potassium-[tetrakis-4-chlorophenyl]-borate is employed most often.

The interaction between the analyte ion and the ionophore must be relatively week to allow reversible electrode function. Its energy usually is in the energy range of the supramolecular interactions. It is advantageous if the ionophore is a neutral species. The charged ionophores interact with any ion of the opposite charge; therefore, usually lower selectivity can be provided with charged ionophores. One of the best known ionophores that successfully could be employed for the preparation of a highly selective ion-selective membrane electrode is the valinomycin. It provides excellent selectivity for potassium ion measurement in the presence of high sodium ion concentration. The valinomycin-based potassium electrode is broadly used in different areas of chemical analysis like in clinical or environmental chemistry.

Up till now, a very high number of reports have been published about results achieved in preparation and testing of different molecules that were found applicable as ionophore in ISEs. Crown ethers [20], calixarenes [21], cryptands [22], podands [23], or other cavitand host species are among the most often prepared and studied types of molecules in this respect.

7.4.1 Ion-selective Field Effect Transistors

The field effect transistors are well-known micro-sized electronic devices that are broadly used in electronic circuits. They are made of a small swab of the silicon substrate. The substrate is lightly p-doped. Two discrete spots placed in a certain distance from each other are strongly n-doped. One of these spots is called the source, while the other is called the drain. If voltage is applied between the source and the drain, then no current flows since the p-doped substrate acts like an insulator between the two n-doped spots. The area above the substrate between the drain and the source is prepared close to the surface but electrically insulated. It is called the Gate.

The electric field applied on the gate can change the charge-carrier distribution in the surface layer. Positive field repulsing the positive holes can make inversion. This means that in the surface layer of the substrate, an n-type conductive channel is formed. In ion-selective field effect transistors [24, 25] (ISFET), the gate is coated with an ion-sensitive layer and an appropriate electrode body is applied (Figure 7.1).

If the ISFET is introduced into a test solution together with a reference electrode, then the voltage between them is influenced by the ion activity. Therefore, the drain current (I_D), that is, the current between the source and the drain, depends on the gate voltage (V_G). No current flows through the gate. Therefore, the current reflects the emf of the measuring cell. The ISFETS can be well used in potentiometric analysis. Since the gate voltage is reflected by the drain current, the dependence cannot be directly related to the Nernst equation. However, most instruments used together with ISFET detection contain an operational amplifier circuitry that keeps the I_D constant by controlling the bias voltage. The change of the bias voltage in this case is equal to the change of the emf in the cell. ISFET probes are applied in conventional size pH sensors as well as in special miniaturized analytical devices.

7.4.2 Severinghaus-type Probes

Selective potentiometric sensors are used as base sensor elements in modified probes. Two types of these are mentioned here briefly: the potentiometric gas sensors and the potentiometric enzyme electrodes.

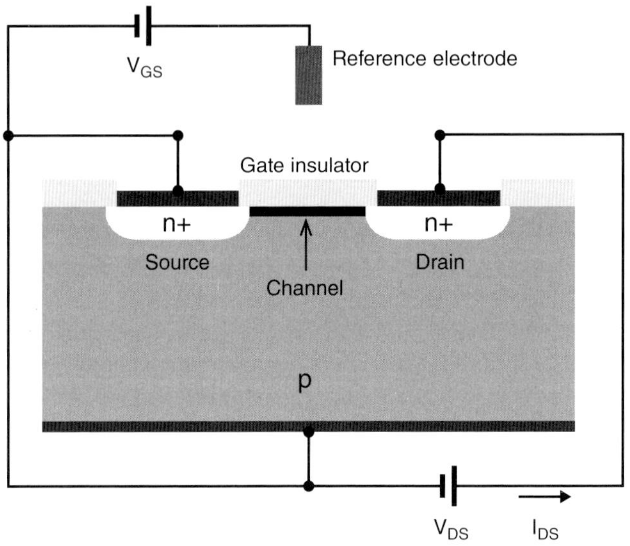

Figure 7.1 *Structure of ISFET [25]*

Selective potentiometric gas sensing probe was first reported by Severinghaus [26, 27]. The probe was made of a pH-selective glass electrode and a reference one set inside the lumen of a cylindrical glass body. The end of the probe was coated with a gas permeable thin membrane that separates the sample and the measuring cell. Inside of the cell, a bicarbonate buffer keeps the pH constant. The surface of the pH-sensitive glass electrode is facing the gas permeable membrane, only a thin buffer film is between. When the probe is in contact with CO_2-containing gas phase or solution, the gas can pass through the membrane and changes the pH.

As the dissociation equilibrium is achieved, a constant pH and cell voltage are achieved. Ionic species will not interfere with the electrode function that is not able to passing through the membrane. Volatile acids can interfere [28]. The pH-sensitive glass electrode can be replaced by the more robust IrO_X sensor [29]. A very similar ammonia gas sensor can be fabricated.

7.4.3 Potentiometric Enzyme Electrodes

The potentiometric enzyme electrodes are made from a basic sensing electrode which is modified with a selective biocatalytic layer coating their measuring surface. The layer of their surface catalyzes a reaction of the analyte. The local concentration change resulted by the reaction is detected by the potentiometric sensor. The first potentiometric biosensor has been reported by Guilbault and coworkers [17, 30]. From that time very high numbers of enzyme electrodes or electrometric biosensors have been reported. Potentiometric detection, however, is less frequently employed in their case comparing to the amperometric one.

In the early times of biosensor research, enzyme sensors were prepared for application in clinical diagnosis. The urea enzyme electrode of Guilbault [17] was built on an ammonium ion-selective membrane electrode prepared with nonactin ionophore. The surface of the electrode was coated with a urease enzyme containing a catalytic reaction layer. If the sensor was introduced into a urea-containing solution, then the urease-catalyzed hydrolysis of the urea took place in the reaction layer. The ammonium ion-selective-base sensor detects the local activity of the liberated ammonium ions:

$$(NH_2)_2CO + 2H_2O + H^+ \xrightarrow{\text{urease}} 2NH_4^+ + HCO_3^-$$

The original electrode was developed for detecting urea in blood and urine samples. An enzyme sensor for measuring urea in milk has been reported by an Indian group [31]. The working principle is identical to that of Guilbault's sensor. The sensor prepared for milk analysis, however, is a flat cell fabricated with screen-printed technology.

7.5 Special Tasks

7.5.1 pH Measurements

The most frequently applied potentiometric measurements involve the determination and adjustment of the pH of different solutions, samples, reagents, reaction media, or buffers.

As is well known, pH is used to characterize the acid base property of a sample. It is defined as the negative logarithm of the hydrogen ion activity (a_H+). Accordingly, it is a dimensionless number:

$$\text{pH} = -\log_{10}(a_H^+)$$

Since the potential of selective potentiometric sensors is in a well-defined relationship with the activity of the detected ion, it is obvious that potentiometric electrodes can be used for pH measurements. The Nernst equation for the hydrogen ion is

$$E = E^0 + \frac{RT}{F}\ln(a_{H^+}) = E^0 - \frac{2.303RT}{F}\text{pH}$$

This means that the linear relationship exists between the electrode potential and the pH. Therefore, if we can find an electrode that shows Nernstian response to hydrogen ion activity, then measuring the E value and having reliable calibration data we can get the hydrogen ion activity and the pH of the tested solution. If E_s and E_x are the electrode potentials in calibrating the standard and the sample, and p_{Hs} and p_{Hx} are the pH of the standard and that of the sample, respectively, then we can easily get p_{Hx}, the unknown p_H of the sample solution,

$$E_s = E^0 - \frac{2.303RT}{F}p_{Hs} \qquad E_x = E^0 - \frac{2.303RT}{F}p_{Hx}$$

$$E_s - E_x = -\frac{2.303RT}{F}(p_{Hs} - p_{Hx}) \qquad p_{Hx} = p_{Hs} + (E_s - E_x)\frac{F}{2.303RT}$$

$$p_{Hx} = p_{Hs} + (E_s - E_x)\frac{F}{2.303RT} = p_{Hs} + \frac{(E_s - E_x)}{S}$$

where S is the slope of the E–p_H dependence. Theoretically, $S = 2.303RT/F$, it is called the "Nerntian" slope. Its value is 59.1 mV/p_H at 25 °C.

This looks simple; however, in practice, there are several obstacles that make this approach not so straightforward.

One thing that needs special care is obtaining a reliable value for the electrode potential. In the potentiometric practice, the cell voltage at $i = 0$, that is, the emf is measured. The emf is a sum of several potential differences. Changes of temperature, convection intensity, and ionic strength can affect these. Unknown changes of the different interfacial or diffusion potentials in the measuring line can bring in changes in the emf, and this brings uncertainty in the experimentally obtained E values.

Deviation of the electrode function from the theoretical one is the other source of error. In connection with this, the well-known acidic and basic error of pH measuring glass electrodes should be mentioned.

The main obstacle, however, is finding calibrating standards and appropriate measuring cell. Unfortunately, there are no solutions available with accurately known hydrogen ion activity. Similarly, there are no procedures for preparing them. Therefore, the actual, traceable pH values must be based on a generally accepted convention. This means that the standard solutions of the known composition are proposed, the pH of which should be accepted. They are called primary standard solutions. The proposed pH values of them are based on the Bates–Guggenheim convention [32] and on potentiometric data experimentally obtained applying the Harned cell. The Harned cell [33] is a cell without transference; therefore, in this case, no diffusion potential is involved in the emf measured. It is defined by the cell symbol

$$\mathrm{Pt}|\mathrm{H_2}|\text{buffer solution, } \mathrm{Cl^-}|\mathrm{AgCl}|\mathrm{Ag}$$

The emf value (E_I) is measured between the two electrodes considering 1 atm partial pressure of the hydrogen gas:

$$E_\mathrm{I} = E^0 - \frac{RT}{F} 2.302 \lg \left[\frac{m_\mathrm{H}\gamma_\mathrm{H}}{m^0} \frac{m_\mathrm{Cl}\gamma_\mathrm{Cl}}{m^0} \right]$$

$$a_\mathrm{H} = \frac{m_\mathrm{H}\gamma_\mathrm{H}}{m^0} = -\lg(a_\mathrm{H}\gamma_\mathrm{Cl}) = \frac{E_I - E^0}{\frac{RT}{F} 2.302} + \lg \frac{m_\mathrm{Cl}}{m^0}$$

where E^0 is the standard emf of the cell, while γ is the activity coefficient of chloride or hydrogen ions.

If in the Harned cell the fixed molality of HCl is applied, then

$$E_\mathrm{Ia} = E^0 - \frac{RT}{F} 2.302 \lg \left[\frac{m_\mathrm{HCl}\gamma_{\pm\mathrm{HCl}}}{m^0} \right]$$

where $\gamma\pm$ is the main activity coefficient.

In order to obtain the pH of a solution from $\lg(a_\mathrm{H}\gamma_\mathrm{Cl})$, an independent way for getting the value of γ_Cl is needed. It is done by extrapolating $\lg(a_\mathrm{H}\gamma_\mathrm{Cl})$ to zero chloride molality ($\lg(a_\mathrm{H}\gamma_\mathrm{Cl})^0$) and using the Bates–Guggenheim convention [34]. The convention proposes that the trace activity coefficient of the chloride ions (γ^0) can be calculated on the basis of Debye–Hückel theory, that is,

$$\lg \gamma_\mathrm{Cl}^0 = \frac{-AI^{1/2}}{1 + \mathrm{Ba}I^{1/2}}$$

where I is the ionic strength, A is a temperature-dependent constant, and $\mathrm{Ba} = 1.5$.

Accordingly, the pH of the proposed primary solutions can be calculated.

$$\mathrm{pH} = \lim m_{\mathrm{Cl}\to 0}\left[\frac{E_I - E^0}{\frac{RT}{F}2.302}\right] + \lg\frac{m_{\mathrm{Cl}}}{m^0} - \frac{AI^{1/2}}{I + 1.5(I/m^0)^{1/2}}$$

The properties of the proposed primary standard solutions are listed in Table 7.1. The table is taken from the report of IUPAC 2002 [35].

In practice, the pH measurements of the so-called secondary standard solutions are often used for calibrating the measuring electrodes. The pH values of these are established in comparison with the primary standards. Their pH values must be consistent with the pH values of the primary standards. The measurements are carried out with the hydrogen electrode and in cells without transference in determination of the pH of the secondary standard solution, if it is possible. The calibrating standards are commercially available, and their pH values at different temperatures are certified by the manufacturer. In Table 7.2, the pH values of a few secondary standard solutions are listed. They were determined with the Harned cell for different temperatures. The table is taken from the report of IUPAC 2002 [35].

The pH values of the calibrating standards are based on the convention but they are more or less consistent.

The application of hydrogen electrodes in the everyday routine of pH determination is not practical. Most often pH sensitive glass electrodes and Ag/AgCl references are used in the cell as shown in the following:

Ag|AgCl|KCl (constant concentration)||standard or sample solution|glass electrode

The emf is measured with the sample in contact with the glass electrode, and the pH of the sample is determined employing the calibration data.

Unfortunately, this cell contains a current bridge; therefore, the E_j junction potential brings in some uncertainty.

If the slope of the calibration curve is known, then one standard can be used for the calibration. However, more reliable pH value of the sample is obtained if the calibration data obtained with several standards are used.

The slope (S) is obtained as $S = \frac{E_{\mathrm{s}2}-E_{\mathrm{s}1}}{p_{\mathrm{Hs}2}-p_{\mathrm{Hs}1}}$, and the unknown $\mathrm{pH_x}$ will be

$$\mathrm{pH_x} = \mathrm{pH_s} - [E_\mathrm{x} - E_\mathrm{s}]/S - [E_{\mathrm{j}2} - E_{\mathrm{j}1}]/S$$

where indexes x, s, j, 1, and 2 stand for the unknown sample, standard, junction, and the numbers of half-cells, respectively.

The pH meters usually have a scale that upon proper adjustment directly displays the pH value of the sample solution. The manuals of the meters usually describe the way of setting the scale. One standard calibration can be employed if the electrode has the Nernstian response slope and the temperature is known. More reliable scale setting can be done employing two or more standards. Usually, the meters provide the slope and bias settings, manual or automatic ones. Advanced pH meters give directions for calibration and scale settings. Some of them recognize the buffers employed and do the scale setting automatically. It is wise, however, if sometimes the functions of the electrode and the meter are checked manually with buffers of known pH, observing the voltage change in millivolts scale.

Table 7.1 *pH values of primary standard solution at different temperature*

Primary standards (PS)	Temperature (°C)										
	0	5	10	15	20	25	30	35	37	40	50
Saturated potassium hydrogen tartrate (at 25 °C)	–	–	–	–	–	3.557	3.552	3.549	3.548	3.547	3.549
0.05 mol kg^{-1} potassium dihydrogen citrate	3.863	3.840	3.820	3.802	3.788	3.776	3.766	3.759	3.756	3.754	3.749
0.05 mol kg^{-1} potassium hydrogen phthalate	4.000	3.998	3.997	3.998	4.000	4.005	4.011	4.018	4.022	4.027	4.050
0.025 mol kg^{-1} disodium hydrogen phosphate + 0.025 mol kg^{-1} potassium dihydrogen phosphate	6.984	6.951	6.923	6.900	6.881	6.865	6.853	6.844	6.841	6.838	6.833
0.03043 mol kg^{-1} disodium hydrogen phosphate + 0.008695 mol kg^{-1} potassium dihydrogen phosphate	7.534	7.500	7.472	7.448	7.429	7.413	7.400	7.389	7.386	7.380	7.367
0.01 mol kg^{-1} disodium tetraborate	9.464	9.395	9.332	9.276	9.225	9.180	9.139	9.102	9.088	9.068	9.011
0.025 mol kg^{-1} sodium hydrogen carbonate + 0.025 mol kg^{-1} sodium carbonate	10.317	10.245	10.179	10.118	10.062	10.012	9.966	9.926	9.910	9.889	9.828

Table 7.2 *pH values of secondary standard solution at different temperature*

Secondary standards	Temperature (°C)									
	0	5	10	15	20	25	30	37	40	50
0.05 mol kg^{-1} potassium tetroxalate[a] [5, 17]	–	1.67	1.67	1.67	1.68	1.68	1.68	1.69	1.69	1.71
0.05 mol kg^{-1} sodium hydrogen diglycolate[b] [31]	–	3.47	3.47	3.48	3.48	3.49	3.50	3.52	3.53	3.56
0.1 mol dm^{-3} acetic acid + 0.1 mol dm^{-3} sodium acetate [25]	4.68	4.67	4.67	4.66	4.66	4.65	4.65	4.66	4.66	4.68
0.1 mol dm^{-3} acetic acid + 0.1 mol dm^{-3} sodium acetate [25]	4.74	4.73	4.73	4.72	4.72	4.72	4.72	4.73	4.73	4.75
0.02 mol kg^{-1} piperazine phosphate[c] [32]	6.58	6.51	6.45	6.39	6.34	6.29	6.24	6.16	6.14	6.06
0.05 mol kg^{-1} tris hydrochloride + 0.01667 mol kg^{-1} tris[c] [5]	8.47	8.30	8.14	7.99	7.84	7.70	7.56	7.38	7.31	7.07
0.05 mol kg^{-1} disodium tetraborate	9.51	9.43	9.36	9.30	9.25	9.19	9.15	9.09	9.07	9.01
Saturated (at 25 °C) calcium hydroxide [5]	13.42	13.21	13.00	12.81	12.63	12.45	12.29	12.07	11.98	11.71

[a] Potassium trihydrogen dioxalate ($KH_3C_4O_8$).
[b] Sodium hydrogen 2,2′-oxydiacetate.
[c] 2-amino-2-(hydroxymethyl)-1,3 propanediol or tris(hydroxymethyl)aminomethane.

For testing the pH of agricultural or food products not very high accuracy is needed. The Harned cell and hydrogen electrode are not being used here. The special electrodes for application in this field are designed for easy and safe operation. Some of the electrodes made with a form that they can be pocked into meat samples, soil, fruits, or other kinds of samples. It is important to avoid broken glass getting slipped in processed food. This limits the applicability of conventional glass electrodes in certain process-controlling units. ISFET sensors, antimony electrodes, or special robust glass are employed in these measurements. Some of the important properties of pH sensors concerning their application in certain areas of food industry are easy way of cleaning, operation at elevated temperature, all solid state construction, option of being sterilized, and so on. The instrument-manufacturing companies commercialize special sensors to satisfy the needs.

7.5.2 Miscellaneous Analytical Tasks by Potentiometric Methods

One of the main advantages of potentiometric methods is that the analysis can be done in line with sampling, sample preparation, and treatments. However, when soaking transducers in samples of high protein, lipid, or other bio matter content, electrode fouling caused by adsorbed biofilm can be expected [36]. The biofilm adsorbed can slow down the response time of the electrodes or it can make local pH changes.

The film can also be adsorbed on the surface of the current bridge, influencing the diffusion potential. The extent of all these interfering effects has been investigated by Upreti and coworkers [37] making pH measurements in cheese processing and milk samples. The performances of glass electrodes and PVC matrix-based membrane electrodes were compared. Similar to the earlier findings [38], the glass electrode performed well in real samples, while the function of the plasticized membrane electrode was affected. Therefore, the adsorbed matter should be periodically removed. Electrode cleaning solutions containing detergents and in some cases digestive enzymes are commercialized. On the other hand, research has been in progress for finding other plastic matrices that avoid or lessen adsorption from those samples. Usually, short-time exposure does not bring in considerable error. Using the fragile glass electrodes in food-processing lines has its potential risk.

Mapping pH and concentration of main ionic components of soil is an obvious step in trying to improve the productivity of different crop production together with limiting the negative environmental impacts of fertilization and soil treatments.

In this, the simple and cost-effective nature of direct potentiometric analysis makes it attractive [39]. Especially, it is so if the analysis can be rightly done at the sample side. Adsett and coworkers [40] developed an automatic sampler that could deliver freshly taken soil samples to the nitrate analyzer. Birrell and Hummel [41] reported on the development and use of a potentiometric real-time flow injection analysis (FIA) soil analyzer. An automatic sampler–analyzer system has been developed originally for soil pH mapping [42]. This apparatus takes the soil sample at different spots, brings into contact of the sample the pH electrode and the current bridge, takes the voltage reading for pH determination, and washes the system to pretreat it for the next measurement. The location of the sampling spot is determined by global positioning system (GPS). In this way the "on-the-go" mapping of the field can be carried out with a good lateral resolution on a cost-effective manner.

This sampling-analyzing apparatus was used by Adamchuk and coworkers [43] in their comparative study conducted to explore the feasibility of application pH, K^+, NO_3^-, and

Na^+ ISEs in field mapping with potentiometric direct soil measurements (DSM). In these works, samples were taken at different locations and analyzed in the laboratory by the apparatus. Reference measurements also were conducted in commercial soil laboratories by conventional laboratory methods. Three pH electrodes, two K^+, two NO_3^-, and one Na^+ electrodes were selected for the experiments.

The accuracy of direct soil measurements (DMS) assessed while comparing against reference measurements was acceptable for the pH test ($R^2 \approx 0.93-0.96$). However, the regressions were found much worse for the other components ($R^2 \approx 0.61-0.62$ for K^+, $R^2 \approx 0.41-0.51$ for NO_3, and $R^2 \approx 0.11$ for Na^+). The response time of the electrodes was short enough to support on the go mapping; however, the tested K^+ and NO_3^- electrodes were only able to distinguish between very high and very small concentrations, while the Na^+ electrode in this measurement was not found applicable.

7.6 Application of Potentiometric Measurements for Anions

7.6.1 Determination of Chloride Ion Concentration

The chloride ion concentration in food and in agricultural samples is relatively high. Assessment and control of chloride ion concentration can be important concerning health, environmental, or technological reasons. Chloride ISEs prepared with silver chloride poly crystalline measuring membrane can be used for analyzing different samples. The electrodes – in the case of several samples – have high enough selectivity enabling the application of direct potentiometric methods. The electrodes are commercially available. High concentration of common cations such as silver, mercury, and lead would interfere with the measurements. The solubility of silver bromide and silver iodide is orders of magnitude smaller than that of silver chloride. Therefore, if bromide, cyanide, and iodide ions are present, they give serious interference with the measurements. Soaking the electrode in a solution with high concentration, these anions can irreversibly damage the measuring surface. Fortunately, in real samples to be analyzed for chloride, only traces of the interfering ions are usually present.

Br^-, I^-, and CN^- ions, if present, can be eliminated by adding a high concentration of the sodium bromate oxidizing agent.

It is obvious that reference electrodes with the potassium chloride filling solution cannot be used since the ions from the filling can contaminate the samples diffusing through the current bridge. Therefore, using the double junction reference electrode is highly recommended. The outer filling solution can be aqueous 0.1 M potassium nitrate or 0.1 M lithium acetate. For samples with ionic strength higher than 0.01 M ionic strength adjusting buffer (ISAB), for example, 5 M $NaNO_3$ has to be added to compensate the main activity coefficient differences between samples and standards. Usually, 2 cm^3 ISAB for the 100 cm^3 sample or standard solution is recommended for problem-free aqueous solution samples.

Before and after sample measurements, the potentiometric measuring cell is calibrated with standard solutions in the range of 3×10^{-5}–0.1 M. The slope of the response curve should be about 54 ± 5 mV/decade at 25 °C. Between measurements, the sample half-cell must be carefully washed and dried to avoid cross contamination. The cell voltage measured in the sample solution must be in the voltage range obtained for calibrating standards.

For analyzing different samples, the sample preparation procedure needed to be carefully worked out.

7.6.1.1 Measurement of Cl^- Concentration in Milk

The high casein content of milk can be adsorbed on the electrode and on the current bridge. This adsorbed protein layer can break down the potentiometric signal. Therefore, the dilution of the sample as well as the standards is necessary. For the milk samples, the addition of the ISAB solution, in this case, 2 volume of the diluted aqueous HNO_3 solution to 1 volume of milk or standard is recommended, that is, the 2:1 ratio is recommended. The ISAB is prepared by diluting 60 cm^3 concentrated HNO_3 to 1 l with de-ionized water.

7.6.1.2 Measurement of Cl^- Concentration in Meat and Meat Products

One gram of a meat sample is sonicated in 100 cm^3 water. The obtained dispersion is filtered. Twenty-five cubic centimeters of the filtrate is mixed in the 1:1 volume ratio with 0.5 M nitric acid used as an ISAB solution. The same ratio of the ISAB is added to the calibrating standards. The usual direct potentiometric method gives the chloride ion concentration of the extracting solution. The sample concentration is given as micrograms per gram of meat.

7.6.1.3 Measurement of Cl^- Concentration in Butter

The chloride content of the butter is extracted with dilute nitric acid. In this procedure, accurately weighted 10 g butter is placed in a 150 cm^3 beaker. Hundred cubic centimeters of dilute (60 cm^3 cc HNO_3 diluted to 1000 cm^3) HNO_3 is added to it and slowly heated while slow magnetic stirring is employed. The butter is melted, but boiling and emulsification are avoided. When complete extraction is achieved, it is allowed to cool down. The water immiscible fat solidifies on the surface. Twenty-five cubic centimeters from the aqueous solution is pipetted out and mixed with double distilled water in a volumetric ratio of 1:1. This solution is used as a sample. To the calibration solutions, the 1:1 ratio of dilute HNO_3 is added as ISAB. Calibration and direct potentiometric measurements are performed in the usual manner. The chloride ion concentration in the butter is usually given in μg/g (= ppm = mg/kg).

7.6.1.4 Measurement of Cl^- Concentration in Mayonnaise

For the preparation of the sample solution, 1 g mayonnaise is introduced into a 1 l volumetric flask. Five hundred cubic centimeters of 1 M nitric acid is added and thoroughly shaken to extract the chloride ions into the aqueous phase. After achieving complete extraction, the flask is filled up to the mark with double distilled water. The solution is filtered and analyzed by the direct potentiometric method. The standards are mixed with 1 M nitric acid ISAB in proportion of 1:1.

7.6.1.5 Measurement of Cl^- Concentration in Soil Samples

Air dry, ground soil sample is selected. Four grams of accurately weighted part of it is introduced into a bottle supplied with an appropriate glass stopper. Fifty cubic centimeters of water is added and the bottle is fixed on a mechanic shaker and 30–90 min shaking is employed for extracting the chloride content. After this, the suspension is filtered and washed. The unified filtrate and washing solution is introduced into a 100 cm^3 volumetric flask and it is filled up to the mark. Two cubic centimeters of the ISAB (5 M $NaNO_3$) is added to the sample before direct potentiometric analysis. Potassium chloride is used for making the calibrating standard solutions. Before reading is taken for calibration, 2 ml ISAB solution must be added to each 100 ml standard and homogenized thoroughly to compensate for different activity coefficients between samples and standards.

7.6.1.6 Chloride Ion Determination in Fruit Juice

The matrix is complicated and the ionic strength in it is high. In order to avoid errors from these uncertainties, the use of standard addition or sample addition is recommended instead of direct potentiometry based on the calibration data.

In a sample addition method, a high, known volume (V_c) standard solution of appropriate, known concentration (C_c) is selected. The cell voltage is measured (E_c) putting it in the sample compartment. After that, an appropriate small known volume (V_s) of the sample is added and the cell voltage is measured (E_s). It is expected that the small volume of the sample does not bring in the noticeable change in the matrix. Therefore, knowing the mentioned voltage, volume, slope, and concentration values, the unknown sample concentration can be calculated.

In order to select an optimal concentration value for the standard, however, a rough direct potentiometric measurement of the chloride concentration of the juice sample is recommended. After calibration, the slope value in the necessary range can also be obtained.

So the procedure is the following:

Using calibration with standards and direct potentiometric measurement a rough estimation of the sample concentration is done and also the slope of the electrode response (S) is determined.

The standard of 50 cm^3 C_c concentration is introduced into the sample half-cell and the steady cell voltage (E_c) is measured. After V_s volume (no more than 2 cm^3) of the fruit juice sample is added, the solution is homogenized and steady reading of the cell voltage (E_s) is taken.

From these values, the C_s sample concentration is calculated.

7.6.2 Determination of Fluoride Ion Concentration

The limited number of well functioning, classical or spectrophotometric methods is available for measuring fluoride ion concentration in different samples. Therefore, after the invention of lanthanum fluoride crystal-based ISE [15], its use as a detector in standardized methods becomes almost general. For example, the Environmental Protection Agency (EPA) METHOD 9214 [44] is for measuring the concentration of fluoride ions in water samples as well as in soil extracts. It is a direct potentiometric method using the ion-selective fluoride electrode and the conventional or double junction reference electrode. The lower limit of detection is 0.025 mg dm^{-3}. Fluoride concentrations from 0.025 to 500 mg dm^{-3} can be measured.

The hydroxide ions seriously interfere with the fluoride measuring function of the electrode; therefore, the pH of the samples and standard solutions must be below 5.5. Decreasing the pH further is not advisable because at low pH values, fluoride forms bifluoride (HF_2) that is not detected by the electrode. Also, in the presence of Al^{3+} or Fe^{3+} ions, fluoride complexes are formed decreasing the detectable fluoride concentration. In order to avoid this interference, a strong complexing agent, like 1,2-cyclohexanedinitrilo-tetraacetic acid ($C_6H_{10}[N(CH_2CO_2H)_2]_2{\cdot}H_2O$), is added to the samples and standard solutions to form complexes with the metals liberating the fluoride ions. For eliminating the interferences and adjusting the ionic strengths, the samples and the calibrating standards are mixed in the 1:1 volumetric ratio with TISAB. The TISAB solution is commercially available; however, for cost-efficiency reason, laboratories often prepare their own one following the procedure as given below.

Preparation of the TISAB solution: Mixture of 500 cm^3 of distilled water 57.0 cm^3 of glacial acetic acid, 58.0 g of sodium chloride, and 4.00 g 1,2-cyclohexanedinitrilo-tetraacetic acid (CDTA) ($C_6H_{10}[N(CH_2CO_2H)_2]_2{\cdot}H_2O$) are mixed and intensively stirred. When a homogeneous solution is obtained, it is cooled to room temperature and the pH is adjusted between 5.0 and 5.5 with 5 M NaOH. It is transferred to a 1 dm^3 volumetric flask and filled up to the mark with distilled water. The ready-made TISAB solution is stored in a polyethylene bottle.

In some cases, tea plants accumulate fluoride ions from the soil. To investigate the risk of drinking too much tea, the Slovenian research team [45] measured the fluoride ion concentration of teas cooked from different tea blends using the simple potentiometric method.

7.6.2.1 *Fluoride Content of Wines*

Fluoride content of wines and musts also drew attention. The first study about potentiometric fluoride measurement in wines was published [46] soon after the appearance of the ion-selective fluoride electrode. A few years later, the methods for fluoride determination have already been extended [47, 48] by applying sample dilution in the 1:1 ratio by the TISAB solution. Hidalgo *et al.* [49] suggested that for achieving reliability, ethanol should be given to the standard solution used for calibrating the fluoride electrode that serves as measuring probe in wine analysis. Martinez and coworkers [50] compared standard addition methods with calibration-based direct potentiometry. In both cases, the standards were made with the same ethanol content as that of the sample wines. The standard addition method appeared more reliable. Studies were conducted [51] for checking the interfering effect of different organic acids usually present in musts and wines, such as oxalic-, citric-, tartaric-, lactic, malic acids. Fortunately, they have not shown the considerable effect on the fluoride electrode function. The official OIV (International Organization of Vine and Wine) method OIV-MA-AS321-03 employs the direct potentiometric procedure with an ion-selective fluoride electrode for measuring fluoride ion concentrations of wines and musts.

7.6.3 Applications of EPA Methods for Anion and Cation Analysis

EPA METHOD 9212 [52] for measuring the chloride ion concentration in water samples as well as in soil extracts. In this direct potentiometric method, the ion-selective chloride electrode and double junction reference electrodes are used. The dynamic range of the analysis is between 2 and 1000 mg/dm^{-3}.

High concentrations of bromide, iodide, cyanide, or sulfide ions interfere with the determination. They can damage the electrode. Free ammonia would also damage the measuring surface of the electrode dissolving the silver chloride crystals.

Sulfide, cyanide, and ammonia can be removed by acidifying the sample to a pH of 4 with dilute sulfuric acid. Bromide and iodide can be removed by treating the acidified sample with potassium bromate (1% solution), which oxidizes the ions to bromine and iodate. Similarly, as in the case of fluoride, the presence of Al^{3+} or Fe^{3+} ions can interfere with the determination forming not detectable complexes with the analyte ions. Their interference can be eliminated by applying a strong complexing agent ethylenediaminetetraacetate (EDTA), disodium salt added in the form of a 2% solution. Before potentiometric measurements, the sample solutions as well as the calibrating standards are diluted in the 50:1 volumetric ratio with 5 M sodium nitrate ionic strength adjusting reagent.

7.6.4 Determination of Potassium Ion Concentration

In order to determine wither potassium fertilization is needed or not, it is important to determine the exchangeable potassium in soil samples [53]. The ion-selective potassium electrode proved applicable for these measurements [54]. A potentiometric flow injection manifold and method was worked out for solving that task by Almeida and coworkers [55]. They used the ion-selective potassium electrode built in a flow through cell.

Grygolowicz-Pawlak and coworkers [56] fabricated two different types of potassium electrodes using valinomycin as ionophore. One was a plasticized PVC matrix-based medium-sized electrode with hydroxyethylenecellulose (HEC)-based gel as an inner electrolyte and a silver/silver chloride inner electrode.

The other was a polyacrylate membrane-based back-side contact (BSC) flat sensor with gold as an inner electrode and redox-active self-assembled monolayer (SAM) as a solid phase. Both electrodes were found to work with good selectivity over calcium, magnesium, and sodium ions present in soils in large amounts. The selectivity was very good toward Ba. $BaCl_2$ applied to extract potassium from soil samples did not affect potassium response in a wide concentration range. The results obtained in soil-extract analysis showed good reproducibility and precision.

7.6.5 Determination of Nitrate Ion Concentration

Nitrate nitrogen can accumulate in leaves of different plants [57]. Eating those plants can be harmful, especially for small children. As is well known, due to high nitrate concentration in their diet young babies may suffer from infantile methemoglobinemia. In that condition, nitrite is substituted for oxygen in hemoglobin. Ion-selective nitrate electrodes have been developed long time ago. Several methods have been worked out [58–60] for using them as indicator electrodes in the analysis of nitrate nitrogen contents of agricultural or food samples. The comparison of the direct potentiometric method with optical measurements [61] proved the advantages of the ion-selective method.

7.6.5.1 Nitrate Contain of Vegetables

In the work of Consalteri *et al.* [62] the samples of six different vegetable species, carrot, wild endive, celery, chicory, spinach, and parsley, were analyzed.

One kilogram of each species was crushed and homogenized after washing. Ten grams from the obtained puree was extracted with the 50 cm^3 and 10 mM $CuSO_4$ solution after intensive shaking for 30 min. After that, the extracts were filtered, filled up to 100 cm^3, and analyzed with direct potentiometry using the commercial nitrate-selective electrode. The simple method was found sufficiently accurate and reliable.

Iren [63] measured the nitrate ion content of potato varieties after harvesting at different times. As sample preparation, the potatoes were sliced and dried at 65 °C to constant weight. One gram of the ground dried matter was shaken with 200 cm^3 0.04 M $(NH_4)_2SO_4$ for 30 min, and analyzed using commercial ion-selective nitrate electrode in a potentiometric cell.

The alcoholic content of the samples affects the potential of the ion-selective nitrate electrodes. Therefore, for analyzing the nitrate content of wine samples, the multiple standard addition method gives more accurate results than calibration curve-based potentiometric measurements.

The flow injection measuring technique combined with the detection with the ion-selective nitrate electrode proved very advantageous in the studies of Andac and coworkers [64].

Figure 7.2 shows the applied FIA manifold. The concentration range determined was between 0.1 and 100 mmol dm^{-3} with the detection limit of 0.066 mmol dm^{-3}. The calibration slope appeared to be 52.96 ± 1.02 mV/decade in the linear range. With the FIA technique, the analysis speed as high as 90 sample h^{-1} could be achieved. The nitrate concentration in fresh leafy vegetables analyzed ranged from 97 to 1861 mg kg^{-1}.

The ion-selective nitrate electrode was made of a graphite epoxy conductive substrate cast inside the end of a copper tube. The measuring surface was formed by coating this substrate with an ionophore containing plasticized PVC film. For the preparation of this measuring film, a solution was made dissolving 4% (w/w) nitrate ionophore (tetradodecyl ammonium nitrate), 30% (w/w) PVC, 66% (w/w) dibutyl phthalate (DBP) in 3 cm^3 THF. The ion-selective cocktail solution was spread over the substrate. The electrode was ready upon the evaporation of the THF.

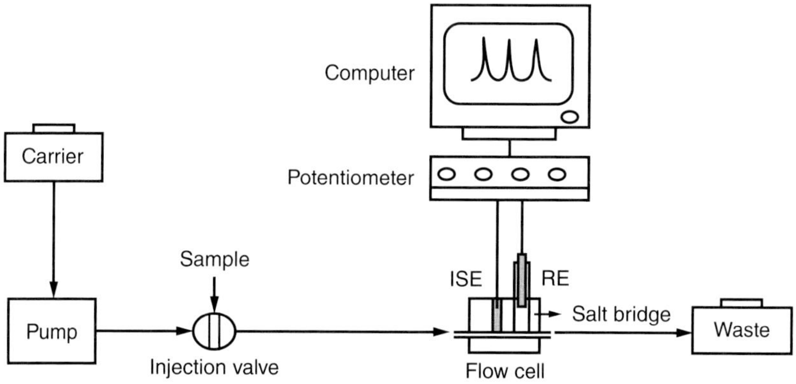

Figure 7.2 *The FIA manifold used by Andac and coworkers [64]*

7.6.6 Determination of Calcium Ion Concentration

The electrode potential of ISEs depends on the activity of the detected ion. If the adsorption or complex formation of the ion decreases the activity of the free ion, then the potentiometric method will give a smaller concentration value than another method that measures the total ion concentration.

Calcium ion concentration is an important property of apples. It seriously affects the post-harvest marketing and storage value of it. Usually, the measurement of the calcium concentration in fruits like apple is made by drying and burning the fruit and analyzing the ashes with atomic absorption spectroscopy. Knee and coworkers [65] observed that the considerable part of calcium ions in the apple is bound to cell walls.

Gao *et al.* [66] investigated the accuracy of calcium measurements in simulated milk ultrafiltrate and skim milk samples. They realized that the free calcium ion concentration is smaller in the sample than the total one owing to the ionic complex formation by phosphate, chloride, and citrate ions. Therefore, they recommend new standard solutions for calibration with ionic profile nearly the same as samples. Using those calibrating solutions, the systematic error could be substantially reduced.

7.6.7 Determination of Sweetening Additive Concentration

Potentiometric biosensors for the analysis of the sweetening additive used in pharmaceutical and food industry have been fabricated by several groups. The basic sensing element in the work of Campanella *et al.* [67] was a potentiometric ammonia gas sensor. The gas permeable membrane of the cell was coated with an immobilized L-aspartase enzyme containing layer. In the substrate-containing solution, the enzyme catalyzes a reaction that results in ammonia gas production. The ammonia passing through the gas permeable membrane changes the pH of the buffer at the pH-sensitive glass membrane surface.

A similar biosensor was made by Nikolelis and Krull [68]; however, they used a Severinghaus-type carbon-dioxide-based sensing element. They tried to find the gas-permeable membrane with optimal carbon dioxide transport properties. In their paper, a kinetic reaction-rate method for the determination of aspartame in dietary foodstuffs was proposed as a rapid and inexpensive alternative to a classical high-performance liquid chromatographic method.

7.6.8 Determination of Fumaric Acid Concentration

Fumaric acid is a food additive (E297) used as an acidity regulator in processed food products such as gelatin dessert powders, instant pudding powders, and ice cream powders. For safety and quality checking, the analysis of fumarate in different food products has special importance.

7.6.8.1 Measuring the Fumarate Content in Gelatin

Recently, Santini *et al.* [69] prepared a fumarate-selective electrode and proved its applicability in the analysis of several dietetic products. The electrode was made of a mercury fumarate ($Hg_2C_4H_2O_4$) containing a pellet. Main parameters were: 0.8 g weight, 6 mm thickness, and about 12 mm diameter. The pellet was glued on a tubular glass body. The internal contact was made through a mercury pool introduced into the lumen of the tube.

For preparation of the pellet, solid $Hg_2C_4H_2O_4$ and liquid mercury were homogenized in a ratio of 1.9:0.3. The obtained powder was mixed with the equal weight of graphite powder and in a press mould 9 tons pressure was employed.

The electrode showed good potential response in the 0.5 M $NaNO_3$ ionic strength adjusting electrolyte. The slope of the calibration curve was -29.2 ± 0.6 mV/decade, almost theoretical for a double charged ion. With good stability, acceptable selectivity, and with its broad linear response range (7.5×10^{-7}–1.0×10^{-2}) and the 6.0–9.0 pH range of applicability, the electrode proved a good analytical tool for food analysis. The multiple standard addition method was applied for analyzing the fumarate content of real gelatin dessert powders, instant pudding powder, and ice cream powders.

7.6.9 Quantification of Food Preservatives

7.6.9.1 *Quantification of Sorbate*

For the quantification of food-preservative sorbate ions in soft drinks, a pressed pellet sorbate-sensitive electrode was prepared [70]. The measuring pellet of it was made of 1.9 g mercury sorbate ($Hg_2(SOB)_2$), 0.3 g metal mercury, and 1.3 g graphite powder homogenized. Part of the powder obtained was transferred to a press mould and 9 ton pressure was employed for 9 min. The pellet was glued on a glass tube electrode body. Mercury and a platinum wire provided the internal contact. The slope of the calibration curve in the range of $5.0 \times 10^{-7} - 1.0 \times 10^{-2}$ M was found almost theoretical (-58.6 ± 1.3) mV/decade with a detection limit of 4.3×10^{-7} M. The proposed sensor displayed good selectivity over a variety of other anions (carboxylates and inorganic anions). The potentiometric sensor was successfully applied to the determination of sorbic acid in real food samples, that is, soft drinks, skim yogurts, jams, and sauces after a special sample treatment. As a first step of the analysis, 10 cm^3 of a certain soft drink sample, 1.0 cm^3 of 0.1 M HNO_3, and 2 cm^3 of the saturated potassium perchlorate were added. The sample was then extracted twice with 10 ml of ethyl ether and vortex mixed for 2 min for each extraction. The solvent from the collected organic phase was then evaporated under reduced pressure. The resulting residue was dissolved in 10 cm^3 of 0.5 M KNO_3 ($pH = 7.5 \pm 0.1$) and an aliquot of 8 cm^3 of the solution was employed for analysis with the sorbate-sensitive electrode using the multiple standard addition method. Similar extraction, solvent evaporation, and dissolution sample pretreatment procedures were used when other kinds of samples such as skim yogurt, jams, and sauces were analyzed.

7.6.9.2 *Quantification of Benzoate*

Benzoic acids (BAs), and their respective sodium, potassium, and calcium salts are the most commonly used preservatives in foodstuffs. Most often high-pressure liquid chromatography (HPLC) methods [71] are used for the determination of benzoates in food and beverage products. Several attempts have been made [72, 73] for developing the benzoate-selective electrode as a sensor applicable in a simple cost-effective measuring procedure. Alizadeh *et al.* [74] developed a solid contact electrode on a platinum-based sensing element. For the preparation of the selective benzoate measuring film, pyrrole was electropolymerized in the presence of sodium benzoate and sodium dodecyl benzenesulfonate. The slope of the calibration curve of the electrode was 55.4 mV/decade in the benzoate concentration range of $9 \times 10^{-5} - 2 \times 10^{-2}$ M. The standard addition method was used for analyzing several commercial soft drink brands. pH adjustment was the only sample preparation step needed.

7.6.10 Determination of Aluminum Ion Concentration

It is believed that aluminum plays important roles in the pathology of Parkinson's disease, Alzheimer's disease, and diseases related to dialysis [75–77]. Therefore, a simple, well working direct potentiometric method could gain large-scale application in food analysis. For these, a selective electrode is needed. Being a three-valence ion, relatively less sensitivity is expected for a direct potentiometric aluminum measuring method. Unfortunately, a relatively small number of successful attempts have been reported about the fabrication of appropriate ion-selective aluminum electrodes.

As a promising step in food analysis, post-synthesis aluminated mezoporous zeolite molecular sieve, AlMCM-41, was used as neutral ionophore for the preparation of aluminum ISEs by Arvand and coworkers [78]. They found that the best performing measuring membrane contained 30% PVC, 67% aminophen as a plasticizer, and 3% ionophore. The slope of the linear range of the electrode response was 19.5 mV/decade, while the dynamic range reached from 10^{-7} to 10^{-1} M. Methods were worked out and tested for measuring the Al^{3+} content of rice, mushroom, potato, and Mg–Al syrup.

7.6.11 Methods for Detecting Histamine

Histamine [2-(4-imidazolyl)-ethylamine] is an important molecule taking part in different biochemical processes. It is also a by-product of bacterial action in food products. Therefore, histamine concentration is commonly used as an indicator of food deterioration. According to the US FDA level of 30 mg kg^{-1} histamine indicates already the bacterial deterioration of fish samples. Regulation of the European Union determines the level of beginning the health hazard range for fishes as 100–200 mg kg^{-1} [79]. Several groups reported the development of more or less selective potentiometric electrodes for detecting histamine released from living cells [80, 81].

Charged ionophore, a thiopyrilium derivative, was employed for preparing an ion-selective histamine membrane electrode by Javanbakht and coworkers [82], while Amini *et al.* [83] tested porphyrin ionophores for the same purposes.

α-cyclodextrin as histamine ionophore was used for preparing the potentiometric electrode applicable in fish freshness test in the work of Amorim [84] and coworkers. Cyclodextrins establishes the lipophylic interaction with histamine form inclusion complexes that are reflected by the potentiometric response. The measuring membrane contained potassium tetra-kis(potassium-chlorophenyl) borate (KTpClPB) as an anionic additive in two different concentrations and 2-fluorophenyl 2-nitrophenyl ether plasticizer (FNDPE), and PVC or PVC-COOH as a supporting polymer matrix. The miniaturized version of the electrode could be used in continuous flow systems.

A matrix cleaning step to eliminate interferences was found important to increase the reliability of histamine measurement in fish samples.

During storage periods, putrefaction processes can proceed in different raw food samples resulting in nucleoside degradation, pH, and redox potential changes. Barat and coworkers [85] investigated the chance of working out fast method for checking fish freshness by applying potentiometric measurement. They used two unselective electrodes, one gold and one silver electrode, and compared the electrode potential changes by storage times with other methods such as microbial test, pH measurement, and ATP released compound concentration. The good correlation validates that the simple potentiometric method can function as an *in situ* freshness test.

Pt, CuS, and Ag_2S electrodes were used in the work of Kaneki and coworkers [86] investigating the applicability of potentiometric measurements for detecting pork freshness. The outputs of these electrodes have been analyzed by principal component analysis (PCA) and multiple regression analysis (MRA) in order to find the correlation with the results of viable bacterial counts. By using the potentiometric sensors, the pork freshness was evaluated and the PCA and MRA corresponded to the degree of bacterial increases more simply and rapidly than other methods.

Artificial noise systems, some of them made of an array of unselective or selective potentiometric sensor spots [87, 88], have already been tested in food analysis, for example, for detecting food quality decline.

References

1. Jagner, D. and Graneli, A. (1976) Potentiometric stripping analysis. *Anal. Chim. Acta*, **83**, 19–26.
2. Blum, R. and Fog, H.M. (1972) Metal buffers as standards in direct potentiometric determination of metal ion activities. *J. Electroanal. Chem. Interfacial Electrochem.*, **34**, 485–488.
3. Sokalski, T., Ceresa, A., Zwickl, T. and Pretsch, E. (1997) Large improvement of the lower detection limit of ion-selective polymer membrane electrodes. *J. Am. Chem. Soc.*, **119**, 11347–11348.
4. Mathison, S. and Bakker, E. (1998) Effect of transmembrane electrolyte diffusion on the detection limit of carrier-based potentiometric ion-based sensors. *Anal. Chem.*, **70**, 303–309.
5. Bakker, E. and Pretsch, E. (2005) Potentiometric sensors for trace-level analysis. *Trends Analyt. Chem.*, **24** (3), 199–207.
6. Eisenman, G., Rudin, D.O. and Casby, J.U. (1957) Glass electrode for measuring sodium ion. *Science*, **126**, 831–834.
7. Nicolsky, B.P. (1937) Theory of the glass electrode. I. Theoretical. *Zh. Fiz. Khim.*, **10**, 495–503.
8. Nägele, M., Bakker, E. and Pretsch, E. (1999) General description of the simultaneous response of potentiometric ionophore-based sensors to ions of different charge. *Anal. Chem.*, **71**, 1041–1048.
9. Gadzekpo, V.P.Y. and Christian, G.D. (1984) Determination of selectivity coefficients of ion-selective electrodes by a matched-potential method. *Anal. Chim. Acta*, **164**, 279–282.
10. Cremer, M. (1906) Über die Ursache der elektromotorischen Eigenschaften der Gewebe, zugleich ein Beitrag zur Lehre von Polyphasischen Elektrolytketten. *Z. Biol.*, **47**, 562–608.

11. Wipf, D.O., Ge, F., Spaine, T.W. and Baur, J.E. (2000) Microscopic measurement of pH with iridium oxide microelectrodes. *Anal. Chem.*, **72** (20), 4921–4927.
12. Boksay, Z. and Lengyel, B. (1974) Vacancy type mechanism of the electrical relaxation processes in glass. *J. Non-Cryst. Solids*, **14**, 79–87.
13. Csákvári, B., Boksay, Z. and Bouquet, G. (1971) Investigation of surface layers on electrode glasses for pH measurement. *Anal. Chim. Acta*, **56**, 279–284.
14. Pungor, A.E. and Hollós-Rokosinyi, E. (1961) On the application of membrane electrodes for the assay of ion concentrations. *Acta Chim. Acad. Sci. Hung.*, **27**, 63.
15. Frant, M.S. and Ross, J.W. (1966) Electrode for sensing fluoride ion activity in solution. *Science*, **154**, 1553–1554.
16. Pick, J., Tóth, K., Pungor, E. *et al.* (1973) A potassium-selective silicone–rubber membrane electrode based on a neutral carrier. *Anal. Chim. Acta*, **64**, 477–480.
17. Guilbault, G.G. and Nagy, G. (1973) Improved urea electrode. *Anal. Chem.*, **45**, 417–419.
18. Bloch, R., Shatkay, A. and Saroff, H.A. (1967) Fabrication and evaluation of membranes bas specific electrodes for calcium ions. *Biophys. J.*, **7**, 865–877.
19. Keil, L., Moody, G.J. and Thomas, J.D.R. (1978) An evaluation of PVC matrix membrane calcium-selective electrodes based on nitrated (octylphenyl)phosphate sensors and phosphonate mediators. *Anal. Chim. Acta*, **96**, 171–175.
20. Lindner, E., Tóth, K., Jeney, J. *et al.* (1990) Novel bis(crown ether)-based potassium sensor for biological applications. *Microchim. Acta*, **100**, 157–168.
21. Kim, Y.H., Cha, N.R. and Chang, S.-K. (2002) A new fluorogenic benzothiazolyl ionophore based upon calix[4]arene-crown-5 ether for calcium determination in aqueous media. *Tetrahedron Lett.*, **43**, 3883–3886.
22. Marchand, A.P., Alihodžić, S., McKim, A.S. *et al.* (1998) Synthesis and alkali metal picrate extraction capabilities of a 4-oxahexacyclo[5.4.1.$0^{2,6}$.$0^{3,10}$.$0^{5,9}$.$0^{8,11}$]dodecane-derived cryptand. A new ionophore for selective ion complexation. *Tetrahedron Lett.*, **39**, 1861–1864.
23. Iimori, T., Erickson, S.D., Rheingold, A.L. and Still, W.C. (1989) Enantioselective complexation with a conformationally homogeneous C_2 podand ionophore. *Tetrahedron Lett.*, **30**, 6947–6950.
24. Bergveld, P. (1970) Development of an ion-sensitive solid-state device for neurophysiological measurements. *IEEE Trans. Biomed. Eng.*, **17**, 70–71.
25. Lee, C.-S., Kim, S.K. and Kim, M. (2009) Ion-sensitive field-effect transistor for biological sensing. *Sensors*, **9**, 7111–7131.
26. Severinghaus, J.W. and Bradley, A.F. (1958) Electrodes for blood pO_2 and pCO_2 analysis. *Appl. Physiol.*, **13**, 515–520.
27. Severinghaus, J.W. and Astrup, P.B. (1986) History of blood gas analysis. III. Carbon dioxide tension. *J. Clin. Monit.*, **2**, 60–73.
28. Neethirajan, S., Jayas, D.S. and Sadistap, S. (2009) Carbon dioxide (CO_2) sensors for the agri-food industry – a review. *Food Bioprocess Technol.*, **2**, 215–221.
29. Yao, S. and Wang, M. (2002) Electrochemical sensor for dissolved carbon dioxide measurement. *J. Electrochem. Soc.*, **149**, H28–H32.
30. Guilbault, G.G. and Montalvo, J. (1970) Enzyme electrode for the substrate urea. *J. Am. Chem. Soc.*, **92**, 2533–2538.

31. Trivedi, U.B., Lakshminarayana, D., Kothari, I.L. *et al.* (2009) Potentiometric biosensor for urea determination in milk. *Sens. Actuators, B*, **140**, 260–266.
32. Bates, R.G. (1973) *Determination of pH*, John Wiley & Sons, Inc., New York.
33. Harned, H.S. and Owen, B.B. (1958) *The Physical Chemistry of Electrolytic Solutions*, Chapter 14, 3rd edn, Reinhold, New York.
34. Bates, R.G. and Guggenheim, E.A. (1960) Report on the standardization of pH and related terminology. *Pure Appl. Chem.*, **1**, 163–168.
35. Buck, R.P., Rondinini, S. and Covington, A.K. (2002) Measurement of pH. Definition, standards, and procedures. *Pure Appl. Chem.*, **74** (11), 2169–2200, ©2002 IUPAC.
36. Kaneda, Y., Kanamori, T. and Iwaida, M. (1977) Nitrate determination in cheese by use of ion selective electrode. *J. Hyg. Chem.*, **23**, 301–306.
37. Upreti, P., Metzger, L.E. and Bühlmann, P. (2004) Glass and polymeric membrane electrodes for the measurement of pH in milk and cheese. *Talanta*, **63**, 139–148.
38. Meyerhoff, M. (1990) New in vitro analytical approaches for clinical chemistry measurements in critical care. *Clin. Chem.*, **36**, 1567–1572.
39. Brouder, S.M., Thom, M., Adamchuk, V.I. and Morgan, M.T. (2003) Potential uses of ion-selective K electrodes in soil fertility management. *Commun. Soils Sci. Plant Anal.*, **34**, 2699–2726.
40. Adsett, J.F., Thottan, J.A. and Sibley, K.J. (1999) Development of an automated on-the-go soil nitrate monitoring system. *Appl. Eng. Agric.*, **15**, 351–356.
41. Birrell, S.J. and Hummel, J.W. (2001) Real-time multi-ISFET/FIA soil analysis system with automatic sample extraction. *Comput. Electron. Agric.*, **32**, 45–67.
42. Adamchuk, V.I., Morgan, M.T. and Ess, D.R. (1999) An automated sampling system for measuring soil pH. *Trans. ASAE*, **42**, 885–891.
43. Adamchuk, V.I., Lund, E.D., Sethuramasamyraja, B. *et al.* (2005) Direct measurement of soil chemical properties on-the-go using ion-selective electrodes. *Comput. Electron. Agric.*, **48**, 272–294.
44. EPA http://www.epa.gov/osw/hazard/testmethods/sw846/pdfs/9214.pd (accessed 11 December 2014).
45. Koblar, A., Tavčar, G. and Ponikvar-Svet, M. (2012) Fluoride in teas of different types and forms and the exposure of humans to fluoride with tea and diet. *Food Chem.*, **130**, 286–290.
46. Martin, C. and Brun, S. (1969) Le dosage du fluor dans les vins. *Trav. Soc. Phann. de Montpellier*, **29**, 161–167.
47. Moreno, F., De la Torre, M.C. and Serrat, M. (1971) Un electrodo especifico para la determinación de pequenas cantidades de ion fluoruro. *Cire. Fann.*, **29**, 315–331.
48. De Baenstg, G., Meriens, J., Vandenwinkel, P. and Massart, D.L. (1973) Détennination de l'ion fluorure dans le vin, par potentiométrie. *Pharm. Belg.*, **28**, 188–194.
49. Hidalgo, J.L., Pastor, M. and Pérezbustamante, A. (1983) Detenninacionde fluoruro en vinos deI Marco de Jerez. *Anal. Bromatol.*, **35**, 67–77.
50. Martinez Rincon, M.C. and Abun Cabeza, L.M. (1990) Flúor en vinos de denominación deorigemméntrida. *An. RealAcad. Fann.*, **56**, 279–282.
51. DelNozal, M.J., Bernal, L. and Martin, M. (1984) Detenninacion espectrofotometrica y potendometrica de elementos minerales en mosto de uva. *Anales Bromatol.*, **35**, 51–65.

52. EPA http://www.epa.gov/osw/hazard/testmethods/sw846/pdfs/9212 (accessed 11 December 2014).
53. Johnston, A.E. (2003) *Understanding Potassium and its Use in Agriculture*, European Fertilizer Manufacturers Association, Brussels, pp. 28–30.
54. Wang, J. and Scott, A.D. (2001) Determination of exchangeable potassium in soil using ion-selective electrodes in soil suspensions. *Eur. J. Soil Sci.*, **52**, 143–150.
55. Almeida, M.I.G.S., Segundo, M.A., Lima, J.L.F.C. and Rangel, A.O.S.S. (2006) Potentiometric multi-syringe flow injection system for determination of exchangeable potassium in soils with in-line extraction. *Microchem. J.*, **83**, 75–80.
56. Grygolowicz-Pawlak, E., Pachecka, K.Á., Wolanin, B. and Malinowska, E. (2006) Towards miniaturized sensors for determination of exchangeable potassium, in soil samples. *Int. Agrophys.*, **20**, 101–105.
57. Blom-Zandstra, M. (1989) Nitrate accumulation in vegetables and its relationship to quality. *Ann. Appl. Biol.*, **115**, 553–561.
58. Baker, A.S. and Smith, R. (1969) Extracting solution for potentiometric determination of nitrate in plant tissue. *J. Agric. Food Chem.*, **17**, 1284–1287.
59. Davenport, J.R. and Jabro, J.D. (2001) Assessment of hand held ion selective electrode technology for direct measurements of soil chemical properties. *Commun. Soil Sci. Plant Anal.*, **32**, 3077–3085.
60. Thottan, J., Adsett, J.F., Sibley, K.J. and MacLeod, C.M. (1994) Laboratory evaluation of the ion selective electrode for use in an automated soil nitrate monitoring system. *Commun. Soil Sci. Plant Anal.*, **25**, 3025–3034.
61. Ott-Borrelli, K.A., Koenig, R.T. and Miles, C.A. (2009) A comparison of rapid potentiometric and colorimetric methods for measuring tissue nitrate concentrations in leafy green vegetables. *HortTechnology*, **19**, 439–444.
62. Consalteri, A., Rigato, A., Clamor, L. and Giandon, P. (1992) Determination of nitrate in vegetables using an ion-selective electrode. *J. Food Compos. Anal.*, **5**, 252–256.
63. Iern, A. (2009) Influence of harvest date on nitrate contents of three potato varieties for off-season production. *J. Food Compos. Anal.*, **22**, 51–55.
64. Andac, M., Eren, H. and Coldur, F. (2011) Determination of nitrate in leafy vegetables by flow injection analysis with potentiometric detection. *J. Food Compos. Anal.*, **19**, 457–462.
65. Knee, M. and Srivastava, P. (1995) Binding of calcium by cell walls and estimation of calcium in apple fruit tissue with an ion selective electrode. *Postharvest Biol. Technol.*, **5**, 19–27.
66. Gao, R., van Leeuwen, H.P., van Valenberg, H.J.F. and van Boek, M.A.J.S. (2011) Accurate determination of the Ca^{2+} activity in milk-based systems by Ca-ISE: effects of ionic composition on the single Ca^{2+} activity coefficient and liquid, junction potentials. *Food Chem.*, **129**, 619–623.
67. Campanella, L., Aturki, Z., Sammartino, M.P. and Tomassetti, M. (1995) Aspartate analysis in formulations using a new enzyme sensor. *J. Pharm. Biomed. Anal.*, **13**, 439–447.
68. Nikolelis, D.P. and Krull, U.J. (1990) Dynamic response characteristics of the potentiometric carbon dioxide sensor for the determination of aspartame. *Analyst*, **115**, 883–888.

69. Santini, A.O., Pezza, H.R. and Pezza, L. (2012) Analytical methods development of a sensitive potentiometric sensor for determination of fumaric acid in powdered food products. *Food Chem.*, **134**, 483–487.
70. Santini, A.O., Pezza, H.R., Carloni-Filho, J. *et al.* (2009) Potentiometric sensor for sorbic acid determination in food products. *Food Chem.*, **115**, 1563–1567.
71. Saad, B., Bari, M.F., Saleh, M.I. *et al.* (2005) Simultaneous determination of preservatives (benzoic acid, sorbic acid, methylparaben and propylparaben) in foodstuffs using high-performance liquid chromatography. *J. Chromatogr. A*, **1073**, 393–397.
72. Sun, C.J., Sun, X.X. and Aboul-Enein, H.Y. (2010) Polymer membrane electrode based on a zinc(II) Schiff base complex for the selective determination of sodium benzoate. *Anal. Lett.*, **43**, 1989–1998.
73. Pezza, L., Santini, A.O., Pezza, H.R. *et al.* (2001) Benzoate ion determination in beverages by using a potentiometric sensor immobilized in a graphite matrix. *Anal. Chim. Acta*, **433**, 281–288.
74. Alizadeh, N., Saburi, N. and Hosseini, S.E. (2012) Rapid determination of benzoate in soft drinks by solid-state benzoate ion selective sensor based on conducting polypyrrole. *Food Control*, **28**, 315–320.
75. Abel-Ghany, M., Ei-Sebae, A.K. and Shalloway, D. (1993) Aluminum-induced nonenzymatic phospho-incorporation into human tau and other proteins. *J. Biol. Chem.*, **268**, 11976–11981.
76. Paik, S.R., Lee, J.H., Kim, D.H. *et al.* (1997) Aluminum-induced structural alterations of the precursor of the non-A beta component of Alzheimer's disease amyloid. *Arch. Biochem. Biophys.*, **344**, 325–334.
77. Kawahara, M., Muramoto, K., Kobayashi, K. *et al.* (1994) Aluminium promotes the aggregation of Alzheimer's amyloid β-protein in vitro. *Biochem. Biophys. Res. Commun.*, **198**, 531–535.
78. Arvand, M., Kermanian, M. and Zanjanchi, M.A. (2010) Direct determination of aluminium in foods and pharmaceutical preparations by potentiometry using an AlMCM-41 modified polymeric membrane sensor. *Electrochim. Acta*, **55**, 6946–6952.
79. European Union (2005) On microbiological criteria for foodstuffs 2073/2005. *Off. J. Eur. Union*, 26.
80. Katsu, T., Kanamitsu, M., Hirota, T. *et al.* (1986) In situ detection of histamine released from mast-cells by using a histamine-sensitive membrane-electrode. *Chem. Pharm. Bull.*, **34**, 3968–3970.
81. Katsu, T. and Hirodo, H. (1999) Determination of histamine release from mast cells using a histamine-sensitive membrane electrode. *Anal. Chim. Acta*, **396**, 189–193.
82. Javanbakht, M., Ganjali, M.R., Norouzi, P. *et al.* (2008) Development of polymeric membrane sensor for the determination of histamine: experimental and theoretical study. *Anal. Lett.*, **41**, 619–639.
83. Amini, M.K., Shahrokhian, S. and Tangestaninejad, S. (1999) Porphyrins as carriers in poly(vinyl chloride)-based membrane potentiometric sensors for histamine. *Analyst*, **124**, 1319–1322.
84. Amorim, G., Souza, R.C., Araújo, A.N. *et al.* (2010) SI lab-on-valve analysis of histamine using potentiometric detection for food quality control. *Food Chem.*, **122**, 871–876.

85. Barat, J.M., Gil, L., Garcia-Breijo, E. *et al.* (2008) Freshness monitoring of sea bream (*Sparus aurata*) with a potentiometric sensor. *Food Chem.*, **108**, 681–688.
86. Kaneki, N., Miura, T., Shimada, K. *et al.* (2004) Measurement of pork freshness using potentiometric sensor. *Talanta*, **62**, 217–221.
87. Bratov, A., Abramova, N. and Ipato, A. (2010) Recent trends in potentiometric sensor arrays – a review. *Anal. Chim. Acta*, **678**, 149–159.
88. Gallardo, J., Alegret, S. and del Valle, M. (2005) Application of a potentiometric electronic tongue as a classification tool in food analysis. *Talanta*, **66**, 1303–1309.

8

Electrochemical Enzyme Biosensors

Ilaria Palchetti and Marco Mascini
Dipartimento di Chimica Ugo Schiff, Università degli Studi di Firenze, Via della Lastruccia 3, 50019, Sesto Fiorentino, Italy

8.1 Introduction

Enzymes were historically the first molecular recognition elements included in biosensors, and continue to be the basis for a significant number of publications in this field. All began in 1962 when Professor Leland C. Clark, in a Conference at a Symposium in the New York Academy of Sciences, described how "to make electrochemical sensors (pH, polarographic, potentiometric, or conductometric) more intelligent," that is by adding "enzyme"[1]. The first example was illustrated by entrapping the enzyme Glucose Oxidase in a dialysis membrane over an oxygen probe. The addition of glucose determined the decrease in oxygen concentration in proportional relation. For this first biosensor was coined the term "enzyme electrode"[2]. Then subsequently in 1967 Updike and Hicks use the same term "enzyme electrode" to describe a similar device where again the enzyme glucose oxidase was immobilized in a polyacrylamide gel onto a surface of an oxygen electrode for the rapid and quantitative determination of glucose [3]. Besides amperometry Guilbault and Montalvo in 1969 use glass electrodes coupled with urease to measure urea concentration by potentiometric measurement [4]. Starting from these first examples, electrochemical transducers combined with an enzyme as the biochemical component form, nowadays, the largest category of biosensors becoming an important tool for detection of chemical and biological components for clinical, food, and environmental monitoring.

Enzymes (and all biological elements, such as tissues, cells, microorganisms, which contain enzymes) represent the class of what are called catalytic elements. Enzyme biosensors have several advantages. These include the catalytic amplification of the biosensor response by modulation of the enzyme activity with respect to the target analyte; a stable source of

Agricultural and Food Electroanalysis, First Edition.
Edited by Alberto Escarpa, María Cristina González and Miguel Ángel López.

material (primarily through biorenewable sources); a very wide range of enzymes available commercially, usually with well-defined and assayed characteristic; the possibility of modifying the catalytic properties or substrate specificity by means of genetic engineering. However, enzyme-based biosensors have also some limitations for use in analytical applications. These include the limited number of substrates for which enzymes have been evolved, the limited interaction between environmental pollutants or food contaminants and specific enzymes, and in the case of inhibitor formats, the lack of specificity in differentiating among compounds of similar classes such as organophosphate and carbamate pesticides. Moreover, stability of the enzymes and the capacity to maintain enzymatic activity over a long period of times continue to be problematic.

Among the enzymes that are commercially available, the most often used in biosensors are oxidoreductase and hydrolase, like glucose oxidase, horseradish peroxidase, and alkaline phosphatase. In particular, glucose oxidase continues to be the most stable and specific enzyme that can be easily obtained in large quantities.

As already mentioned, genetic engineering helps in the careful selection of the location and type of mutations giving rise to enzymes with enhanced or particular properties, such as higher affinity toward specific analytes, higher stability, higher electron transfer rates, and residues able to provide an oriented or more stable immobilization. These improvements have already resulted in biosensors with enhanced performance. Recent progress with respect to genetically modified enzyme biosensors has been well reviewed [5–8].

An emerging field of catalytic biorecognition elements is so-called whole-cell systems. They have, recently, benefited enormously from the recent improvements in recombinant DNA technology [9] and there is renewed interest in their use in monitoring environmental pollution, toxicity, and food contaminations. Whole-cell systems are based on complex cellular functions, among which enzyme-catalyzed reactions play an important role. The biosensors are constructed by the fusion of promoters (responsive to the relevant environmental conditions) to easily monitored reporter genes. Depending on the choice of reporter gene, expression can be monitored by the production of color, light, fluorescence, or electrochemical reactions. Although there are numerous examples of genetic modification involving bacteria, yeast, algae, and tissue culture cells, genetically engineered microorganism (GEM) are most often reported in cell-based biosensors.

In principle, enzyme-based biosensors have potential application in the agro-food analysis, within three main areas that are food safety, food quality, and process control. The term food safety involves the concept of the production and marketing of harmless food, monitoring the presence of contaminants, such as residues of pesticides, fertilizers, heavy metals, and other toxic organic compounds also used as additives. Food quality is related not only to safety but mainly to nutritional value and acceptability. Thus, in this context freshness, appearance, flavor, texture, and composition are food characteristics that have to be controlled. Moreover, enzyme biosensors allow the determination and quantification, on-line, of compounds of interest in process control, such as fermentation, sugar, alcohols contents, and so on.

Typical examples of enzymes involved in food applications are cholinesterase for organophosphorous and carbamate pesticide analysis; tyrosinase or laccase for analysis of phenols, quinones, and related compounds; glucose oxidase for sugar content analysis, carboxyl esterase, alcohol oxidase, carboxypeptidase, L-aspartase, peptidase, aspartate aminotransferase for aspartame, xanthine oxidase to detect the level of freshness in fish, glucose oxidase, β-galactosidase, fructose dehydrogenase, and other dehydrogenase to detect glucose in fruit juice, lactose, or lactulose in milk, and so on. In Table 8.1 are reported some examples of biosensors used in the detection of contaminants, antinutrients, and analytes to evaluate food quality, freshness, and shelf-life as developed in the author's lab. Obviously, many other examples can be found in literature.

8.2 General Features of Enzyme Biosensors

Enzyme-based biosensors may be classified according to the analytes or reactions that they monitor, for example, direct monitoring of analyte concentration of reactions producing or consuming such analytes is the direct mode. Some environmental or food contaminants selectively inhibit the activity of certain enzymes, thus the resulting product concentration is affected. This inhibition is analytically useful and has been used advantageously in the development of many biosensing devices. The monitoring of inhibitor or activator of the enzyme is considered the indirect mode.

In enzyme-based biosensors, the enzyme reacts selectively with its substrate. The detailed mode of action of enzymes can be found in any standard biochemistry book, and here we will remind only the basic enzyme catalysis mechanism, which is as follows [20, 21]:

$$S + E \underset{k_{-1}}{\overset{k_1}{\leftrightarrow}} ES \xrightarrow{k_2} E + P$$

where S is the substrate, E is the enzyme, ES is the enzyme–substrate complex, and P is the product. Considering the kinetic theory, at the steady state, the concentration of the enzyme–substrate complex is constant, so the rate of formation of the complex from its component is balanced by the rate of its breakdown back to enzyme and forward to its products. The overall rate of the reaction (rate of formation of products) is given by the Michaelis–Menten equation, as follows:

$$v = \frac{d[P]}{dt} = \frac{-d[S]}{dt} = k_2[ES] = k_2[E_0]\frac{[S]}{K_m + [S]}$$

where K_m is the Michaelis–Menten constant $[K_m = (k_{-1} + k_2)/k_1]$.

When $[S] \gg K_m$, a maximum value of the rate constant, V_{max}, is reached, so that $V_{max} = k_2[E_0]$, with $[E_0]$ as the initial concentration of the enzyme, and when $[S] = K_m$, $v = V_{max}/2$.

For a given enzyme, the rate of reaction v is a function of the ratio $[S]/K_m$, and:

when $[S]/K_m \rightarrow 0$, $[S] \ll K_m, v \rightarrow (V_{max}/K_m)[S]$ (first-order kinetics with respect to $[S]$),
when $[S]/K_m$ becomes large, $[S] \gg K_m, v \rightarrow V_{max}$ (zeroth-order kinetics with respect to $[S]$).

Now, if we consider the response of an enzyme-based biosensor to the addition of the enzymatic substrate, this response is determined by the concentration of the product (P) of the enzymatic reaction on the surface of the electrode (Figure 8.1). This reaction is controlled by the rate of two simultaneous processes, that is, the enzymatic conversion of the substrate (S) and the diffusion of the product from the enzyme layer.

Table 8.1 Examples of enzyme-based biosensors developed in the author's lab for evaluation of food quality, safety, and process control

Analyte	Matrix	Recognition element	Transduction system	References
Food safety				
Pesticides				
Phosdrin; DDVP; Fenthion; Diazinon; Malathion; Carbofuran; Propoxur	Buffer	Choline oxidase + AChE[a]	Amperometric	[10]
Carbofuran	Fruits and vegetables	Choline oxidase + AChE	Amperometric	[11]
Carbofuran	Water	AChE + TCNQ	Amperometric	[12]
Carbofuran	Water, fruit, and vegetable	AChE + TCNQ	Amperometric	[13]
Carbofuran	Buffer	AChE + cobalt(II) phthalocyanine	Amperometric	[14]
Food quality and food process control				
Anti-nutrient and toxic components				
Amygdalin (bitter almond)	Buffer	*β*-glucosidase	Potentiometric	[15]
Antioxidant properties				
Polyphenols	Olive oils	Tyrosinase	Amperometric	[16]
Freshness				
Biogenic amines	Fish	Xanthine oxidase	Amperometric	[17]
Process control				
Lactate	Milk	l- and d-lactate dehydrogenase	Amperometric	[18]
Lactose,glucose	Milk	*β*-galactosidase and glucose oxidase	Amperometric	[19]

[a] Ache: Acetylcholinesterase.

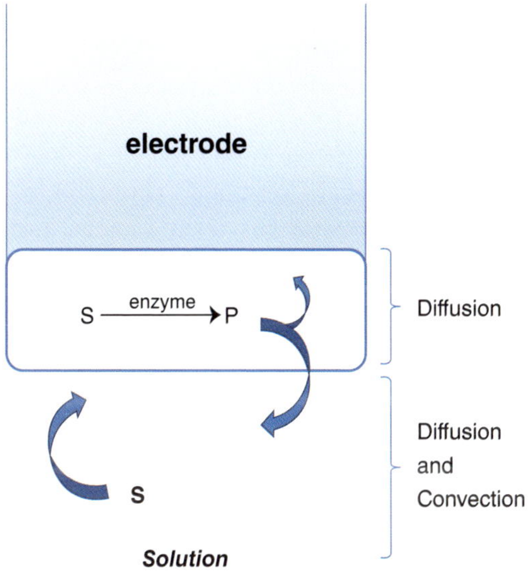

Figure 8.1 Schematic of an enzyme-electrode: diffusion of the substrate and the product in the enzyme layer on an electrode

In particular, an enzyme electrode operates via a five-step process:(1) the substrate must be transported from the bulk of solution toward the enzyme layer. (2) The substrate must diffuse through the layer to the enzyme active site. (3) Reaction occurs at the active site. (4) Product formed in the enzymatic reaction is transported through the layer to the surface of the electrode. (5) Product is measured at the electrode surface.

The first step, transport of the substrate, is most critically dependent on the stirring rate of the solution, so that rapid stirring will bring the substrate very rapidly to the electrode surface. If the layer is kept very thin using highly active enzymes then steps (2) and (4) are eliminated or minimized; since step (3) is very fast, the response of an enzyme electrode should theoretically approach the response time of the transducer. Many researchers have shown experimentally that one can approach this behavior using a thin enzyme layer and rapid stirring.

Moreover, if there is a high enzyme activity, the decrease in the substrate concentration is not totally compensated by the transfer from the bulk solution due to the diffusion limitation, and because of this, only a fraction of the enzyme active centers is involved in the interaction with a substrate. In this case (diffusion control of the response), the sensitivity of the immobilized enzyme toward inactivation, (inhibitor effect), is lower than that for the enzyme used in solution.

These considerations can be explained from a theoretical point of view as follows considering the response in the transient time and at the steady state. The concentration of the substrate and product at the transducer depends on the enzyme Michaelis constant K_m, the activity V_{max} of the enzymatic layer, the thickness of the layer, and the diffusion coefficients of the substrate and of the product. The product concentration at the external surface of the electrode is assumed to be zero because the diffusion of the species occurs in the active layer. The temperature, V_{max}, K_m, and coefficient of diffusion D_s and D_p (diffusion coefficient of the substrate and of the product at the interior of the active layer) are all considered constant throughout the entire layer. It is also assumed that there is no consumption of the substrate or the product by the transducer. If e is the thickness of the layer, the ratio e^2/D_s determines the response time of the biosensor because it expresses the time taken for the enzymatic layer to reach the steady state [21]. This response time can be reduced by modifying the layer so that it is extremely permeable to the substrate (increase D_s) or very thin (decrease e). Reduction in enzymatic layer thicknessis the most effective method because this value is squared in the expression for response time. In reality, an excessive decrease in the thickness of the membrane will also affect its mechanical properties. Moreover, the response time of the biosensor can never be less than the response time of transducer.

The response of an enzyme sensor in the steady state depends largely on the ratio of the substrate concentration $[S]$ to the enzyme Michaelis constant K_m. When $[S]/K_m$ is large, the reaction rate reaches a maximal value V_{max}, which is proportional to the number of active sites of the immobilized enzyme. The reaction rate is independent of the substrate concentration, and the product concentration at the contact with the electrode is the same for all high substrate concentration. The quantity of enzyme in the layer determines the linear zone in the response to the substrate concentration. This zone corresponds to first-order kinetics with respect to substrate concentration, whereas the region with a plateau has zeroth-order kinetic. When the substrate concentration is very high($[S] \gg K_m$), the biosensor is no longer capable of determining the substrate but may determine inhibitors which affect the minimal rate of the enzymatic reaction V_{max}.

8.2.1 Enzyme Biosensor Assembly

To prepare the biosensor, the enzyme has to be immobilized in a thin layer at the surface of the transducer. Among all the several methods available, the following are the most generally employed:

1. Physical entrapment within an inert polymeric membrane (in this case the enzyme is mixed with a monomer solution, which is then polymerized to a gel – polyacrylamide gel, starch, agar gel, and so on, thus trapping the enzyme) or behind a membrane (in this case the enzyme solution is simply confined by an analyte permeable membrane, such as a dialysis membrane, as a thin film covering the indicator electrode; this method is also called microencapsulation);
2. Adsorption is the simplest and fastest way to prepare immobilized enzymes. Adsorption can roughly be divided into two classes: physical adsorption and chemical adsorption. Physical adsorption is weak and occurs mainly via van der Waals interactions. Chemical adsorption is stronger and involves the formation of covalent bonds.
3. Covalent bonding of the enzymes on membranes or surfaces activated by means of bifunctional groups or spacers, such as gluteraldehyde, carbodiimide, and so on (when the enzyme is bound to the solid support using a bi- or multifunctional agent, such as gluteraldehyde, which forms a bridge between biocatalytic species or proteins, the technique is also called cross-linking or reticulation). Several companies put on the market pre-activated membranes suitable for the immediate preparation of any bioactive

membrane and this appeared as a real improvement at least for the easy use of enzyme sensors (like Immunodyne produced by Pall Industries, NY).

Enzyme immobilization is one of the most important steps involved in the biosensor design. The choice of the technique used for connecting the biological component (enzyme) to the electrode is crucial, since the stability and the sensitivity largely depend on enzymatic layer configuration. Generally the technique of chemical immobilization (covalent bonding) is preferred. This procedure improves the enzyme stability and defines a diffusion layer where the product, formed by the catalytic action of the enzyme, diffuses partly to the electrode surface and partly back to the bulk of the solution; therefore, by defining the steady state one obtains a reproducible signal value. However, the real advantage of the enzyme immobilization through chemical bonding is the lifetime of the final probe, which can reach 6–12 months.

Several enzymes can be immobilized within the same reaction layer to increase the range of possible biosensor analytes, or to provide efficient regeneration of enzyme co-substrates, or to improve the biosensor selectivity by decreasing the local concentration of electrochemical interfering substances.

8.2.2 Enzyme Biosensor Stability

Enzymes are proteins. They are sensitive to denaturation by pH, temperature, or aging. Enzymes have an optimal pH range in which their activity is maximal; this pH range should be compatible with the transducer. Moreover, most of the biological systems have a very narrow range of temperature (15–40 °C). The most important problem and main drawback for industrial exploitation is the short lifetime associated with the biological elements.

The gradual loss in activity generates a decrease in V_{max}. Stability of enzyme biosensor is related to enzyme loading, and in general, if the membrane contains more active enzyme, the biosensor is more stable and has a longer lifetime. However it is not always possible to satisfy this condition and a compromise must be adopted [20, 21].

The removal of interferences has also been the other important aspect for the wide use of biosensors for industrial processes. This problem could be solved by using multilayer membranes. The main role of the membrane is to prevent interferences from passing into the bioactive layer. Cellulose acetate membrane allows only molecule of the size of hydrogen peroxide to cross and contact the platinum anode, thus preventing interference from ascorbic acid or uric acid, for example, at the fixed potential in the case of a glucose biosensor.

8.2.3 Biosensors Based on the Principle of Enzyme Inhibition

Biosensors based on the principle of enzyme inhibition have by now been applied for a wide range of significant analytes such as organophosphorous pesticide (OP), organochlorine pesticides, derivatives of insecticides, heavy metals,and so on. The choice of enzyme/analyte system is based on the fact that these toxic analytes inhibit normal enzyme function. In general, the development of these biosensing systems relies on a quantitative measurement of the enzyme activity before and after exposure to a target analyte. Typically the percentage of inhibited enzyme (I%) that results after exposure to the inhibitor is quantitatively related to the inhibitor (i.e., analyte) concentration and the incubation time. Consequently, the residual enzyme activity is inversely related to the

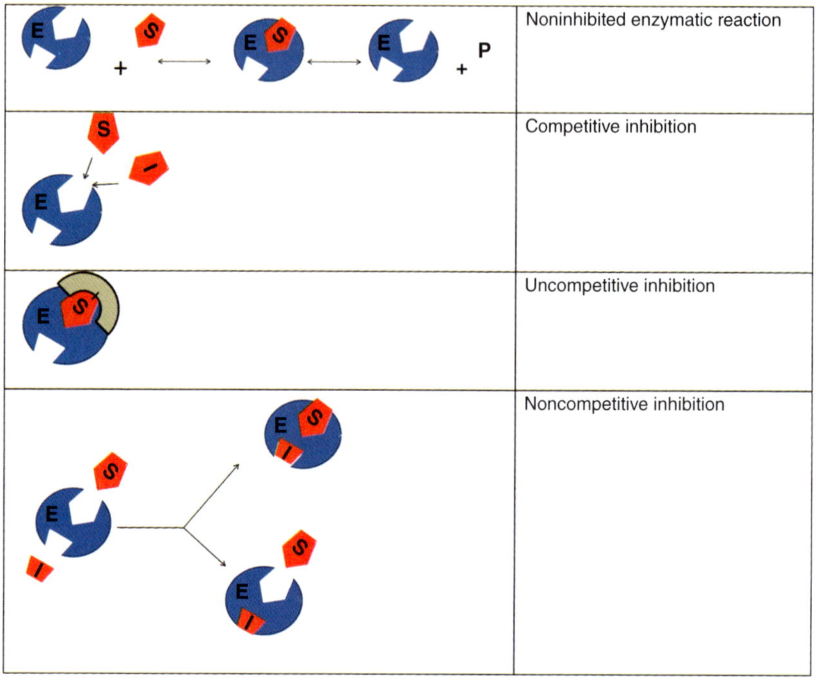

Figure 8.2 *Schematic of the principal modes of inhibition*

inhibitor concentration. The inhibition can be either reversible or result in an irreversible inactivation of the enzyme [21].

As reported in Figure 8.2, different modes of inhibition are possible.

Inhibitors structurally related to the substrate may be bound to the enzyme active center and compete with the substrate (competitive inhibition). If the inhibitor is not only bound to the enzyme but also to the enzyme–substrate complex, the active center is usually deformed and its function is thus impaired. In this case the substrate and the inhibitor do not compete with each other (noncompetitive inhibition). Competitive and noncompetitive inhibitions affect the enzyme kinetics differently. A competitive inhibitor does not change V_{max} but increases the K_m; on the contrary, a noncompetitive inhibition results in an unchanged K_m and in a decrease in V_{max}. In the case of mixed inhibition, the inhibitor binds the enzyme and the enzyme–substrate complex with a different affinity. For uncompetitive inhibition, the inhibitor binds only when the enzyme–substrate complex is formed [21].

8.3 Analytical Features of Enzyme Based Biosensors

8.3.1 Biosensor Calibration

The maximum concentration of substrate that can be measured with an enzyme electrode is related to K_m. When the concentration is near to K_m the linear relation between substrate concentration and the extent of enzyme reaction fails and a limiting value is reached. The difficulty of measuring high concentration is usually overcome by diluting the sample. This procedure has the advantage of the ability to control pH and ionic strength by the use of suitable buffers, so that samples and standards are measured under identical conditions.

Immobilization of the enzyme is thought to increase the K_m value, the increase being related to change in substrate, diffusion effects, or changes in enzyme structure.

The minimum concentration is often related to the minimum amount of electroactive substance that can be measured with the electrochemical transducer.

The calibration graphs are therefore often S-shaped, leveling off at high concentration due to the K_m or K_m-related maximum concentration measurable and at low values by the performances of the transducer. The useful part of the curve is the linear zone where a variation in substrate concentration gives a variation in the signal of the transducer. The slope of the curve in this zone corresponds to the sensitivity of the biosensor because it expresses a variation in the signal obtained as a function of the analyte concentration.

One of the most important parameters concerning the concentration range is the activity of the enzyme in the membrane layer. As a general rule, the slope of the sensor will increase with increasing amount of enzyme up to a limiting value. The activity of the enzyme determines the percentage of reaction of the substrate, and hence the amount of electroactive species. However, with increasing amounts of enzyme loading, the percentage of substrate reacting reaches a limit that matches its complete transformation.

8.3.2 Biosensor Calibration for Determination of Inhibitors

When an enzyme sensor is intended to determine a species that inhibits an enzymatic reaction, the substrate concentration is fixed at a given value, and the inhibitor concentration is varied. In this case, it is possible to represent both the variation in potential and the variation corresponding to the percentage inhibition ($I\%$):

$$I\% = \frac{S_0 - S_I}{S_0} \times 100$$

where S_0 and S_I are the signal of the biosensor in the presence and in absence of inhibitors, for the same concentration of substrate. The advantage of using $I\%$ is its dimensionless character [21].

The true value of limit of detection (LOD) for biosensor based on inhibition can be defined as the concentration of the inhibitor where the confidence interval does not overlap that of the zero concentration of the inhibitor standard [8].

8.3.3 Practical Aspects

8.3.3.1 Response Time

Mathematical models describing response times have been considered in the literature. This is the essential feature of a biosensor, and depends on the procedure of assembling the biosensor, that is the method of immobilization, the electrochemical transducer employed, the geometry of the measuring cell, and the kinetics of the enzymatic reaction.

For fast response times and steady states, a fast but constant rate of stirring of the substrate solution is recommended.

8.3.3.2 *Thickness of the Enzyme Layer*

The time required to reach steady-state potential reading is dependent on the enzyme layer thickness because of the diffusion parameter for the substrate to reach the active sites of the enzyme and of the electroactive species to diffuse through the membrane to the sensor. A mathematical model relating the thickness of the membrane, *d*, the diffusion coefficient, *D*, the Michaelis constant, K_m, and the maximum velocity of the enzyme reaction, V_{max}, has been developed:

$$V_{max}\ d^2/DK_m = V$$

where *V* compares the rate of chemical reaction in the membrane with the rate of diffusion through the membrane. The larger *V* is, the faster is the enzyme catalysis relative to the diffusion process.

Membranes as thin as possible are recommended for the best results, and this is helped by the use of highly active enzyme.

There is a twofold effect on increasing the enzyme activity in the layer that is in close proximity to the electrode surface. First, a complete conversion of the substrate into products will be ensured; secondly, the response time of the electrode will be affected. The increase in the amount of enzyme affects the thickness of the membrane. This results in an increase in the time required for the substrate to diffuse through the membrane. Hence, for the best results, it is recommended that an enzyme of as high an activity as possible is used to ensure rapid kinetics by achieving the thinnest possible membrane.

8.3.3.3 *Effect of Additional Membranes*

Besides the enzymatic reacting layer, many biosensors, especially designed for biological or clinical applications, incorporate one or more membranes, which serve mainly two important functions: (i) protective barrier – to prevent the interference of molecules, such as proteins or cells of biological samples, with the reaction layer or to decrease the influence of electrochemical species detected by the transducer. It also reduces leakage of the reacting layer components into the sample solution. (ii) Diffusional outer barrier for the substrate: the linear dynamic ranges may be large if the sensor response is controlled by the substrate diffusion through the membrane and not by the enzyme kinetics. This control is achieved by placing a thin outer membrane over a highly active enzyme layer. Generally, thinner are the membranes, shorter is the biosensor response time.

8.4 Examples of Electrochemical Enzymatic Biosensors for Food Analysis

Enzyme-based biosensors have been developed both with potentiometric as well as amperometric transducers. Different potentiometric transducers have been used including the glass electrode for the measurement of pH or monovalent ions, specific electrodes sensitive to anions or cations, and gas electrodes such as pCO_2 o pNH_3 electrodes. If an amperometric electrode is used as the transducer of a biosensor, a consumption of reaction products is involved; this is the major difference from a potentiometric electrode. Different types of electrodes can be used in amperometric biosensor such as oxygen sensor, gold, platinum, or graphite electrodes.

Sometimes the three modes of oxidation reactions that occur in biosensors are referred to as first-, second-, and third-generation, as follows:[20]

First generation – oxygen sensor based systems,
Second generation – mediator-based systems,
Third generation – directly coupled enzyme electrodes.

Nowadays some innovative techniques for electrode preparation have been proposed that are characterized by the possibility of mass-production and high reproducibility. Among these, the equipment needed for screen-printing technology is less complex and less costly and thus this is one of the most used techniques for sensor production. Screen-printing technology consists of depositing inks on a substrate in a film of controlled pattern and thickness. The inks may be printed on several kinds of supports like glass, ceramic, or plastic. Many different types of ink are, now, commercially available, differing in composition and electrical behavior. The interest in screen-printed sensors as electrochemical transducers in biosensor production is due to the possibility of making them disposable; this characteristic arises from the low cost and the mass-production of these systems. In electrochemistry, a disposable sensor offers the advantage of not suffering from the electrode fouling that can result in loss of sensitivity and reproducibility. The single-use sensors have other important advantages, especially working in the field of clinical analysis, such as avoidance of contamination among samples. Moreover, the dimensions of these screen-printed devices are important to satisfy the needs of decentralized testing. The high degree of reproducibility that is possible for these onetime use electrodes eliminates the cumbersome requirement for repeated calibration. The type of instrument used for these measurements is also very easy to obtain and can be inexpensive and compact; this allows for the possibility of *in situ* measurements.

In the following section an example of the use of disposable graphite sensor based for food analysis will be described. In particular, the use of these sensors to develop enzymatic biosensors for pesticide detection based on AChE (acetylcholinesterase) enzyme inhibition will be described.

8.4.1 Detection of Pesticides

The anticholinesterase inhibitors (organophosphorous and carbamate), have come into widespread use in the last decades, because they are less persistent in the environment than other pesticides, such as organochlorine. However their presence in water and food is a potential hazard to human health and there is a growing interest in their rapid and accurate determination. Standard methods, based on gas chromatography (GC), are very reliable but there is the need for fast and innovative methods. The use of enzymatic biosensors, and especially electrochemical biosensors, for organophosphorus and carbamate pesticides detection has been reported by many authors [22–26].

These are based mainly on the use of enzyme AChE; it hydrolyzes the acetylcholine to choline and acetic acid which are not electroactive; therefore another enzyme can be added in the reaction scheme such as choline oxidase which oxidizes the choline to produce hydrogen peroxide easily detected by amperometry [10, 11]. Examples of these approaches are reported in Figure 8.3, where calibration curves of different pesticides are shown.

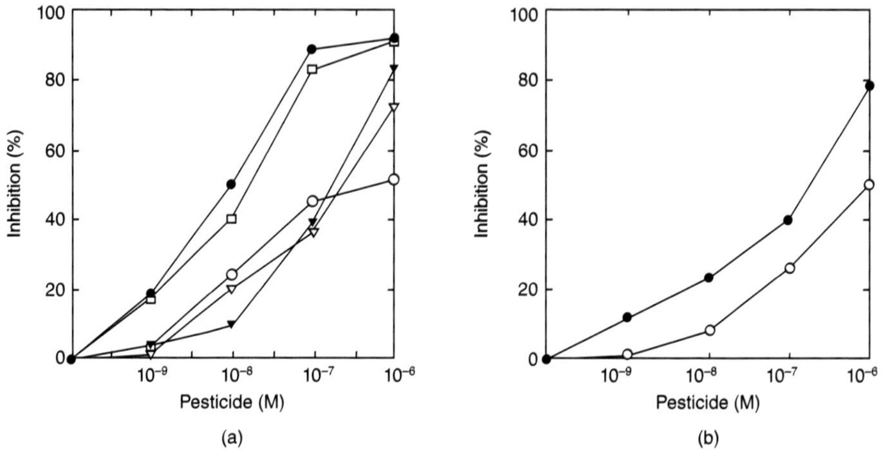

Figure 8.3 *(a) Inhibition effect of different organophosphorus compounds on a pesticide biosensor: Phosdrin; DDVP; Fenthion; Diazinon; Malathion. (b) Inhibition effect of different carbamates on a pesticide biosensor: Carbofuran; Propoxur. Incubation time: 10 min. Acetylcholine concentration: 0.5 mM, 0.1 M borate buffer pH 9.0 with 0.1 M KCl. Applied potential: +700 mV vs. screen-printed electrode (SCE). Reprinted with permission from Ref. [10]*

Some other authors report the use of acetylthiocholine as substrate, which product (thiocholine) can be oxidized directly at the electrode surface [27]. The major problem associated with the use the acetylthiocholine as electroactive substrate is the high potential required to oxidize the thiocholine. This can be overcome using an electron transfer mediator. Our group developed two different methods based on the use of two different mediators:

1. Disposable TCNQ-Nafion modified screen-printed electrode coupled to differential pulse voltammetry (DPV)[12] and
2. Disposable cobalt(II) phthalocyanine (CoPC) modified screen-printed electrodes coupled to chronoamperometry [14] .

CoPC was indicated as one of the most suitable for the detection of thiol-containing molecules and the resulting oxidation signals occur at lower voltages, thus limiting the electrochemical interference of other oxidable compounds. Using these modified screen-printed carbon electrodes (SPCEs), under optimized chronoamperometric conditions, it was possible to detect AChE inhibitors, such as carbofuran, through the study of the AChE activity. Actually, the AChE free in solution was incubated with the pesticide. The inhibitory effect of the pesticide determines a decrease in the catalytic activity of AChE. As a consequence, less thiocholine is produced and thus a current value lower than that recorded in a blank solution was obtained. This current decrease was correlated with the pesticide concentration. A detection limit of 2.0×10^{-10} M for carbofuran was found in an analysis time of 15 min. To increase the versatility of the device, a reproducible and reliable immobilization strategy of AChE onto the SPCE surface was studied. The AChE was immobilized by cross-linking with glutaraldehyde, bovine serum albumin (BSA) and Nafion® onto the surface of the modified SPCE. The composition of the surface

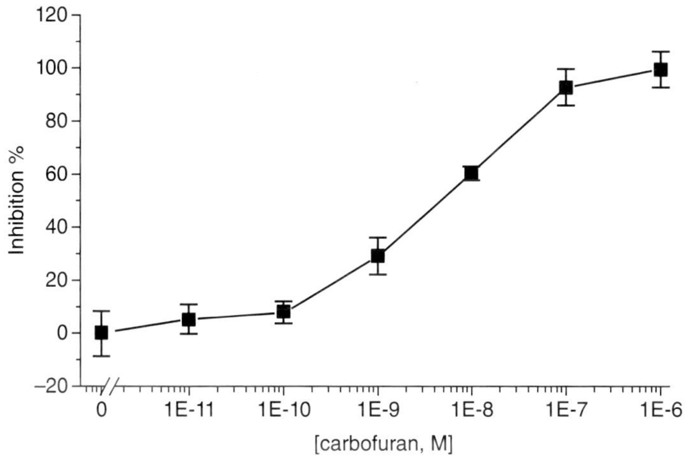

Figure 8.4 *Inhibition plot of carbofuran onto AChE-based biosensor. Reprinted with permission from Ref. [14]*

protein layer (enzyme units, glutaraldehyde, BSA, and Nafion® amounts) was optimized to obtain high and reliable response toward the substrate and AChE inhibitors. In the optimized conditions, the dynamic range for carbofuran detection was 10^{-10} to 10^{-7} M with a detection limit of 4.9×10^{-10} M, for an analysis time of 15 min (Figure 8.4). This is an important feature, considering that the immobilization can determine a loss of the activity of the enzyme that influences the sensitivity as well as dynamic range of the pesticide detection. Moreover, the proposed method was less prone to electrochemical interferences since the incubation and measurement were performed in two separate steps. The modification of a SPCE with CoPC provided significant improvement in the performance of the AChE biosensor. This was revealed in the decrease of the working potential of the TCh oxidation and in a fast and reliable response toward specific substrate, ATCh. The limits of detection obtained are small enough to detect trace amounts of anticholinesterase compounds. Taking into account the simplicity of the proposed method, the device offers good opportunities for direct and inexpensive detection of pesticides.

8.5 Conclusion

Enzymatic biosensors represent powerful tools for analysis in the agro-food industry. In recent years, the number of scientific investigations and reviews on biosensors has been very high, which reflects the considerable interest in this theme.

To be competitive with other existing technologies, enzymatic biosensor must be easy to use, portable, and inexpensive to deploy. In our opinion, all these aspects are within the reach of existing manufacturing technologies and immediate development. The real challenge, nowadays, is the possibility of obtaining stable enzyme with improved specificity and response time. Recent progress with respect to genetically modified enzyme biosensors opens new possibilities.

However, the lag between the high level of scientific and technological development and the limited use of these devices in the agro-food sector could be due also to different reasons. Among these, difficulties to change well-established methods and legislation are playing an important role.

References

1. Clark, L.C. (1956) Monitor and control of blood and tissue oxygenation. *Trans. Am. Soc. Artif. Int. Organs*, **2**, 41–48.
2. Clark, L.C. and Lyons, C. (1962) Electrode systems for continuous monitoring cardiovascular surgery. *Ann. N.Y. Acad. Sci.*, **102**, 29–45.
3. Updike, S.J. and Hicks, G.P. (1967) The enzyme electrode. *Nature*, **214**, 986–988.
4. Guilbault, G.G. and Montalvo, J. (1969) A urea specific enzyme electrode. *J. Am. Chem. Soc.*, **91**, 2164–2169.
5. Rogers, K.R. (2006) Recent advances in biosensor techniques for environmental monitoring. *Anal. Chim. Acta*, **568**, 222.
6. Campas, M., Prieto-Simonc, B., and Marty, J.L. (2009) A review of the use of genetically engineered enzymes in electrochemical biosensors. *Semin. Cell Dev. Biol.*, **20**, 3.
7. Tudorache, M. and Bala, C. (2007) Biosensors based on screen-printing technology, and their applications in environmental and food analysis. *Anal. Bioanal. Chem.*, **388**, 565.
8. Amine, A., Mohammadi, H., Bourais, I., and Palleschi, G. (2006) Enzyme inhibition-based biosensors for food safety and environmental monitoring. *Biosens. Bioelectron.*, **21**, 1405.
9. Ron, E.Z. (2007) Biosensing environmental pollution. *Curr. Opin. Biotechnol.*, **18**, 252.
10. Cagnini, A., Palchetti, I., Lionti, I., Mascini, M., and Turner, A.P.F. (1995) Disposable ruthenized screen-printed biosensors for pesticides monitoring. *Sens.Actuators, B*, **24–25**, 85–89.
11. Palchetti, I., Cagnini, A., Del Carlo, M., Coppi, C., Mascini, M., and Turner, A.P.F. (1997) Determination of anticholinesterase pesticides in real samples using a disposable biosensor. *Anal. Chim. Acta*, **337**, 315–321.
12. Hernandez, S., Palchetti, I., and Mascini, M. (2000) Determination on anticholinesterase activity for pesticides monitoring using acetylthiocholine sensor. *Int. J. Environ. Anal. Chem.*, **78**, 3–4.
13. Mascini, M. and Palchetti, I. (2001) Electrochemical biosensor for evaluation of contaminants in food for quality improvement. *Arch.Ind. Hyg. Toxicol.*, **52**, 49–59.
14. Laschi, S., Ogończyk, D., Palchetti, I., and Mascini, M. (2007) Evaluation of pesticide-induced acetylcholinesterase inhibition by means of disposable carbon-modified electrochemical biosensors. *Enzyme Microb. Technol.*, **40**, 485–489.
15. Mascini, M. and Liberti, A. (1974) An enzyme-coupled cyanide solid-state electrode. *Anal. Chim. Acta*, **68** (I), 177–184.
16. Capannesi, C., Palchetti, I., Mascini, M., and Parenti, A. (2000) Electrochemical sensor and biosensor for polyphenol detection in olive oils. *Food Chem.*, **71** (4), 553–562.

17. Volpe, G. and Mascini, M. (1996) Enzyme sensors for determination of fish freshness. *Talanta*, **43**, 283–289.
18. Marrazza, G., Cagnini, A., and Mascini, M. (1994) l- and d-Lactate assay in real milk samples with immobilized enzyme reactors and graphite electrode. *Talanta*, **41**, 1007–1014.
19. Mascini, M. and Pilotton, R. (1990) Flow analysis of lactose and glucose in milk, an improved electrochemical biosensor. *Food Chem.*, **36**, 213–222.
20. Eggins, B.R. (2002) *Chemical Sensors and Biosensors*, John Wiley & Sons, Ltd, Chichester.
21. Minh Canh, T. (1993) *Biosensors*, Chapman & Hall, London.
22. Kulys, J. and D'Costa, E.J. (1991) Printed amperometric sensor based on TCNQ and cholinesterase. *Biosens. Bioelectron.*, **6**, 109–115.
23. Kumaran, S. and Morita, M. (1995) Application of a cholinesterase biosensor to screen for organophosphorus pesticides extracted from soil. *Talanta*, **42** (4), 649–655.
24. Cagnini, A., Palchetti, I., Mascini, M., and Turner, A.P.F. (1995) Ruthenised screen-printed choline oxidase-based biosensors for measurement of anticholinesterase activity. *Mikrochim. Acta*, **121**, 155–166.
25. Bernabei, B., Chiavarini, S., Cremisini, C., and Palleschi, G. (1993) Anticholinesterase activity measurement by a choline biosensor: application in water analysis. *Biosens. Bioelectron.*, **8**, 265.
26. Tran-Minh, C. and Pandey, P.C. (1990) Insecticide determination with enzyme electrodes using different enzyme immobilization techniques. *Biosens. Bioelectron.*, **5**, 461.
27. Mionetto, N., Marty, J.L., and Karube, I. (1994) Acetylcholinesterase in organic solvents for the detection of pesticides: biosensor application. *Biosens. Bioelecton.*, **9**, 463–470.

9

Electrochemical Immunosensors

M. Teresa Fernández-Abedul, M. Begoña González-García and Agustín Costa-García
Departamento de Química Física y Analítica, Facultad de Química, Universidad de Oviedo, 33006, Oviedo, Julián Clavería, 8, Spain

9.1 Introduction

Various definitions have been given to the word "biosensor." The journal *Biosensors and Bioelectronics* defines a biosensor as an "analytical device incorporating a biological material, a biologically derived material, or a biomimic, intimately associated with or integrated within a physicochemical transducer or transducing microsystem. . . ." It is also said that a biosensor harnesses the exquisite sensitivity and specificity of biology in conjunction with physicochemical transducers to deliver complex bioanalytical measurements with simple, easy-to-use formats [1]. Biosensors may be classified according to the specific biological activity mechanism or, alternatively, to the mode of physicochemical signal transduction. In the case of immunosensors, the biological element is an immunoreagent, that is to say, an antibody (or a fragment) or an antigen. An electrochemical principle of transduction (commonly amperometric or impedimetric) is employed in those named electrochemical immunosensors. In this context, food electrochemical immunosensors are a developed research field that is continuously growing, due to the increasingly more stringent food quality and safety requirements. However, commercial development is far from being in accordance to this explosion of knowledge.

This increase in the research related to this field can be observed in Figure 9.1 where a graph represents the number of publications appeared in 2001, 2003, and the period comprised between 2005 and 2013 (last entry February 2014). Data were gathered from the databases Scopus and Google Scholar. According to the latest source, the number of publications for 2006 is only a 30% of this for 2013 (1490 publications). Without considering

Agricultural and Food Electroanalysis, First Edition.
Edited by Alberto Escarpa, María Cristina González and Miguel Ángel López.

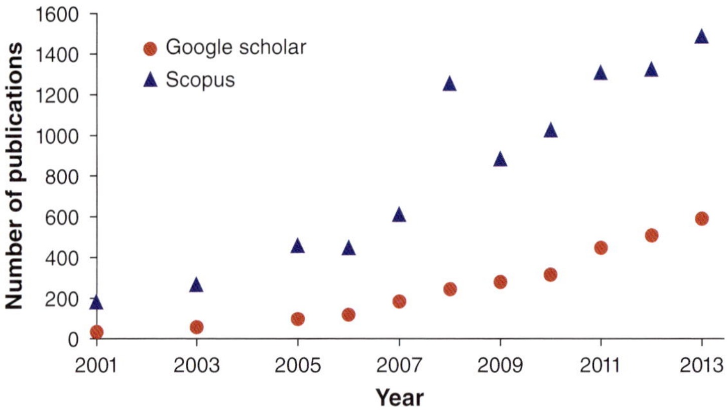

Figure 9.1 *Number of publications for "food electrochemical immunosensor" in different databases; February 2014*

special peaks (such as this in 2008), a rate of 151 and 67 publications per year is observed for Scopus and Google Scholar, respectively.

As it can be seen, the number of publications is enormous. The aim of this chapter, rather than attempting a systematic review of the field, is to show in a didactic manner the possibilities of electrochemical immunosensing for food analysis. There are some parts that could be found in other chapters of this book, such as, for example, the use of miniaturized electrodes, nanomaterials, or electrochemical techniques. However, a focus on the knowledge required for understanding their application to food analysis is always made, leaving a deeper treatment for those chapters. On the other hand, the reader can find many important references to the field in recent reviews, from the most general to the particular ones. Thus, recent trends in antibody-based sensors [2] (immunosensors) or smart electrochemical biosensors (from advanced materials to ultrasensitive devices) [3] include the application of the new classes of electrode nanomaterials as do those reviews on electrochemical (bio) sensing based on carbon nanomaterials (nanotubes [4] or graphene [5]), or metal nanoparticles (NPs) [6, 7]. Very useful is the review on the current/experimental aspects of (electrochemical) immunosensors [8, 9] where several issues including the use of magnetic beads (MBs) are considered. This special solid support has been the subject of reviews on the detection and quantification of bioterrorism/biohazard agents [10] and its coupling with electrochemical arrays [11]. Special emphasis is made on the use of screen-printed electrodes (SPEs) (commented in Section 9.1.1) for biosensing in microbiology [12]. Referring to the biorecognition events, the important issue of antibody orientation on biosensor surfaces has been also reviewed [13]. More related to the food analysis, a revision of nanobiotechnologies for the detection and reduction of pathogens [14] comments the problem of food related diseases, being *Campylobacter jejuni, Escherichia coli* O157:H7, *Listeria monocytogenes* and *Salmonella enterica serovar typhimurium,* the four major food-borne pathogens. Since the number of cases of infections increases, a need to monitor them throughout the food chain from production, processing, and distribution to the point-of-sale is claimed. Critical issues in sensor science to aid food and water safety [15] consider important points like the role of nanotechnology, the use of conducting polymers as emerging nanomaterials, and the integration of micro/nanofluidics as well as dealing with the diversity of target analytes or the complexity of food matrices and potential recognition technologies. More specialized reviews on microarray technology for major chemical contaminants (mycotoxins, biotoxins, pesticide, and pharmaceutical residues) analysis in food [16], on methods of analysis for the determination of trichothecene mycotoxins [17] or ochratoxin-A [18, 19] in food, on electrochemical biosensors for the detection of mycotoxins [20] and pesticides [21] or on the analytical control of gluten-free food [22] can be found in the bibliography. A very interesting review on healthy and adverse effects of plant-derived functional metabolites considers the need of revealing their content and bioactivity in complex food matrices [23].

Due to the complex nature of food samples, there are often strong matrix interferences that can have a huge effect on the results reported. Electroactive interferences are relatively rare, suggesting that electrochemical approaches might be of broad utility for monitoring Ag/Ab interactions, with many advantages over other principles of detection. The low cost of instrumentation, the possibility of performing *in situ* analysis, or the insensitivity to turbid samples are among them [17]. In a review that considers various techniques used for the detection of mycotoxins [19], electrochemical biosensing is seen as rapid and cost-effective, with good selectivity and sensitivity, portable, and with no expertise required. Moreover, it is useful for performing multiplex detection, suitable for screening, and open to commercial issues. According to recent scientific literature, the biosensors dominating the food sector of the market in the last 3 years are related to electrochemical transduction systems [23]. However, the selection of the immobilizing matrix and the technique is crucial to achieve enhanced sensing performance.

9.2 Defining the Problem: The Targets

Analytical chemistry is a well-established discipline but also a very vivid one. The analytical process aims to solve problems by translating socioeconomic or scientific-technological statements into analytical problems related to some of the four known questions: What and how much? How structured? How bound? and/or how distributed?[24] The main goals of analytical chemistry are, therefore, to obtain more (bio)chemical information with better quality, increasingly using less materials, time, cost, and risk for personnel and environment. This is summarized in three basic, interconnected and continuous trends, namely, automation, simplification, and miniaturization [25, 26]. A more recent trend that derives also from these is the significant reduction in the cost of analytical devices. Electrochemical immunosensing fits properly with these aims and is tremendously adequate for miniaturized devices. On the other hand, immunoassays are the basis of important modules of clinical automated analyzers and many developments in this field can be easily transferred to food analysis. Electrochemical immunosensors are a good example of simple devices that try to solve real-life problems. Referring to the cost, and in the special case of immunosensing, although reagents are expensive, miniaturization of devices allows small reagent consumption as well as does a careful design of the assays. In this context, electrochemical immunosensors reveal themselves as a very appropriate alternative for food analysis, especially when *in situ* or field analysis is required. By comparison with the clinical field where point-of-care devices are required, food analysis will have soon competitive point-of-production, point-of-distribution, or point-of-sale analytical devices.

The targets for which analysis is being required in food by immunosensing can be classified by the problems they try to solve. The diversity of species to be detected in food is enormous. Different types can be found and in Figure 9.2, those related to food and water safety are sorted. Main classification is made by their biological (prionic, viral, and bacterial) or chemical (toxins and pesticides) nature. Toxins, which are chemicals produced by metabolic processes in microorganisms are divided by size, being proteins those of higher size. This, as it will be commented later, is going to condition the format of the immunoassay, for example, sandwich or competitive. The pesticides are classified according to the chemical nature into organic (possessing chloride, carbamate, or phosphate groups) or inorganic. Apart from these, other chemicals that can be toxic such as polycyclic aromatic hydrocarbons (PAHs) or veterinary pharmaceuticals have to be considered. Moreover, it has to be taken into account that analytes could be found all along the food chain, from the farms and plantations to the primary and secondary food industry or even the stores and restaurants. In this section, examples of electrochemical immunosensors for different

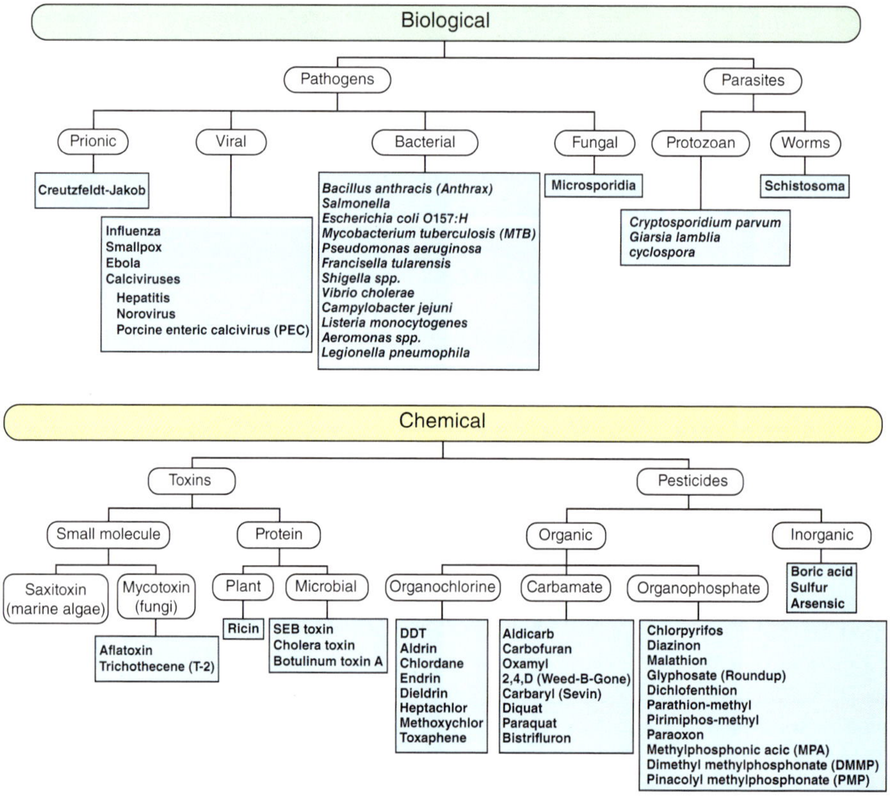

Figure 9.2 *Hierarchical categorization of priority target analytes for food and water. The rectangular boxes contain specific examples of pathogens, parasites, toxins, and pesticides (Reproduced with permission from [15]. Copyright © 2012, American Chemical Society)*

targets, including also pharmaceuticals or special analytes such as those related to food allergies, are given.

Related to the first step in the manufacturing process, common targets in agriculture are *herbicides* and *pesticides,* for weed or pest control. Over the past few decades, the use of chemicals within agriculture has widespread and the effect on the ecosystems has become public concern. Apart from possible environmental effects, potential contamination of water supplies and foodstuffs can have harmful results for the population. Therefore, stringent legislation for control of use and release of these contaminants has been introduced and electrochemical immunosensors have been developed for their analysis. For example, the determination of chlorinated herbicides such as picloram [27] or atrazine [28], the latter being the most commonly used in the United States and probably in the world, and found to be a persistent environmental contaminant, is of high interest. Paraquat, a rapid-acting and broad-spectrum quaternary nitrogen herbicide, is excluded from the routine multiresidue monitoring programs due to the permanent ionic character and high hydrophilicity that require different sample pretreatments. In this context, electrochemical magnetoimmunosensing is an excellent option that allows its direct detection and accurate quantification in potato extracts [29]. Also public concern justifies the determination of those belonging to the carbamate family such as isoproturon [30], employed to control various grasses and included in the list of substances of major concern in European waters, or molinate [31], used in rice production. Similar analytes for which immunosensors have been developed are pesticides (also from the carbamate family) such as carbofuran [32–35] or endosulfan [36], which can exist in field water samples somewhat longer than in pure water by binding to sediments and soil particles. The insecticide chlorpyrifos, which belongs to organophosphate family, has also been determined in food samples such as cabbage or lettuce [37].

Other toxic chemicals that could appear in food and require control are *PAHs*. They can be produced naturally during thermal geologic production and burning of vegetation, but also in cigarettes, vehicle exhaust, and broiled or smoked food. Concerning to immunosensing, it is difficult to produce antibodies that are specific for only one PAH because of the similarities in the molecular structure, electron density, and molecular weight, and most antibodies have been raised against benzo [a] pyrene, phenanthrene, and anthracene. The first is one of the most characteristic markers, and electrochemical immunosensors for its determination have been developed [38, 39]. Since its concentration correlates well with the total PAH content, it is considered the representative constituent of PAHs.

In relation also to the first step in the food industry, common analytes that are found in animal care in farms are *pharmaceuticals,* and examples of electrochemical immunosensors for their determination are reported. Monitoring, in most of the cases, is aimed to avoid their illegal use. Antibiotics have been widely used in food-producing animals for prevention and treatment of various infections. The United States and the European Union have established maximum residue limits (MRLs) for antibacterial agents that are authorized for use as veterinary drugs in food-producing animals to protect human health from the harmful residues. One of the oldest antibacterial agents is chloramphenicol, first isolated from *Streptomyces venezuelae* and now produced through chemical synthesis. The use is restricted to poultry due to its toxic effects. However, illegal use because of its accessibility and low cost requires sensitive analytical methods for food screening [40]. On the other hand, tetracyclines refer to several antibiotics with the same basic chemical structure and good activity against both Gram-negative and positive microorganisms. The analysis is useful not only from the point of view of health but also because it may lead to monetary losses in dairy industry by inhibiting starter cultures in food technological processes [41]. Other antibacterial agents are nitrofurans, used in gastrointestinal infections. Also the illegal use of these compounds needs an effective control and analysis by immunosensing of metabolites AMOZ (5-morpholino-3-amino-2-oxazolidone) [42] and AOZ [43] was suggested. Similarly, the sulfonamide family includes a large spectrum of synthetic bacteriostatic antibiotics used for therapeutic and prophylactic purposes. Improper administration leaves residues in food products and immunosensing has been proposed for determination in honey [44] or milk [45]. Neomycin is an aminoglycoside that was used for treating bacterial infections in veterinary medicine but ototoxicity and nephrotoxicity have been found. Immunosensors for sensitive detection of residues in foodstuffs from animals are, therefore, justified [46, 47], similarly to what happens for kanamycin [48, 49]. Another veterinary drug for which immunosensors have been proposed is enrofloxacin (from the fluoroquinolones family, with an MRL in milk for it and its primary metabolite, ciprofloxacin [50]). Salbutamol is one of the molecules that has been used in human and veterinary medicine for the treatment of pulmonary disorders and can be misused in farm animals, where a high dose gives rise to a preferential muscle-to-fat ratio, resulting in financial gain for the farmer. Illegal use in several countries requires ultrasensitive methodologies to monitor therapeutic use as to control illegal one by biosensing [51, 52]. Related to animal care, it has to be taken under consideration that many substances have been used not only for disease prevention or treatment but also for improving characteristics such as growth rate, reduced fat deposition, or increase in protein gain. This is the case of clenbuterol that has been also illegally used. Its usage for improving the production of lean meat can imply large amounts of residues in the produced meat with serious side effects and food poisoning. Since several countries have forbidden its use, control is required [53–55] and determinations are even performed in swine urine [56] or bovine hair [57]. Continuous surveillance of abuse is necessary and biosensing is one of the most appropriate alternatives [58–60].

On the other hand, methods for the detection of *microbial species* have attracted great interest because of the need to detect disease-causing microbes arising from contamination in water and food sources. Every year, pathogens in water and food sources cause 1.5 millions of deaths [61]. Among the microorganisms present in food, beneficial (e.g. those needed for cheese or yoghourt production), indicators (advisers of quality deterioration, contamination, and possible existence of pathogenic species), and pathogens (causing severe health damage to the consumers) have to be differentiated. Therefore, the detection of microorganisms is very essential and can be performed by electrochemical immunosensing [62]. One highly relevant and commonly targeted indicator for routine analysis of contaminated water and food sources is the bacterium *E. coli.* Although most strains of *E. coli* are harmless, their presence could be indicative of harmful effects and thus their analysis is of much interest [63]. Moreover, some of them are pathogenic as in the case of, for example, *E. coli* O157:H7, making its determination of high interest [64–70]. One of the most frequently reported causes of food-borne illness in developed countries is *Salmonella* and immunosensors based on specific antibodies are commonly employed [71, 72] for analysis. Cooking generally destroys most harmful bacteria, but undercooked foods, processed ready-to-eat foods, and minimally processed foods can contain harmful bacteria that are serious health threats [10]. Similarly, *Shigella* is one of the major causes of human infectious diseases, usually acquired by drinking contaminated water or eating uncooked food, and is responsible for millions of cases of diarrhea worldwide. One of the serogroups, *Shigella flexneri*, shows strong capacity for being acid resistant and salt tolerant, surviving at infective levels in water and food throughout the storage period, being of relevance its determination [73]. Other microorganisms for which detection by electrochemical immunosensing has been developed are *Staphylococcus aureus* [74], *Enterobacter sakazakii* [70, 75, 76], or the fungus *Phakopsora pachyrhizi* [77], the last one being an important pathogen that worries farmers involved in the crop culture since it can quickly defoliate plants (e.g., soybean), reducing production and quality, and leading to serious economic losses as well as spreading to neighboring farming. The development of multiplexed detection of the most common food-borne pathogens is also very interesting since screening of food should be possible in a rapid way [70, 78]. On the other hand, the ability to endure usual sanitation procedures and pasteurization conditions makes detection of *Bacillus cereus* of interest [79] in several products such as vanilla sauce, custards, soups, rice, pasta, meat, or dairy products. Finally, the detection of other microorganisms such as viruses is also relevant, as in the case of the subgroup J of avian leucosis viruses [80, 81].

Related also to microbiological activity, one of the reasons of their unbeneficial effects is the production of *toxins*. Fumonisins are mycotoxins produced by a variety of fungi of the *Fusarium* genus, considered natural contaminants of cereal grains worldwide. A maximum level of fumonisins has been recommended by the FDA and they are considered as carcinogens by the EPA. Therefore, biosensing of FB1, the most abundant and toxic of a family of related compounds [82] and deoxynivalenol (DON, a group B trichothecene) [83] or zearalenone (ZEA, a nonsteroidal estrogen) [84–86] as well as a resorcyl lactone derivative [87] is of much interest. Similarly, cyanotoxins (e.g., microcystin, MC) are produced by cyanobacteria and represent a significant hazard to public health (by liver damage) due to the release to water supplies. MC-LR is the most toxic species with WHO limits for drinking water, and is the target of several immunosensors [88–90]. On the other hand, staphylococcal enterotoxin B (SEB) is produced by *S. aureus* and is one of the toxins responsible for staphylococcal poisoning in humans, being produced by some countries as a biological weapon [91]. Need for rapid and effective screening justifies the development of immunosensors [92–94]. Similarly, rigorous control of ochratoxin A (OTA), a toxin produced from *Aspergillus ochraceus* and *Penicillium verrucosum* that contaminates cereals, coffee beans, beans, grapes, and dried fruits, led to the development of many immunosensors [95–102]. They have also been developed for other frequent toxins derived from fungi *Aspergillus flavus* and *Aspergillus parasiticus,* named aflatoxins, B1 being the most abundant and carcinogenic [103] with many immunosensors reported [104–107]. When aflatoxin B1 (AFB1) is ingested by cows through contaminated feed, it is transformed into aflatoxin M1 (AFM1) through enzymatic hydroxylation. Therefore, its determination in milk is also relevant [108] since it becomes secreted by the mammary gland of dairy cows. One of the first mycotoxins isolated was citrinin, a secondary metabolite produced by fungi of the *Aspergillus, Monascus,* and *Penicillium* genera. In this case, it is also necessary to establish maximum permitted levels for protecting public health and determination in rice samples has been proposed [109].

Special attention among toxins has to be paid to those that are active in acidic media. Actually, the FDA classifies acidic foods into low acid and acid. Although the acidic environment is regarded as microbiologically safe, some biological toxins maintain their toxicity (e.g., ricin that has been used as biological warfare). Immunointeraction at acidic pH has an extremely low rate and adjustment is almost always necessary. Therefore, the determination of toxins at these pH values is a challenge [110]. Finally, special toxins are those present in marine media such as okadaic acid, a lipophilic biotoxin produced by *Dinophysis* and *Prorocentrum* dinoflagellates, mainly accumulated in the hepatopancreas of bivalves. The need for analysis is clear [111–115] because it is considered as the most concerning diarrheic shellfish poisoning for human health worldwide. Other marine toxins are cyclic polyether neurotoxins produced by *Karenia brevis* (brevetoxins) causing neurological shellfish poisoning. Exposure in this case may occur during red tide with compounds aerosolized by wind and surf and can be found in seafood, being biosensing adequate for food control [116].

Additives are the other type of analytes related to food immunosensing, although they are not as common as the other ones mentioned above. Colorants, such as Sudan azo dyes, are synthetic chemicals employed in industrial products but also found as additives in foods, such as hot chili powder, relishes, chutneys, seasonings, sauces, and ready meals because of their bright and vivid red color. However, Sudan I was recognized as a potential carcinogen to both human beings and animals and its use is forbidden all over the world, which justifies its immunonsensing [117].

A very important group of biosensors in the food field is devoted to *food allergy or intolerance,* an abnormal (immunological or not) response due to sensitivity to a food component. It represents an important health problem, especially in industrialized countries where food allergy has been estimated to affect around 1–2% of the adult population and up to 8% in children under 3 years [118]. Cow's milk allergy is one of the most common food allergies in childhood with two major allergens: casein and β-lactoglobulin. Interest, therefore, is focused on the development of biosensors for milk allergen detection [119, 120]. Critical health problem is also caused by allergies to peanuts and tree nuts, especially among Americans. The incidence has so increased that they have been called epidemic[121]. These allergies can be revealed by detecting IgE antibodies to the allergens in patient serum [122]. Related to this is the celiac disease, an autoimmune-mediated disorder that primarily affects the gastrointestinal tract characterized by inflammation due to the ingestion of gluten, a protein made up of gliadins and glutenins found in cereals. Gliadin is deaminated by tissue transglutaminase (tTG), the detection of anti-tTG autoantibodies being the basis of the celiac disease diagnosis by immunosensing [123, 124]. Due to this, the inclusion of gluten in food labeling has been required to prevent harmful effects in celiac patients. The FDA stated in 2008 that foods labeled with the term gluten-free may not exceed gluten content of 20 mg l^{-1} and the same regulation has been taken in 2009 by EC. Therefore, it is important to have methods for the detection of gliadin, not only for celiac patients but also for industries generating gluten-free foodstuffs for rapid testing of incoming raw materials or for checking gluten contamination throughout the food production process [125].

Related also to food analysis is the immunosensing of leptin, a proteic *hormone* that plays an important role in regulating food intake and body composition, since its binding to brain ventromedial nucleus gives the brain a signal of satiety sensation. In healthy individuals, a decrease in leptin increases neuropeptide production in the hypothalamus, which in turn increases appetite and decreases energy expenditure. Immunosensing analytical methodologies for this obesity-related hormone are under development [126].

Finally, an important group of analytes is related to *functional or nutraceutical food.* Attempts to improve human health and to satisfy people's desire for healthcare without intake of pharmaceuticals have led the food industry to focus attention on nutraceutical food. Although they are thought to promote good health, sometimes undesirable effects are also reported. Furthermore, several products present on the market show few benefits and sometimes even reverse unhealthy effects could be produced. The efficacy is often unconvincing and epidemiological studies are necessary to prove the truth of their claims. Therefore, there is a need for reliable analytical control systems to measure the bioactivity, content, and quality of these additives in the complex food matrix. Biosensors are promising candidates for routine control of functional foods [23]. However, only very few works deal with supplemented food. A biosensor is proposed for determination of biotin (vitamin H) on fortified commercial dietary supplement and infant formula. In this case, a very well-known high-affinity ligand–protein interaction (biotin-streptavidin) is employed in a similar way to antigen–antibody interactions [127].

9.3 Recognizing the Target

Immunosensing is based on the specific molecular recognition of an antigen by an antibody. The use of antibodies as recognition molecules is one of the many successful examples of man's exploitation of a biological phenomenon. Although the use of the immunological interaction was recognized as long ago as the 1890s, development as quantitative analytical technique dates only from 1960s. Immunosensor development is continuously advancing by the multidisciplinary cooperation of biology, chemistry, and physics but there is still a great deal on the improvement of devices. Therefore, knowing about the characteristics of the antibody, the recognition molecule that is generated by an antigen, is of extreme importance for achieving a competitive and useful analytical device. In this section, important issues referring to immunoreagents, antibodies, and antigens will be considered.

9.3.1 Antibodies

Antibodies (Abs) are the molecules in charge of the recognition event in immunosensing. They are glycoproteins that are produced in mammals as an immune response to extraneous molecules. These are the product of the humoral response of the organism (opposite to the cellular answer) to fight against possible pathogens. Special types of lymphocytes, the cells of the lymphatic system, are responsible for the guard of the organism. Among them, lymphocytes B that maturate in the bone marrow are the producers of antibodies. These extraordinary proteins can be envisaged through a basic structure that is shown in Figure 9.3a.

Antibody molecules are also called immunoglobulins, since they are globulins in charge of the immune response. This is the reason why they are simplified as "Ig" followed by the class in capital roman letters and the subclass in subscripted numbers, as in, for example, "IgG_1." The polypeptidic structure consists of four chains, equal in pairs. There are two long (H or heavy) and two short (L or light) chains. The aminoacidic residues are grouped in globular regions stabilized by disulfide bonds, called domains. Two and four domains constitute light and heavy chains, respectively. Apart from these intrachain bonds, there are interchain disulfide bonds that join both heavy chains and also light to heavy ones. Looking

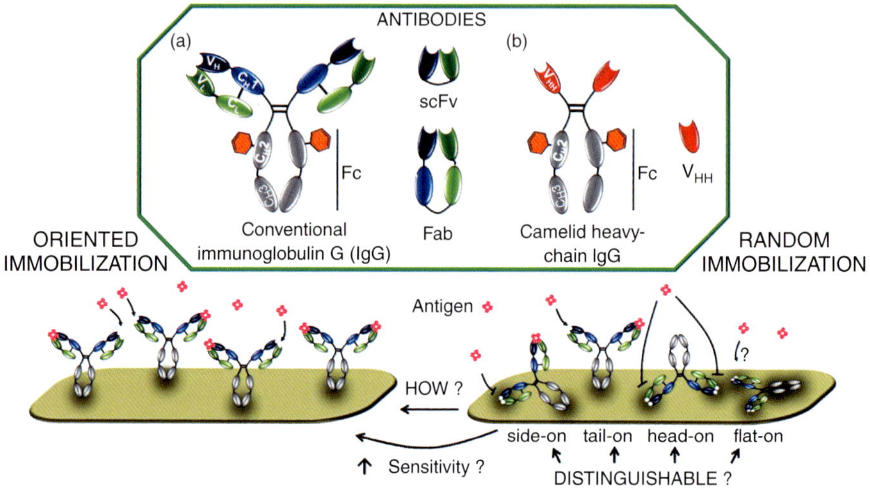

Figure 9.3 Schematic depiction of (a) natural antibodies with single-chain variable fragment, antigen-binding Ig fragment, and (b) camelid heavy-chain IgG with the variable fragment HH. Random or oriented immobilization is represented at the bottom of the figure (Reprinted with permission from [13], published by The Royal Society of Chemistry)

into the sequence of antibody aminoacids, it was observed that the variability (ratio between the number of different aminoacids in one position *vs.* the frequency of the most common aminoacid in this position) was very low in most of the domains (constant domains, C_L and C_H; that are called C_H1, C_H2, and C_H3, starting the 3 in the carboxy terminus of the protein). On the contrary, domains that are at the amino terminus are more variable (V_L and V_H). If variability is evaluated more carefully, three hypervariable (HV) regions, called complementarity determining regions (CDRs), are found. These regions, which are situated among framework regions (FRs), form the antigen-binding site or paratope, the "nest" for the antigen. The antibody molecule is bivalent since it includes two binding sites for the antigen, which are equal in naturally occurring molecules (bivalent but monospecific antibodies). There is a flexible region among C_H1 and C_H2 that allows movement of the "arms" of the antibody to best bind the antigen, called the hinge region. Finally, it has to be considered that antibodies are not just proteins but glycoproteins, and therefore they are associated to carbohydrates most commonly found in the C_H2 domain.

Antibodies can also act as antigens and can be classified according to the antigenic determinant sites (epitopes), which can group types of molecules into classes (isotypes). Number of domains in the heavy chains, percentage of carbohydrates, number of disulfide bonds, and so on, are differences between classes of immunoglobulins. For example, human immunoglobulins (hIgs) can be grouped depending on their heavy chain, distinguishing five different chains, named in Greek letters: γ, μ, α, δ, ε that correspond to immunoglobulins IgG, IgM, IgA, IgD, and IgE, ordered by abundance in serum. Subclasses can be also found (e.g., IgG_1, IgG_2, and IgG_3).

The basic structure of an antibody is a monomer although molecules could associate to form oligomers, which can help in the immunoglobulin-specific function. This is the case of the secretory IgA, a dimmer in which two molecules are grouped by a joining aminoacid chain (J-chain) and stabilized by other called the secretory component. Taking into account that IgA is the immunoglobulin present in all the secretions as a first barrier, this dimerization (tetravalent) can help in stabilizing it. Similarly, the IgM, the immunoglobulin in charge of the first answer to the pathogens, needs antigen-binding sites for a first neutralization. Therefore, it exists as a pentamer, where five molecules form a star with all the 10 antigen-binding sites (decavalent) directed outside. The characteristics of these immunoglobulins, functions, and other details, can be found in general immunology textbooks [128–130].

When the antibody is considered an analytical reagent, several characteristics have to be taken into account, mainly title, affinity, specificity, and cross reactivity. The *titer* of the antibody is a measurement of its concentration. Usually serial dilutions are made for interaction with antigen, the titer being defined as the maximum dilution of antibody in which a specified signal is observed (e.g., a detectable signal, one unit of absorbance, etc.). Therefore, titer depends on the technique, mainly due to the different sensitivity. When two antibodies are compared, this with a higher titer for the same technique (1 : 5000 *vs.* 1 : 200) possesses higher activity and therefore higher dilution can be made for obtaining similar signals.

Affinity has also to be taken into consideration. It relates to the bond strength between both binding sites (epitope and paratope for antigen and antibody, respectively). On the other hand, when designing an immunoassay one of the first selections that has to be made is the type of antibody, especillay in which *specificity* is referred: polyclonal or monoclonal. The first type comes from the purification of the serum of an animal that has been immunized with the antigen (antisera). Different lymphocytes produce different antibodies (in response to different epitopes in the antigen). In the case of monoclonal antibodies, the immune cells producing antibodies have been made immortal (by hybridoma technology) and have been isolated in wells in a cell *per* well way. Therefore, antibodies are made by identical immune cells that are clones of a unique parent cell. The procedure for obtaining monoclonal antibodies takes usually several months and therefore their price is incremented when compared to polyclonal reagents; but on the other hand, more specific antibodies are obtained. Since several possibilities (of affinity and specificity) exist, a careful selection has to be made in the design of the immunoassay. This is crucial to obtain a selective and sensitive analytical device. Related to this is the term *cross-reactivity*, which refers to the response of the antibody to an antigen, different (but usually structurally related) to this that has generated it. It is calculated as the ratio of the signals of the immunological interaction for the cross-reagent and for the specific antigen, usually expressed as a percentage.

Talking about *dimensions,* and now that total immersion on nanotechnological approaches is occurring, could be interesting comment that a basic antibody molecule approximates 10 nm in length. Therefore, these molecules are very appropriate tools for constructing "bottom-up" nanotechnological approaches.

In most of the immunosensors, the whole molecule of the antibody is used. However, the use of *fragments* is increasing due to their advantages. The antibody molecule has two different functionalities: antigen binding and effector functions, both of which happen at different parts of the molecule (amino and carboxy parts, respectively). In cases where only antigen binding is required, fragments that conserve this function could be used and the whole molecule is not needed. Reduced size or avoided adsorptions are some of the advantages obtained.

Antibody can be lyzed into different constitutive fragments using the proteolytic enzymes pepsin or papain, two enzymes that helped to understand the structure of these molecules. In the first case, the molecule is divided above the disulfide bridges of the hinge region and therefore two separate fragments with antigen-binding availability are obtained (Fab, each of them monovalent, see Figure 9.3a). The other part is crystallizable and named Fc. In the case of papain, since the break happens below the disulfide bridges, the fragment with antigen-binding capacity is divalent (Fab'_2) and the other part breaks into small peptides and the fragment pFc′. Apart from enzymatic fragmentation, the molecule can be vertically broken by chemical reduction of the disulfide bonds that keep together the heavy chains. In this way two monovalent fragments are obtained, usually utilized for immobilization or labeling purposes. One of the advantages that can be of interest for immunosensing is the reduction in the nonspecific binding resulting from Fc interactions and a potential higher sensitivity in antigen detection as a result of reduced steric hindrance from large protein epitopes. Although not common in food analysis, the use of a recombinant Fab-fragment has been reported for deoxynivalenol detection in food samples [83]. Fab fragments could also be combined to form dimers and trimers.

Achieving smaller molecules with high binding capacity has been performed by using only variable domains, constituting the fragment Fv, usually joined by a single aminoacidic chain (scFv, Figures 9.3a and 9.4). This is obtained by spontaneous association of V_H and V_L domains generated by recombinant techniques (genetic engineering) that are linked through a chain of 15 amino acids. A disulfide bridge, instead, will give rise to a dsFv fragment. Further size reduction into monomeric single domain antibodies (sdAbs), V_H or V_L, resulted in loss of affinity toward the antigen [132]. Although very scarce in the bibliography, some works can be found on the use of these fragments, such as the usage of scFv for detection of the food pathogen *Listeria monocytogenes* [133].

As commented earlier, natural antibodies are monospecific but *multispecific* antibodies would be of relevance in several applications, most of them related to therapy. Bispecific antibodies can be generated chemically (by combination of two reduced fragments) or biologically (through hybridoma technology). However, a large number of nonfunctional molecules are obtained. Recombinant techniques can produce bivalent molecules (mono or bispecific) by dimerization of two V_H-V_L fragments (Figure 9.4). To avoid the generation of Fv fragments (by the association between the V_H and V_L domains of the same fragment, generating the monovalent aminoacidic chain), the linker is reduced to approximately five amino acids. In this way, one V_H has to associate to a V_L different to this in the same fragment. Summarizing, two fragments assemble into a dimeric molecule (diabody). Further reduction in the linker length to 0–2 aminoacids results in generation of triabodies (trimeric) or tetrabodies (tetrameric). In the case of bispecific diabodies, recombinant techniques produce a fragment ($V_HA-V_LB-V_HB-V_LA$) that reorganizes for generating the epitopes for binding antigens A and B. In a similar way, fragments can be produced to generate trivalent mono- or trispecific triabodies as well as tetravalent mono- or tetraspecific tetrabodies.

Antibodies can be obtained from *different species* and in this way not only rabbit, mouse, goat, dog, sheep, and so on, but antibodies are also found from kangaroo, wild pig, or camelids. This last case deserves special consideration. These antibodies lack the light chain and the first constant domain of the heavy chain, leaving one single domain for effective antigen binding, known as V_{HH} or nanobody[134] (Figure 9.3b). Another novel Ig with

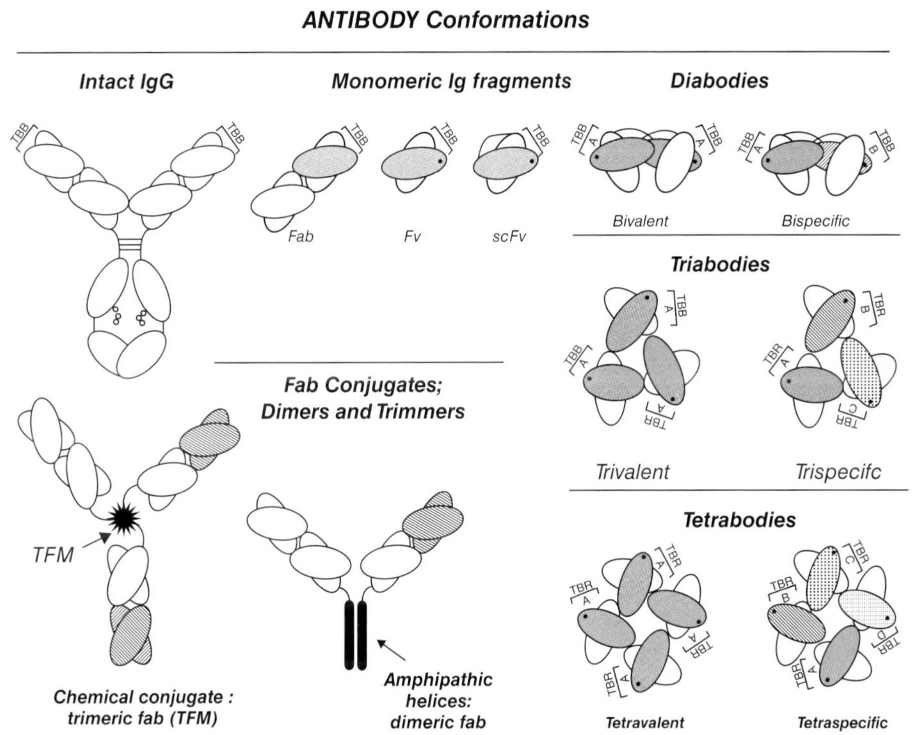

Figure 9.4 *Schematic representation of intact Ig (bivalent) alongside monomeric Ig fragments (Fab, Fv, scFv) and diabodies, triabodies, and tetrabodies. V-domains and C-domains are represented by ovals. The different target-binding regions (TBRs) of Fv modules are labeled TBR-A, TBR-B, and so on. The different patterns of the V_H domains are meant to represent different target specificity. For size comparison, chemically conjugated Fab trimers, and helix-conjugated Fab dimers are also shown (Reprinted from [131] with permission from Elsevier)*

one variable domain, called novel antigen receptor (V_{NAR}), was discovered in cartilaginous fish, such as sharks [135]. Both are small, highly soluble, possess superior stability, and seem very adequate for biosensing purposes.

Since antibodies are used as therapeutic molecules that are produced in species different than human, and to reduce the risk of immunogenicity, *humanization* is a common procedure. This is very frequent in monoclonal antibodies produced in mice. A mouse–human antibody chimera can be generated by substituting a murine by a human Fc fragment. This chimeric antibody is then humanized at 75%. Higher percentages (99%) are obtained when only murine CDRs are maintained (mosaic antibodies). Beyond this, changes in the aminoacid sequence of the HV regions can improve the affinity toward the antigen. All these possibilities remain unexplored in the biosensing area and could be of much interest in a near future.

Finally, and in which refers to the recognition element, it can be taken into account that there are some approaches that use synthetic antibodies that mimic the natural ones. This is the case of molecularly imprinted polymers (MIPs), where the imprinting effect is achieved by polymerization of functional monomers and cross-linkers in the presence of template molecules, either the analyte or analogs. After removing the templates from the resultant network, a polymer with template-selective binding capacity is obtained. This principle has been used for the determination of residues of the antibiotic streptomycin in milk and honey [136]. Similarly, aptamers, functional nucleic acids selected from combinatorial oligonucleotide libraries to bind specific target molecules, are seen as synthetic antibodies that can follow the same formats as immunosensors (aptasensors) for determination of, for example, tetracycline [137] or OTA [138].

9.3.2 Antigens

The antigen (Ag) is the molecule that enters an organism and produces immune response. It possesses immunogenicity (capacity for producing the immune response) and antigenicity (capacity of interacting with the antibodies that has generated). There are small molecules that are not immunogenic and require conjugation to a carrier molecule for producing antibodies. However, they can interact with the generated antibodies. Since they lack immunogenicity and possess antigenicity, they are called incomplete antigens or *haptens*. The difference between complete and incomplete antigens is important from the point of view of immunoassay design. For example, high-size antigens have many binding sites or epitopes and sandwich-type assays could be performed. On the other hand, special detection methodologies could be based on the different sizes between antigen (small hapten) and immunological complex (that includes the antibody, 150 kDa approximately). Apart from this, not only the size but also the chemical nature, conformation, number of epitopes, or accessibility, among other, has to be considered. Small molecules are a challenge for the direct sensing, since a change in the analytical signal has to be detected and this is usually faint. Therefore, labeled methods are commonly preferred.

It also has to be taken into account that antibodies are proteins, and therefore, as was commented before, they can be antigens as well, interacting with anti-antibodies that have been generated by immune response. However, in which refers to the specific application of food analysis, antigens (without antibody activity) are the most common analytes. In clinical samples, detection of antibodies, generated after contact with an external pathogen or nonaccepted molecule, is indicative of pathogen presence. In this clinical context, but also related to food analysis, food allergies (commented in Section 9.2) can be included. For example, in peanut allergy, three proteins produce the majority of specific immunoglobulin E (the one that assesses the existence of an allergic process) and point-of-care diagnostic tools of IgE-mediated peanut allergy are related to the analysis of this protein [122]. Similarly, in the case of gluten sensitivity, the detection of gliadin is performed in food [125] and antibodies are detected in the serum of celiac people [124].

9.4 Immunosensing Architectures

A crucial requirement of an immunosensor is to immobilize the biomolecules onto the sensor surface in a reproducible way while keeping their biological activity and controlling spatial distribution. Although antibodies have been increasingly used as recognition elements in biosensors, a key challenge remains: the perfect immobilization on the adequate substrate. Antibodies with better affinities and higher stabilities are preferred to improve biosensor performance. Therefore, research works are many times associated to antibody production.

One of the characteristics that mostly represent the evolution of immunosensing is the architecture of the recognition layer or the surface engineering, concepts aimed to mean the diversity of the interactions and the number of different functional components that are arranged together. Sensor optimization is directed towards surface preparation aiming to promote specific binding and suppressing the nonspecific one. Strategies for immobilization may result in *specific or random orientation* of antibodies. The orientation is dependent on the self-organizing capacity of the antibodies, which may be steered by specific reactive groups on the surface, on the antibody, or on both. Specific orientation of immobilized antibodies is not easily achieved, since they usually carry several copies of reactive groups. Four exemplary molecular orientations can be adopted by a basic immunoglobulin molecule (Figure 9.3b): side-on (one Fc and one Fab attached to the surface), tail-on (Fc attached to the surface), head-on (both Fabs attached to the surface), or flat-on (all three fragments attached to the surface) [13]. For the highest analyte binding, antibodies should display free antigen-binding regions after immobilization. Controlling the orientation will, therefore, lead to better analyte binding, resulting in improved biosensor sensitivity. Possibility of attachment through carbohydrates or thiol groups (in chemically reduced fragments) is not common in food analysis. Some cases of immobilizations through Protein A [34, 44] or Protein G [31, 45, 77, 84, 102, 108, 113], both binding the Fc portion of the antibody leaving free the binding sites, are reported. An ending histidine tag in the Fc fragment (by recombinant methodologies) that completes the coordination sites of an immobilized metal cation (e.g., Cu^{2+} or Ni^{2+} through a complex with NTA (N_α, N_α-bis (carboxymethyl)-L-lysine hydrate) chelator) could also be used [139]. This strategy raises the possibility of binding reversibly the antibody by using a complexing agent of the metal without denaturation of the antibody (Figure 9.5).

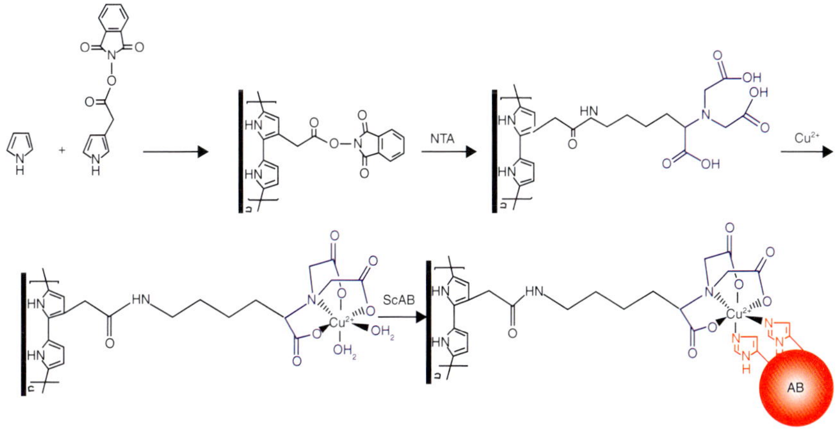

Figure 9.5 *Schematic representation of the fabrication process of the multilayer system of an immunosensor on a gold electrode (Reproduced from [139], with permission from Elsevier)*

Even though and knowing all the advantages of orientation, a random immobilization is the most common approach for the sake of simplicity, especially in the case of non-regenerable sensors. It is also reported that orientation of the capture molecule hardly affects high-affinity interactions, while it leads to strong improvements in sensitivity for low-affinity interactions [140].

Immobilization possibilities are almost infinite. Since it is essential to immobilize antibodies on surfaces without changing their binding activity and specificity, strategies should be mild. Surfaces can be used either directly or functionalized, and used either as two-dimensional surfaces or three-dimensional matrices. Noncovalent and covalent immobilization strategies have been reviewed very recently [13]. Options are numerous and it is not possible to report all of them here. Generally, many different approaches that combine nanostructures (either only one or hybrid approaches), polymers with cationic/anionic properties, proteins, cross-linkers, or other molecules with various functionalities can be found in the bibliography. Typically the *layer design* is written as a succession of abbreviations separated by hyphens (components included in the same step) and forward slashes (different steps of the procedure) being indifferent to start or finish with the outer or inner layer. In the cases where some of the reagents are first included in one of the components, they could be named with the symbol "at" (@) such in the case of CD@Fe_3O_4 that are cyclodextrines on Fe_3O_4 nanospheres [80] or Pd@SBA-15 that stands for Pd deposited from a solution of H_2PdCl_4 on mesoporous silica with uniform tubular channels (SBA-15) [59]. Some examples corresponding to amperometric detection using labels are reported in Table 9.1, meanwhile in Table 9.2 are included those label-free approaches, most of them impedimetric. The functions of the components are diverse and some important related points are commented in the sections below.

There is an increasing interest in developing label-free immunosensors to overcome several drawbacks and simplify the assays. Direct readout is possible with fewer reagents and lower cost assays. As will be commented in the label-free methods section, the electron transfer reaction of a probe at an immunosensor is commonly studied (Faradaic methods). Upon specific molecular recognition of the antigen (or antibody) by the immobilized capture antibody (or antigen), there will be changes in the interfacial charge, capacitance, resistance, or mass, which in turn have a direct effect on the electron-transfer reaction of the probe at the immunosensor–electrolyte solution interface. In Table 9.2, a compendium of surface engineering for label-free food immunosensors is reported, as well as the corresponding analytes and samples.

9.4.1 Components of the Sensing Layer

Since the detection principle of the immunological event is electrochemical, an electrodic surface, which can be considered as the primary transducer, is involved. This is commonly a glassy carbon electrode (GCE), a SPE, a gold disk, and so on. Issues related to this and the detection principle involved will be commented below in Section 9.6. However, the generation of the sensor requires the integration of the recognition element, and usually different steps, components, or layers are included. As can be seen in Tables 9.1 and 9.2, the possibilities are enormous, and sometimes the complexity of the architecture is considerable. Very interesting is the possibility of adding beneficial properties by including more materials, by functionalization (covalent or noncovalent) or by using hybrid structures.

Table 9.1 Examples of electrochemical immunosensors related to food analysis with amperometric detection using labels

Sensing layer	Transduction approach	Electrochemical technique	Analyte	Sample	Comment	Reference
Au-BSA-Ag	Ab-HRP	Amperometry in TMB/H_2O_2 solution	Clenbuterol	Livestock urine	Competitive assay Sixteen electrodes	[55]
Au/CHIT/Fe_3O_4/PAMAM/AuNPs/Ag/HRP	HRP-MWCNTs-Ab	CV in H_2O_2	Salbutamol	–	Competitive assay	[52]
Au/AuNPs-PAAD/BSA-Ag	Ab-HRP	DPV in o-PD and H_2O_2	Brevetoxins	Molluscs	Au-PAADs are generated by chemical reduction of $HAuCl_4$-dendrimer (generation 4.0) Competitive assay	[116]
Au/DT2/tTG	Ab_2-HRP	CV/EIS in $[Fe(CN)_6]^{3-/4-}$ amperometry (TMB and H_2O_2)	Anti-tTG	Blood serum samples of celiac patients	DT2 and Ab by EDC/NHS chemistry DT2 is a long-chain, carboxylic-terminated bipodal alkanethiol PEG groups in alkanethiol chain were effective in eliminating NSB	[123]
SPGE/CMD/OTA-BSA	Ab-HRP and OTA	Chronoamperometry in TMB/H_2O_2	OTA	Wine	Competitive assay CMD was passive and covalently adsorbed	[101]
SPGE-Anti Ab/Ab	Ag-HRP	Chronoamperometry in TMB/H_2O_2	Fumonisin	Corn	Competitive assay	[82]

(continued overleaf)

Table 9.1 *(continued)*

Sensing layer	Transduction approach	Electrochemical technique	Analyte	Sample	Comment	Reference
MB-ST-B-OA graphite electrode	Ab/antiAb-AP	DPV in pNPP solution	Okadaic acid	Mussels	Sensitivity was increased by ED when compared to colorimetric assays Competitive assay	[113]
GEC (graphite epoxy composite)/ MB-Ag	Ab_2-HRP	Amperometry in H_2O_2 and HQ (mediator)	Gliadin	Gluten-free foodstuff	Tosyl activated MBs Competitive assay	[125]
PG (pyrolytic graphite)/PDDA/ GSH-AuNPs/Ag	Ab_2-HRP	EIS, TMB/H_2O_2 biocatalyzed formation of insoluble product	Ab to peanut allergen Ara h2	–	Faradaic and non-Faradaic EIS, and amperometry were compared. Non-Faradaic EIS provided the best detection limit and linear range	[122]
GCE-GS-CD	GOD-Ab_2/Fc-CD@Fe_3O_4	DPV	Subgroup J of avian leucosis viruses	Avian serum	CDs impede agglomeration of GS GOD is immobilized on nanospheres Carboxymethyl CDs are bond through EDC/NHS as coupling agents and EP as cross-linker	[80]
GCE/SDBS-GS-Ab_1	Pd@SBA-15/Ab_2/ BMIMBr	Amperometry after addition of H_2O_2	Salbutamol	Serum	Acyl chloride linking reaction GA is employed as cross-linker	[59]
GCE/PB/ MWCNTs/Ab_1	GOD/GO/ AuNPs-Ag	DPV after addition of glucose	Clenbuterol	Feed sample	Competitive assay Layer by layer methodology Optimization by EIS	[53]

GCE/CHIT/GSE	HRP-CNS-Ab	DPV in solution containing H_2O_2	Microcystin-LR	Water	GSs are prepared from GO by reduction with hydrazine Competitive assay HRP and Ab are conjugated by EDC/NHS chemistry	[89]
GCE/CHIT/ AuNP/Ab_1	Ab_2-cathecol-Au-Fe_3O_4 (magnetic nanoparticle)	DPV	AFB1	–	CHIT gives -NH_2 groups Ab is covalently bound by EDC/NHS chemistry EIS and CV are employed for characterization	[105]
GCE-NPG-Ab_1	Na-Mont-THI-HRP-Ab	Amperometry in H_2O_2 solution	Zeranol	Beef liver	GA is used for incorporating both HRP and Ab EIS for characterization	[87]
GCE-MWCNT-Ab	AgNP-GO-Ag	DPV (oxidation to Ag^+)	Clenbuterol	Urine	Competitive assay CV for characterization with $[Fe(CN)_6]^{3-/4-}$	[56]
GCE/Au nanoclusters/ Ag-BSA	Ab/Ab-HRP	Amperometry in presence of H_2O_2 and HQ	Picloram	Peach extract	Competitive assay EIS for characterization Nanoclusters are electrodeposited on a PC template, later solved with chloroform	[27]

(continued overleaf)

Table 9.1 *(continued)*

Sensing layer	Transduction approach	Electrochemical technique	Analyte	Sample	Comment	Reference
GCE/CHIT/AuNPs/Ab/THI-CHIT/AuNPs-HRP	HRP	Chronoamperometry in H_2O_2/CV in $[Fe(CN)_6]^{3-/4-}$	*Bacillus cereus*	Milk	The resulting immune complex hinders the electron transfer due to steric hindrance	[79]
GCE/dendritic-SiO_2NPs/Ag-BSA	Ab_1/Ab_2-HRP	CV/EIS CV in HQ/H_2O_2	Benzo[a]pyrene	Water	Ethylenediamine is electropolymerized on GCE	[38]
					SiO_2NPs contains carboxyl groups from PAA. EDC/NHS chemistry is employed Competitive assay between free and immobilized Ag for Ab_1	
GCE/SWNHs/Ag	Ab-HRP	DPV in *o*-PD and H_2O_2	Microcystin-LR	Water	EDC/NHS for Ag binding	[88]
					Competitive assay Dahlia flowerlike spherical aggregates composed of thousands of graphitic tubule closed ends with cone shaped horns are employed	[141]
GCE/AuNPs/pDPB/Ab_1	Hyd-MWCNTs (AuNPs)-Ab_2	CV for characterizing chronoamperometry	Neomycin	Beef, pork, chicken	Electrodeposited AuNPs	[46]
					Ab immobilization through EDC/NHS chemistry pDPB generated electrochemically from the monomer MWCNT (AuNPs) were combined to Hyd and Ab_2 through EDC/NHS AuNPs were chemically generated in a dispersion of MWCNTs	

GCE/AuNP/Ab_1	$NiCo_2O_4$-Nf-THI-Au-Ab_2	EIS in $[Fe(CN)_6]^{3-/4-}$ DPV in H_2O_2	Proposed for micotoxins	Food samples	Sandwich type immunoassay Demonstrated for CEA	[142]
GCE/AuNP/PB/AuNP/PANI (PAA)/Ag	Ab_2-HRP-AuNPs	CV in H_2O_2	Salbutamol	Swine feed and pork	Competitive assay	[51]
GCE/AuNPs/poly-TTCA/PAMAM/CdS/Ab	Ag-Hyd	CV and chronoamperometry in H_2O_2	Chloramphenicol	Beef, chicken, and pork meat	Covalent immobilization between carboxyl groups of CdS and amino groups of antibodies	[40]
SPCE/MB-Protein G-Ab_1	Ag-phage-borne peptides /antipeptide-HRP	SWV pyrocatechol as substrate	Molinate	River water	Competitive assay	[31]
SPCE/CMC/sodium alginate/MWCNTs/Ab-HRP	HRP incorporated with one antibody	CV-change in charge transfer in THI and H_2O_2	*E. coli* O157:H7 *Enterobacter sakazakii*	–	Multiplexed	[70]
SPCE/MBs/Protein G/Ab_1	Ab_2-HRP	Amperometry in HQ and H_2O_2	Tetracycline antibiotics residues	Milk	Competitive assay	[45]
SPCE/ERGO/Ab-HRP	HRP in electrode	CV in THI and H_2O_2	*Enterobacter sakazakii*	–	Label is immobilized on the electrode	[75]
tTG-SPCE-MWCNT-AuNPs	Ab_2-AP	Anodic stripping CV after addition of 3-IP/Ag^+	Tissue transglutaminase	Human sera	AuNPs facilitate the reduction of silver	[124]
SPCEs/MBs-Ab_1	Ab_2/AuNPs	DPV	*Salmonella*	Skimmed milk	–	[72]

(continued overleaf)

Table 9.1 *(continued)*

Sensing layer	Transduction approach	Electrochemical technique	Analyte	Sample	Comment	Reference
SPCE-OA	Ab/antiAb-AP	DPV in NP substrate	Okadaic acid	Mussels	CV and EIS for control of layers	[111]
					Immobilization of antigen through a diazotization-coupling reaction Competitive assay	
SPCE-MB-Prot G-Ab	OTA-HRP	SWV in pyro-catechol/ H_2O_2	OTA	Red wine	–	[102]
SPCE/MB-Prot G-Ab_1	AFM1-HRP	Amperometry in H_2O_2/MPMS solution	AFM1	Milk	–	[108]
SPCE-MB-Prot G-Ab_1	Ab_2/Ab_3-AP	DPV in NP	*Phakopsora pachyrhizi*	–	To be developed for detection of soybean rust	[77]
SPCE/MWCNTs/ AuNP	Ab-AP	CV for silver oxidation 3-IP and silver nitrate	Celiac disease (antigliadin and antitTG)	Serum	Dual electrode	[143]
					In situ electrodeposited AuNPs Silver-enhancement Multiplexed (spatial separation)	
SPCE/PAH-MWCNT	CuS, CdS, or PbS nanocrystals are tagged to the three specific Abs	SWASV in a buffer containing Hg^{2+}	*E.coli, Salmonella, Campylobacter*	Milk	Multiplexed detection	[78]
					Polyaniline interacts electrostatically with CNTs GA bind amino groups and Ag PAH avoids CNTs aggregation	

SPCE/MWCNT-SA/Ab-HRP	HRP in electrode	CV in THI and H_2O_2	*Shigella flexneri*	–	SA improves the solubility of MWCNTs and acts as chemical scaffold for biomolecule immobilization	[73]
SPCE/MBs-ST/B-Ab	OTA-HRP	DPV in HQ/H_2O_2	OTA	Red wines	Competitive assay	[97]
SPCE-MBs-Prot G-Ab	Ag-HRP	SWV in pyrocatechol and H_2O_2	OTA	Red wine	Competitive assay Four-parameters logistic regression	[102]
SPCE-membrane-Ag-OVA	Ag and Ab-HRP	Amperometrý (CV) in *o*-phenylene-diamine and H_2O_2	Isoproturon	Ground soil	Three designs are evaluated, including one for integration with LFA and FIA	[30]
SPCE/MWNTs-Anti Ab/Ab	Ag-HRP	Cronoampero-metry TMB/H_2O_2	Clenbuterol	Pig urine samples	They can work well without loss of activity in 100 days Competitive assay	[58]
SPCE/MB-(Ag-DSC-cHSA)	Ag and Fab-B/ST/B-HRP	Chronoampero-metry in TMB/H_2O_2	DONl	Wheat, breakfast cereals, and baby food	Four-parameter logistic regression Eight-channel multiplexed Fab was site-specifically biotinylated from single engineered cysteine	[83]
SPCE/AuNPs-BSA-OTA	Ab-B/ea-HRP	DPV in HQ and H_2O_2	OTA	Wheat	ea-AP was also checked with NPO Four-parameter logistic regression	[96]

(continued overleaf)

Table 9.1 *(continued)*

Sensing layer	Transduction approach	Electrochemical technique	Analyte	Sample	Comment	Reference
SPCE-MB-ProtG-Ab_1	Ag-HRP	DPV (HQ and H_2O_2)	Zearalenone	Baby food samples	Competitive assay MBs are used on electrodes for the transduction step, whereas the affinity reaction is performed in a different support. Matrix effect is minimized by improved washing and separation steps	[84]
SPCE/THI/AuNP/Ab	HRP Si CNT/Ab	DPV In H_2O_2	SEB	(Soy)milk, fruit juices, pork food	HRP and TEOS can diffuse inside CNTs HRP-doped silica nanoparticles could be grown inside the CNTs	[92]
SPCE/MB-Protein A/Ab	Ag-$(CH_2)_5$-AP	DPV (in NP)	Antimicrobial sulfonamide	Honey	Eight-electrodes AP array Four-parameters logistic regression	[44]
SPCE/calomel	Fe_3O_4 nanoparticles-Ag	Stripping voltammetry in acidic solutions	*Salmonella typhimurium*	Excrements of infected hens	Acidic decomposition of the immunocomplex	[71]
SPCE-Ab	Ag-HRP	CV, EIS for characterizing	Sulfonamide antibiotics	Milk	Covalent coupling with 4-aminobenzoic acid grafted on the electrode and EDC/NHS Competitive assay Specificity for sulfonamide antibiotic residues Dual electrodes	[45]

PAMAM = poly(amido amine);PAAD = poli (amido amine) dendrimer; SPGE = screen-printed gold electrode; CMD = carboxymethylated dextran; GSH = glutation; SDBS = sodium dodecylbenzene sulfonate; THI = thionine; PAA = polyacrylic acid; NSB = nonspecific binding; SWNH = single-walled nanohorn; pDPB = poly-2,5-di-(2-thienyl)-1*H*-pyrrole-1-(*p*-benzoic acid); CV = cyclic voltammetry; THI = thionine; PANI = polyaniline; TTCA = 5,2′,5′,2″-terthiophene-3′-carboxyl acid; CMC = carboxymethyl chitosan; HQ = hydroquinone; OVA = ovalbumin; MWNT = multi-wall nanotube; DSC = N,N_2-disuccinimidyl carbonated; cHSA = carbamylated human serum albumin; ST = streptavidin; TEOS = tetraethoxysilane; SA = sodium alginate.

Table 9.2 *Examples of electrochemical immunosensors related to food analysis with label-free approaches*

Sensing layer	Electrochemical technique	Analyte	Sample	Comment	Reference
Au/dithiol/AuNPs/Ab	Impedimetric	AMOZ	Pork, shrimp, honey, egg, chicken muscle	Au-S bond is employed	[42]
Au/MBA/Ab	CV/EIS in $[Fe(CN)_6]^{3-/4-}$	Sudan I	Chili powder samples	Specific monoclonal antibodies were prepared	[117]
Au/cysteine/AuNPs/Ab	CV/EIS	Aflatoxigenic *Aspergilli*	Soybean	–	[34]
Au{GNPs/CHIT-MWNTs-SiO_2@THI}n	CV	*E.coli* O157:H7	Milk and water	CHIT-MWNTs enhances the stability of THI After four layers current does not increase	[66]
Au/PANI-AuNG/[D(n-C4)[m][PF_6]]/magnetosomes-Ab	EIS/CV	SEB	Milk	IL used as electrolyte due to its intrinsic conductivity and wide electrochemical window Immobilization is performed by LBL Magnetosomes are produced by magnetotactic bacterias	[94]
Au-HA-Ab	EIS in $[Fe(CN)_6]^{3-/4-}$	*E.coli* O157:H7	–	Ab immobilized by EDC/NHS chemistry HA immobilized through 11-mercapto-1-undecanol and epichlorohydrin Flow cell for sample inlet and outlet	[64]
Au-MB-Prot G-Ab	DPV	Okadaic acid	–	–	[113]
Au/{AuNPs/DMDPSE}n	CV EIS	Carbofuran	Lettuce, cabbage, green peppers, tomato, strawberries	Charge transfer depends on the number of layers, that are generated by LBL	[35]

(*continued overleaf*)

Table 9.2 *(continued)*

Sensing layer	Electrochemical technique	Analyte	Sample	Comment	Reference
Au/MPA/Ab	CV EIS	*Staphylococcus aureus*	–	MPA forms SAMs EDC/NHS chemistry is employed	[74]
Au/MA/Ab	EIS in $[Fe(CN)_6]^{3-/4-}$	Furazolidone	Crab, pork	Immobilization of Ab by EDC/NHS chemistry	[43]
Sputtered gold (300 nm)/MNPs /Ab	EIS in $[Fe(CN)_6]^{3-/4-}$ SPR	OTA	Wine	MNPs-Ab using carbodiimide chemistry Glutaraldehyde links amino group of thiolamine Ethanolamine was used for deactivating remaining succinimide groups	[95]
Quartz crystal Au/MPA/Ab	C_{dl} is proportional to concentration	*E. coli* O157:H7	Pasteurized milk, ground beef, fresh spinach	Capacitive immunosensors based on quartz crystal Au electrode	[69]
Pt-nanoporous alumina membrane-Ab	CV of ferrocene methanol	*E.coli*	–	Aluminum is sputtered, anodized, and etched Viability of cells is evaluated by addition of glucose	[63]
Pt/PANI/PV-SO_3^-/Ab	EIS	OTA	Corn, wheat, coffee	Electropolymerization of doped aniline in acidic media gives a conducting protonated form of PANI referred as an emeraldine salt Antibody is electrostatically immobilized	[99]
GCE/MWCNTs/PTH/AuNPs/Ab/BSA	CV and DPV	Salbutamol	Swine feed samples	Competition between Ag and Ag-BSA	[60]

GCE/MWCNTs/P-L-Arg/AuNPs/Ab	CV/DPV in $[Fe(CN)_6]^{3-/4-}$	Casein	Cheese	Milk allergies	[119]
				AuNPs by chemisorption on NH_2	
GCE/silica gel-ionic liquid-Ab	EIS in$[Fe(CN)_6]^{3-/4-}$	AFB1	Bee pollen	Ionic liquid enhances electron conduction pathway due to good conductivity	[103]
GCE/SiSG-Ab	CV/EIS $[Fe(CN)_6]^{3-/4-}$	Carbofuran	Cabbage and lettuce	Sol–gel entrapment of Ab strategy	[32]
GCE/MW-OEG/FDMA/epitope	SWV	Enrofloxacin	Skimmed milk	Displacement assay	[50]
			Stream water	Molecular wire is oligo (phenylethynylene) with ferrocene	
GCE/4-aminophenyl and phenyl/SWNTs/FDMA/Ag	SWV	Endosulfan	Water	The rigidity of the SWNTs is essential to allow access of the antibody to the surface epitope without hindrance from the surface	[36]
GCE/GS-Nf/THI/PtNPs/Ab	CV EIS	Kanamycin	Animal derived foods	Electroactivity of THI is greatly enhanced with GS and Pt LBL immobilization methodology	[49]
GCE/AuNP/TU/AuNP/Ab	CV EIS	Carbofuran	Cabbage	Immobilization through Au-S bonds and the Ab by LBL	[33]
GCE/AuNPs/PB-MWCNTs-CTS-Prot A-Ab	CV EIS	Carbofuran	Cabbage and lettuce	Noncovalent immobilization	[34]
Graphene/poly DPB functional conducting polymer/AuNPs/ionic liquid/GCE	DPV in $[Fe(CN)_6]^{3-/4-}$	Microcystin-LR	Water	AuNPs are electrodeposited on the graphene modified electrode	[90]

(continued overleaf)

Table 9.2 *(continued)*

Sensing layer	Electrochemical technique	Analyte	Sample	Comment	Reference
GCE-polymer-Ab	SWV	Atrazine	–	Monomer has hydroxyl (electropolymerization), quinone (transduction), and hydroxyatrazine (bioreceptor) groups	[28]
GCE/PTBA-AuNPs-GR/Ab	EIS in $[Fe(CN)_6]^{3-/4-}$	Avian leucosis viruses	Avian serum	Covalent bond between the boric acid group of PTBA and the glycosil of Ab	[81]
GCE/MWCNTs-THI-CHIT/GA/Ab	CV EIS	Chlorpyrifos	Cabbage, lettuce, Chinese chives	CHIT has excellent film forming and adhesion ability and can dissolve MWCNTs adequately	[37]
WGS/PB-CHIT/NPG/GCE	CV, DPV, EIS in $[Fe(CN)_6]^{3-/4-}$	Kanamycin	Pork meat	Ultrathin nanoporous gold membranes with infinite amount of pore channels are employed	[48]
GCE Ph-NH_2/SWNTs/PEG/FDMA/endosulfan	SWV	Endosulfan	Environmental water	GC was derivatized with mixed layers of 4-aminophenyl/phenyl PEG aims to avoid protein adsorption Endosulfan is attached to ferrocene by amide coupling Vertically aligned CNTs are employed Displacement assay	[36]

Ab/4-AP/graphene-SPE	DPV in $[Fe(CN)_6]^{3-/4-}$	ß-lactoglobulin	Cake, cheese snacks, sweet biscuit	Graphene modified SPEs were directly purchased Grafting is performed on graphene by reduction of 4-nitrophenyldiazonium cations and nitro groups to amines. Activation using GA	[120]
HRP-Ab-MWCNT-[BMiM]PF_6-NaDDBS/THI-SPCE	CV-change in charge transfer	*Enterobacter sakazakii*	Proposed for infant formula	THI is chemisorbed by carboxylic MWCNTs	[76]
SPCE/Ab	CV EIS	Okadaic acid	Mussels	Ab is attached through EDC/NHS to a 4-carboxyphenyl film generated from the diazonium salt	[112]
Carbon paste electrode-PNTs-Ab	CV in $[Fe(CN)_6]^{3-/4-}$	*E. coli* O157:H7	–	Carbon paste is made by sol–gel methodologies	[68]
Ab-sol–gel derived nano-ZnO/ITO	CV, DPV, EIS $[Fe(CN)_6]^{3-/4-}$	OTA	–	ZnO increases electroactive surface of the electrode	[100]
GR/conducting polymer/AuNPs/IL	CV, EIS in $[Fe(CN)_6]^{3-/4-}$	AFB1	Food samples	Several ionic liquids were evaluated 1,3-di(isobutyl) imidazolium bis(trifluoromethanesulfonyl) imide is dispersed in chitosan	[104]
Paper/Ab-CNTs (in PSS-water solution)	Chronoamperometry in $[Fe(CN)_6]^{3-/4-}$	Neomycine	Milk	Filtration paper strips were dipped into coating solution (LBL procedure) and then freeze-dried to minimize denaturation of the antibody	[47]

4-AP = 4-aminophenol (also pAP); GSH = glutation; MBA = mercaptobenzoic acid; DMDPSE = 4,4′-thiobisbenzenethiol; MPA = 3-mercaptopropionic acid; PTH = polythionine; MNP = magnetic nanoparticles; MPA = 3-mercaptopropionic acid.

Among the components used, some are added to improve the detection and generate a secondary transducer (e.g., conductive polymers or nanomaterials on primary transducers). Sometimes they can have an important role in both the detection and the immobilization of the recognition element. For example, nanomaterials have adsorptive and electrocatalytic properties that are very advantageous for the development of immunosensors [144], and many are simultaneously employed for improving the immobilization of antibodies (taking advantage of the high surface-to-volume ratio) as well as for improving the electron transfer. In many cases they are functionalized to incorporate active groups that allow proper immobilization. This is the case, for example, of carbon nanotubes (CNTs), which can improve the electron transfer and at the same time favor immunoreagent immobilization by adsorption. In this section, those cases where modification of the transducer is made will be considered, with special focus on nanomaterials.

Since the explosion of research in nanotechnology, the incorporation of *nanomaterials* to the sensing element is very common. Depending on the assay format, sensitivity commonly needs a high loading of immunoreagents. To obtain more amount of antibody matrix on electrode surface, covering with high surface-to-volume ratio materials is chosen by many researchers. A great diversity of nanomaterials can be used nowadays, including 0D materials such as metallic (commonly gold) or magnetic nanoparticles, 1D as in the case of nanotubes, nanofibers, or nanowires, CNTs being the most common, or more recently a 2D carbon material (graphene). Although carbon can also be found as a 0D material, as in the case of fullerene, its use has not widespread in immunosensing. CNTs, first synthesized by Iijima [145], who prepared carbon structures consisting of needle-like tubes, each needle comprising coaxial tubes of graphitic sheets, have found great applicability in immunosensing.

As a novel class of nanomaterial, graphene, prepared by Novoselov and Geim [146], and its derivatives have attracted increasing attention, being regarded as the next generation because of remarkable electronic, optical, thermal, chemical, and mechanical properties. Graphene is the constituent of fullerenes (by wrapping up), CNTs (by rolling), or graphite (by stacking). The most widely accepted structure is that of an extended graphene sheet (GS) decorated with hydroxyl and epoxy functional groups on the basal plane with carboxyl and carbonyl groups at the edges. This GS, a monolayer of carbon atoms packed into a close, honeycomb two-dimensional lattice, also tends to form irreversible agglomerates or even restacks to form graphite through strong $\pi-\pi$ stacking and van der Waals interactions [147]. Graphene-modified electrodes usually have been obtained by reduction of either chemical (chemically reduced graphene oxide (CRGO)) or electrochemical (electrochemically reduced graphene oxide (ERGO)) oxides. The last one features controllability, speed, and green nature. An example of graphene modification is shown in Figure 9.6, where a triple amplification assay is depicted. The amplification agents are: (i) graphene that improves the electron transfer and increases surface area for immobilizing capture antibody, (ii) CD@Fe_3O_4 that has host–guest capability and larger free room for secondary antibodies, and (iii) GOD (glucose oxidase) that produces enzymatic amplification.

It has to be taken into consideration that a *great variety* of materials is included under the same name. For example, CNTs can be present in different forms and therefore properties and applications could be diverse [144]. Apart from variations in diameter or length, two main classes can be distinguished: multiple-wall carbon nanotubes (MWCNTs) and

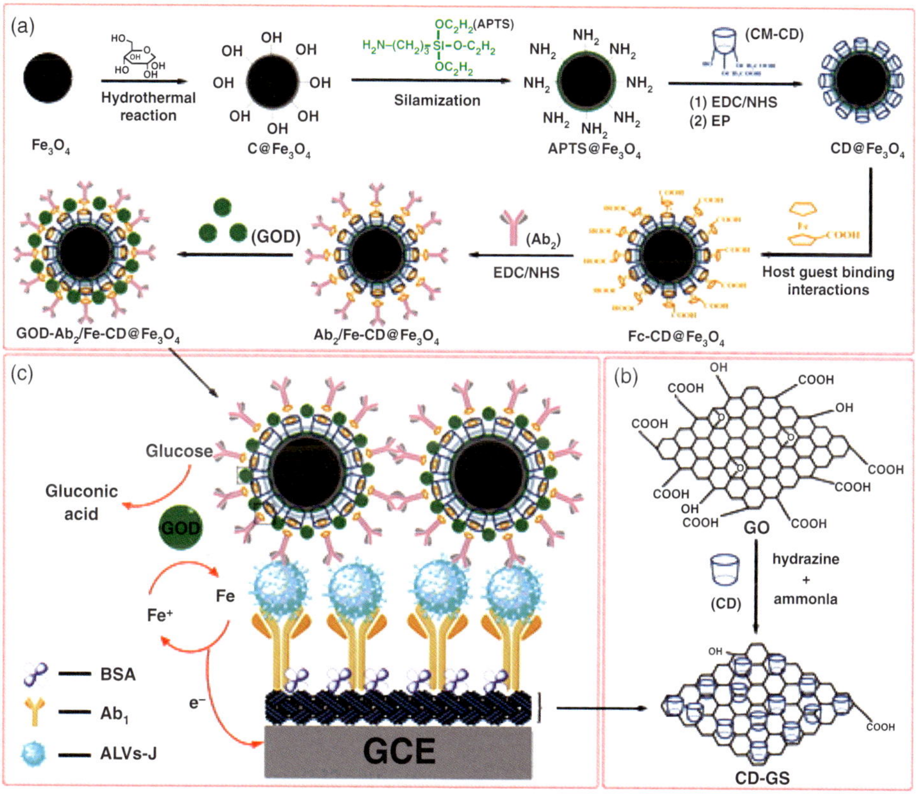

Figure 9.6 *Schematic representation of the preparation of GOD-Ab$_2$/Fc-CD@Fe$_3$O$_4$ (a), CD-GS (b), and (c) schematic description of electrochemical response mechanism (Reproduced from [80], with permission from Elsevier)*

single-wall carbon nanotubes (SWCNTs) depending on the existence of several concentric tubes or only one graphite sheet rolled up, respectively. On the other hand, electronic properties depend on the structure of SWCNTs, mainly diameter and chirality. They can be then classified into two categories: metallic (armchair) and semiconducting (zigzag or chiral) [148]. In the case of MWCNTs, "hollow-tube," "herringbone," or "bamboo" morphological variations can be found [149]. Moreover, closed- or open-ended CNTs are possible. Chemically functionalized CNTs with groups such as -COOH, -OH, -SH, or -NH_2 are available and other functionalizations are also possible. Summarizing, even when they have a very simple chemical composition and atomic-bond configuration, they exhibit the most extreme diversity in structure and in turn in properties and behavior, similarly to what happens to the rest of nanomaterials. A homogeneous reagent is not available yet and fabrication methodologies have to be directed to the consecution of this homogeneity. Depending on the procedures employed for the fabrication, oriented (forest-like or vertically aligned CNTs) or randomly distributed (spaghetti-like) nanotubes could be generated, with notorious differences in their electroactivity [150]. In some applications, specific nanotubes are used, such as in the case of vertically aligned CNTs for impedimetric determination of endosulfan [36]. Similarly, peptide nanotubes (PNTs) are tubular structures that are assembled *via* intermolecular hydrogen bonds between carboxylic acid and amido groups. They can also be modified since they already have functional groups by themselves on their surface. In addition, PNTs can be produced under milder conditions and have been proposed for use on carbon paste electrodes [68]. In the case of metallic structures, the variety is also high and nanorods, nanoparticles, nanoclusters, nanocrystals, and so on, are found in the bibliography.

Apart from the diversity of materials, their *dispersion* has to be carefully issued. The insolubility of MWCNTs in most common solvents is the major barrier for the development of well-defined MWCNT-modified electrodes. It has to be cautiously considered since the dispersing agent could affect the electrochemical response. Following with the example of CNTs, several reports deal with this issue [151, 152]. The immobilization layer is very important since a thicker composite film increases electron transfer resistance and slows down substrate diffusion. This is the reason why in many of the immunosensors revised, nanosized materials (mainly NPs and nanotubes) are included. In which refers to NPs, the use of *gold* NP*s* exceeds that of the rest. AuNPs (colloidal gold) that have been used for long also as electrochemical labels [153] result very conveniently because they can be generated either by electrodeposition on the electrode surface or alternatively, by chemical reduction from $HAuCl_4$ solutions. However, other interesting NPs such as magnetic (increasingly used), lipid, or dendritic nanoparticles could also be employed.

Binary compounds (M_xO_y), especially *magnetic nanomaterials,* are of intense interest. Since magnetic particles, not only nanosized but also at the microscale, can be used as an easy renewable surface (see also Section 9.4.3), much effort is being made in their synthesis, characterization, modification, and application. Among these nanomaterials, the cobalt-containing spinel oxides MCo_2O_4 (M = Cu, Mn, Ni, Zn, Mg, etc.) have drawn considerable attention by virtue of their physicochemical properties and potential for many technological applications such as catalysts, electrode materials, and so on. Mesoporous spinel cobaltites with large surface-to-volume ratios and uniform pore diameters are desirable especially for processes taking place at the phase boundary between solid particles and a liquid or gas phase. The conductivity of magnetic porous nanomaterials can be improved by doping noble metallic NPs into the pores, which facilitates electron communication as in the case of Ni [136], approach employed for including labels. Another inorganic material employed for including labels is the nanomontmorillonite, aluminosilicate clay mineral, abundant and very low in price [87]. In which refers to the transduction event, nano-ZnO has also been considered [27].

A special case of polymers are *dendrimers,* spheroidal nanostructures that are engineered to carry molecules encapsulated in their interior void spaces or attached to the surface. Size, shape, and reactivity are determined by generation (shells) and chemical composition of the core, interior branching, and surface functionalities. Conjugation to the dendrimer surface can give detecting agents or affinity ligands [154]. One dendrimer has several possible sites to couple active species. Since biomolecules has to be electrically connected to the base electrode, if the conductivity of the dendrimer is not good it can be increased by metal–dendrimer composites such as in the case of AuNPs. Several references to the use of dendrimers can be found in Tables 9.1 and 9.2 [40, 52, 116]. SiO_2 particles with polymer in their surface, which possesses functional groups for immobilization purposes, have also been called dendritic due to the similarities with these branched structures [38].

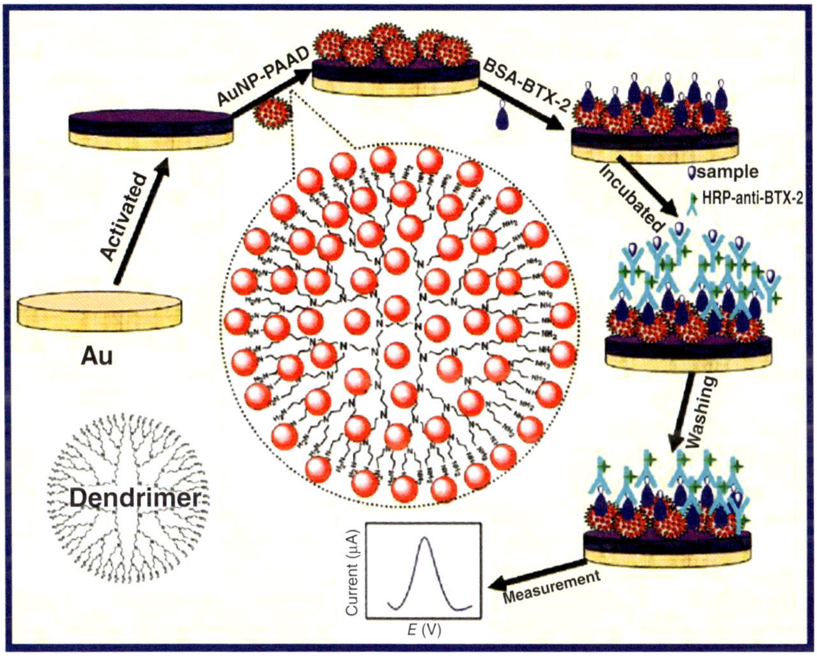

Figure 9.7 Schematic representation of the fabrication and measurement process of the competitive-type electrochemical immunosensor (Reproduced with permission from [121]. Copyright © Cambridge University Press, 2007)

In Figure 9.7, a special case of elegant branched polymers with functional groups in the surface is depicted.

On the other hand, composite materials not only possess the advantages from the component nanomaterials but also have more excellent properties due to a synergic action. They have been exploited for electrochemical sensing applications since the combination of materials enhances the individual properties. Therefore, in many cases *hybrid structures* with, for example, NPs and nanotubes or graphene, and so on, can be found. Since another chapter of this book revises the role of nanomaterials in food electroanalysis, they are not dealt with very deeply here but illustrative examples of this important trend can be found in Tables 9.1 and 9.2.

Not only nanometric structures such as tubes or particles are important but also the application of nanoporous materials such as the case of *nanoporous* gold films [48, 87] (NPG) that could be prepared by selective dissolution of silver form Au/Ag alloys under free corrosion conditions. In this case, stability of the electrode can be increased and biomolecule immobilization could happen inside the tunable pores. Due to their well-defined nanopores, anodized alumina membranes have found many applications, among them, impedimetric immunosensing [63].

Many other components are included for *stabilizing or dispersing nanomaterials,* or *providing groups for covalent immobilization* of immunoreagents. Chitosan (CHIT) is a derivative of the natural polysaccharide chitin, and has excellent film-forming and adhesion properties, together with nontoxicity and good biocompatibility [37]. Association with MWCNTs (which are adequately solved on CHIT without assembling into bundles) provides appropriate microenvironment for immobilizing molecules and promoting electron transfer. Similarly, hyaluronic acid is a linear anionic hydrophilic biopolymer (polysaccharide) commonly found in many animal tissues. Its carboxyl groups facilitate cross-linking to solid surfaces and can be used as a matrix for antibody immobilization [64]. Many other polymers such as polyarginine [119] or conductive polyaniline [78, 99] also provide groups for further immobilization procedures. Stability of the components of the final layer is very important and ionic liquids are being introduced in sensing methodologies [76, 90, 94, 103] with this aim, since they generate a favorable microenvironment. Due to their conductivity, they help in promoting electron transfer processes [59].

Other components that are found in the sensing layer are the *redox probes or mediators* of enzymatic reactions. In this way, thionine [37, 66, 76, 79, 92], which provides a large amount of hydrophilic amino groups with planar aromatic structure, allows strong interaction with MWCNTs through $\pi - \pi$ stacking forces, enhancing the electroactivity of CNTs. Mediators such as Prussian blue [53] or ferrocene [36] could also be employed.

9.4.2 Surface Engineering Procedures

Not only the components of the sensing layer are important but also the strategies of integration between them and with the primary transducer are of paramount importance. Passive adsorption is one of the simplest and most frequently used immobilization techniques, based mainly on weak noncovalent bindings. However, random protein orientation can lead to the obstruction of functional binding sites and can result in the loss of affinity or activity. Thus, covalent attachment of a protein layer on a chemically functionalized surface produces a more stable layer with correct orientation. This influences sensitivity and specificity of the immunoassay as density of immobilized protein can be better controlled and nonspecific adsorption may be decreased [155]. On the other hand, covalent procedures are usually longer and more tedious and are less justified when disposable surfaces can be used.

Among those procedures based on weak noncovalent binding, it has to be highlighted this based on *layer-by-layer* (LBL) generation. The LBL method is seen as a very promising tool for the fabrication of nanostructured films with high organization at the nanoscale level. In most cases, the adsorption is based on the attraction of opposite ionic charges located on polyelectrolyte molecular structures (Figure 9.8). For the growth of multilayer films, the adsorption occurred as a result of alternating the immersion process of a solid substrate in cationic and anionic polyelectrolyte solutions. The excess of weakly adsorbed polyelectrolyte is washed in with a solvent solution and dried. Also biomolecules are used as components for film growth, and different roughness, thickness, or porosity can be controlled by changing experimental conditions [156]. This methodology is used in many of the examples given in Tables 9.1 and 9.2 [33, 35, 47, 49, 94]. For example, combination of AuNPs/GO (graphene oxide) performed to increase conductivity and biocompatibility is carried out by electrostatic interaction using positive poly-diallyldimethylammonium chloride (PDDA) as a linker [53].

Affinity interactions are also employed, especially through Protein A [34, 44] or Protein G [31, 45, 77, 84, 102, 108, 113]. Anti-anticapture antibodies [58, 82] or the known

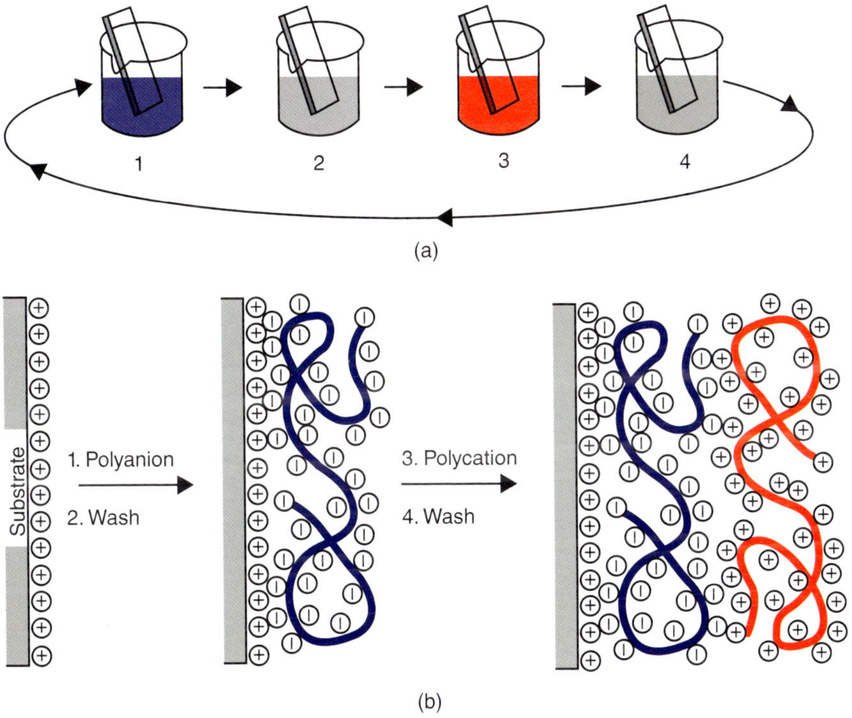

Figure 9.8 *Scheme of the film deposition process using slides and beakers. Polyanion and polycation deposition (1 and 3, respectively) are followed by washing steps (2 and 4). First two adsorption steps of $(A/B)_n$ procedure (Reproduced with permission from [156]. The Royal Society of Chemistry)*

interaction avinid–biotin [97] are used. Although noncovalent in nature, multiple bonding gives high strength to this type of interaction.

Bioactive molecules could be entrapped on *sol–gel* matrices [27, 32, 103] and also in or on electrically *conductive polymers,* approach that has been extensively explored in an effort to provide intimate contact between elements [28, 38, 46, 99]. The latest fits perfectly with electrochemical detection since the electrode generates polymer (that in turn is conductive) from the respective monomers. Immobilization can be achieved by three methods [139]: (i) entrapping biomolecules in an electropolymerized film formed with a mixture of monomers and biomolecules, (ii) electrostatic binding of biomolecules directly onto specific groups generated on the substrate surface, and (iii) covalent attachment of the biomolecules to an electropolymerized polymer film carrying pendant-reactive groups such as amino or carboxyl group. In this case, minimum physisorption, reducing the likelihood of leaching of the captured biomolecules, is produced.

An important approach that should be considered is that of *self-assembled monolayers* (SAMs), developed by Bain and Whitesides [157], very common between gold substrates and sulfur compounds. The Au-S bond could be considered almost covalent. Thiols having functionalities at the opposite end can be self-assembled for allowing further biomolecule immobilization [42, 123] (see examples in Figures 9.11 and 9.12). One of

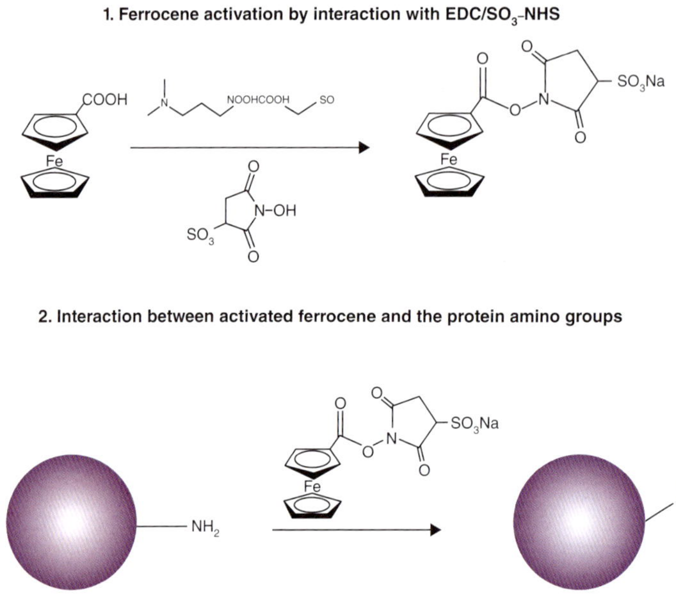

Figure 9.9 *Schematic representation of the bioconjugation following the EDC/NHS chemistry exemplified through the labeling with ferrocene*

the main applications of SAMs is their usage as blocking agents when gold electrodes are used. They occupy free sites in an ordered manner avoiding nonspecific adsorptions, but the length should be carefully selected for allowing electron transfer.

The *covalent* attachment is widely used in immunosensors, being the most common this based on the EDC/NHS (1-ethyl-3-(3-dimethylaminopropyl)-carbodiimide/N-hydroxysulfosuccinimide sodium salt) chemistry (examples reported in Tables 9.1 and 9.2 [38, 43, 45, 46, 64, 74, 80, 88, 89, 105, 112, 123]), the chemistry of carbodiimide. This is used for binding molecules with carboxy and amino terminal groups, and proteins possess both. As can be seen in the scheme of Figure 9.9, where an example of bioconjugation with ferrocene is represented, EDC activates the carboxylated molecule, which in this case is ferrocene (through the derivative ferrocene monocarboxylic acid) but any carboxylated substrate is also suitable. The succinimide ester (in this case the soluble form, sulfosuccinimide) favors the interaction with amino groups forming an amide bond. In the context of covalent attachment, the use of cross-linkers such as glutaraldehyde [59, 78, 87] also has to be considered, where two ending aldehyde groups allows acting as a bridge molecule.

Apart from the specific recognition, *nonspecific adsorption* is an important factor that should be considered when engineering electrode surface since it increases considerably background signal. As was commented before, SAMs were used as blocking agents but also other molecules, commonly proteins, are included in the immunological procedures. The most common is bovine serum albumin (BSA) but others such as casein, gelatin, or more complex matrices such as skimmed milk or newborn calf serum are also employed.

9.4.3 Renewable Solid Surfaces

A special and important trend in immunoassays is the usage of renewable solid surfaces. Magnetic (micro or nano) particles or beads (MBs) are becoming more used (some examples can be revised in Tables 9.1 and 9.2). This is mainly due to the possibility of using a different surface for interaction and transduction. The last one can be reusable meanwhile the solid surface where immunointeraction occurs can be disposed. Magnetic particles provide a high area for interaction with enhanced kinetics since stirring is possible. Once the interaction took place, they can be removed from the media with the help of a magnet. Furthermore, transfer to the conductive surface (in case the magnet is not located under the electrode) allows transduction of the immunological event. Once measurement has been recorded, they can be released by separating the magnet. Even when they are not real sensors since different surfaces are employed for recognition and transduction, scientists tend to call them magnetoimmunosensors. Their use has been well accepted in the scientific community and the number of publications that employ them increases continuously.

This special surface consists of a paramagnetic or superparamagnetic (depending on the size: μm or nm) core (being magnetite, Fe_2O_3, the most common) that is surrounded by a polymeric outer layer suitable for functionalization. Their advantages can be summarized as follows: (i) they are similar in size to that of biomolecules and they can serve as solid surface for recognition events, (ii) they possess a high surface-to-volume ratio for immobilization of different biomolecules, (iii) the kinetic of interactions is faster than that interactions are faster than those that take place in common solid surfaces since they can be stirred, (iv) they can be grouped under an external magnetic force in a fast and simple way without requiring centrifugation, (v) washings and other steps of immunoassays can be easily performed by cycles of suspension and agglomeration, (vi) they can be moved from one place to the other, either in a drop or by means of a flow following magnetic forces, (vii) they can be placed on the working electrode for performing transduction of the biological event, (viii) the possible hindrance of electron transfer by the immobilization of immunoreagents is avoided, and (ix) they can be easily removed to regenerate the transducer surface and increasing the precision. In Figure 9.10, a schematic representation of an electrochemical immunosensor in which MBs are fixed on a SPE surface after performing a competitive enzymatic (HRP (horseradish peroxidase)-based) assay for zearalenone is shown. A similar system is included in a microfluidic system with a gold working electrode [86].

MBs can suffer some modification for possessing an entity on the surface that can immobilize the immunoreagent. An affinity interaction such as this of Protein G [52, 77, 84, 102, 108] or Protein A [44] with the Fc fraction of antibodies is common but also the useful streptavidin–biotin interaction is employed [113]. Apart from this modification, magnetic particles have to be commonly treated for avoiding aggregation. To overcome this inconvenience, magnetosomes can be produced by magnetotactic bacteria [94].

A different renewable solid surface is this employed in membrane-based assays where the membrane allows the covalent immobilization of the reagents, giving rise to a more stable and reproducible assay. The membrane is brought into contact with a SPE so that the immunoassay can be directly transduced [30].

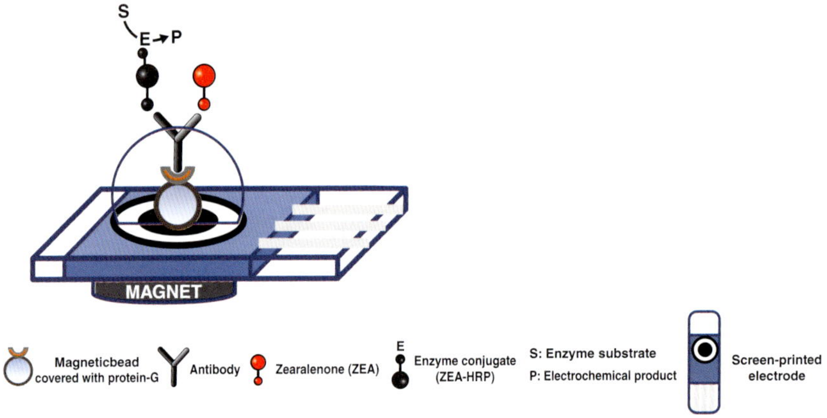

Figure 9.10 *Schematic representation of the electrochemical immunosensor strategy for the detection of zearalenone (Reproduced from [83], with permission from Elsevier)*

9.5 Performing Affinity Interactions for Molecular Recognition

Once one of the immunoreagents has been immobilized and the surface has been blocked in order to avoid nonspecific adsorptions, the immunological interaction occurs. Different formats are possible. All of them are heterogeneous in nature since a solid phase (this of the immunosensor) is used for selective analyte recognition, which is in solution. In the case where the immunological interaction produces a change in some observable property, labels are not needed and simple formats (antibody immobilized and interaction with antigen or *vice versa*) are common. In many cases, a redox probe is employed (either in solution or immobilized on the surface) for evaluating the change in its electrochemical behavior after immunointeraction. Competitive or sandwich formats (see examples in Tables 9.1 and 9.2) and even *displacement* assays [50] could be employed. The last one (example of the Figure 9.11) is based on the interaction between an antigen (that shares an epitope with the analyte) and an analyte-specific antibody. This is a case where a commonly considered negative property (cross-reactivity between structurally similar molecules) is positively used. Furthermore, interaction with the analyte (antigen that has generated the specific antibody) displaces the antibody producing a change in the response (electron transfer of an immobilized redox probe). In the example, ciprofloxacin and enrofloxacin are antigen and analyte, respectively, with similar molecular structures. The antibody is a voluminous antienrofloxacin IgM.

When a label is needed for providing the antigenic complex of an observable property that can be measured and related to the concentration of analyte (antigen or antibody), either sandwich or competitive formats are followed. In the *sandwich format* (many of the assays in Table 9.1, and the example of Figure 9.6) for determining antigens, at least two different antibodies are used: the capture antibody (immobilized on the solid surface) and the detection antibody that possess a detectable molecule. Therefore, when more antigen is present more signal coming from the label is obtained and the calibration curve (dose–response curve) shows a positive slope. Both antibodies participate in excess *versus* the antigen.

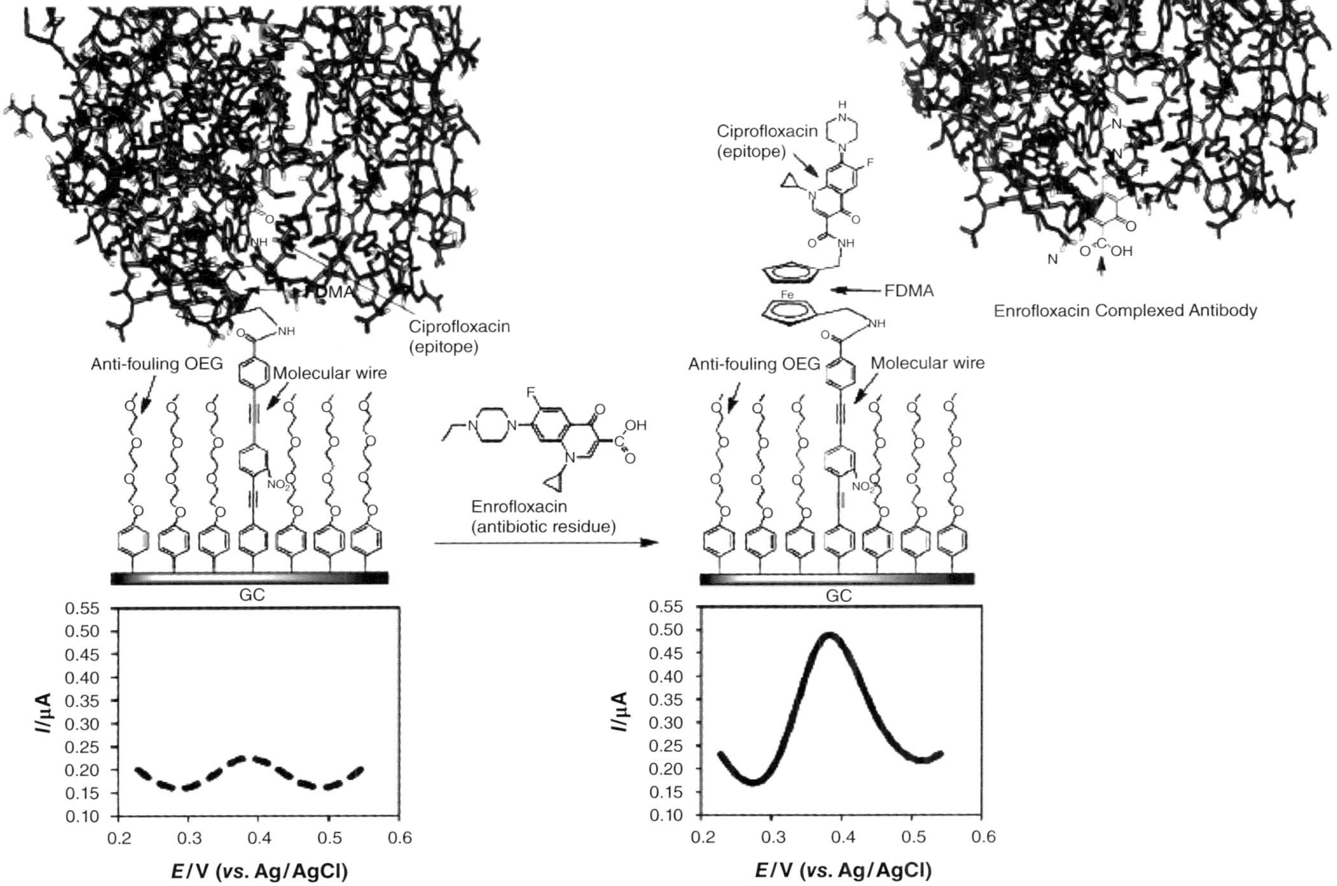

Figure 9.11 *Schematic of an immunobiosensor. The left hand side shows the electrode surface with anti-enrofloxacin IgM bound before the detection of enrofloxacin, with the corresponding electrochemical signal. The right hand side shows the detection after the antibody dissociates from the surface as a result of free enrofloxacin in solution with the expected increase in electrochemistry (Reproduced with permission from [50], copyright © 2011 WILEY-VCH Verlag GmbH & Co. KGaA, Weinheim)*

Sometimes, the detection antibody does not bring a label. In this case, this is provided by a third antibody that recognizes an epitope of the second antibody (usually an idiotype, common to a class of immunoglobulins, e.g., anti-IgG). This approach allows the use of the same labeled antibody for different immunoassays, which is very advantageous from the commercial point of view. With the same reagent a very broad spectrum of immunodeterminations can be fulfilled. This is also the case where the analyte is the antibody and the capture element is the antigen (see the example of Figure 9.12). The detection antibody is in this case specific to human antibodies (that in turn are specific to tTG). Opposite to this is the trend to simplify procedures, performing in many cases one-step (or one-pot) approaches.

The *competitive format* is preferred in many cases, especially when the antigen is small and binding with two antibodies is not possible. Two different configurations are found in the references: direct [31, 45, 53, 56, 58, 82, 97, 102, 113] and indirect [27, 51, 52, 55, 84, 88, 89, 101, 111, 116]. The requirement for a competition is that one of the reagents (e.g., the antibody) is limited and free and labeled antigens compete by the limited amount of antibody (example of Figure 9.10). Another possibility is immobilization of the antigen (example of Figure 9.7) and competition between this and the free antigen for the labeled antibody that is also in solution. In many cases, the molecule that is labeled is not "the

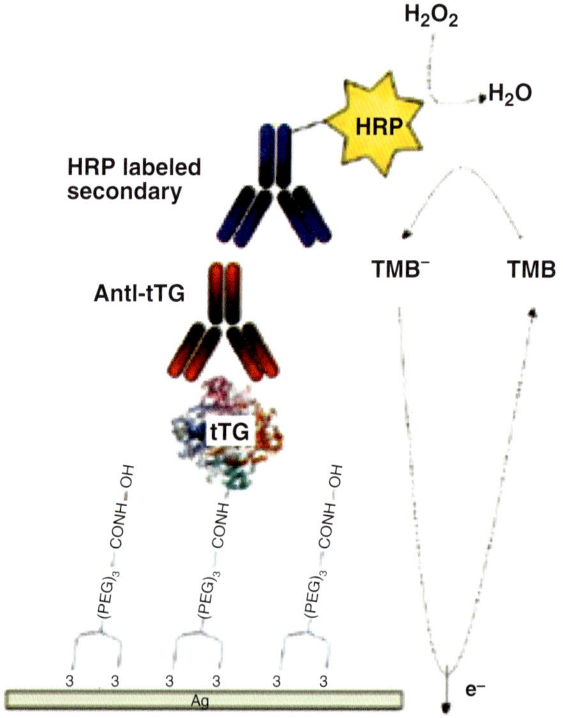

Figure 9.12 *Schematic of the electrochemical immunosensor assay architecture with 22-(3,5-bis((6-mercaptohexyl)oxy)phenyl)-3,6,9,12,15,18,21-heptaoxadocosanoic acid dithiol PEG-6 carboxylate (DT2) (Reproduced from [123], with permission from Elsevier)*

same" analyte molecule but a "similar" one, such as in the case of benzo[a]pyrene (antigen) and aminopyrene (that is the one immobilized) [38]. Competition for a limited amount of labeled antibody is then established between the immobilized and free antigen. When the analyte concentration increases, less labeled bioreagent (either the antigen in the first or the antibody in the second case) is conjugated to the solid phase and less signal is obtained. In both cases, direct and indirect formats, the signal decreases with analyte concentration and the dose–response curve presents a negative slope.

9.6 Transducing Immunological Events

After the immunological event has taken part, it has to be transduced. As was commented in Section 9.1, a sensor consists of two parts: recognition and transduction, usually very selective and sensitive. Both are commonly combined in a single device (except in the case of the so-called magnetoimmunosensors or automatized devices) with complicated architectures. Surface is carefully engineered to get better performance and the best analytical features. Therefore, the primary transducer (e.g., a SPE) is converted into a different one that could be called the secondary transducer. Connection with the primary transducer is often achieved with the help of conductive nanomaterials (e.g., CNTs, molecular wires, etc.) or polymers (e.g., polyaniline). In this section that is devoted to the important transduction of the immunological events, the main electrode materials and types, the approaches for the detection, the use of labels and label-free formats, as well as the main techniques are summarized.

9.6.1 Electrode Materials and Types

The choice of the electrode that is the final surface for transduction in electrochemical immunosensing is crucial, not only for being the foundation of the device but also for aspects such as the sensitivity of the method, the possibility of adopting different immobilization procedures, or the cost of the assay. This subject is excellently commented in an above cited review [8]. Electrodes used for this task are commonly made of metals such as gold or platinum and different forms of carbon. Among the reported examples in Table 9.1, only four electrodes are of gold [55, 116, 123], one of them is screen-printed [82]. However, in combination with label-free approaches (Table 9.2), gold is more common [34, 35, 42, 43, 52, 64, 74, 94, 113, 117]. A sputtered thin-film gold layer [95] and gold electrodes on quartz crystal [69] were also employed and some examples with platinum working electrodes [63, 99] were found. In many cases, gold is used for immobilizing thiol-containing molecules (especially those forming SAMs) taking advantage of the Au-S bond, such as in the case of DT2 (22-(3,5-bis((6 mercaptohexytl)oxyl)phenyl)-3,6,9,12,15,18,21-heptaoxadocosanoic acid) [123], cysteine [34], or a dithiol [42] as in Figure 9.12. Gold as transducer material has been known for its stability, very good electrical conductivity, chemical inertness, corrosion resistance, and also easy chemical modification for covalent molecular immobilization. However, carbon, especially in the form of GCE, is the most common one. Some sensor was constructed on graphite epoxy composite [125] or pyrolitic graphite [122] and a label-free approach on carbon paste working electrode was also reported [68]. The use of nonconventional substrates that can incorporate conductive materials together with the employ of nanomaterials opens the field to very interesting approaches, like in the case of CNTs on paper [47]. In any case, using either conventional or advanced electrodes, in labeled or label-free approaches, nanomaterials are widely employed for different purposes. CNTs, first synthesized by Iijima [145], gold NPs (known for long as colloidal gold), or more recently GSs are common components of immunosensors. For example, the electrochemical features of CNTs make them suitable for use in Faradaic processes, due to the fast electron-transfer kinetics, by the presence of edge plane graphite sites within the walls and at the ends of CNTs [158] or in non-Faradaic processes such as the large changes in conductance when CNTs, due to the presence of the cloud of electrons surrounding their walls, can accept or withdraw charges from or to molecules in their nearest chemical environment [4].

The important trend of assay parallelization (similarly to the ELISAs (enzyme linked immunosorbent assays) performed in microtiter plates) has increased the use of miniaturized electrodes and, film electrodes, especially those fabricated by screen-printed technology are widespreading. By their importance, these will be discussed in a separate section.

9.6.1.1 Film Electrodes

Conventional solid electrodes are usually included in a polymer cover (e.g. Teflon) with one end for connection to the instrument and the other as sensing surface. Although these sensors are widely used in electrochemical laboratories and in many applications that are here reported, they are not the best ones for immunosensor applications. This is because an electrochemical immunoassay usually requires several steps and various calibration solutions/samples have to be measured. This would result in the need of a high number of such electrodes or in the reuse of the same immunosensor for several times. The price of gold electrodes as reported in ref. [8] (surprisingly similar for carbon electrodes) ranges from 60 to 300€. Moreover, as these are not intended for single use, and often before each use they must be washed and cleaned very carefully, they are time-consuming. An additional drawback is that they are not very amenable to high-throughput modification as they need to be immersed in a quite large volume of solution, not less than 1 ml and they need an external reference electrode, and often a counter electrode. However, they commonly appear in food analysis bibliography because the main goal is obtaining a proof-of-concept of different immunosensing architectures, with a more specific academic interest. They should be limited to this, to evaluate new immunological approaches since they do not possess high practical or commercial use. This is a typical example where end-users, academic researchers, and product manufacturers/sellers are disconnected [159] and solutions for approaching them have to be found.

The use of films as electrodes makes possible numerous experiments that would be difficult or impractical to implement with conventional bulk electrodes. Conductive films employed as electrodes are usually classified as "thin" (thickness in the nanometer range) or "thick" (in the micrometer scale). There are numerous film fabrication methods available, depending on the material, and are included in "thin-film" and "thick-film" technologies. The most common materials for thin-film electrodes are gold and platinum metals and are deposited as a continuous film by sputtering or vacuum evaporation, commonly on an insulating substrate such as quartz, fused silica, glass, or polymeric materials. In many cases, adhesion is improved by a thin layer of an intermediate material as titanium or chromium.

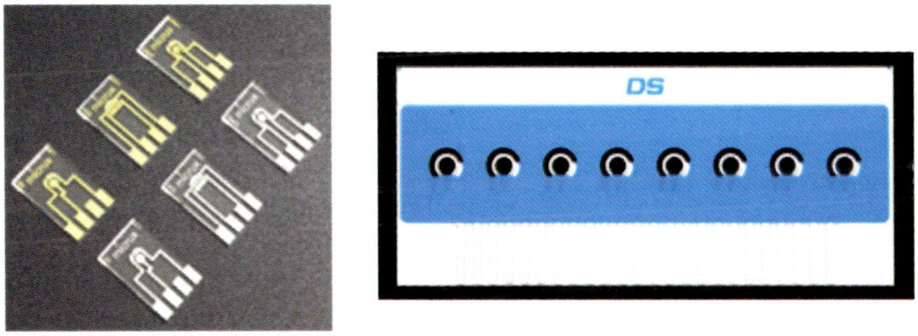

(a) (b)

Figure 9.13 Examples of thin-film (a) and thick-film (b) electrodes (Reproduced with permission from Micrux Fluidic (www.micruxfluidic.com) and DropSens (www.dropsens.com))

To obtain discontinuous films with different geometries (e.g., interdigitated arrays) or electrodes with special geometry it is necessary to combine these methods with lithography. Gold and platinum 150 nm thin-film electrodes included with pseudoreference and counter-electrodes in an electrochemical cell with a volume as small as 1–5 μl are commercially available as well as interdigitated thin-film arrays [160] (Figure 9.13a). Moreover, they can be homemade fabricated by using a sputtering facility that usually is found associated to scanning electron microscopy. However, although they constitute an inexpensive alternative, their use in food analysis applications is still very scarce [95].

Concerning to thick-film electrodes, screen-printing is the most common method used to fabricate electrochemical disposable devices. The electrodes obtained with this technology are usually called SPEs. This technology has become increasingly popular for the fabrication of electrodes and complete cells in applications where a disposable, single-use electrochemical device is desired. This avoids problems that usually arise from electrode fouling. In addition, the use of SPEs also permits handling a very small sample volume, which is always wanted when extraction procedures are employed for sample treatment [17]. Masks with desired patterns are applied to a fine screen whose mesh allows transfer the ink or other printable material that is pressed through the mesh with a squeegee blade. Layers could be deposited successively to define the required electrode. The finished assembly is then treated according to the type of electrode desired. In some cases (metal electrodes) the inks are fired at high or low temperature and in other cases (carbon films), the electrodes are ready as soon as the solvent has evaporated and the film has dried. The most common substrates used to fabricate SPEs are ceramic aluminum, polyester flexible film, and polycarbonate sheet. The usage of polymeric materials is becoming very interesting for generating flexible sensors, which are very important in diagnostics [162] and will probably be widely developed for food packaging. Metal (e.g., gold, platinum, silver, copper, titanium, or bismuth) and carbon as single material (including mesoporous carbon) or modified with mediators (Co-phthalocyanine, Meldola's blue, Prussian blue, ferrocyanide, bismuth oxide, nickel oxide, polyaniline, or nanomaterials such as CNTs and metal NPs) SPEs are available in a single or multielectrode configuration (Figure 9.13b) [161]. A pseudoreference and auxiliary electrodes are commonly screen-printed too and an insulating ink delimitates the electrochemical cell. They can also be homemade fabricated with the employ of a screen-printer machine. However, precise interelectrode measurements are required when disposable electrodes are employed and this is best achieved through mass-production that in turn allows manufacture of reproducible and inexpensive electrodes as well as mechanically robust. Other important features these electrodes exhibit are related to the miniaturization of the corresponding device along with their ease of handling and manipulation. The planar nature of the SPE allows the easy modification of the working electrode and performing the measurements in low-volume drops (e.g., 30 μl) that can be decreased to, for example, 8 μl with the employ of a special mesh [161]. Reduced reagent and sample volumes also decrease the cost of the analysis and facilitate the waste disposal. In Tables 9.1 and 9.2 many examples on their use can be found. It is expected that with the employ of more flexible substrates as paper, printed electrodes with low-cost facilities are used not only for diagnostics [163] but also for food analysis.

Multielectrode configuration allows parallelization of the assays. In this way, eight-electrode arrays are available, in a way that material employed in conventional ELISA microtiter plates (8×12 wells) such as eight-channel micropipettes could be used. This format was employed for detection of antimicrobial sulfonamides in honey [44] or clenbuterol in livestock urine [55] with 8 or 16-electrode arrays, respectively. Apart from these arrays, dual-electrode formats in which both working electrodes share the reference and auxiliary electrodes can be employed for single [45] or multianalyte determinations [143], similarly to what happens with a four-channel SPCE (screen-printed carbon electrode) design for simultaneous bianalyte determination [70] (see also Section 9.7.1).

9.6.2 Electrochemical Detection Methodologies

One of the most common and known immunoassays are ELISA tests. They are widespread due to the enzymatic amplification step, the easiness of use, the employ of 96-well plate suitable for high-throughput and multianalyte tests, the flexibility in format assay and the possibility of determination of both, antigens and antibodies. Even being very sensitive, chromogenic, fluorogenic, or electrochemiluminescent approaches suffer from several drawbacks associated with these types of measurement: requirement for generally bulky and power-intensive, light sources, detectors, and monochromators as well as false signals arising from complex colored samples. Moreover, in the case of optical methods, the well-known Lambert–Beer law implies that a minimum sample volume and path length is required to achieve certain performances. In this context, electrochemical methods appear as the most promising alternative to optical approaches for immunosensing [8]. The increasing miniaturization of electrochemical instrumentation converted the methodologies based on this principle in very appropriate ones for field detection since really portable equipment is available. Low energy consumption (devices that can be worked with the employ of batteries) and wireless connection to a computer are required for performing real *in situ* analysis. Inexpensive, hand-held devices that couple the most common forms of electrochemical analysis directly to "the cloud" using any mobile phone (allowing on-board mixing of samples, operating with a wide range of electrode formats, and transmitting data over voice) have been proposed for diagnostics use in resource-limited settings [164] and could be applied to field food analysis.

Different strategies are possible in electrochemical detection, mainly labeled (Figure 9.14a) or label-free (Figure 9.14b).

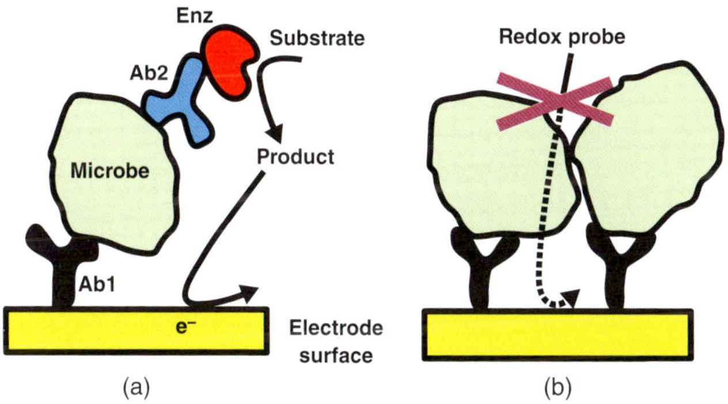

Figure 9.14 *Examples of electrochemical immunosensor formats: (a) sandwich assay (enzymatic label), and (b) direct assay (Reprinted with permission from [62], Copyright © 2013 by ESG)*

In the first approach, the antibody Ab_1 captures the target microbe, which is interrogated using the antibody labeled with enzyme. The substrate becomes converted to an electrochemically active product measured at the electrode. In the second approach, the captured microbe impedes the redox probe to access the electrode surface. Impedance (or capacitance, or decrease in the electron transfer) can be directly measured. Label-free approaches are simpler, involve a lower number of steps, and decrease commonly both, analysis time and cost. However, sensitivity usually becomes enhanced when a label (especially when amplification occurs) is involved. In next subsections, both approaches are commented in more detail.

9.6.2.1 *Label-Free Approaches*

When no label is employed, a bottleneck is how to achieve direct electrochemical transduction when there is no intrinsic charge transfer reaction following molecular recognition. There are two types of label-free strategies: *Faradaic* and *non-Faradaic*. The first one consists of measuring the electroactivity of a redox probe, either in solution or directly immobilized on the sensor surface, in such a way that its behavior could be influenced by steric hindrance of heavy molecules. The detection of the target is then performed simply by recording the redox current before and after recognition [165]. Any electrochemical technique can be used for monitoring the response, but commonly cyclic, differential pulse or square-wave voltammetries as well as chronoamperometry are employed. The ferri/ferrocyanide pair and also ferrocene/ferricinium ion are useful redox couples commonly employed for monitoring the behavior of the electrode interface. The presence of species such as antibodies or interaction products led to the formation of layers on the electrode surface. The increase in the steric hindrance, and the decrease in the electrode active sites, impede the charge transfer. On the other hand, instead of measuring the current generated by the electrochemical process of the redox probe, the resistance to the charge transfer could also be measured by electrochemical impedance spectroscopy (EIS).

Faradaic and non-Faradaic impedance spectroscopies are effective methods to probe the interfacial electron-transfer resistance at a functionalized electrode. In Faradaic impedance sensors, the interaction between a biological receptor and a target species (e.g., antigen–antibody) recruited from the solution causes a change in interfacial electron transfer kinetics between a redox probe in solution and electrode sites. This change is detectable by monitoring the charge-transfer resistance (R_{CT}), which increases as molecules bind to the electrode surface. Usually the variation of ΔR_{CT} is percentually measured with respect to a reference resistance value. Non-Faradaic impedance employs no redox probes, and impedance changes arise primarily from displacement of water and ions by molecules binding to the sensor surface. The advantages of the non-Faradaic include the elimination of the redox probe and rapid direct measurement. One potential disadvantage of impedance measurement is the higher influence of nonspecific binding. In Table 9.2, several examples of label-free approaches, most of them based on the measurement of the Faradaic impedance (evaluating the resistance to the charge transfer of the pair ferro/ferricyanide), are reported.

In EIS, a low-amplitude (5–10 mV) sinusoidal potential is superimposed on a fixed DC potential that promotes the redox reaction of the probe (Faradaic format). The sinusoidal current obtained is then used to calculate the impedance over a wide frequency range. A Nyquist plot (imaginary (Z'') *vs.* real (Z') components of impedance being represented as a function of decreasing frequency) often reveals different features depending on the nature of the electrode reaction. If electron-transfer reaction of the probe is the rate-determining electrode reaction step, the Nyquist plot will show a semicircle, and if diffusion of the probe is rate determining, a linear plot is obtained. In the former, the high-frequency region of the semicircle is dominated by double-layer capacitance and this region often shows an offset on the x-axis by a value corresponding to the magnitude of the electrolyte solution resistance (R_s). Meanwhile, the intercept of the low-frequency region of the semicircle on the x-axis represents the sum of R_s and the electron transfer resistance (R_{et}, also named R_{CT}). The last one is conveniently estimated from the diameter of the semicircle on the x-axis. When progressively more antigens bind the immunoreagent, this resistance increases, exhibiting a linear relationship with analyte concentration, often being exploited to establish calibration plots. However, several other factors could cause false-positive increase and care should be taken when applying EIS as the detection technique [166]. Computer simulations based on an equivalent circuit, usually the Randles cell (Figure 9.15), which represents the immunosensor–eletrolyte solution interface, are then performed to obtain fitted Nyquist plot and extract electrical parameters responsible for the impedimetric change.

The components of the circuit are as follows: R_s, Z_w (Warburg impedance resulting from the diffusion of ions from the bulk electrolyte to the electrode), C_{dl} (double-layer capacitance, mainly dependent on the electrode area), and R_{et}. Since R_s and Z_W represent bulk properties of the electrolyte solution and diffusion features of the redox probe in solution, they are not affected by physicochemical transformation occurring at the surface of electrode. Only C_{dl} and R_{et}, which can be estimated, depend on the surface transformations.

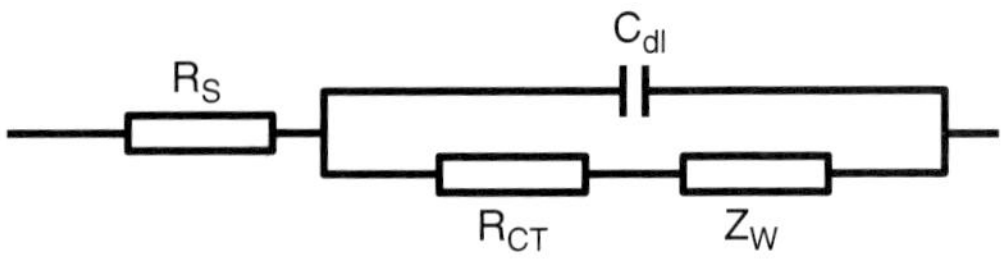

Figure 9.15 *Equivalent circuit corresponding to the Randles cell*

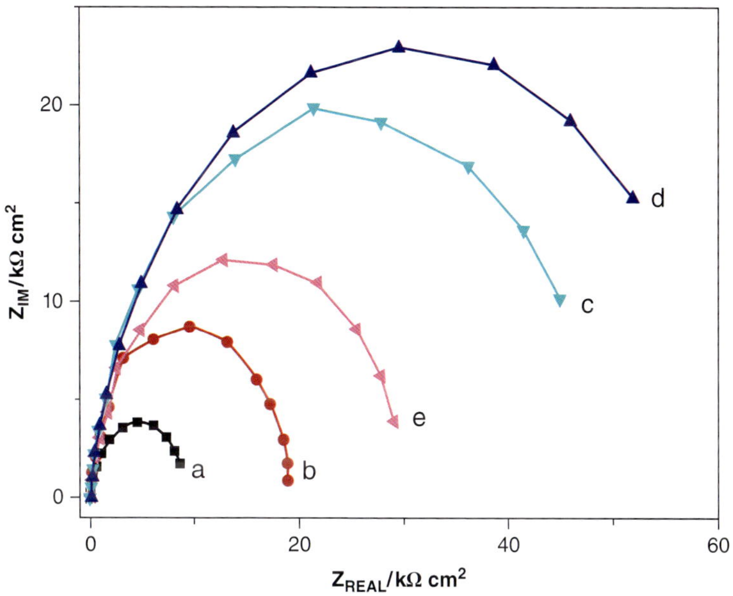

Figure 9.16 *Nyquist diagram for the Faradaic impedance measurements corresponding to the electrodes: (a) Au; (b) Au/TA; (c) Au/TA/GA; (d) Au/TA/GA/BSA; and (e) Au/TA/GA/BSA/Ab-MNP (Reproduced from [94], with permission from Elsevier)*

Figure 9.16 shows the Nyquist plot for the determination of OTA based on the use of functionalized magnetic NPs [95].

Similar to EIS, SWV (square-wave voltammetry) is another frequency-dependent electrochemical technique that could also be used in label-free Faradaic immunosensing [167]. In this case, a train of potential pulses is superimposed on a staircase potential signal with the latter centered between a cathodic pulse and an anodic pulse of the same amplitude. During each cathodic pulse, the analyte diffuses to the electrode surface and it is immediately reduced. During the anodic pulse, analyte that was just reduced is reoxidized. The current is sampled just before and at the end of each pulse and the current difference between these two points is then plotted against the staircase potential in a SW voltammogram. A linear potential scan in SWV is faster than EIS record and a familiar peak-shaped signal is more easily interpreted.

Capacitive immunosensors [69] (ECS, electrochemical capacitance spectroscopy) are based on the measurement of capacitance, instead of resistance. An electrode immersed in an electrolyte solution can generally be described as a capacitor due to its ability to store charges. For a given potential the electrode will possess a charge q_m and the solution charge q_s, where q_m is equal to $-q_s$. Charged species and dipoles will be oriented at the electrode/solution interface, generating the electrical double layer, with a capacitance, C_{dl}. It can be said that EIS and ECS are sustained on the same mathematical principle, measurement of interfacial electrical transfer function $Z(\omega)$ (where ω is the angular frequency) [168]. However, in analyzing the corresponding transfer function, the signal is mathematically converted to $C(\omega)$. The electroanalytical signal is then capacitive-based (charge storage) and not resistive-based (redox impediment). In capacitance measurements, this is plotted against frequency. A change in the double-layer capacitance is produced by the immunological interaction and can be the basis of determinations.

Apart from the possibilities of label-free determination, impedance and capacitance measurements are usually employed for characterization purposes of surfaces later applied in labeled amperometric approaches (see comments inTable 9.1).

Some special devices are constructed using more advanced strategies such as the generation of a PDMS chip that integrates a nanoporous alumina membrane in which antibodies against *E. coli* O157:H7 are immobilized. MBs are preconcentrated on this area with the aid of a magnet and after interaction, magnetic field is off. The impedance changes between two platinum electrodes situated at both sides of the chamber. The sensing surface in this case does not contact the electrodes directly, they are in close proximity [67].

9.6.2.2 *Amperometric Response to Labels*

Amperometry relies on the measurement of the current generated by the reduction/oxidation of a molecule on an electrode surface at an applied potential. This can be maintained at a fixed value (amperometry, current plotted *vs.* time) or scanned with time (voltammetry, current plotted *vs.* potential). Cyclic voltammetry is mainly used for characterization, while differential pulse voltammetry (DPV) or square-wave voltammetry (SWV) are usually employed to sensitively quantify analytes. Chronoamperometry is becoming increasingly used due to the instrumental simplicity added. A fixed potential at which oxidation or reduction occurs is applied to the electrode and potential scan is not required. As antibody and antigen are intrinsically unable to act as redox partners, for being electrochemically inert in most of the cases, a label that aids in generating an electrochemical signal is often conjugated to either the antibody or the antigen. The signal is then quantitatively related to the analyte after performing the immunointeraction.

Both antibodies and antigens can be labeled. In case the analyte is a small molecule that cannot interact with more than one antibody (due to the presence of only one epitope or to steric hindrance) the antigen (analyte) is labeled and a competitive assay in which both free and labeled antigen compete by a limited amount of antibody is performed. When the antibody is labeled, sandwich or competitive (free and labeled antibody compete by a limited amount of antigen) approaches are commonly employed. In these cases, the antigen is higher in size and possesses more than one binding site.

The labeling of an immunoreagent, antibody, or antigen, with species of powerful analytical properties, such as enzymes and nanomaterials, is a common strategy to produce and amplify signal readout in immunoassay. The labeling protocols in immunoassay can be divided into three [169]: (i) the detection antibody is linked to one label to form a stable molecular conjugate. This protocol is simple and well established. However, the label load is rather limited. Covalent and noncovalent binding is possible, although higher stability of the conjugate is obtained with the covalent approach; (ii) The label and the detection antibody are coimmobilized on the surface of a performed nanosized matrix, for example, metal NPs, CNTs, and so on. This protocol improves the label load. One of the advantages of high surface-to-volume ratio materials is that they allow multipoint attachment of biomolecules to surfaces and reduces protein unfolding resulting in enhanced stability of the biomolecule; (iii) The label is entrapped or encapsulated in a porous or hollow matrix: silica NPs, liposome, and apoferritin. The large interior space of these matrices allows immobilization of a large amount of labels.

Referring to the use of labels, there are two main types of assays: direct or indirect. In the first one, an electroactive molecule is conjugated to the immunoreagent meanwhile in the second one an *enzyme* is attached. This protein converts a substrate into an electroactive product that subsequently undergoes a redox reaction upon applying a suitable potential. The use of enzymes has enormously widespread (especially horseradish peroxidase and alkaline phosphatase (AP/ALP)), mainly due to the sensitivity improvement by the inherent amplification: one enzyme molecule can convert many molecules of substrate into product. HRP catalyzes the reduction of inorganic peroxides, commonly H_2O_2. Transfer of electrons occurs from a substrate (cofactor) that in turn is converted into the oxidized form. The reduction current of this product can be quantitatively related to the amount of enzyme. Since it is conjugated to an immunoreagent, it can be finally related to the analyte. Many molecules have been employed as substrates: 3,3′,5,5′-tetramethylbenzidine (TMB) [55, 83, 101, 122, 123], pyrocatechol [31, 102], *o*-phenylenediamine (*o*-PD) [58, 88, 116], 5-methylphenazinium methylsulfate (MPMS) [108], 4-*tert*-butylcatechol (4-TBC) [86], hydroquinone [27, 38, 59, 84, 96, 97], or thionine [70, 73, 75]; these are added to the solution of detection or are coimmobilized in the sensing layer. The back electrochemical reduction is in some cases detected at very advantageous low potentials such as 4-TBC at 0.0 V on gold electrode [86]. On the other hand, AP or ALP converts orthophosphoric monoesters into alcohols. The enzymatic product is electroactive and the current of the electrochemical process can be related to the analyte. Various phosphate esters can act as substrates: mainly α-NP [44, 77, 111] or *p*-nitrophenylphosphate (pNPP) [113], and also 3-indoxyl phosphate (3-IP) together with silver cations [124, 143, 170]. In this case, AP hydrolyzes 3-IP resulting in an indoxyl intermediate that will reduce silver ions present in solution. Metallic silver and indigo blue are produced. Thus, the silver enzymatically deposited on the electrode surface (SPCE with a nanohybrid structure of MWCNTs and electrodeposited AuNPs) can be detected through the oxidation peak of silver when an anodic scan is carried out. An example of an immunoassay that follows this principle is shown in Figure 9.17. In this case, the electrochemical signal of the enzymatic product is not recorded but the oxidation of the electrodeposited silver, which enhances the sensitivity. A special case is the immobilization of a labeled (HRP) antibody, producing the immunological interaction the obstruction of a channel between the active center of the immobilized enzyme and the substrate (thionine) in solution [75].

Although any enzyme that generates an electroactive product could be used, those (HRP and AP) are the most common and only some sporadic cases on the use of other enzymes, such as GOD [80] (glucose oxidase) that oxidizes glucose to gluconic acid, are found (see an example in Figure 9.6). In this case, the electrochemical reduction of the mediator, included in a complex architecture that uses magnetic particles and host–guest interactions, constitutes the analytical signal.

Among the *electroactive entities,* much effort has been made for finding an appropriate electrochemical labeling system, similarly to what happens for other principles of detection, such as fluorescence. The number of commercial markers for fluorescent methodologies is enormous; however, providers that offer kits for electrochemical labeling were not found. Electrochemical labels could be classified depending on their nature as organic

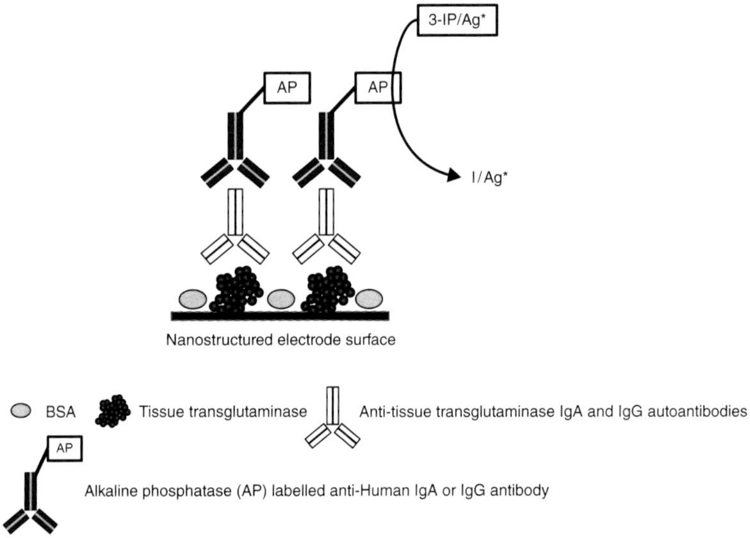

Figure 9.17 *Schematic representation of an immunosensor architecture with AP enzymatic labeling (Reproduced from [124], with permission from Elsevier)*

or inorganic, with organometallic compounds in between. Attempts have been made for conjugating the antibody to different organic molecules looking for the highest sensitivity. There are two main functions of this label: reporter (species that suffer an electrochemical process that can be related to analyte) or catalyst (molecule that produces a catalytic current from a redox probe, usually added to the detection solution). In the last case, the labeling with molecules such as hydrazine has to be taken into consideration, which serves as a catalyst for the reduction of hydrogen peroxide. The catalytic current monitored could be related to the analyte, for example, chloramphenicol [40] or neomycin [46]. A similar approach is carried out employing Pd NPs as catalyst labels for H_2O_2 reduction in the determination of salbutamol [59]. On the other hand, many electroactive organic molecules were covalently bound to immunoreagents to follow their electrochemical processes but application to food analysis is not very relevant. Although scarce, some examples, such as this in Figure 9.18, which presents an interesting immunoassay architecture ending with catechol as label [105], can be found. Oxidation to *o*-quinone is the basis of the analytical signal.

Among the organometallic compounds, ferrocene stands as one of the most common redox probes for conjugation, employing different derivatives, such as these with hydroxyl or carboxyl functionalities. As commented before, it could also be used as enzyme mediator.

Concerning to inorganic markers, first approaches were introduced by Heineman's group using cations such as Bi^{3+} and In^{3+} as labels of antibodies. Conjugation was made through complexation and anodic stripping voltammetric detection was performed after they were released from the corresponding labeled proteins [171]. Related to metals, colloidal gold (today referred as gold nanoparticles) has been introduced for electrochemical immunosensing some years ago [157]. Moreover, the sensitivity could be enhanced by taking advantage of the ability of reducing silver and generating a shell that can be

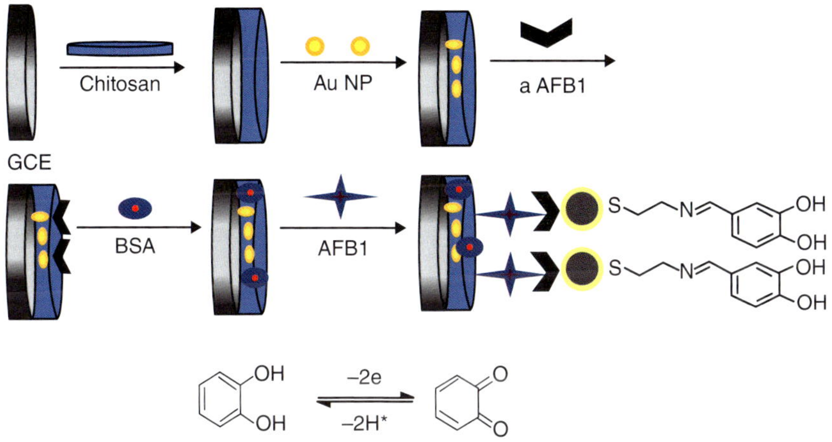

Figure 9.18 Fabrication of GCE/chitosan/AuNP/a AFB1-AFB1-a AFB1-catechol-Au-Fe_3O_4 immunoelectrode (Reproduced from [105], with permission from Elsevier)

anodically stripped off. Nanomaterials have gained enormous interest, not only as appropriate surfaces for immobilization and transduction but also as electrochemical labels (see scheme of Figure 9.19).

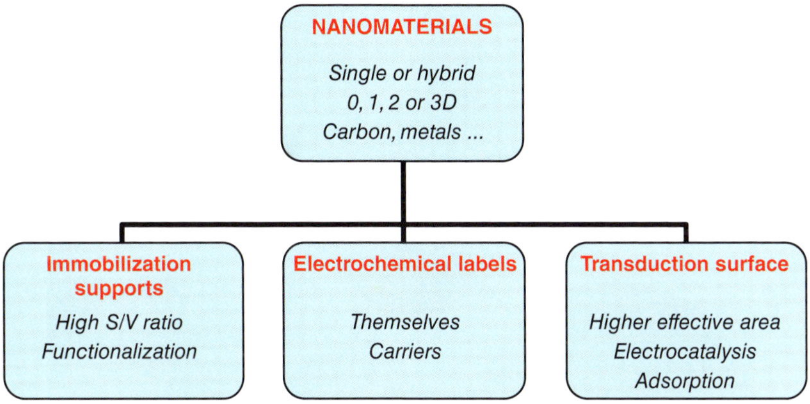

Figure 9.19 Role that nanomaterials play in food immunosensing

Related to food analysis, most of the nanosized labels are NPs, not only gold but also silver [56] and quantum dots [78], which can also be accumulated on the transducer for further stripping. CuS, CdS, and PbS have been electrodeposited after adding Hg^{2+} to the solution for recording later the square-wave anodic stripping voltammogram. The use of different labels is one of the strategies for performing multiplexed analysis (see Section 9.7.1). They can be used as carriers, providing a higher amount of label per binding event, such as in the case of carbon nanospheres for immobilizing both, the antibody and enzyme (HRP) [89], achieving biomolecules enhanced stability.

Magnetic particles are employed not only as renewable surfaces but also for introducing labels [71, 80, 142]. In conjugation procedures, several steps are usually required and purification of the bioconjugate is needed. In this context, MBs provide a perfect surface for separation of free and labeled fractions of the immunoreagent that is going to be used in one of the steps of the whole immunoassay procedures. Particles of Fe_3O_4 are the most common but other types such as magnetic mesoporous $NiCo_2O_4$ nanosheets are also prepared [142].

9.7 Advancing in Real Immunosensing

Demand of analytical information is increasing continuously as well as improvements in quality and quantity. Characteristics that have to be kept in mind are depicted in Figure 9.20. They are at the same time achieved goals and never-end objectives. Determining selective and sensitively, in a precise and accurate way, a number of analytes (multiplexed) as fast as possible with low-cost, simple, and stable miniaturized devices that are easy-to-use, with low-energy consumption requires continuous research.

On the other hand, even when field analysis is necessary in many steps of the food chain (farms, crops, etc.) and small devices with portable instrumentation are required, food processes in industries need in many cases automatization, especially when the number of samples is high (e.g., food quality). Moving the approaches from off- to at-, on-, and in-line analysis is always desirable, but all the possibilities have to be considered and all of them need innovative approaches for generating the ideal food analysis platform. In this section, some issues related to multianalyte determination and automatization are considered.

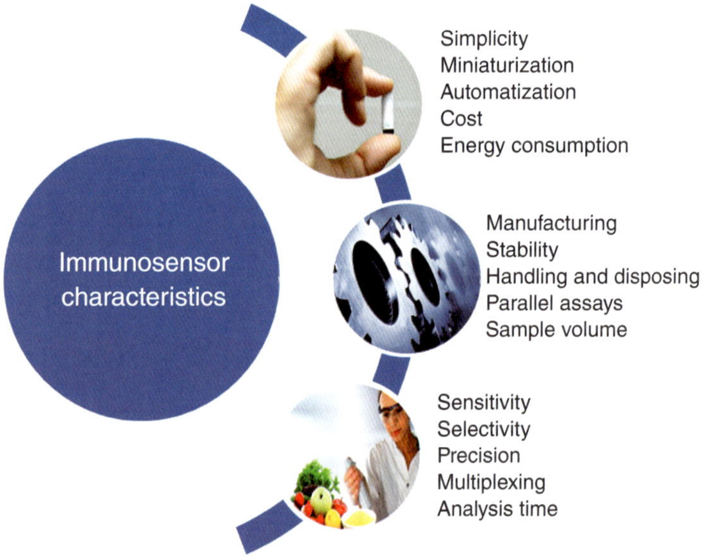

Figure 9.20 *Characteristics of an ideal immunosensor in which refers to the device, the performance, and the analytical features*

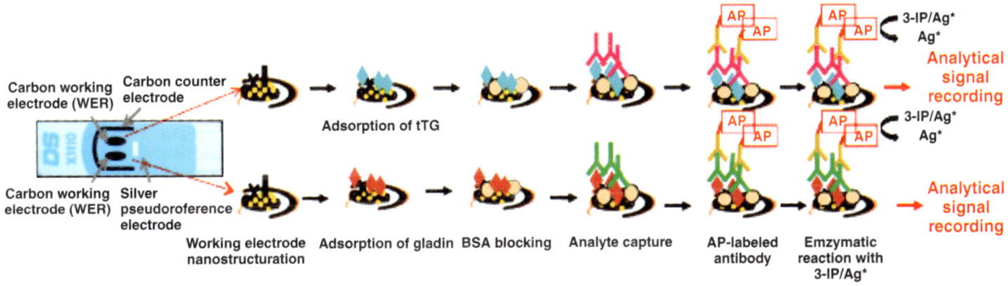

Figure 9.21 *Schematic representation of the immunosensing strategy followed for the analytical detection of celiac disease serological markers (Reproduced from [122], with permission from Elsevier)*

9.7.1 Multiplexed Determinations

The possibility of determining different analytes in the same sample (multiplexing) is of paramount importance nowadays. When dealing with multianalyte determination, different strategies could be followed [172]:

1. Determination of the *analytes as a whole,* without differentiation among them. This is the simplest assay, since single label, "one pot" assay is carried out. If there are two analytes, A and B, a positive response may indicate that there is A, or B, or both, but without differentiation between them. This is very important in the case of determining the presence of pathogens in food, where the presence of only one of them implies a health hazard. It does not matter if it is A or B; the food has to be disposed or treated in a different manner.
2. *Spatially resolved approaches,* in which the use of single labels is allowed. The simplest approach is parallelizing single analyte assays with individual reaction areas. In the case of analytes anti-A and anti-B, the same label could be employed but the capture reagents (A and B in this case) are spatially immobilized. Therefore, the signal for one analyte is independent to the signal for the other. Some of the steps (blocking, washings, incubation with secondary antibody, etc.) can be simultaneously performed. This is the strategy employed for the determination of two microorganisms, *E. coli* O157:H7 and *Enterobacter sakazakii,* whose HRP-labeled antibodies are immobilized in different modified SPCEs of an array of four [75]. In Figure 9.21, a scheme of the bianalyte immunoassay for the determination of autoantibodies for tTG and gliadin [143], both involved in the celiac disorder is shown.
   The use of SPEs, which include two different working electrodes sharing reference and auxiliary electrodes, is very advantageous because most of the steps are common for both and the same drop of solution can be employed, decreasing analysis time and costs. In case an enzyme is used, measurements have to be made before the product diffuses from one electrode to the other. In the approach of Figure 9.21, silver is generated and deposited on gold NPs that nanostructure the electrode. Therefore, cross-talking between electrodes is avoided. On the other hand, a multipotentiostat is useful for simultaneous measurement and decreasing analysis time. This strategy seems to be very interesting since designing sensor arrays is one of the actual immunosensing trends.

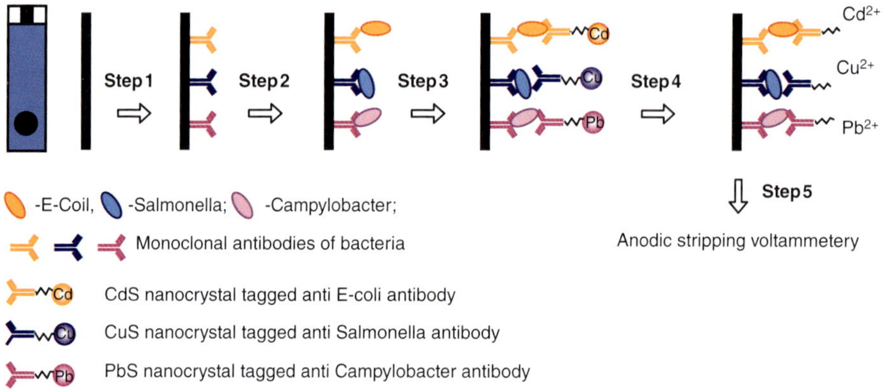

Figure 9.22 *Multiplexed detection of pathogens using NC antibody conjugates and MWCNT-PAH/SPE. Step 1: antibody immobilization; Step 2: immunocapture; Step 3: NC-antibody conjugates immunobinding; Step 4: dissolution of metal ions from NC; and Step 5: SWSV analysis (Reproduced from [143], with permission from Elsevier)*

3. It is also possible to perform multiplexed analysis with a *sequential detection,* that is to say, separation in time of the measurements. In this case, the same surface (multispecific surface) is employed for the immunointeractions of analytes, for example, A and B with their respective antibodies. However, since the same label is used, measurement for A is first taken and then a second measurement corresponding to A and B is made. The difference between both is related to analyte B. With this strategy, since two different steps have to be performed for final interactions, analysis time is increased when compared with the spatial separation approach, especially when more than two analytes are determined.
4. *Multiple labels* can also be employed for multiplexing purposes. Possibilities are enormous and new attempts are continuously made. The most appropriate is to look for labels that could be detected simultaneously and the assay complicates as the number of analytes increases. When enzymes are involved, care should be taken to maintain the activities of the molecules since detection conditions (composition or pH of the medium) as well as substrates are different. Therefore, simultaneous detection is almost impossible and enzymatic interactions then take place sequentially, increasing analysis time. One of the possibilities (Figure 9.22) is to use quantum dots as labels. CdS, CuS, and PbS are labels of antibodies specific for *E. coli, Samonella,* and *Campylobacter.* In this assay there is no spatial or timely separation and all the three labels are simultaneously measured in the same anodic stripping square-wave voltammogram, after acidic release from the biomolecule.

There is another possibility for performing multiplexed analysis. This consists of using separation-based methods. In this case, incubation of A and B with anti-A and anti-B is performed all at once, using later a separation technique. However, the incorporation of this new operation is out of the scope of the biosensors where the recognition and transduction surfaces are commonly the same or are in close proximity.

9.7.2 Automation of Immunosensors

One of the challenges of immunosensors is their automation, especially relevant in food quality, when high number of samples, or continuous food processes are involved. Even when an in-line determination is more adequate, immunosensing is still being performed, in most of the cases, as an at-line strategy. Flow injection analysis allows introducing automatically samples in a flow that includes all the steps needed for the determination. The versatility of flow analysis makes easy the adaptation to each of the immunoassay procedures. The flow is employed for both, for presenting different reagents to the immunosensor and for helping in a regeneration step. Depending on this, immunosensors can be classified as disposable or reusable. Although disposable devices are very advantageous, mass production of low cost and precise devices is required. In case this is not possible, and traditional electrodes are employed, a regeneration step, commonly made with a low-pH solution (e.g., glycine pH 2.0) that turns back immunological interactions by reversing the equilibrium, is commonly employed. This can be performed by flowing a desorption buffer on the immunosensor. Some examples are reported in the bibliography, as this of a competitive indirect assay in which free and immobilized citrinins compete for a specific antibody [109]. The electrode is glassy carbon (conventional), and the surface is nanostructured with electrodeposited gold and activated with cysteamine that allows the immobilization of citrinin labeled with chicken egg ovalbumin with the help of glutaraldehyde as cross-linker. Immobilization procedures are in many cases long and tedious and regeneration seems a solution for maintaining a complex architecture, but even when automatization makes it affordable, parallelization of assays in disposable devices reduces time considerably.

Flow cells were used for amperometric immunosensors applied to detection of microrganisms such as *Francisella, Salmonella,* or *Escherichia* in bioaerosols [62] or zearalenone in corn silage [85]. On-line determinations are also aimed through automated flow-through amperometric immunosensors such as this for okadaic acid in mussel samples employing MBs as renewable solid surface [115]. Impedimetric measurements could also be performed in flow cells such as detection of *Escherichia coli* O157:H7 [64].

Flow systems are being displaced by microfluidics [173], in which fluids are manipulated in channels with dimensions of tens of micrometers. Immunoassays can greatly be benefited from miniaturization in microfluidic systems [174]. At these small scales, fluids exhibit laminar flow, that is, fluidic streams flow parallel to each other, and mixing occurs only by diffusion. Advantages over conventional methods [175] are as follows: (i) increased surface area to volume ratio that speeds up antibody–antigen interaction, (ii) smaller dimensions that reduce the consumption of expensive reagents and samples, and (iii) automated fluid handling that can improve reproducibility and throughput. Magnetic particles have also been added to a microfluidic chip–SPCE for the detection of clenbuterol in bovine hair samples [57], OTA in apples [98], zearalenone in feedstuffs [86], or *Botrytis cinerea* in tissue of fruits [176] through a competitive direct assay between free and labeled (with HRP the three first, or AP the last one) antigens, respectively, by the immobilized antibody. Immunosensors could be then included in a microfluidic system for the development of advantageous integrated analytical immunosystems. The great advance that the development of paper microfluidics is suffering has to be taken into account, especially in which refers to versatility and low-cost applications. Although food applications are really scarce, it is expected that they will increase considerably. Since electrochemical sensors in microfluidic systems are the subject of another chapter, they will not be discussed in detail here.

9.8 Processing Data

There are some differences between the rest of analytical methodologies and immunoassay [130]: (i) there is a nonlinear relationship between response and analyte concentration so that a more sophisticated method than drawing a straight line must be used, (ii) no single unique line fits the data, so a choice must be made, which introduces the risk of bias, and (iii) relatively large assay errors occur; these errors are not constant in every region of the assay working range and in every batch, so the curve must be drawn anew for every batch and with some uncertainty. Many of these problems could be reduced by an increase in the number of calibration points and replicates. However, economics makes this decision a compromise. In curve fitting, linear interpolation is preferred, and sometimes, logarithm transformation is employed. The logit function is also commonly found in processing immunosensing data. In most of the cases, the regression model produces two parameters (slope and intercept). The four parameter logistic equation is a nonlinear regression model that introduces two more parameters: one for the assay at zero concentration and other for the nonspecific binding:

$$y = D + (A - D) \, / \, (1 + (x/C)^B)$$

where A is the response value at zero standard concentration, B is Hill's slope, which refers to the steepness of the curve (negative or positive), and C is the inflection point (since a sigmoidal curve is obtained). This corresponds to the concentration of analyte where $y = (D - A)/2$, and D is the response value for infinite standard concentration. Several examples of curve fitting employing this model are found in Table 9.1 [44, 83, 96, 102].

9.9 Conclusions

Food analysis is a very important field that involves very different applications such as *quality food processing, food and health safety, or development of new functional food among other.* Therefore, there are many possible analytes all along the food chain, from the farm to the table. It can be cited as most common residues of herbicides/pesticides or veterinary pharmaceuticals, pathogenic microorganisms, and their toxins, toxic substances such as PAHs, additives, and substances related to food allergies, intolerances as well as controllers of food uptake.

Electrochemical immunosensors are very adequate devices for food analysis, since the *enormous selectivity of the antibody–antigen system* is combined to *highly sensitive detection techniques.* Today that *bioinspired materials and devices* are increasingly intended due to the degree of perfection that millions of years have achieved, the immunological approach results very appropriate for detection of analytes in a complex matrix such as food. Moreover, immunoreagents, especially antibodies, are improving and fractions, multispecific or humanized antibodies are available. Since more attention is paid to genetic engineering and bioinspiration fields, new interesting reagents are envisaged in a near future.

The whole process of immunosensing is being improved step by step, not only choosing the best reagents but also in which refers to the *immobilization, interaction, and transduction stages.* Since the number of components used for increasing the benefits of the final immunosensor rises notoriously, as does the complexity and variety of transducer layers, a real *surface engineering or complex architecture* is required. In this context, a shared trend is the capitalization of nanotechnology and emerging nanomaterials in the majority of the new developments. A great variety of *nanomaterials* can be employed from simple one-material platforms to hybrid materials or composites that combine the properties of several. Since they are beneficial not only for immobilization of the recognition element but also for transduction (as support or labels) purposes, their use has enormously widespread. The high surface-to-volume ratio coupled with their electron transport properties are their most appreciated properties. On the other hand, due to the fact that the dimensions of nanomaterials are on the same scale as biomolecules, this emerging research is unveiling exciting possibilities for studying their interactions with biological species.

Electroanalytical methodologies allow *very different transduction possibilities. Label-free* approaches are increasing due to the inherent simplicity, but higher sensitivity and selectivity is usually achieved with *labeled strategies.* Therefore, the transduction event can be direct or indirect depending on if the event is directly transduced by a change in a property that can be measured by the transducer or indirectly with the help of a label. Amperometric detection in which a Faradaic current is measured after completion of the immunological interaction is commonly employed. In label-free approaches, differences in the electron transfer of a redox probe are commonly followed. On the other hand, since molecules involved in the immunointeraction are electrochemically inert, a label is required for converting them into electroactive entities and detecting if a binding event has occurred. Enzymes are the most common labels due to the inherent amplification but direct labeling systems are also possible. In this context, nanomaterials are increasingly employed. Enzymatic amplification-free approaches are aimed to reduce analysis time and complexity of the protocols. In this context, other sensitivity enhancer particles that can carry multilabels, such as dendrimers or nanovesicles (e.g., liposomes [177]), will probably be widely applied in the near future. On the other hand, there are some nonelectrochemical principles of detection that are based on the employ of conductive layers, and therefore they could be employed for electrochemical sensing. This is the case of surface plasmon resonance detection of ractopamine residues in pork [178].

Although conventional electrodes are usual in academic papers (e.g., GCEs), the use of *disposable electrodes,* which allow parallelization of assays, is increasing and is the way to convert academic research tools into commercial sensors. A technology that will allow a rapid production of large numbers of high-quality and inexpensive sensors is required. One potential technology is that of SPEs which can be mass-produced at low cost using existing technology, allowing their disposability. Although some of them are considered reusable, antigen and antibody have high affinities ($10^5 - 10^{11}$ M) and present multivalent interactions of high avidity making a challenge to reverse this strong interaction without rendering the device unusable by removing or denaturing the biological recognition element. Moreover, analysis time is increased by sometimes inaccurate regeneration steps performed under extremely harsh conditions. A renewable sensing component of an immunosensor is highly desirable. In this context, the employ of *MBs* results very interesting since they are an adequate substrate for heterogeneous immunointeractions and they can be disposed and renewed after transduction. *Arrays of electrodes* are very useful for making parallelization of assays easy and also for performing *multiplexed analysis. Automation of assays* is especially wanted in food process and when a high number of samples is involved.

Commercial devices are being developed at lower rate than those made for research purposes, mainly due to additional stringent requirements such as, for example, damage resistance, lower cost, or being noninvasive as well as other requirements for being approved by regulatory agencies. Moreover, real-world conditions present a number of obstacles for reliable and rapid assays of food. For example, the complex matrix (proteins, fats, carbohydrates, preservatives, and other constituents) of raw, processed, or precooked food. Furthermore, the distribution of biological and chemical contaminants are not homogeneous in food and water and unfavorable sensing environments of farms, factories, and transportation containers with variation in temperature, pressure, or mechanical, and so on, make food analysis a more complicate challenge [15]. However, the ultimate goal and challenge of food sensors is to function in field conditions, whether in agricultural settings, food processing facilities, or consumer venues. Even though these difficulties, the number of papers on food sensors is increasing, showing the *degree of maturity of these devices and their possibilities for real implementation.* It has to be taken into account that entering the market without a reasonable niche volume is difficult. Market drivers are the detection of bacterial and viral pathogens or biological weapons [2]. As toxin contamination usually occurs in trace amounts per gram of foodstuff, sensitive and accurate analytical methods for determination are highly desirable. Apart from this, their development is supposedly going to find *new and interesting applications fields,* such as this of nutraceutical and functional food analysis. The growing interest of consumers in the role nutrition plays on health is the primary incentive behind the success of healthy diets [23].

Food immunosensing is undoubtedly a growing field that will continue incorporating interesting approaches. With all the effort is being made, the future will undoubtedly bring multianalyte miniaturized immunosensor platforms that perform all testing in a fast one device-one place way with improved characteristics.

Abbreviations

Ab	antibody
Ag	antigen (also silver)
AFB1	aflatoxin B1
AFM1	aflatoxin M1
ALP	alkaline phosphatase (also AP)
AMOZ	5-morpholino-3-amino-2-oxazolidone
AOZ	3-amino-2-oxazolidinone
AP	alkaline phosphatase (also ALP)
4-AP	4-aminophenol (also pAP)
BSA	bovine serum albumin
CD	cyclodextrin
CEA	carcinoembryonic antigen
cHSA	carbamylated human serum albumin

CHIT or CTS	chitosan
CMC	carboxymethyl chitosan
CMD	carboxymethylated dextran
CNS	carbon nanosphere
CNTs	carbon nanotubes
CRGO	chemically reduced graphene oxide
CTS or CHIT	chitosan
CV	cyclic voltammetry
DDBS	dodecylbenzene sulfonate
DMDPSE	4,4′-thiobisbenzenethiol
DON	deoxynivalenol
DSC	N,N_2-disuccinimidyl carbonated
DPV	differential pulse voltammetry
DT2	22-(3,5-bis((6 mercaptohexytl)oxyl)phenyl)-3,6,9,12,15,18,21-heptaoxadocosanoic acid
ea	extravidin
EC	European community
ECS	electrochemical capacitance spectroscopy
ED	electrochemical detection
EDC	(3-dimethylaminopropyl)-*N′*-ethylcarbodiimide hydrochloride
EIS	electrochemical impedance spectroscopy
ELISA	enzyme-linked immunosorbent assay
EPA	Environmental Protection Agency
ERGO	electrochemically reduced graphene oxide
FB1	fumonisin B1
FDA	food and drug administration
FDMA	ferrocenedimethylamine
FIA	flow injection analysis
GA	glutaraldehyde
GCE	glassy carbon electrode
GEC	graphite epoxy composite
GNP	gold nanoparticle
GO	graphene oxide
GOD	glucose oxidase
GR	graphene
GS	graphene sheet
GSH	glutation
HQ	hydroquinone
HRP	horseradish peroxidase
Hyd	hydrazine
IL	ionic liquid
3-IP	3-indoxyl phosphate
ITO	indium tin oxide
LFA	lateral flow assay
MA	mercaptoacetic acid
MBA	mercaptobenzoic acid

MIP	molecularly imprinted polymer
MNPs	magnetic nanoparticles
MPA	3-mercaptopropionic acid
MPMS	5-methylphenazinium methyl sulfate
MW	molecular wire
MWCNTs	multiwall carbon nanotubes
MWNTs	multiwall nanotubes
NC	nanocrystals
Nf	Nafion
NHS	*N*-hydroxysulfosuccinimide sodium salt
NP	nanoparticle/naphthyl phosphate
NPG	nanoporous gold
NSB	nonspecific binding
NTA	N_α, N_α-bis (carboxymethyl)-L-lysine hydrate
OEG	oligo (ethylene glycol)
o-PD	*o*-phenylenediamine
OVA	ovalbumin
PAA	polyacrylic acid
PAAD	poly(amido amine) dendrimer
PAH	polycyclicaromatichydrocarbons
PAMAM	poly(amido amine)
PANI	polyaniline
PB	Prussian blue
PC	polycarbonate
PDMS	polydimethylsiloxane
pDPB	poly-2, 5-di-(2-thienyl)-1*H*-pyrrole-1-(*p*-benzoic acid)
PDB	2, 5-di-(2-thienyl)-1*H*-pyrrole-1-(*p*-benzoic acid)
PDDA	poly (diallyldimethylammonium chloride)
PEG	polyethylenglycol
PG	pyrolytic graphite
PSS	polystyrene sulfonate
PTBA	poly (3-thiophene boronic acid)
PTH	polythionine
$PV\text{-}SO_3^-$	poly(vinyl sulfonic acid)
SA	sodium alginate
SDBS	sodium dodecylbenzene sulfonate
SEB	staphylococcal enterotoxin
SiSG	silica gel sol gel
SPCE	screen-printed carbon electrode
SPE	screen-printed electrode
SPGE	screen-printed gold electrode
SPR	surface plasmon resonance
ST	streptavidin
SWASV	square-wave anodic stripping voltammetry
SWCNTs	single-walled carbon nanotubes
SWNHs	single-walled nanohorns

SWSV	square-wave stripping voltammetry
SWV	square-wave voltammetry
TA	thiolamine
4-TBC	4-*tert*-butylcatechol
TBR	target binding regions
TEOS	tetraethoxysilane
THI	thionine
TMB	3,3′,5,5′-tetramethylbenzidine
TTCA	5,2′,5′,2″-terthiophene-3′-carboxyl acid
TU	thiourea
WHO	World Health Organisation
WGS	water-soluble graphene sheet
ZEA	zearalenone

References

1. Turner, A.P.F. (2013) Biosensors: sense and sensibility. *Chem. Soc. Rev.*, **42**, 3184–3196.
2. Holford, T.R.J., Davis, F. and Higson, S.P.J. (2012) Recent trends in antibody based sensors. *Biosens. Bioelectron.*, **34**, 12–24.
3. Sadik, O.A. and Mwilu, S.K. (2010) Smart electrochemical biosensors: from advanced materials to ultrasensitive devices. *Electrochim. Acta*, **55**, 4287–4295.
4. Yáñez-Sedeño, P., Riu, J., Pingarrón, J.M. and Rius, F.X. (2010) Electrochemical sensing based on carbon nanotubes. *Trends Anal. Chem.*, **29**, 939–953.
5. Gan, T. and Hu, S. (2011) Electrochemical sensors based on graphene mater. *Microchim. Acta*, **175**, 1–19.
6. Siangproh, W., Dungchai, W., Rattanarat, P. and Chailakapul, O. (2011) Nanoparticle-based electrochemical detection in conventional and miniaturized systems and their bioanalytical applications: a review. *Anal. Chim. Acta*, **690**, 10–25.
7. Wang, J. (2012) Electrochemical biosensing based on noble metal nanoparticles. *Microchim. Acta*, **177**, 245–270.
8. Gopinath, S.C.B., Tang, T.-H., Citartan, M. *et al.* (2014) Current aspects in immunosensors. *Biosens. Bioelectron.*, **57**, 292–302.
9. Ricci, F., Adornetto, G. and Palleschi, G. (2012) A review of experimental aspects of electrochemical immunosensors. *Electrochim. Acta*, **84**, 74–83.
10. Pedrero, M., Campuzano, S. and Pingarrón, J.M. (2012) Magnetic beads based electrochemical sensors applied to the detection and quantification of bioterrorism/biohazard agents. *Electroanalysis*, **24**, 470–482.
11. Laschi, S., Centi, S. and Mascini, M. (2011) Electrochemical arrays coupled with magnetic separators for immunochemistry. *Bioanal. Rev.*, **3**, 11–25.
12. Alonso-Lomillo, M.A., Domínguez-Renedo, O. and Arcos-Martínez, M.J. (2010) Screen-printed biosensors in microbiology: a review. *Talanta*, **82**, 1629–1636.
13. Trilling, A.K., Beekwilder, J. and Zuilhof, H. (2013) Antibody orientation on biosensor surfaces: a minireview. *Analyst*, **138**, 1619–1627.
14. Gilmartin, N. and Kennedy, R.O'. (2012) Nanobiotechnologies for the detection and reduction of pathogens. *Enzyme Microb. Technol.*, **50**, 87–95.

15. Farahi, R.H., Passian, A., Tetard, L. and Thundat, T. (2012) Critical issues in sensor science to aid food and water safety. *ACS Nano*, **6**, 4548–4556.
16. Zhang, Z., Li, P., Hu, X. *et al.* (2012) Microarray technology for major chemical contaminants analysis in food: current status and prospects. *Sensors*, **12**, 9234–9252.
17. Meneely, J.P., Ricci, F., van Egmond, H.P. and Elliot, C.T. (2011) Current methods of analysis for the determination of trichothecene mycotoxins in food. *Trends Anal. Chem.*, **30**, 192–203.
18. Hayat, A., Paniel, N., Rhouati, A. *et al.* (2012) Recent advances in ochratoxin A-producing fungi detection based on PCR methods and ochratoxin A analysis in food matrices. *Food Control*, **26**, 401–415.
19. Kaushik, A., Arya, S.K., Vasudev, A. and Bhansali, S. (2013) Recent advances in detection of ochratoxin-A. *Open J. Appl. Biosens.*, **2**, 1–11.
20. Vidal, J.C., Bonel, L., Ezquerra, A. *et al.* (2013) Electrochemical affinity biosensors for detection of mycotoxins: a review. *Biosens. Bioelectron.*, **49**, 146–158.
21. Mostafa, G.A.E. (2010) Electrochemical biosensors for the detection of pesticides. *Open Electrochem. J.*, **2**, 22–42.
22. Pereira da Silva Neves, M.M., González-García, M.B., Nouws, H.P.A. *et al.* (2010) Celiac disease diagnosis and gluten-free food analytical control. *Anal. Bioanal. Chem.*, **397**, 1743–1753.
23. Lavecchia, T., Rea, G., Antonacci, A. and Giardi, M.T. (2013) Healthy and adverse effects of plant-derived functional metabolites: the need of revealing their content and bioactivity in a complex food matrix. *Crit.Rev. Food Sci. Nutr.*, **53**, 198–213.
24. Ortner, H.M. (1992) Analytical Chemistry – today's definition and interpretation. *Fresenius J. Anal. Chem.*, **343**, 825–826.
25. Valcárcel, M. (1997) A modern definition of analytical chemistry. *Trends Anal. Chem.*, **16**, 124–131.
26. Escarpa, A. (2005) *Miniaturisation of Analytical Techniques*, John Wiley & Sons, Inc., Hoboken, NJ.
27. Chen, L., Zeng, G., Zang, Y. *et al.* (2010) Trace detection of picloram using an electrochemical immunosensor based on three-dimensional gold nanoclusters. *Anal. Biochem.*, **407**, 172–179.
28. Tran, H.V., Yougnia, R., Reisberg, S. *et al.* (2012) A label-free electrochemical immunosensor for direct, signal-on and sensitive pesticide detection. *Biosens. Bioelectron.*, **31**, 62–68.
29. Garcia-Febrero, R., Valera, E., Muriano, A. *et al.* (2013) An electrochemical magneto immunosensor (EMIS) for the determination of paraquat residues in potato samples. *Anal. Bioanal. Chem.*, **405**, 7841–7849.
30. Baskeyfield, D.E.H., Davis, F., Magan, N. and Tothill, I.E. (2011) A membrane-based immunosensor for the analysis of the herbicide isoproturon. *Anal. Chim. Acta*, **699**, 223–231.
31. Arévalo, F.J., González-Techera, A., Zon, M.A. *et al.* (2012) Ultra-sensitive electrochemical immunosensor using analyte peptidomimetics selected from phage display peptide libraries. *Biosens. Bioelectron.*, **32**, 231–237.
32. Sun, X., Du, S., Wang, X. *et al.* (2011) A label-free electrochemical immunosensor for carbofuran detection based on a sol–gel entrapped antibody. *Sensors*, **11**, 9520–9531.

33. Sun, X., Li, Q. and Wang, X. (2012) *IEEE Sens. J.*, **12**, 2071–2076.
34. Sun, X., Du, S. and Wang, X. (2012) Amperometric immunosensor for carbofuran detection based on gold nanoparticles and PB-MWCNTs-CTS composite film. *Eur. Food Res. Technol.*, **235**, 469–477.
35. Sun, X., Zhu, Y. and Wang, X. (2012) Amperometri immunosensor based on deposited gold nanocrystals/4,4′-thiobisbenzenethiol for determination of carbofuran *Food Control*, **28**, 184–191.
36. Liu, G., Wang, S., Liu, J. and Song, D. (2012) An electrochemical immunosensor based on chemical assembly of vertically aligned carbon nanotubes on carbon substrates for direct detection of the pesticide endosulfan in environmental water. *Anal. Chem.*, **84**, 3921–3928.
37. Sun, X., Cao, Y., Gong, Z. *et al.* (2012) An amperometric immunosensor based on multi-walled carbon nanotubes-thionine-chitosan nanocomposite film for chlorpyrifos detection *Sensors*, **12**, 17247–17261.
38. Wang, C., Lin, M., Liu, Y. and Lei, H. (2011) *Electrochim. Acta*, **56**, 1988–1994.
39. Lin, M., Liu, Y., Sun, Z. *et al.* (2012) Electrochemical immunoassay of benzo[a]pyrene based on dual amplification strategy of electron-accelerated Fe_3O_4/polyaniline platform and multi-enzyme-functionalized carbon sphere label. *Anal. Chim. Acta*, **722**, 100–106.
40. Kim, D.-M. (2010) Md.A. Rahman, M.H. Do, C. Ban, Y.-B. Shim, An amperometric chloramphenicol immunosensor based on cadmium sulfide nanoparticles modified-dendrimer bonded conducting polymer. *Biosens. Bioelectron.*, **25**, 1781–1788.
41. Conzuelo, F., Gamella, M., Campuzano, S. *et al.* (2012) Disposable amperometric magneto-immunosensor for direct detection of tetracyclines antibiotics residues in milk. *Anal. Chim. Acta*, **737**, 29–36.
42. Jin, W., Yang, G., Wu, L. *et al.* (2011) *Food Control*, **22**, 1609–1616.
43. Yang, G., Jin, W., Wu, L. *et al.* (2011) Development of an impedimetric immunosensor for the determination of 3-amino-2-oxazolidone residue in food samples. *Anal. Chim. Acta*, **706**, 120–127.
44. Centi, S., Stoica, A.I., Laschi, S. and Mascini, M. (2010) Development of an electrochemical immunoassay based on the use of an eight-electrodes screen-printed array coupled with magnetic beads for the detection of antimicrobial sulfonamides in honey *Electroanalysis*, **22**, 1881–1888.
45. Conzuelo, F., Gamella, M., Campuzano, S. *et al.* (2012) Disposable and integrated amperometric immunosensor for direct determination of sulfonamide antibiotics in milk. *Biosens. Bioelectron.*, **36**, 81–88.
46. Zhu, Y., Son, J.I. and Shim, Y.-B. (2010) Amplification strategy based on gold nanoparticle-decorated carbon nanotubes for neomycin immunosensors. *Biosens. Bioelectron.*, **26**, 1002–1008.
47. Wu, X., Kuang, H., Hao, C. *et al.* (2012) Paper supported immunosensor for detection of antibiotics. *Biosens. Bioelectron.*, **33**, 309–312.
48. Zhao, B.Y., Wei, Q., Xu, C. *et al.* (2011) Label-free electrochemical immunosensor for sensitive detection of kanamycin. *Sens. Actuators, B*, **155**, 618–625.
49. Wei, Q., Zhao, Y., Du, B. *et al.* (2012) Ultrasensitive detection of kanamycin in animal derived foods by label-free electrochemical immunosensor. *Food Chem.*, **134**, 1601–1606.

50. Khor, S.M., Liu, G., Peterson, J.R. *et al.* (2011) An electrochemical immunobiosensor for direct detection of veterinary drug residues in undiluted complex matrices *Electroanalysis*, **23**, 1797–1804.
51. Huang, J., Lin, Q., Zhang, X. *et al.* (2011) *Food Res. Int.*, **44** (92–97).
52. Liu, S., Lin, Q., Zhang, X. *et al.* (2011) *Sens. Actuators, B*, **156**, 71–78.
53. Lai, Y., Bai, J., Shi, X. *et al.* (2013) Graphene oxide as nanocarrier for sensitive electrochemical immunoassay of clenbuterol based on labeling amplification strategy. *Talanta*, **107**, 176–182.
54. Li, Z., Wang, Y., Kong, W. *et al.* (2012) *Sens. Actuators, B*, **174**, 355–358.
55. Li, S., Zhang, J., Deng, W. *et al.* (2012) *Electroanalysis*, **24**, 1–7.
56. Bai, J., Lai, Y., Jiang, D. *et al.* (2012) Ultrasensitive electrochemical immunoassay based on graphene oxide–Ag composites for rapid determination of clenbuterol *Analyst*, **137**, 4349–4355.
57. Regiart, M., Fernández-Baldo, M.A., Spotorno, V.G. *et al.* (2013) Ultra sensitive microfluidic immunosensor for determination of clenbuterol in bovine hair samples using electrodeposited gold nanoparticles and magnetic micro particles as bio-affinity platform. *Biosens. Bioelectron.*, **41**, 211–217.
58. Liu, G., Chen, H., Peng, H. *et al.* (2011) A carbon nanotube-based high-sensitivity electrochemical immunosensor for rapid and portable detection of clenbuterol. *Biosens. Bioelectron.*, **28**, 308–313.
59. Cui, Z., Cai, Y., Wu, D. *et al.* (2012) *Electrochim. Acta*, **69**, 79–85.
60. Mi, Q., Wang, Z.W., Chai, C.Y. *et al.* (2011) Multilayer structured immunosensor based on a glassy carbon electrode modified with multi-wall carbon nanotubes, polythionine, and gold nanoparticles. *Microchim. Acta*, **173**, 459–467.
61. Prüstün, A., Bos, R., Gore, F. and Bartram, J. (2008) *Safer Water, Better Health: Costs, Benefits and Sustainability of Interventions to Protect and Promote Health*, World Health Organization, Geneva.
62. Skládal, P., Kovár, D., Krajicek, V. *et al.* (2013) *Int. J. Electrochem. Sci.*, **8**, 1635–1649.
63. Cheng, M.S., Lau, S.H., Chow, V.T. and Toh, C.-S. (2011) Membrane-based electrochemical nanobiosensor for *Escherichia coli* detection and analysis of cells viability *Environ. Sci. Technol.*, **45**, 6453–6459.
64. Joung, C.-K., Kim, H.-N., Im, H.-C. *et al.* (2012) Ultra-sensitive detection of pathogenic microorganism using surface-engineered impedimetric immunosensor. *Sens. Actuators, B*, **161**, 824–831.
65. Chan, K.Y., Ye, W.W., Zhang, Y. *et al.* (2013) Ultrasensitive detection of *E. coli* O157:H7 with biofunctional magnetic bead concentration via nanoporous membrane based electrochemical immunosensor. *Biosens. Bioelectron.*, **41**, 532–537.
66. Li, Y., Cheng, P., Gong, J. *et al.* (2012) Amperometric immunosensor for the detection of *Escherichia coli* O157:H7 in food specimens *Anal. Biochem.*, **421**, 227–233.
67. Joung, C.-K., Kim, H.-N., Lim, M.-C. *et al.* (2013) A nanoporous membrane-based impedimetric immunosensor for label-free detection of pathogenic bacteria in whole milk. *Biosens. Bioelectron.*, **44**, 210–215.
68. Cho, E.C., Choi, J.-W., Lee, M. and Koo, K.-K. (2008) Fabrication of an electrochemical immunosensor with self-assembled peptide nanotubes. *Colloids Surf. A*, **313–314** (95-99).

69. Li, D., Feng, Y., Zhou, L. *et al.* (2011) Label-free capacitive immunosensor based on quartz crystal Au electrode for rapid and sensitive detection of *Escherichia coli* O157:H7 *Anal. Chim. Acta*, **687**, 89–96.
70. Dou, W., Tang, W. and Zhao, G. (2013) A disposable electrochemical immunosensor arrays using 4-channel screen-printed carbon electrode for simultaneous detection of *Escherichia coli* O157:H7 and *Enterobacter sakazakii*. *Electrochim. Acta*, **97**, 79–85.
71. Brainina, K.Z., Kozitsina, A.N. and Glazyrina, Y.A. (2010) *IEEE Sens. J.*, **10**, 1699–1704.
72. Afonso, A.S., Pérez-López, B., Faria, R.C. *et al.* (2013) Electrochemical detection of Salmonella using gold nanoparticles. *Biosens. Bioelectron.*, **40**, 121–126.
73. Zhao, G., Zhan, X. and Dou, W. (2013) *Anal. Biochem.*, **408**, 53–58.
74. Braiek, M., Rokbani, K.B., Chrouda, A. *et al.* (2012) An electrochemical immunosensor for detection of *Staphylococcus aureus* bacteria based on immobilization of antibodies on self-assembled monolayers-functionalized gold electrode *Biosensors*, **2**, 417-426.
75. Hu, X., Dou, W., Fu, L. and Zhao, G. (2013) A disposable immunosensor for *Enterobacter sakazakii* based on an electrochemically reduced graphene oxide-modified electrode *Anal. Biochem.*, **434**, 218–220.
76. Zhang, X., Dou, W., Zhan, X. and Zhao, G. (2012) A novel immunosensor for *Enterobacter sakazakii* based on multiwalled carbon nanotube/ionic liquid/thionine modified electrode *Electrochim. Acta*, **61**, 73–77.
77. Mendes, R.K., Laschi, S., Stach-Machado, D.R. *et al.* (2012) A disposable voltammetric immunosensor based on magnetic beads for early diagnosis of soybean rust. *Sens. Actuators, B*, **166–167**, 135–140.
78. Viswanathan, S., Rani, C. and Ho, J.A. (2012) Electrochemical immunosensor for multiplexed detection of food-borne pathogens using nanocrystal bioconjugates and MWCNT screen-printed electrode. *Talanta*, **94**, 315–319.
79. Kang, X., Pang, G., Chen, Q. and Liang, X. (2013) Fabrication of *Bacillus cereus* electrochemical immunosensor based on double-layer gold nanoparticles and chitosan *Sens. Actuators, B*, **177**, 1010–1016.
80. Shang, K., Wang, X., Sun, B. *et al.* (2013) β-cyclodextrin-ferrocene host-guest complex multifunctional labeling triple amplification strategy for electrochemical immunoassay of subgroup J of avian leukosis viruses. *Biosens. Bioelectron*, **45**, 40–45.
81. Wang, Z., Shang, K., Dong, J. *et al.* (2012) Electrochemical immunoassay for subgroup J of avian leukosis viruses using a glassy carbon electrode modified with a film of poly(3-thiophene boronic acid), gold nanoparticles, graphene and immobilized antibody. *Microchim. Acta*, **179**, 227–234.
82. Kadir, M.K.A. and Tothill, I.W. (2010) Development of an electrochemical immunosensor for fumonisins detection in foods *Toxins*, **2**, 382–398.
83. Romanazzo, D., Ricci, F., Volpe, G. *et al.* (2010) H. vanEgmond, M. Vehniäinen, G. Palleschi, Development of a recombinant Fab-fragment based electrochemical immunosensor for deoxynivalenol detection in food samples. *Biosens. Bioelectron.*, **25**, 2615–2621.
84. Hervás, M., López, M.A. and Escarpa, A. (2010) Simplified calibration and analysis on screen-printed disposable platforms for electrochemical magnetic bead-based

immunosensing of zearalenone in baby food samples. *Biosens. Bioelectron.*, **25**, 1755–1760.
85. Panini, N.V., Bertolino, F.A., Salinas, E. *et al.* (2010) *Biochem. Eng. J.*, **51**, 7–13.
86. Panini, N.V., Salinas, E., Messina, G.A. and Raba, J. (2011) Modified paramagnetic beads in a microfluidic system for the determination of zearalenone in feedstuffs samples. *Food Chem.*, **125**, 791–796.
87. Feng, R., Zhang, Y., Li, H. *et al.* (2013) Ultrasensitive electrochemical immunosensor for zeranol detection based on signal amplification strategy of nanoporous gold films and nano-montmorillonite as labels. *Anal. Chim. Acta*, **758**, 72–79.
88. Zhang, J., Lei, J., Xu, C. *et al.* (2010) Carbon nanohorn sensitized electrochemical immunosensor for rapid detection of microcystin-LR. *Anal. Chem.*, **82** (1117–1122).
89. Zhao, H., Tian, J. and Quan, X. (2013) A graphene and multienzyme functionalized carbon nanosphere-based electrochemical immunosensor for microcystin-LR detection. *Colloids Surf. B*, **103** (38–44).
90. Ruiyi, L., Qianfang, X., Zaijun, L. *et al.* (2013) Electrochemical immunosensor for ultrasensitive detection of microcystin-LR based on graphene–gold nanocomposite/functional conducting polymer/gold nanoparticle/ionic liquid composite film with electrodeposition. *Biosens. Bioelectron.*, **44**, 235–240.
91. Sabransky, V., Pinelis, D., Korolev, S. *et al.* (2004) Immunogenicity of the histidine-to-tyrosine staphylococcal enterotoxin B mutant protein in C3H/HeJ mice *Toxicon*, **43**, 433–438.
92. Tang, D., Tang, J., Su, B. and Chen, G. (2010) Ultrasensitive electrochemical immunoassay of staphylococcal enterotoxin B in food using enzyme-nanosilica-doped carbon nanotubes for signal amplification. *J. Agric. Food Chem.*, **58**, 10824–10830.
93. Chai, C. and Takhistov, P. (2010) Label-free toxin detection by means of time-resolved electrochemical impedance spectroscopy *Sensors*, **10**, 655–669.
94. Wu, L., Gao, B., Zhang, F. *et al.* (2013) A novel electrochemical immunosensor based on magnetosomes for detection of staphylococcal enterotoxin B in milk *Talanta*, **106**, 360–366.
95. Zamfir, L.-G., Geana, I., Bourigua, S. *et al.* (2011) Highly sensitive label-free immunosensor for ochratoxin A based on functionalized magnetic nanoparticles and EIS/SPR detection. *Sens. Actuators, B*, **159**, 178–184.
96. Vidal, J.C., Bonel, L., Duato, P. and Castillo, J.R. (2011) Improved electrochemical competitive immunosensor for ochratoxin A with a biotinylated monoclonal antibody capture probe and colloidal gold nanostructuring. *Anal. Methods*, **3**, 977–984.
97. Vidal, J.C., Bonel, L., Ezquerra, A. *et al.* (2012) An electrochemical immunosensor for ochratoxin A determination in wines based on a monoclonal antibody and paramagnetic microbeads. *Anal. Bioanal. Chem.*, **403**, 1585–1593.
98. Fernández-Baldo, M.A., Bertolino, F.A., Fernández, G. *et al.* (2011) Determination of ochratoxin A in apples contaminated with *Aspergillus ochraceus* by using a microfluidic competitive immunosensor with magnetic nanoparticles *Analyst*, **136**, 2756–2762.
99. Muchindu, M., Iwuoha, E., Pool, E. *et al.* (2011) *Electroanalysis*, **23**, 122–128.
100. Ansari, A.A., Kaushik, A., Solanki, P.R. and Malhotra, B.D. (2010) Nanostructured zinc oxide platform for mycotoxin detection *Bioelectrochemistry*, **77**, 75–81.

101. Heurich, M., Kadir, M.K.A. and Tothill, I.E. (2011) An electrochemical sensor based on carboxymethylated dextran modified gold surface for ochratoxin A analysis. *Sens. Actuators, B*, **156**, 162–168.
102. Perrotta, P.R., Arévalo, F.J., Bettorazzi, N.R. *et al.* (2012) Development of a very sensitive electrochemical magneto immunosensor for the direct determination of ochratoxin A in red wine. *Sens. Actuators, B*, **162**, 327–333.
103. Zaijun, L., Zhongyun, W., Xiulan, S. *et al.* (2010) A sensitive and highly stable electrochemical impedance immunosensor based on the formation of silica gel-ionic liquid biocompatible film on the glassy carbon electrode for the determination of aflatoxin B1 in bee pollen *Talanta*, **80**, 1632–1637.
104. Linting, Z., Ruiyi, L., Zaijun, L. *et al.* (2012) An immunosensor for ultrasensitive detection of aflatoxin B1 with an enhanced electrochemical performance based on graphene/conducting polymer/gold nanoparticles/the ionic liquid composite film on modified gold electrode with electrodeposition. *Sens. Actuators, B*, **174**, 359–365.
105. Masoomi, L., Sadeghi, O., Banitaba, M.H. *et al.* (2013) A non-enzymatic nanomagnetic electro-immunosensor for determination of Aflatoxin B_1 as a model antigen. *Sens. Actuators, B*, **177**, 1122–1127.
106. Sun, X., Yan, L., Tang, Y. and Zhang, Y. (2012) A rapid and specific immunosensor for the detection of aflatoxigenic Aspergilli. *Eur. Food Res. Technol.*, **234**, 1013–1021.
107. Srivastava, S., Kumar, V., Ali, M.A. *et al.* (2013) Electrophoretically deposited reduced graphene oxide platform for food toxin detection *Nanoscale*, **5**, 3043–3051.
108. Paniel, N., Radoi, A. and Marty, J.-L. (2010) Development of an electrochemical biosensor for the detection of aflatoxin M1 in milk *Sensors*, **10**, 9439–9448.
109. Arévalo, F.J., Granero, A.M., Fernández, H. and Raba, J. (2011) *Talanta*, **83**, 966–973.
110. Chai, C., Lee, J. and Takhistov, P. (2010) Direct detection of the biological toxin in acidic environment by electrochemical impedimetric immunosensor *Sensors*, **10**, 11414–11427.
111. Hayat, A., Barthelmebs, L., Sassolas, A. and Marty, J.-L. (2011) An electrochemical immunosensor based on covalent immobilization of okadaic acid onto screen printed carbon electrode via diazotization-coupling reaction *Talanta*, **85**, 513–518.
112. Hayat, A., Barthelmebs, L. and Marty, J.-L. (2012) Electrochemical impedimetric immunosensor for the detection of okadaic acid in mussel sample. *Sens. Actuators, B*, **171–172**, 810–815.
113. Hayat, A., Barthelmebs, L., Sassolas, A. and Marty, J.-L. (2012) Development of a novel label-free amperometric immunosensor for the detection of okadaic acid. *Anal. Chim. Acta*, **724**, 92–97.
114. Hayat, A., Barthelmebs, L. and Marty, J.-L. (2011) Enzyme-linked immunosensor based on super paramagnetic nanobeads for easy and rapid detection of okadaic acid. *Anal. Chim. Acta*, **690**, 248–252.
115. Dominguez, R.B., Hayat, A., Sassolas, A. *et al.* (2012) Automated flow-through amperometric immunosensor for highly sensitive and on-line detection of okadaic acid in mussel sample *Talanta*, **99**, 232–237.
116. Tang, D., Tang, J., Su, B. and Chen, G. (2011) Gold nanoparticles-decorated amine-terminated poly(amidoamine) dendrimer for sensitive electrochemical immunoassay of brevetoxins in food samples. *Biosens. Bioelectron.*, **26**, 2090–2096.

117. Xiao, F., Zhang, N., Gu, H. *et al.* (2011) A monoclonal antibody-based immunosensor for detection of Sudan I using electrochemical impedance spectroscopy *Talanta*, **84**, 204–211.
118. Helm, R.M. and Burks, A.W. (2000) Mechanisms of food allergy. *Curr. Opin. Immunol.*, **12**, 647–653.
119. Cao, Q., Zhao, H., Yang, Y. *et al.* (2011) Electrochemical immunosensor for casein based on gold nanoparticles and poly(L-Arginine)/multi-walled carbon nanotubes composite film functionalized interface. *Biosens. Bioelectron.*, **26**, 3469–3474.
120. Eissa, S., Tlili, C., L'Hocine, L. and Zourob, M. (2012) Electrochemical immunosensor for the milk allergen β-lactoglobulin based on electrografting of organic film on graphene modified screen-printed carbon electrodes. *Biosens.Bioelectron.*, **38**, 308–313.
121. deLeón, M.P., Rolland, J.M. and O'Hehir, R.E. (2007) The peanut allergy epidemic: allergen molecular characterisation and prospects for specific therapy. *Expert Rev. Mol. Med.*, **9**, 1–18.
122. Liu, H., Malhotra, R., Peczuh, M.W. and Rusling, J.F. (2010) Electrochemical immunosensors for antibodies to peanut allergen ara h2 using gold nanoparticle-peptide films. *Anal. Chem.*, **82**, 5865–5871.
123. Dulay, S., Lozano-Sánchez, P., Iwoha, E. *et al.* (2011) Electrochemical detection of celiac disease-related anti-tissue transglutaminase antibodies using thiol based surface chemistry. *Biosens. Bioelectron.*, **26**, 3852–3856.
124. Neves, M.M.P.S., González-García, M.B., Nouws, H.P.A. and Costa-García, A. (2012) Celiac disease detection using a transglutaminase electrochemical immunosensor fabricated on nanohybrid screen-printed carbon electrodes. *Biosens. Bioelectron.*, **31**, 95–100.
125. Laube, T., Kergaravat, S.V., Fabiano, S.N. *et al.* (2011) Magneto immunosensor for gliadin detection in gluten-free foodstuff: towards food safety for celiac patients. *Biosens. Bioelectron.*, **27**, 46–52.
126. Chen, W., Lei, Y. and Li, M. (2010) *Electroanalysis*, **22**, 1078–1083.
127. Kergaravat, S.V., Gómez, G.A., Fabiano, S.N. *et al.* (2012) Biotin determination in food supplements by an electrochemical magneto biosensor *Talanta*, **97**, 484-490.
128. Golub, E.S. (1987) *Immunology: A Synthesis*, Sinauer Associates, Sunderland, MA.
129. Diamandis, E.P. and Christopoulos, T.K. (1996) *Immunoassay*, Academic Press, San Diego, CA.
130. Price, C.P. and Newman, D.J. (1993) *Principles and Practice of Immunoassay*, MacMillan, New York.
131. Todorovska, A., Roovers, R.C., Dolezal, O. *et al.* (2001) Design and application of diabodies, triabodies and tetrabodies for cancer targeting. *J. Immunol. Methods*, **248**, 47–66.
132. Borrebaeck, C.A., Malmborg, A.C., Furebring, C. *et al.* (1992) Kinetic analysis of recombinant antibody–antigen interactions: relation between structural domains and antigen binding *Biotechnology*, **10**, 697–698.
133. Tully, E., Higson, S.P. and Kennedy, R.O'. (2008) The development of a "labeless" immunosensor for the detection of Listeria monocytogenes cell surface protein, Internalin B. *Biosens. Bioelectron.*, **23**, 906–912.

134. Hamers-Casterman, T.A.C., Muyldermans, S., Robinson, G. *et al.* (1993) Naturally occurring antibodies devoid of light chains. *Nature*, **363**, 446–448.
135. Greenberg, A.S., Avila, D., Hughes, M. *et al.* (1995) A new antigen receptor gene family that undergoes rearrangement and extensive somatic diversification in sharks. *Nature*, **374**, 168–173.
136. Liu, B., Tang, D., Zhang, B. *et al.* (2013) Au(III)-promoted magnetic molecularly imprinted polymer nanospheres for electrochemical determination of streptomycin residues in food. *Biosens. Bioelectron.*, **41**, 551–556.
137. Zhou, L., Li, D.-J., Gai, L. *et al.* (2012) Electrochemical aptasensor for the detection of tetracycline with multi-walled carbon nanotubes amplification. *Sens. Actuators, B*, **162**, 201–208.
138. Barthelmebs, L., Hayat, A., Limiadi, A.W. *et al.* (2011) Electrochemical DNA aptamer-based biosensor for OTA detection, using superparamagnetic nanoparticles. *Sens. Actuators, B*, **156**, 932–937.
139. Hafaid, I., Chebil, S., Korri-Youssoufi, H. *et al.* (2010) Effect of electrical conditions on an impedimetric immunosensor based on a modified conducting polypyrrole. *Sens. Actuators, B*, **144**, 323–331.
140. Thilling, A.K., Harmsen, M.M., Ruigrok, V.J.B. *et al.* (2013) The effect of uniform capture molecule orientation on biosensor sensitivity: dependence on analyte properties. *Biosens. Bioelectron.*, **40**, 219–226.
141. Yuge, R., Ichihashi, T., Shimakawa, Y. *et al.* (2004) Preferential deposition of pt nanoparticles inside single-walled carbon nanohorns. *Adv. Mater.*, **16**, 1420–1423.
142. Li, Q., Zeng, L., Wang, J. *et al.* (2011) Magnetic mesoporous organic–inorganic $NiCo_2O_4$ hybrid nanomaterials for electrochemical immunosensors. *Appl. Mater. Interfaces*, **3**, 1366–1373.
143. Neves, M.M.P.S., González-García, M.B., Delerue-Matos, C. and Costa-García, A. (2012) Multiplexed electrochemical immunosensor for detection of celiac disease serological markers. *Sens. Actuators, B*, **187**, 33–39.
144. Fernández-Abedul, M.T. and Costa-García, A. (2008) Carbon nanotubes (CNTs)-based electroanalysis. *Anal. Bioanal. Chem.*, **390**, 293–298.
145. Iijima, S. (1991) Helical microtubules of graphitic carbon. *Nature*, **354**, 56–58.
146. Geim, A.K. and Novoselov, K.S. (2007) The rise of graphene. *Nat. Mater.*, **6**, 183–191.
147. Li, D., Müller, M.B., Gilje, S. *et al.* (2008) Processable aqueous dispersions of graphene nanosheets. *Nat. Nanotechnol.*, **3**, 101–105.
148. Dai, H. (2002) Carbon nanotubes: opportunities and challenges. *Surf. Sci.*, **500**, 218–241.
149. Banks, C.E. and Compton, R.G. (2006) New electrodes for old: from carbon nanotubes to edge plane pyrolytic graphite. *Analyst*, **131**, 15–21.
150. I. Álvarez-Martos, A. Fernández-Gavela, J. Rodríguez-García, N. Campos-Alfaraz, A.V. García-Delgado, D. Gómez-Plaza, A. Costa-García, M.T. Fernández-Abedul, *Sens. Actuators, B*, **192**, 253–260 (2014).
151. González-García, R., Fernández-la-Villa, A., Costa-García, A. and Fernández-Abedul, M.T. (2013) Dispersion studies of carboxyl, amine and thiol-functionalized carbon nanotubes for improving the electrochemical behavior of screen printed electrodes. *Sens. Actuators, B*, **181**, 353–360.

152. Premkumar, T., Mezzenga, R. and Geckeler, K.E. (2012) Carbon nanotubes in the liquid phase: addressing the issue of dispersion *Small*, **8**, 1299-1313.
153. González-García, M.B., Fernández-Sánchez, C. and Costa-García, A. (2000) Colloidal gold as an electrochemical label of streptavidin–biotin interaction. *Biosens. Bioelectron.*, **15**, 315–321.
154. Franc, G. and Kakkar, A. (2010) "Click" methodologies: efficient, simple and greener routes to design dendrimers. *Chem. Soc. Rev.*, **39**, 1536–1544.
155. Bilitewski, U. (2006) Protein-sensing assay formats and devices. *Anal. Chim. Acta*, **568**, 232–247.
156. Iost, R.M. and Crespilho, F.N. (2012) Layer-by-layer self-assembly and electrochemistry: applications in biosensing and bioelectronics. *Biosens. Bioelectron.*, **31**, 1–10.
157. Bain, C.D. and Whitesides, G.M. (1989) Modeling organic surfaces with self-assembled monolayers. *Angew. Chem. Int. Ed. Engl.*, **28**, 506–512.
158. Lawrence, N.S. and Wang, J. (2006) *Electrochem. Commun.*, **8**, 71–76.
159. Whitesides, G.M. (2013) Cool, or simple and cheap? Why not both? *Lab Chip*, **13**, 11–13.
160. Micrux Fluidic, www.micruxfluidic.com (accessed 5 May 2014).
161. Dropsens, www.dropsens.com (accessed 5 May 2014).
162. Windmiller, J.R. and Wang, J. (2013) Wearable electrochemical sensors and biosensors: a review *Electroanalysis*, **25**, 29–46.
163. Martinez, A.W., Phillips, S.T. and Whitesides, G.M. (2010) Diagnostics for the developing world: microfluidic paper-based analytical devices. *Anal. Chem.*, **82**, 3–10.
164. Nemiroski, A., Christodouleas, D.C., Hennek, J.W. *et al.* (2014) Universal mobile electrochemical detector designed for use in resource-limited applications *Proc, Natl. Acad. Sci*, U.S.A.. doi: 10.1073/pnas.1405679111
165. B. Piro, S. Reisberg, G. Anquetin, H.-T. Suc, M.-C. Pham,.Quinone-based polymers for label free reagentless electrochemical immunosensors: applications to proteins, antibodies and pesticide detection. *Biosensors*, **3**, 58–76 (2013).
166. Bogomolova, A., Komarova, E., Reber, K. *et al.* (2009) Challenges of electrochemical impedance spectroscopy in protein biosensing. *Anal. Chem.*, **81**, 3944–3949.
167. Liu, X., Duckworth, P.A. and Wong, D.K.Y. (2010) Square wave voltammetry versus electrochemical impedance spectroscopy as a rapid detection technique at electrochemical immunosensors. *Biosens. Bioelectron.*, **25**, 1467–1473.
168. Fernandes, F.C.B., Santos, A., Martins, D.C. *et al.* (2014) Comparing label free electrochemical impedimetric and capacitive biosensing architectures. *Biosens. Bioelectron.*, **57**, 96–102.
169. Fu, Y., Li, P., Bu, L. *et al.* (2010) Chemical/biochemical preparation of new polymeric bionanocomposites with enzyme labels immobilized at high load and activity for high-performance electrochemical immunoassay. *J. Phys. Chem. C*, **114**, 1472–1480.
170. Neves, M.M.P.S., González-García, M.B., Santos-Silva, A. and Costa-García, A. (2012) Voltammetric immunosensor for the diagnosis of celiac disease based on the quantification of anti-gliadin antibodies. *Sens. Actuators, B*, **163**, 253–259.
171. Hayes, F.J., Halsall, H.B. and Heineman, W.R. (1994) Simultaneous immunoassay using electrochemical detection of metal ion labels. *Anal. Chem.*, **66**, 1860–1865.
172. Brecht, A. and Abuknesha, R. (1995) Multi-analyte immunoassays application to environmental analysis. *Trends Anal. Chem.*, **14**, 361–371.

173. Whitesides, G.M. (2006) The origins and the future of microfluidics *Nature*, **442**, 368-373.
174. Ng, A.H.C., Uddayasankar, U. and Wheeler, A.R. (2010) Immunoassays in microfluidic systems. *Anal. Bioanal. Chem.*, **397**, 991–1007.
175. Bange, A., Halsall, H.B. and Heineman, W.R. (2005) Microfluidic immunosensor systems. *Biosens. Bioelectron.*, **20**, 2488–2503.
176. Fernández-Baldo, M.A., Messina, G.A., Sanz, M.I. and Raba, J. (2010) Microfluidic immunosensor with micromagnetic beads coupled to carbon-based screen-printed electrodes (SPCEs) for determination of *Botrytis cinerea* in tissue of fruits. *J. Agric. Food Chem.*, **58**, 11201–11206.
177. Liu, Q. and Boyd, B.J. (2013) Liposomes in biosensors *Analyst*, **138**, 391–409.
178. Lu, X., Zheng, H., Lli, X.-Q. *et al.* (2012) Detection of ractopamine residues in pork by surface plasmon resonance-based biosensor inhibition immunoassay. *Food Chem.*, **130**, 1061–1065.

10

Electrochemical Genosensors

Briza Pérez-López[1,2] *and Arben Merkoçi*[1,3]

[1]*Nanobioelectronics and Biosensors Group, ICN2 – Institut Catala de Nanociencia i Nanotecnologia, Universitat Autònoma de Barcelona, 08193, Bellaterra, Barcelona, Spain*

[2]*Devices, Design and Engineering, LEITAT Technological Center, C/de la Innovació, 2, 08225, Terrassa, Barcelona, Spain*

[3]*Institució Catalana de Recerca i Estudis Avançats, ICREA, Passeig Lluís Companys, 23, 08010, Barcelona, Spain*

10.1 General Introduction on Electrochemical Genosensors

Genosensors (or DNA biosensors) represent an important type of affinity biosensors in which the biorecognition molecules are oligonucleotides of known sequences and the recognition event is the hybridization with the complementary sequences to the target DNA of interest [1]. DNA is composed of two strands of nucleotide polymers wound around each other to form a double helix. The nucleotide chains are held together by hydrogen bonds between specific pairs of nucleotides. Adenine (A) in one strand binds to thymine (T) in the complementary strand. Similarly, guanine (G) in one strand forms a hydrogen bond with cytosine (C) in the opposite strand [1].

DNA biosensors combine a biological recognition agent (single-stranded DNA, ssDNA, also called a probe) and a transducer where the first provides a good selectivity and the second converts the recognition event into a measurable electronic signal, which can be modified with any nanomaterial or electroactive label to increase the signal sensitivity, improving the performance of the sensor [2]. The major requirements for a successful operation of a DNA biosensor are high specificity (including observation of a change in a single nucleotide) and high sensitivity.

The application of genosensors in food industry could lead to immense improvements in various food industry processes. These devices can be used in the raw material selection,

Agricultural and Food Electroanalysis, First Edition.
Edited by Alberto Escarpa, María Cristina González and Miguel Ángel López.

food processing (quality control), monitoring of storage conditions as well as ensuring food safety and security.

Genosensors enable one to improve the detection of food contaminants, reducing the health risks and medical costs associated with food-borne illness. Some of the contaminants that can be found in food are: bacteria, virus, parasites, seafood toxins, mycotoxins, pesticides, toxic metals, veterinary drug residues, undesirable fermentation products, and so on [3].

As DNA is more thermostable than many proteins, analyses using nucleic acid are less liable to be disrupted by processing of foodstuffs.

There are a variety of methods to detect the DNA content of food, which can be used to unequivocally identify the nature of the product [4]. Among the various systems for nucleic acid detection, electrochemical DNA analysis can involve direct detection based on a guanine signal (label-free) [5] or an electrocatalytic mechanism (label-based). Quantum dots (QDs) [6, 7], metal nanoparticles (NPs) [8, 9], enzymes [10, 11], and metal complexes [12, 13] can be employed as labels. This chapter focuses on electrochemical biosensing systems based on DNA hybridization events, which offer novel routes for food safety and security applications. Particularly, it describes in detail different approaches reported in the latest years on the immobilization of oligonucleotides on electrochemical transducers for sensing of various compounds with interest in food industry. In addition, some interesting applications in other fields that can easily be extended to that of food are also included.

10.1.1 Operation Principles

DNA biosensor technologies are under intense investigation owing to their great promise for the rapid and low-cost detection of specific DNA sequences. In the following sections, various DNA detection strategies based on electrochemical/electrical techniques that also involve nanomaterials and are with interest for food quality control will be described.

10.1.1.1 Label-Free (Direct) Detection

After the discovery of nucleic acid (NA) electroactivity at the end of the 1950s by Paleček [14], there has been much progress in the development of label-free electrochemical DNA detection (or direct detection) [5, 15–23]. Direct methods rely on the intrinsic electrochemical properties of DNA (the oxidation of purine bases, particularly guanine) are based on changes of the interfacial properties after hybridization [24] (see Figure 10.1) being these the most common strategies used in label-free sensors. These changes in resistance or conductivity of an interface modified with duplex DNA as compared to ssDNA have been reported by Paleček and Bartošík [25].

The label-free detections are based on the direct measurement of a phenomenon occurring during the biochemical reactions on a transducer surface as shown in Figure 10.2.

The first report in which guanine oxidation chemistry was used for target detection was described by Wang and coauthors [26]. The authors reported that the chronopotentiometric signal of guanine decreased after the incubation of an oligo d[G]-modified electrode with a guanine-free target.

Guanine as a purine base is the most easily oxidized base in DNA, and its irreversible peak has been found at the 1.0 V vs. Ag/AgCl reference electrode [27]. This peak height

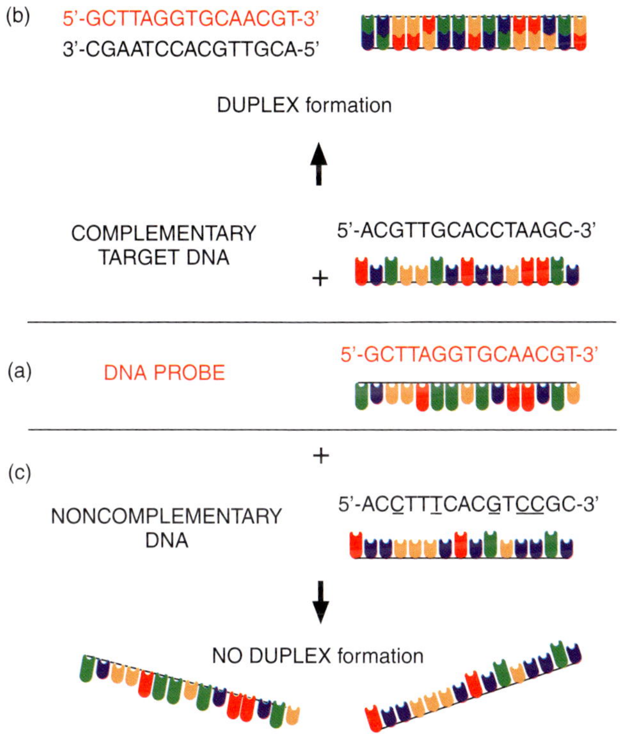

Figure 10.1 *Schematic representation of DNA hybridization. (a) DNA probe 5′-GCTTAGGTGCAACGT-3′. (b) DNA duplex formation from two oligodeoxynucleotide (ODN) strands – DNA probe (5′-GCTTAGGTGCAACGT-3′) and complementary target DNA (3′-CGAATCCACGTTGCA-5′). (c) Noncomplementary DNA. Reproduced with permission from [25]. Copyright © 2012, American Chemical Society*

is dependent on the NA concentration and the guanine base composition in the DNA sample [5].

Electrochemical detection methods are based on the electroactivity of the nucleotide bases [28]. Redox processes of NAs mostly involve the reduction and oxidation of bases. NA oxidation has been analyzed at different surfaces, such as carbon, platinum, gold, silver, and so on, and the NA reduction has been observed only at mercury electrodes. For G being more advantageous to study the oxidation of its reduction product (7,8-dihydroguanine), for example, when the mercury electrode is exposed to highly negative potentials (∼−1.8 V at neutral pH), its residues in an adsorbed NA molecule are reduced. G-reduction product remains at the electrode and is reoxidized back to G in an anodic scan, obtaining an oxidation peak of ∼−0.3 V [25].

The use of oligonucleotides as probe and the appearance of a guanine signal upon hybridization with the target opened a new field in electrochemical research [26]. These methods are promising for the rapid and low-cost detection of specific DNA sequences. Furthermore, their detection protocols eliminate the use of redox indicators minimizing the assay time. Several works related to the electrochemical DNA detection have been reported by using the G signal without any external indicator [29–32].

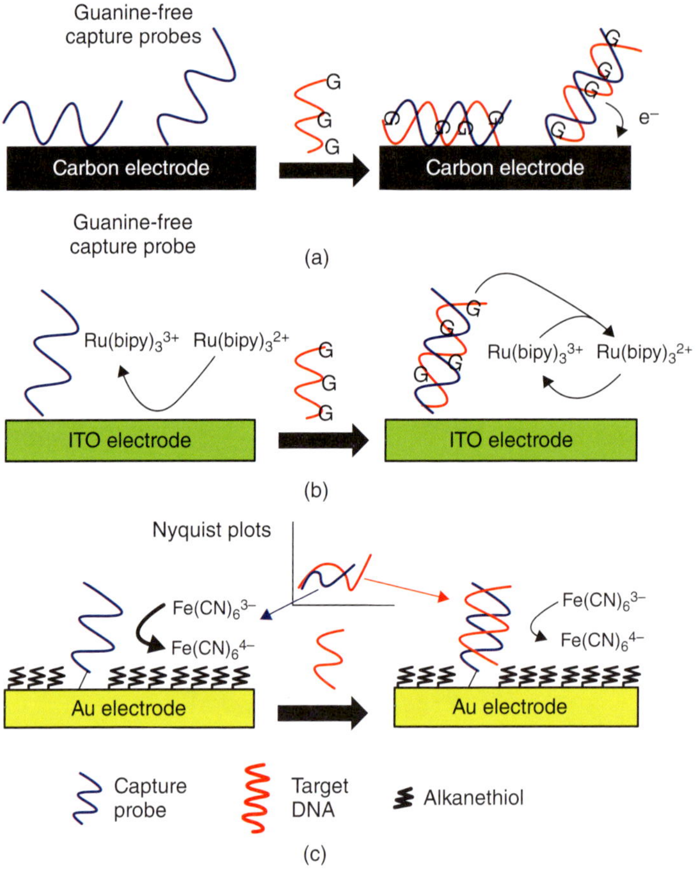

Figure 10.2 *Scheme of label-free hybridization commonly used in electrochemical DNA hybridization sensors. (a) Target DNA guanine bases on carbon electrode after the hybridization. (b) Representation of the electrocatalytic oxidation of guanine on ITO electrode by means of redox mediator. (c) Increase the electron transfer resistance by use an electroactive redox couple and a gold electrode surface. Reproduced from [25], with permission from American Chemical Society*

DNA oxidation at carbon electrodes is associated with the irreversible oxidation of guanine and adenine [33]. For example, the G oxidation signal observed at +1.0 V, without external labels, has been used to monitor telomerase activity by using a carbon graphite electrode (CGE) as an electrochemical transducer [5]. Telomerase activity has been detected in cell extracts containing as low as 100 ng μl^{-1} of protein. This label-free assay is practical in the quantitative determination of telomerase activity providing a cheap and simple detection protocol for the diagnosis of cancer that can also be extended to the analysis of food related to DNA.

A novel label-free electrochemical genosensor for the detection of catechol-*o*-methyltransferase (COMT) Val108/158Met polymorphism based on the guanine signal change by using a disposable CGE in connection with differential pulse voltammetry

(DPV) has been developed by Ozkan-ariksoysal *et al.* [29]. This developed technology was performed by using short synthetic oligonucleotides and polymerase chain reaction (PCR) amplicons of 203 bp length to measure the change of the guanine oxidation signal obtained at ~+1.0 V after DNA hybridization between the probe and the target (synthetic target or denatured PCR samples). COMT-specific oligonucleotides were immobilized onto the carbon surface with a simple adsorption method in two different ways: (i) guanine-containing targets were attached or (ii) inosine-substituted probes were attached onto the electrode. The detection limit ($S/N = 3$) was 2.44 pmol of target sequence in the 30 μl samples.

The development and validation of electrochemical DNA biosensors that incorporate nanomaterials as a signal transducer for the direct detection of analyte have also been developed. The most widely investigated polymeric nanomaterials used for biomolecule immobilization are conducting polymers such as polyaniline (PANI), poly(phenylenevinylene), polypyrrole, polythiophene, polyacetylene, and polyindole [34]. The unique electronic structure of polymeric nanomaterials is responsible for their remarkable high electrical conductivity, ease of processibility, low ionization potentials, good environmental stability, and high electron affinity [35]. For example, the use of *polymeric nanoparticles* for DNA biosensors has been recently reported by Zhang *et al.* [17]. They designed a novel architecture by combining the strong adsorption ability of Fe_2O_3 microspheres to the DNA probes and the excellent conductivity of self-doped PANI nanofibers (copolymer of aniline and *m*-aminobenzenesulfonic acid) on carbon ionic liquid electrode (CILE) for electrochemical impedance sensing of the immobilization of probe ssDNA for the sensitive detection of sequence-specific DNA of phosphoenol-pyruvate carboxylase (PEPCase) gene. $[Fe(CN)_6]^{3-/4-}$ was employed as an internal indicator. The strong adsorption ability of Fe_2O_3 microspheres and the excellent conductivity of self-doped PANI nanofibers enhanced the sensitivity of DNA hybridization recognition. In the same study, DNA hybridization events were monitored by using label-free electrochemical impedance spectroscopy (EIS). The response of the optimized biosensor was measured under the wide concentration range from 1.0×10^{-13} to 1.0×10^{-7} mol l^{-1}, with a detection limit of 2.1×10^{-14} mol l^{-1}.

Other nanomaterials used as modifiers of electrotransducers are one-dimensional *nanowires* (NWs). In recent years, research in gallium nitride nanowires (GaN) has been intensified for designing DNA biosensors [18, 19]. Chen *et al.* [18] have developed GaN NWs for the label-free electrochemical detection of target DNA with an anthrax lethal factor (LF) sequence by using a dual route of EIS and photoluminescence (PL) measurements. The GaN NW biotransducer showed enhanced sensitivity to surface-immobilized DNA molecules as NWs provided high surface binding energies for more binding sites to probe DNA and surface-enhanced charge transfer capability of the analyte. This novel biosensor revealed excellent selectivity and specificity, down to picomolar concentration, as well as high response sensitivity.

Carbon nanotubes (CNTs) are used as carriers. CNTs serve as a part of the transducer that carries DNA probes and, at the same time, promotes electron transfer. Thus, CNTs have brought further advancements into the rapidly evolving field of DNA sensors. The excellent conductivity of CNTs allows their use as highly sensitive DNA sensors [36, 30]. Also, the inorganic oxide nanoparticles (i.e., CeO_2, Fe_2O_3, etc.) are used as immobilizing carriers of the ssDNA probe [17, 37].

Magnetic nanoparticles-based biosensors present certain advantages for DNA detection in terms of sensitivity, specificity, and time- and cost-efficiency. Among label-free assays, the quartz crystal microbalance (QCM) piezoelectric sensor has proved to be a useful platform in the efficient detection of pathogenic microorganisms (including *B. anthracis)* that are of importance to food and environmental safety, biosecurity, and medical diagnostics. Based on this platform, electrically active magnetic (EAM) nanoparticles are being used as a concentrator of DNA targets as well as electrochemical transducers for the detection of the *B. anthracis* protective antigen A (pag A) gene [21].

As mentioned before, several electrochemical biosensors based on direct DNA detection or catalyzed oxidation of DNA G residues require the combination of nanomaterials, DNA-recognition, and electrical detection protocols, allowing them to improve the sensitivity of the devices. Other promising technologies for the analysis of DNA through the use of labels, as described in the following section, have been obtained.

10.1.1.2 Label-Based (Indirect) Detection

The labeling technologies for electrochemical DNA hybridization assays are based on the incorporation of electroactive labels [38]. QDs, metal NPs, enzymes, and metal complexes are offering attractive properties to act as DNA labels [6–13] (as shown in Figure 10.3 for metal NPs). With the use of labels, the detection sensitivity improves due to the high surface area, favorable optical, and electronic properties and electrocatalytic activity as well as good biocompatibility induced by nanometer size and specific physicochemical characteristics. Their high sensitivity has led to extensive applications in electrochemical sensing. The DNA detection systems have special interest for further applications in clinical analysis, food quality, and safety as well as other industrial applications [9].

The method of indirect DNA detection using electroactive indicators offers a very attractive route, which has recently been preferred to convert the hybridization/recognition events into measurable electrochemical signals. The use of sensitive electrochemical indicators is of particular significance for indirect DNA detection. For example, methylene blue (MB) has been widely used as an electrochemical electroactive label for monitoring DNA hybridization reaction due to its various affinities to ssDNA and dsDNA (double stranded deoxyribonucleic acid). MB specifically binds to the guanine bases in ssDNA and a high signal is observed. However, following the hybridization of probe with complementary DNA, a lower signal current is observed due to the less amount of MB that could incorporate into dsDNA. This is due to less or no accessibility of the guanine bases in dsDNA.

On the other hand, organic dyes in combination with CNTs are also used in electrode fabrication and hybridization techniques for developing an ultrasensitive, selective, and miniaturized electrochemical DNA biosensor for quick and reliable DNA sequence analysis. CNTs have good biocompatibility which allows them to increase the attached DNA amount on the substrate surface, thanks to their high surface area, an enhanced electronic conductivity, and a high mechanical resistance in electroanalytical chemistry can accelerate the electron-transfer rate between the redox active ssDNA molecule and electrode.

Li and coauthors [40] have immobilized a chitosan film doped with CNTs onto the graphite electrode and co-immobilized fish sperm DNA for the detection of salmon sperm. They employed MB as a redox-active indicator for the electrochemically quantitative

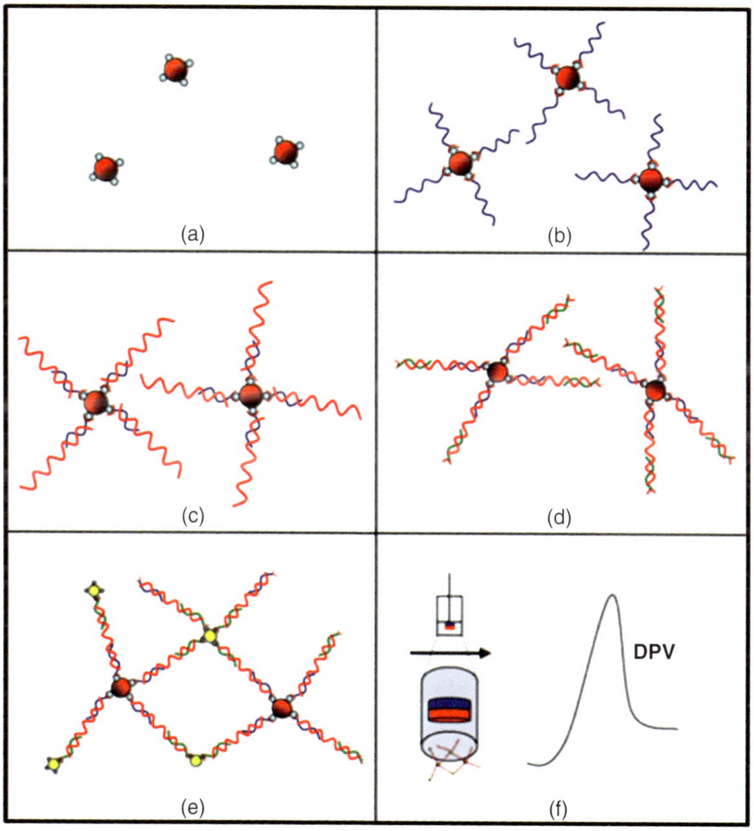

Figure 10.3 *Scheme of DNA hybridization assay based on the incorporation of metal–nanoparticles (not in scale): (a) streptavidin-coated magnetic beads, (b) immobilization of the biotinylated capture DNA probe onto the magnetic beads, (c) addition of target DNA (first hybridization event), (d) addition of biotinylated signaling DNA probe (second hybridization event), (e) tagging by using the streptavidin-AuNPs, and (f) Au-NPs-DNA-magnetic bead conjugate accumulated on the surface of the M-grafito-epoxi composite electrode (GECE) and direct DPV electrochemical detection of Au-NPs tag in the conjugate. Reproduced with permission from Elsevier. Adapted from Ref. [9]*

detection of the DNA hybridization signal, founding that CNTs can enhance the electroactive surface area threefold (0.28 ± 0.03 and 0.093 ± 0.06 cm^2 for chitosan–CNT- and chitosan-modified electrodes, respectively) and can accelerate the rate of electron transfer between the redox-active MB and the electrode. A low detection limit of 0.252 nM fish sperm DNA was achieved, and no interference was found in the presence of human serum albumin. The DPV signal of MB was linear over the fish sperm DNA concentration range of 0.5–20 nM.

An electrochemical DNA biosensor has been reported by Tichoniuk *et al.* [41] for the discrimination between ss- and dsDNA, and the voltammetric detection of target NA fragments typical of the aerolysin gene for the detection of *Aeromonas hydrophila* in food.

The genosensor recognition layer was prepared using mixed self-assembled monolayer (SAM) consisting of thiolated ssDNA probe and diluent molecules – mercaptoalcohol: mercaptoethanol (MCE) or mercaptohexanol (MCH) or mercaptononanol (MCN). Thereby, a detection of specific NA fragments was performed with a simple and reliable hybridization monitoring system based on the electroactive indicator of methylene blue. The biosensor analytical response was adequate to the concentration of target DNA fragments in the hybridization solution and was not interfered with by the presence of noncomplementary NA fragments. This DNA hybridization biosensor displays a promising detection system for the detection of food-borne pathogens.

The *enzymes* have also been widely used as labels for a sensitive DNA quantification [10, 11, 42–45]. For example, Rochelet-Dequaire *et al.* [46] have investigated that the alkaline phosphatase (AP) amplifies the analytical response of human cytomegalovirus DNA sequence (HCMV DNA) by adding the diaphorase (DI) enzymatic amplifier in the solution. The electrochemical detection was performed through arrays of carbon screen-printed DNA biosensors in association with the AP enzyme label and with the *p*-aminophenylphosphate (PAPP) substrate. The amount of hybrids immobilized on the electrode surface was labeled with an anti-digoxigenin AP conjugate and quantified electrochemically by measuring the activity of the AP label through the hydrolysis of the electroinactive PAPP substrate into the *p*-aminophenol (PAP) product. The intensity of the cyclic voltammetric anodic peak current resulting from the oxidation of PAP into *p*-quinoneimine (PQI) was related to the number of viral amplified DNA targets present in the sample. The obtained detection limit (6×10^7 copies of HCMV-amplified DNA fragments per electrochemical cell) has been very competent with other monoenzymatic electrochemical DNA sensors [47].

Metal NPs offer very attractive properties to act as DNA labels [8, 48, 49]. For example, Castañeda and co-authors [9, 50] have used Au NPs as labels for the electronic detection of DNA hybridization improving the sensitivity of bioelectronic assays by several orders of magnitudes. Their protocol relies on capturing the NPs to the hybridized target, followed by the highly sensitive anodic stripping electrochemical measurement of the metal tracer. The sensitivity of the NPs long lifetime along with multiplexing capability has also led to an explosive growth of *nanoparticles/QD*-based DNA electrochemical assays [7, 51–54] in addition to the use of metal complexes to improve the sensitivity for DNA detection [12, 13, 55].

10.2 Detection Methodologies

The electrochemical DNA detection is based on the electrochemical signals due to NA electroactivity, labeling of the target, or the probe with electroactive species, including NPs or changes in various electrochemically detectable DNA properties, which are related to current/potential, redox kinetics, impedance, or changes in the DNA structure owing to conformation changes, mass transportation, van der Waals interactions, and so on.

DNA detection offers several ways of interfacing DNA recognition events with electrochemical signal transduction. Among the many established methods, amplification is one of the most important concepts because it permits a highest analytical sensitivity. Some of the most widely used methods applied in DNA sensors such as a voltammetric, potentiometric, conductometric, impedimetric, and scanning electrochemical microscope will be described in the following sections.

10.2.1 Voltammetric/Stripping

Voltammetry is an electrochemical technique in which a controlled potential is applied to a working electrode, and any current that results from an electron transfer reaction at the electrode is measured. There are different modes of voltammetry depending of the information desired. Cyclic voltammetry (CV) is a technique for the characterization of electroactive species where a solution component is electrolyzed (oxidized or reduced) by placing the solution in contact with an electrode surface, and then making that surface sufficiently positive or negative in voltage to force electron transfer. Solid-state detection of NPs has been used in numerous DNA analysis exploiting the intrinsic electrochemical properties of the metal NPs used as labels. DPV analyses have been performed to quantify the metal NPs collected through the biospecific interaction, and that are related to the target analyte concentration (DNA/protein). For example, the detection of AuNPs by DPV for two Au NPs-based genomagnetic sensors designs for the detection of DNA hybridization based on a magnetically induced the direct electrochemical detection of gold tags on magnetic graphite-epoxy composite electrodes has been studied (including testing for a single and three-base mismatch). The proposed electrochemical detection formats are simple, sensitive enough, have a low cost, fast response time, and are potentially useful for fast clinical screenings [9]. Recently, a novel nanoparticle-based-enhanced methodology for the detection of ssDNA using nanoporous alumina filter membranes, containing pores of 200 nm in diameter, has also been reported [56]. This developed nanochannel/nanoparticle biosensing system would have enormous potential in future miniaturized designs adapted to mass production technologies such as screen-printing technology.

Square wave voltammetry (SWV) is a very sensitive and fast technique (voltammograms can be obtained in a few seconds), therefore, very well suited to study DNA hybridization. A label-free DNA electrochemical sensor based on a peptide nucleic acid (PNA)-functionalized conductive polymer has been presented by Reisberg *et al.* [57]. They used SWV as a technique to detect and quantify PNA–DNA hybridization. PNA was attached covalently onto a quinone-based electroactive polymer and following changes in its flexibility upon hybridization generates electrochemical changes at the polymer–solution interface. An increase in the peak current of quinone was observed upon the hybridization of probes by the target DNAs, whereas no change is observed with noncomplementary sequence. Hybridization was detected by recording the modification of the redox process of the quinone group.

An example based on an electrochemical genomagnetic hybridization assay has been reported by our group [39] by using a stripping technique, which takes advantage of an efficient magnetic separation/mixing process and of the use of cadmium sulfide QDs (CdS QDs) for DNA detection. A cystic-fibrosis-related DNA sequence was sandwiched between the two DNA probes. One DNA probe is linked via biotin–streptavidin bonding with magnetic beads and the other one via thiol groups with the CdS QD used as tags. The electrochemical detection of a CdS QDs–DNA complex was based on the stripping of electrochemically reduced cadmium at the hybridization solution using the SWV technique, giving a well-shaped and sensitive analytical signal (see Figure 10.4). Zhang's

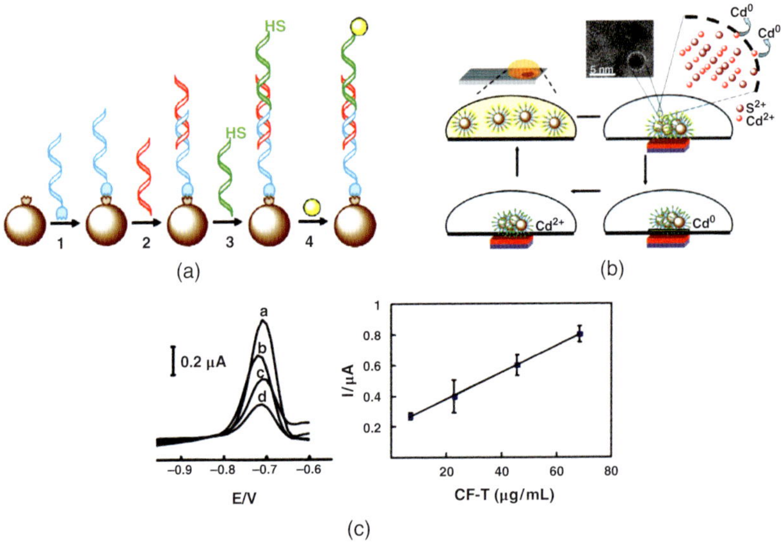

Figure 10.4 *(A) Scheme of genomagnetic hybridization assay (not to scale): (1) immobilization of the biotinylated DNA probe onto the streptavidin-coated magnetic beads, (2) first hybridization between target DNA and capture DNA, (3) second hybridization between target DNA and signaling DNA modified thiol, (4) addition and capture of the CdS QDs. (B) Schematic representation of electrochemical detection of DNA sandwich complex labeled with CdS QDs on screen-printing electrode surface. (C) Square-wave stripping voltammograms after blank subtraction to increase concentrations of target DNA: (a) 68 μg ml^{-1}, (b) 45 μg ml^{-1}, (c) 23 μg ml^{-1}, and (d) 6.8 μg ml^{-1}. Also shown is the corresponding calibration plot (right) over the range 6.8–68 μg ml^{-1} of target DNA. The measuring solution was the hybridization solution. Square-wave voltammetric scan with frequency of 25 Hz, step potential 10 mV and amplitude of 30 mV was performed. The deposition potential of −1.1 V for 120 s was applied. Reproduced by permission of IOP Publishing. [39]*

group have developed an ultrasensitive electrochemical detection of DNA based on the PbS nanoparticle tags and nanoporous gold electrode by the indirect determination of the amount of lead ions through differential pulse anodic stripping voltammetry (DPASV). This protocol could detect the target DNA of as low as femtomolar and exhibited excellent selectivity against one-base mismatched DNA and noncomplementary DNA. Under the optimum conditions, the anodic stripping peak current of lead demonstrated a good linear relationship with the target DNA concentration in the range of 9.0×10^{-16}–7.0×10^{-14}M. A detection limit of 2.6×10^{-16}M of target DNA was achieved [48].

10.2.2 Potentiometric

The signal measured with a potentiometric method is based on the potential difference between the working and reference electrodes. The working electrode's potential depends on the concentration of the analyte, whereas that the reference electrode is used to provide a defined reference potential.

In 1996, Wang's group [28] reported the first use of PNAs as the recognition layer in DNA biosensors. They reported the use of PNA as a recognition probe for the electrochemical detection of the hybridization event using chronopotentiometric measurements. This method consists of four steps: (i) probe (PNA) immobilization onto the transducer surface, (ii) hybridization, (iii) indicator binding ($Co(phen)_3{}^{3+}$), and (iv) chronopotentiometric transduction. Conditions for attaining a full surface coverage were assessed by measuring the dependence of the chronopotentiometric guanine peak area upon the PNA solution concentration or the adsorption time. The accumulated ($Co(phen)_3{}^{3+}$) was measured by using an initial potential of +0.5 V and a constant current of −6.0 μA. The reported chronopotentiometric response represents the difference in indicator peak areas at the hybrid- and probe-coated electrodes. The anodic signal at around +1.0 V, corresponding to the oxidation of guanine, was used as the analytical signal. The measurements were performed by using DNA and PNA at carbon-paste electrode (CPE) or glassy-carbon electrode (GCE) in 0.2 M acetate buffer (pH 5.0).

Nowadays, several works related to potentiometric genosensors by using labeled and label-free biosensing platforms [58–62] with different applications have been reported. Figure 10.5 shows a potentiometric detection of DNA hybridization using as label a Ca^{2+} ion [58]. However, not many potentiometric genosensors have been found for the detection of pathogens. The detection of *Escherichia coli* (*E. coli*) cells in vegetable food using the

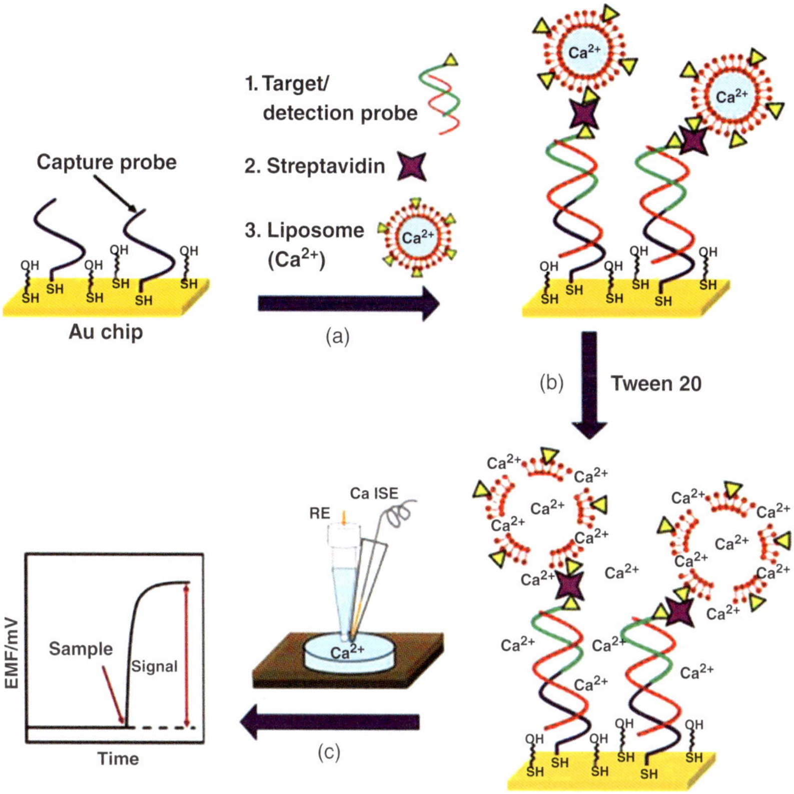

Figure 10.5 *Schematic representation of the potentiometric detection of DNA hybridization based on ion-loaded liposomes and* Ca^{2+}*-ion selective electrodes (ISE). Reproduced from [58] with permission from the Royal Society of Chemistry*

potentiometric alternating biosensing (PAB) system based on the light-addressable potentiometric sensor (LAPS) has been demonstrated by Ercole and coauthors [64]. LAPS was used as a transducing element, for detecting pH variations due to NH_3 production by an urease-*E. coli* antibody conjugate. The liquid phase was analyzed by the PAB system and conventional Colony Forming Unit (CFU) methods. The PAB system appeared to be very sensitive and fast, in comparison with conventional methods, and the concentration of 10 cells ml^{-1} was detected in an assay time of about 1.5 h, which was from 10 to 20 times shorter than the conventional CFU procedure.

10.2.3 Impedimetric

EIS is a very powerful tool for the analysis of interfacial capacitance and resistance changes occurring at conductive or semiconductive surfaces. A small amplitude sinusoidal excitation applied to the system measures the changes in the electrical impedance of the medium. Among the used electrochemical techniques, EIS has been shown very effective and sensitive for the characterization of biomaterial functionalized electrodes and biocatalytic transformation at electrode surfaces. The principle of impedimetric sensing of DNA hybridization is based on the formation of nucleic acid/DNA complexes at an electrode surface. The resulting negatively charged interface electrostatically repels the negatively charged redox indicator $Fe(CN)_6^{3-/4-}$. The electrostatic repulsion of the redox-active molecules leads to an enhanced charge transfer resistance, Rct. Thus, Rct will increase with the increasing amount of hybridized DNA. Bonnani and co-authors have reported studies based on the use of nanomaterials by using the EIS technique for the DNA detection [63, 65, 66]. One of their works is based on chemically modified graphene nanomaterials as sensing platforms (Figure 10.6). They employed a hairpin DNA as a highly selective probe for the detection of single nucleotide polymorphisms (SNPs) correlated with Alzheimer's disease, finding that graphene oxide provides the best ability to discriminate a complementary sequence from one carrying an SNP [63]. Thus, this work could provide a simple, sensitive, low-cost, label-free sensing device for the detection of SNPs correlated with genetic disorders using a graphene-based platform.

The integration of impedance with biological recognition technology has led to the development of impedimetric genosensors finding widespread use in the recent years to detect and/or quantify a variety of food-borne pathogens [68, 69]. Some other works related to DNA detection by using EIS have also been reported [70–72].

10.2.4 Conductometric

Conductometric biosensing systems are based on the specific biological interaction that occurs on a supporting surface where they may vary the resistance or conductance of the substrate. The capacitance or conductivity measurements can be made to determine the number of particles and, therefore, target molecules that fill the gap [73].

Detection of DNA hybridization used in connection with conductometric measurements after labeling with AuNPs has been demonstrated by Mirkin's group [67]. They developed an array based on electrical detection utilizing oligonucleotide functionalized Au NPs and closely spaced interdigitated microelectrodes. The oligonucleotide probe was immobilized in the gap between the two microelectrodes. The hybridization event, thus, localizes Au

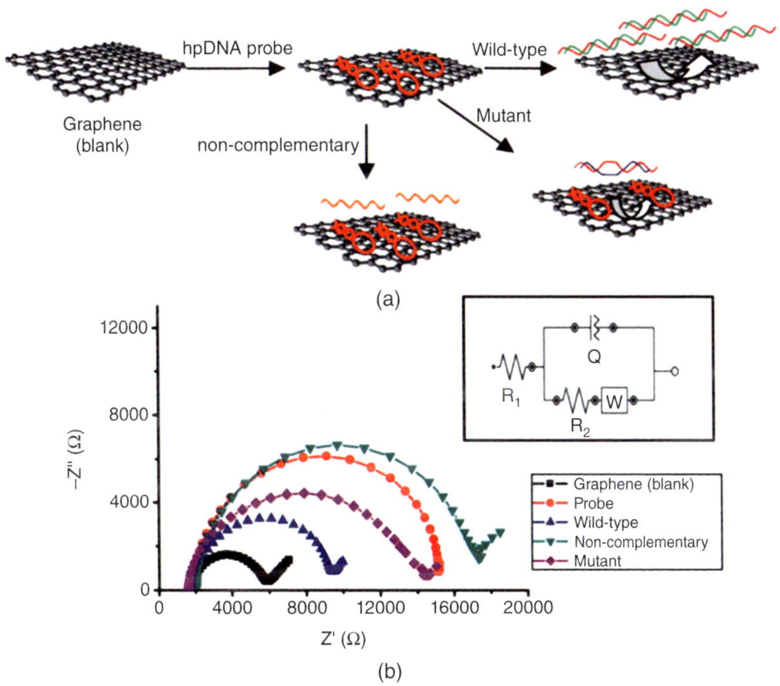

Figure 10.6 *(a) Schematic of the physical adsorption onto the graphene electrode surface to immobilize the hairpin DNA (hpDNA) probes. (b) Nyquist plots of the graphene surface and after modification with the hpDNA probes and different ssDNA target strands. Concentration of probes: 10 μM, concentration of each target: 0.3 μM. (Inset) Randles equivalent circuit used to fit the experimental data. Reproduced from [63] with permission from the Royal Society of Chemistry*

NPs in the electrode gap and along with subsequent silver deposition leads to measurable conductivity signals (see Figure 10.7). Such hybridization-induced conductivity signals, associated with resistance changes across the electrode gap, offer high sensitivity with a 0.5 pM detection limit.

10.3 Applications

A variety of methods to identify the DNA content in food, environmental, and clinical samples have been developed. Various techniques have been described and discussed related to food samples such as meat, fish, and so on. Some reported studies will be described as below.

10.3.1 Species Identification

As DNA is more thermostable than many proteins, analyses using nucleic acid are less liable to be disrupted by processing of foodstuffs. Furthermore, DNA is present in the majority of the cells of an organism, potentially enabling identical information to be

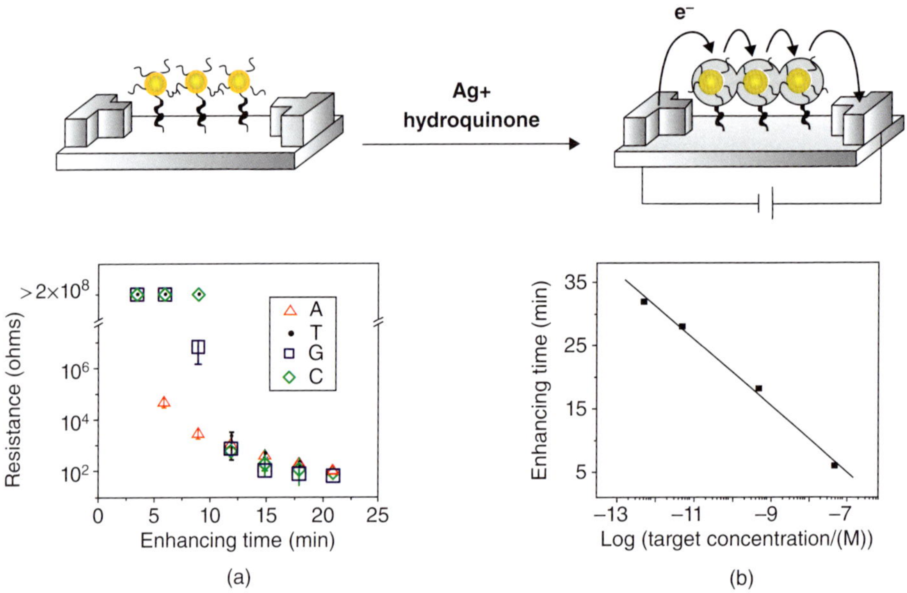

Figure 10.7 (A) Scheme showing a conductivity assay in which gold is accumulated in the gap and later on a silver enhancement procedure in the presence of hydroquinone. (B) (a) Resistance of the electrode arrays measured as a function of increasing silver enhancing time washing with 0.01 M phosphate buffered saline (PBS) at room temperature before silver enhancing. (b) A graph of the silver enhancing time required to reach a resistance value of 100 kΩ as a function of target concentration showing that target can be detected in the 50 nM to 500 fM concentration range by adjusting silver enhancing time. All rights reserved with permission from the Royal Society of Chemistry [67]

obtained from any appropriate sample from the same source, regardless of the tissue of origin [4].

A promising technological platform for the construction of a portable species identification gadget is the development of a new generation of nanomaterial-based electrochemical biosensors. DNA biosensors have been applied for the identification of specific genes in meat [72, 74, 75], which are important from cultural, religious, health, and commercial perspectives. Thus, an easy, rapid, and sensitive method of meat species detection in raw or processed foods has been carried out by Tamiya *et al.* [74]. In this study, they tried to distinguish species-specificity in control and processed pork, chicken, and bovine meats using loop-mediated isothermal amplicons (LAMP) and disposable electrochemical printed (DEP) chips. An electrochemical genosensor with the DEP chips detects the amplicons by linear sweep voltammetry (LSV) observation of the DNA–Hoechst33258 interaction on the chip surface (as shown in Figure 10.8). They showed that this method is more specific and free of unwanted amplifications compared to the multiplexed polymerase chain reaction (M-PCR) method and gave limits of detection of ~20.33 ng/μl (3×10^4 copies/reaction), ~78.68 pg/μl (3×10^2 copies/reaction), and ~23.63 pg/μl (30 copies/reaction) for pork, chicken, and bovine species, respectively.

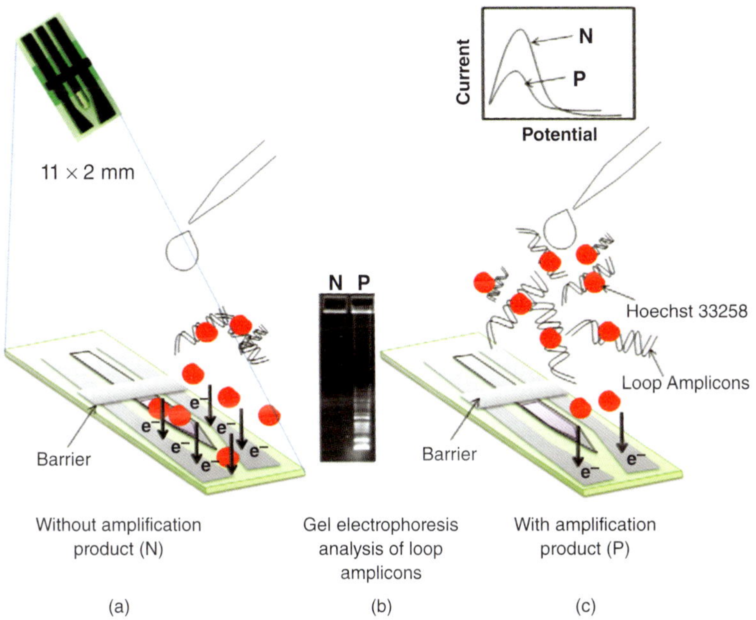

Figure 10.8 *Species-specific identification and detection of loop amplicons based on DNA–H33258 interactionusing the disposable electrochemical printed (DEP) chip. (a) Higher rate of diffusion and oxidation of H33258 molecules on electrode surface in the absence of DNA. (b) Lower oxidation observed due to DNA–H33258 interaction and for the low amount of unbound/free H33258 molecule. (c) Gel electrophoresis analysis of loop amplicons. Schematic representation of the electrochemical response (on the right top). Reproduced from [74] with permission from Elsevier*

Another electrochemical genosensor was developed by Mascini and coauthors [75] for the hybridization detection of real samples of total genomic DNA extracts of mammalian species (bovine). They immobilized two inosine-modified (guanine-free) DNA sequences of 21 and 25 based on the screen-printed electrodes (SPEs) as a capturing agent and the detection of the duplex formation, via guanine oxidation of the target. The duplex formation was detected using the SWV technique. The developed genosensors were applied to identify bovine and porcine DNA extract samples without PCR.

Electrical DNA hybridization biosensors are capable of converting DNA–DNA recognition events into an electronic signal-transduction process and identify different species in food. Further work is needed to realize the full potential of this new class of biosensors for the analysis of large DNA sequences and its application in species identification.

10.3.2 Contaminant Monitoring

DNA sensors related to the rapid identification and quantification of pathogenic bacteria such as *Salmonella* spp., *Lysteria monocytogenes, E. coli 0157:H7, Legionella sp.* have been developed [77–79]. Moreover, these devices have also been extensively exploited in the detection of pathogenic microorganisms relevant to biodefense and environmental contamination applications.

On the other hand, the detection of phenolic contaminants by developing a CNT-based DNA biosensor was reported by Zheng and co-authors [80]. They immobilized DNA on a GCE modified with MWNTs dispersed in Nafion (DNA/MWNTs/GCE). The DNA-modified electrode exhibited two well-defined oxidation peaks corresponding to the guanine and adenine residues of DNA, respectively. Phenol, *m*-cresol, and catechol owing to their interactions with DNA showed noticeable inhibition toward the response of the electrode. These biosensors obtained the detection limits of 16, 1.2, and 0.52 μM for phenol, *m*-cresol, and catechol, respectively.

Environmental contamination is also presented in the nanomaterials owing to its toxic properties. For this reason, Peng *et al.* [76] have developed an electrochemical genosensor that can distinguish single-wall carbon nanotubes (SWCNTs) and MWCNTs in cell extracts, which is based on the capability of SWNTs to specifically induce human telomeric i-motif (a tetrameric DNA structure with protonated cytosine–cytosine base pairs) formation (see Figure 10.9). The application of this sensor using SWNTs has been demonstrated in cancer cell extracts with a good selectivity and a detection limit of 0.2 ppm. The DNA sensor consists of a short single-stranded DNA containing a human telomeric i-motif DNA sequence. This DNA is modified with a redox-active MB and attached to gold electrodes. In the absence of a target, the immobilized 26-mer DNA remains unfolded in the buffer observing a Faradaic current due to the attached MB tag. In the presence of the i-motif

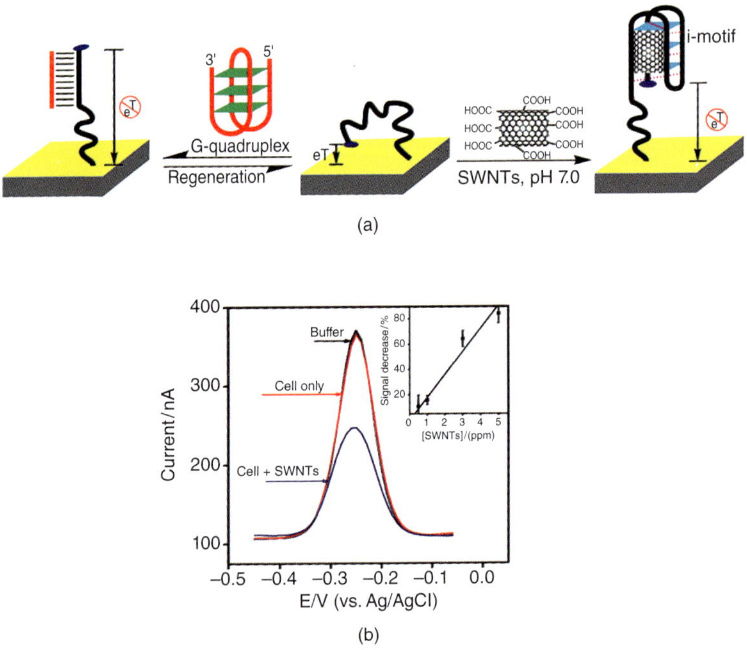

Figure 10.9 *(a) Scheme of thei-Motif telomeric DNA-based electrochemical DNA (E-DNA) sensor. (b) E-DNA sensor that detects intracellular SWNTs in extracts of the human chronic myelogenous leukemia K562 cells. (Inset) Calibration curve measured in* 1×10^4 *cell* ml^{-1} *cell supernatant (cells were cultured in the absence of SWNTs). Reproduced with permission from [76]. Copyright © 2009, American Chemical Society*

cDNA, the Faradaic current decreases, indicating the formation of telomeric DNA duplex, which prevents the MB tag to approach the electrode surface allowing the detection of 0.2 ppm of SWNTs. Furthermore, the decrease in the signal strongly depends on the type of CNT, being possible the discrimination between SWCNTs and MWCNTs. This study provides a new development in the design of a DNA-based biosensor for CNT detection. It could apply to help combat pollution in the environment.

10.4 Conclusions and Future Trends

In this chapter, we have reported some examples related to DNA sensors, especially used in food applications, by employing electrochemical detection techniques. A variety of sensing systems based on label-free techniques utilizing electrochemical and/or surface activity as well as direct methods that rely on the intrinsic electrochemical properties of DNA (the oxidation of purine bases, particularly guanine) have been presented in the different sections of this chapter.

The field of genosensors is an active research area where many formats and designs are proposed in order to achieve better biodetection features. Moreover, scientists still should increase efforts to optimize its use in real samples, as well as to improve its reproducibility and stability. The major requirements for a successful operation of a DNA biosensor are high specificity (including observation of a change in a single nucleotide) and high sensitivity. Nowadays, electrochemical sensing of DNA is one of the most attractive and promising techniques, because it furthermore provides fast, facile, and low-cost detection of small volumes of DNA in miniaturized devices. These biodevices have attracted much attention in a broad range of applications that do not include only food safety and testing but also clinical diagnostics, genetics, pathology, forensics, and many other fields.

ACKNOWLEDGMENTS

ICN2 acknowledges support from the Severo Ochoa Program (MINECO, Grant SEV-2013-0295).

References

1. Payne, W. L., Hill, W. E., Datta, A.R., Feng, P., Lampel, K.A., *Bacteriological Analytical Manual*, AOAC, International, Chapter 24, 8th edn, Revision A, 1998.
2. Pividori, M.I., Merkoçi, A. and Alegret, S. (2000) Electrochemical genosensor design: immobilisation of oligonucleotides onto transducer surfaces and detection methods. *Biosens. Bioelectron.*, **15** (5-6), 291–303.
3. Pividori, M.I. and Alegret, S. (2010) Electrochemical biosensors for food safety. *Contrib. Sci.*, **6** (2), 173–191.
4. Lockley, A. and Bardsley, R. (2000) DNA-based methods for food authentication. *Trends Food Sci. Technol.*, **11** (2), 67–77.
5. Eskiocak, U., Ozkan-ariksoysal, D., Ozsoz, M. and Oktem, H.A. (2007) Label-free detection of telomerase activity using guanine electrochemical oxidation signal. *Anal. Chem.*, **79** (22), 8807–8811.

6. Zhang, C., Xu, J., Zhang, S. *et al.* (2012) One-pot synthesized DNA-CdTe quantum dots applied in a biosensor for the detection of sequence-specific oligonucleotides. *Chemistry (Weinheim an der Bergstrasse, Germany)*, **18** (27), 8296–8300.
7. Baş, D. and Boyaci, I.H. (2011) Photoelectrochemical competitive DNA hybridization assay using semiconductor quantum dot conjugated oligonucleotides. *Anal. Bioanal. Chem.*, **400** (3), 703–707.
8. Kwon, S.J. and Bard, A.J. (2012) DNA analysis by application of Pt nanoparticle electrochemical amplification with single label response. *J. Am. Chem. Soc.*, **134** (26), 10777–10779.
9. Castañeda, M.T., Merkoçi, A., Pumera, M. and Alegret, S. (2007) Electrochemical genosensors for biomedical applications based on gold nanoparticles. *Biosens. Bioelectron.*, **22** (9-10), 1961–1967.
10. Low, K.-F., Chuenrangsikul, K., Rijiravanich, P. *et al.* (2012) Electrochemical genosensor for specific detection of the food-borne pathogen, vibrio cholerae. *World J. Microbiol. Biotechnol.*, **28** (4), 1699–1706.
11. Ferapontova, E.E., Hansen, M.N., Saunders, A.M. *et al.* (2010) Electrochemical DNA sandwich assay with a lipase label for attomole detection of DNA. *Chem. Commun. (Camb.)*, **46** (11), 1836–1838.
12. García, T., Casero, E., Revenga-Parra, M. *et al.* (2008) Architectures based on the use of gold nanoparticles and ruthenium complexes as a new route to improve genosensor sensitivity. *Biosens. Bioelectron.*, **24** (2), 184–190.
13. Niu, S.-Y., Wu, M.-L., Hu, L.-Z. *et al.* (2009) Nucleic acid biosensor for DNA hybridization detection using rutin–Cu as an electrochemical indicator. *Electrochim. Acta*, **54** (5), 1564–1569.
14. Paleček, E. (2002) Past, present and future of nucleic acids electrochemistry. *Talanta*, **56** (5), 809–819.
15. Stempkowska, I., Ligaj, M., Jasnowska, J. *et al.* (2007) Electrochemical response of oligonucleotides on carbon paste electrode. *Bioelectrochemistry (Amsterdam)*, **70** (2), 488–494.
16. Yang, T., Zhou, N., Zhang, Y. *et al.* (2009) Synergistically improved sensitivity for the detection of specific DNA sequences using polyaniline nanofibers and multi-walled carbon nanotubes composites. *Biosens. Bioelectron.*, **24** (7), 2165–2170.
17. Zhang, W., Yang, T., Li, X. *et al.* (2009) Conductive architecture of Fe_2O_3 microspheres/self-doped polyaniline nanofibers on carbon ionic liquid electrode for impedance sensing of DNA hybridization. *Biosens. Bioelectron.*, **25** (2), 428–434.
18. Chen, C.-P., Ganguly, A., Wang, C.-H. *et al.* (2009) Label-free dual sensing of DNA molecules using GaN nanowires. *Anal. Chem.*, **81** (1), 36–42.
19. Chen, C.-P., Ganguly, A., Lu, C.-Y. *et al.* (2011) Ultrasensitive *in situ* label-free DNA detection using a GaN nanowire-based extended-gate field-effect-transistor sensor. *Anal. Chem.*, **83** (6), 1938–1943.
20. Gao, Z., Agarwal, A., Trigg, A.D. *et al.* (2007) Silicon nanowire arrays for label-free detection of DNA. *Anal. Chem.*, **79** (9), 3291–3297.
21. Pal, S. and Alocilja, E.C. (2010) Electrically active magnetic nanoparticles as novel concentrator and electrochemical redox transducer in *Bacillus anthracis* DNA detection. *Biosens. Bioelectron.*, **26** (4), 1624–1630.

22. Fang, B., Jiao, S., Li, M. *et al.* (2008) Label-free electrochemical detection of DNA using ferrocene-containing cationic polythiophene and PNA probes on nanogold modified electrodes. *Biosens. Bioelectron.*, **23** (7), 1175–1179.
23. Ben-Yoav, H., Dykstra, P.H., Bentley, W.E. and Ghodssi, R. (2012) A microfluidic-based electrochemical biochip for label-free diffusion-restricted DNA hybridization analysis. *Biosens. Bioelectron.*, **38** (1), 114–120.
24. Mateo-Martí, E. and Pradier, C.-M., A novel type of nucleic acid-based biosensors: the use of PNA probes, associated with surface science and electrochemical detection techniques, in *Intelligent and Biosensors*, InTech, 2010, 323–344.
25. Paleček, E. and Bartošík, M. (2012) Electrochemistry of nucleic acids. *Chem. Rev.*, **112** (6), 3427–3481.
26. Wang, J., Cai, X., Rivas, G. *et al.* (1997) Recognition and detection at chronopotentiometric DNA chips. *Science*, **12** (7), 587–599.
27. Wang, J. (2002) Electrochemical nucleic acid biosensors. *Anal. Chim. Acta*, **469** (1), 63–71.
28. Wang, J., Paleček, E., Nielsen, P.E. *et al.* (1996) Peptide nucleic acid probes for sequence-specific DNA biosensors. *Biochemistry*, **7863** (10), 7667–7670.
29. Ozkan-ariksoysal, D., Tezcanli, B., Kosova, B. and Ozsoz, M. (2008) Design of electrochemical biosensor systems for the detection of specific DNA sequences in PCR-amplified nucleic acids related to the polymorphism based on intrinsic guanine signal. *Society*, **80** (3), 588–596.
30. Wang, J., Kawde, A.-N. and Musameh, M. (2003) Carbon-nanotube-modified glassy carbon electrodes for amplified label-free electrochemical detection of DNA hybridization. *Analyst*, **128** (7), 912.
31. Hason, S., Pivonkova, H., Vetterl, V. and Fojta, M. (2008) Label-free sequence-specific DNA sensing using copper-enhanced anodic stripping of purine bases at boron-doped diamond electrodes. *Anal. Chem.*, **80** (7), 2391–2399.
32. Tosar, J.P., Keel, K. and Laíz, J. (2009) Two independent label-free detection methods in one electrochemical DNA sensor. *Biosens. Bioelectron.*, **24** (10), 3036–3042.
33. Drummond, T.G., Hill, M.G. and Barton, J.K. (2003) Electrochemical DNA sensors. *Nat. Biotechnol.*, **21** (10), 1192–1199.
34. Malhotra, B.D., Chaubey, A. and Singh, S.P. (2006) Prospects of conducting polymers in biosensors. *Anal. Chim. Acta*, **578** (1), 59–74.
35. Peng, H., Zhang, L., Soeller, C. and Travas-Sejdic, J. (2009) Conducting polymers for electrochemical DNA sensing. *Biomaterials*, **30** (11), 2132–2148.
36. Zhang, Q.D., Piro, B., Noël, V. *et al.* (2011) Applications of carbon nanotubes to electrochemical DNA sensors: a new strategy to make direct and selective hybridization detection from SWNTs. *Adv. Nat. Sci.: Nanosci. Nanotechnol.*, **1** (4), 045011.
37. Zhang, W., Yang, T., Zhuang, X. *et al.* (2009) An ionic liquid supported CeO_2 nanoshuttles-carbon nanotubes composite as a platform for impedance DNA hybridization sensing. *Biosens. Bioelectron.*, **24** (8), 2417–2422.
38. Pournaghi-Azar, M.H., Hejazi, M.S. and Alipour, E. (2006) Developing an electrochemical deoxyribonucleic acid (DNA) biosensor on the basis of human interleukine-2 gene using an electroactive label. *Anal. Chim. Acta*, **570** (2), 144–150.

39. Marin, S. and Merkoçi, A. (2009) Direct electrochemical stripping detection of cystic-fibrosis-related DNA linked through cadmium sulfide quantum dots. *Nanotechnology*, **20** (5), 055101.
40. Li, J., Liu, Q., Liu, Y. *et al.* (2005) DNA biosensor based on chitosan film doped with carbon nanotubes. *Anal. Biochem.*, **346** (1), 107–114.
41. Tichoniuk, M., Gwiazdowska, D., Ligaj, M. and Filipiak, M. (2010) Electrochemical detection of foodborne pathogen aeromonas hydrophila by DNA hybridization biosensor. *Biosens. Bioelectron.*, **26** (4), 1618–1623.
42. Zhang, J., Lao, R., Song, S. *et al.* (2008) Design of an oligonucleotide-incorporated nonfouling surface and its application in electrochemical DNA sensors for highly sensitive and sequence-specific detection of target DNA. *Anal. Chem.*, **80** (23), 9029–9033.
43. Yean, C.Y., Kamarudin, B., Ozkan, D.A. *et al.* (2008) Enzyme-linked amperometric electrochemical genosensor assay for the detection of PCR amplicons on a streptavidin-treated screen-printed carbon electrode. *Anal. Chem.*, **80** (8), 2774–2779.
44. Walter, A., Wu, J., Flechsig, G.-U. *et al.* (2011) Redox cycling amplified electrochemical detection of DNA hybridization: application to pathogen *E. coli* bacterial RNA. *Anal. Chim. Acta*, **689** (1), 29–33.
45. He, Y., Zeng, K., Zhang, X. *et al.* (2010) Electrochemistry communications ultrasensitive electrochemical detection of nucleic acid based on the isothermal strand-displacement polymerase reaction and enzyme dual amplification. *Electrochem. Commun.*, **12** (7), 985–988.
46. Rochelet-Dequaire, M., Djellouli, N., Limoges, B. and Brossier, P. (2009) Bienzymatic-based electrochemical DNA biosensors: a way to lower the detection limit of hybridization assays. *Analyst*, **134** (2), 349–353.
47. Lermo, A., Campoy, S., Barbé, J. *et al.* (2007) *In situ* DNA amplification with magnetic primers for the electrochemical detection of food pathogens. *Biosens. Bioelectron.*, **22** (9-10), 2010–2017.
48. Hu, K., Liu, P., Ye, S. and Zhang, S. (2009) Ultrasensitive electrochemical detection of DNA based on PbS nanoparticle tags and nanoporous gold electrode. *Biosens. Bioelectron.*, **24** (10), 3113–3119.
49. Castañeda, M.T., Alegret, S. and Merkoçi, A. (2007) Electrochemical sensing of DNA using gold nanoparticles. *Electroanalysis*, **19** (7-8), 743–753.
50. Pumera, M., Castañeda, M.T., Pividori, M.I. *et al.* (2005) Magnetically trigged direct electrochemical detection of DNA hybridization using Au_{67} quantum dot as electrical tracer. *Langmuir: ACS J. Surf. Colloids*, **21** (21), 9625–9629.
51. Chen, J., Zhang, J., Yang, H. *et al.* (2010) A strategy for development of electrochemical DNA biosensor based on site-specific DNA cleavage of restriction endonuclease. *Biosens. Bioelectron.*, **26** (1), 144–148.
52. Loaiza, Ó.A., Jubete, E., Ochoteco, E. *et al.* (2011) Gold coated ferric oxide nanoparticles based disposable magnetic genosensors for the detection of DNA hybridization processes. *Biosens. Bioelectron.*, **26** (5), 2194–2200.
53. Li, H., Sun, Z., Zhong, W. *et al.* (2010) Ultrasensitive electrochemical detection for DNA arrays based on silver nanoparticle aggregates. *Anal. Chem.*, **82** (13), 5477–5483.

54. Selvaraju, T., Das, J., Jo, K. *et al.* (2008) Nanocatalyst-based assay using DNA-conjugated Au nanoparticles for electrochemical DNA detection. *Langmuir*, **24** (17), 9883–9888.
55. Zhao, C., Ren, J., Gregolinski, J. *et al.* (2012) Contrasting enantioselective DNA preference: chiral helical macrocyclic lanthanide complex binding to DNA. *Nucleic Acids Res.*, **40** (16), 8186–96.
56. de la Escosura-Muñiz, A. and Mekoçi, A. (2010) Nanoparticle based enhancement of electrochemical DNA hybridization signal using nanoporous electrodes. *Chem. Commun. (Camb.)*, **46** (47), 9007–9009.
57. Reisberg, S., Dang, L.A., Nguyen, Q.A. *et al.* (2008) Label-free DNA electrochemical sensor based on a PNA-functionalized conductive polymer. *Talanta*, **76** (1), 206–210.
58. Chumbimuni-Torres, K.Y., Wu, J., Clawson, C. *et al.* (2010) Amplified potentiometric transduction of DNA hybridization using ion-loaded liposomes. *Analyst*, **135** (7), 1618–1623.
59. Numnuam, A., Chumbimuni-Torres, K.Y., Xiang, Y. *et al.* (2008) Potentiometric detection of DNA hybridization. *J. Am. Chem. Soc.*, **130** (2), 410–411.
60. Wu, J., Chumbimuni-Torres, K.Y., Galik, M. *et al.* (2009) Potentiometric detection of DNA hybridization using enzyme-induced metallization and a silver ion selective electrode. *Anal. Chem.*, **81** (24), 10007–10012.
61. Du, M., Yang, T. and Jiao, K. (2010) Rapid DNA electrochemical biosensing platform for label-free potentiometric detection of DNA hybridization. *Talanta*, **81** (3), 1022–1027.
62. Zhou, Y., Yu, B., Guiseppi-Elie, A. *et al.* (2009) Potentiometric monitoring DNA hybridization. *Biosens. Bioelectron.*, **24** (11), 3275–3280.
63. Giovanni, M., Bonanni, A. and Pumera, M. (2012) Detection of DNA hybridization on chemically modified graphene platforms. *Analyst*, **137** (3), 580–583.
64. Ercole, C., Del Gallo, M., Mosiello, L. *et al.* (2003) *Escherichia coli* detection in vegetable food by a potentiometric biosensor. *Sens. Actuators B*, **91** (1-3), 163–168.
65. Bonanni, A. and Del Valle, M. (2010) Use of nanomaterials for impedimetric DNA sensors: a review. *Anal. Chim. Acta*, **678** (1), 7–17.
66. Bonanni, A., Esplandiu, M.J. and Del Valle, M. (2010) Impedimetric genosensing of DNA polymorphism correlated to cystic fibrosis: a comparison among different protocols and electrode surfaces. *Biosens. Bioelectron.*, **26** (4), 1245–1251.
67. Park, S.-J., Taton, T.A. and Mirkin, C.A. (2002) Array-based electrical detection of DNA with nanoparticle probes. *Science (New York)*, **295** (5559), 1503–1506.
68. Yang, L. and Bashir, R. (2008) Electrical/electrochemical impedance for rapid detection of foodborne pathogenic bacteria. *Biotechnol. Adv.*, **26** (2), 135–150.
69. Wang, Y., Ye, Z. and Ying, Y. (2012) New trends in impedimetric biosensors for the detection of foodborne pathogenic bacteria. *Sensors (Basel)*, **12** (3), 3449–3471.
70. Li, C., Li, X., Liu, X. and Kraatz, H.-B. (2010) Exploiting the interaction of metal ions and peptide nucleic acids-DNA duplexes for the detection of a single nucleotide mismatch by electrochemical impedance spectroscopy. *Anal. Chem.*, **82** (3), 1166–1169.
71. Mohan, S., Nigam, P., Kundu, S. and Prakash, R. (2010) A label-free genosensor for BRCA1 related sequence based on impedance spectroscopy. *Analyst*, **135** (11), 2887–2893.

72. Shamsi, M.H. and Kraatz, H.-B. (2011) Electrochemical identification of artificial oligonucleotides related to bovine species. Potential for identification of species based on mismatches in the mitochondrial cytochrome C1 oxidase gene. *Analyst*, **136** (22), 4724–4731.
73. Ambrosi, A., Merkoci, A. and De la Escosura, A. (2009) Gold nanoparticles: a versatile label for affinity electrochemical biosensors, in *Biosensing Using Nanomaterials*, John Wiley & Sons, Inc., pp. 177–198.
74. Ahmed, M.U., Hasan, Q., Mosharraf Hossain, M. *et al.* (2010) Meat species identification based on the loop mediated isothermal amplification and electrochemical DNA sensor. *Food Control*, **21** (5), 599–605.
75. Mascini, M., Del Carlo, M., Minunni, M. *et al.* (2005) Identification of mammalian species using genosensors. *Bioelectrochemistry (Amsterdam)*, **67** (2), 163–169.
76. Peng, Y., Wang, X., Xiao, Y. *et al.* (2009) i-motif quadruplex DNA-based biosensor for distinguishing single- and multiwalled carbon nanotubes. *J. Am. Chem. Soc.*, **131** (38), 13813–13818.
77. Ligaj, M., Oczkowski, T., Jasnowska, J. *et al.* (2003) Electrochemical genosensors for detection of *L. Monocytogenes* and genetically-modified components in food. *Pol. J. Food Nutr. Sci.*, **12** (48), 61–63.
78. Rai, V., Nyine, Y.T., Hapuarachchi, H.C. *et al.* (2012) Electrochemically amplified molecular beacon biosensor for ultrasensitive DNA sequence-specific detection of *Legionella sp*. *Biosens. Bioelectron.*, **32** (1), 133–140.
79. Farabullini, F., Lucarelli, F., Palchetti, I. *et al.* (2007) Disposable electrochemical genosensor for the simultaneous analysis of different bacterial food contaminants. *Biosens. Bioelectron.*, **22** (7), 1544–1549.
80. Zheng, Y., Yang, C., Pu, W. and Zhang, J. (2009) Carbon nanotube-based DNA biosensor for monitoring phenolic pollutants. *Microchim. Acta*, **166** (1-2), 21–26.

11

Electrochemical Biosensors Based on Nanomaterials

Joseph Wang
Departments of Nanoengineering, University of California, 9500 Gilman Drive, San Diego, CA, 92093, USA

11.1 Why Nanoscale Materials?

Nanotechnology has recently become one of the most exciting forefront fields in electroanalytical chemistry. Nanotechnology is defined as the creation of functional materials, devices, and systems through control of matter at the 1–100 nm scale. A wide variety of nanoscale materials of different sizes, shapes, and compositions are now available [1]. The remarkable interest in nanomaterials is driven by their many desirable properties. In particular, the ability to tailor the size and structure and, hence, the properties of nanoscale materials offers excellent prospects for designing novel sensing systems and improving the performance of bioanalytical assays of foods or beverages and of our surrounding environment. This chapter highlights recent advances in nanomaterials for such electrochemical biosensing devices.

11.2 Nanowires, Nanotubes, and Nanoparticles

Research efforts on metal and metal semiconductor nanoparticles have flourished in recent years [2, 3]. Metal nanoparticles are generally defined as isolable particles between 1 and 50 nm in size that are prevented from agglomerating by protective shells. Owing to their small size, such nanoparticles have physical, electronic, and chemical properties that are different from those of bulk materials. Such properties strongly depend on the number and

Agricultural and Food Electroanalysis, First Edition.
Edited by Alberto Escarpa, María Cristina González and Miguel Ángel López.

kind of atoms that make up the particle. Several reviews have addressed the synthesis and properties of nanoparticles [2, 3]. Typically, such particles are prepared by the chemical reduction of the corresponding transition metal salts in the presence of a stabilizer (capping agent such as citrate or thiol) which binds to their surface to impart high stability and rich linking chemistry and provides the desired charge and solubility properties. Designer particles, including colloidal gold or inorganic nanocrystals, have found broad applications in many forms of biological tagging schemes. For example, colloidal quantum dot nanocrystals have been widely used for optical bioassays because of their size-tunable light-emitting properties [4]. Over the past decades, we have witnessed considerable research efforts aimed at designing powerful electrochemical biosensors based on nanoparticle tags and amplification platforms. A variety of nanomaterials have been shown to impart distinct advantages and improvements of electrochemical biosensing systems [5].

One-dimensional (1D) nanomaterials, such as carbon nanotubes (CNTs) and semiconductor- or conducting-polymer nanowires, are particularly attractive for bioelectronic detection [5]. Because of the high surface-to-volume ratio and novel electron transport properties of these nanostructures, their electronic conductance is strongly influenced by minor surface perturbations (such as those associated with the binding of macromolecules). Such 1D nanomaterials, thus, offer the prospect of rapid (real-time) and sensitive label-free bioelectronic detection, and massive redundancy in nanosensor arrays. The extremely tiny dimensions of these nanomaterials would allow packing a huge number of sensing elements onto a small footprint of an array device. Metal and conducting-polymer nanowires can be readily prepared by a template-directed electrochemical electrodeposition into the pores of a membrane template [6]. CNTs are particularly exciting 1D nanoscale materials that have received considerable attention owing to their unique structure-dependent electronic and mechanical properties [7, 8]. CNT can be divided into single-wall carbon-nanotubes (SWCNTs) and multiwall carbon-nanotubes (MWCNTs). SWCNTs possess a cylindrical nanostructure (with a high aspect ratio), formed by rolling up a single graphite sheet into a tube. SWCNTs can, thus, be viewed as molecular wires with every atom on the surface. MWCNTs are comprised of an array of such nanotubes that are concentrically nested like rings of a tree trunk. Since the discovery of the CNT in their early 1990s, there has been considerable activity exploring their attractive properties for a wide range of sensing applications, in connection with different transduction principles. The attractive properties of CNTs offer the possibility of developing superior electrochemical sensing devices, ranging from amperometric enzyme electrodes to label-free DNA hybridization biosensors [9]. The tailored electronic conductivity of conducting polymers, coupled with their ease of processing/modification and rich chemistry, make them extremely attractive as 1D sensing materials. Recently developed CNT/conducting-polymer nanowire materials [10], based on incorporating oxidized CNT as the charge-balancing dopants within electropolymerized films, should further enhance the sensing capabilities of 1D materials.

In the following sections, I will discuss how the unique properties of nanomaterials can lead to the creation of a new generation of high-performance electrochemical biosensors for diverse applications in food, environmental, and clinical analyses.

11.3 Nanomaterial-based Electrochemical Biosensors

Electrochemical biosensors offer several distinct advantages for food and environmental analyses. In particular, such devices offer attractive avenues for interfacing, at the molecular level, biomolecular recognition events, and electronic signal-transduction processes. In addition, electrochemical devices are extremely suitable for meeting the portability, cost, and power requirements of decentralized testing and indicate great promise for a wide range of food, environmental, and biomedical applications [11, 12]. Nanoscale materials can be used for enormously enhancing a variety of electrochemical biosensing protocols and the present chapter is divided accordingly. The organization of nanomaterials into controlled surface architectures is crucial for the successful realization of these biosensing applications.

11.3.1 Nanomaterial-based Biocatalytic Sensors

Enzyme electrodes have been widely used for monitoring a wide range of substrates of environmental or food significance. An important challenge in amperometric enzyme electrodes is the establishment of satisfactory electrical communication between the active site of the enzyme and the electrode surface [13]. The redox center of most oxidoreductases is electrically insulated by a protein shell. Because of this shell, the enzyme cannot be oxidized or reduced at an electrode at any potential. The possibility of direct electron transfer between enzymes and electrode surfaces could pave the way for superior reagentless biosensing devices, as it obviates the need for co-substrates or mediators and allows efficient transduction of the biorecognition event. "Trees" of aligned CNTs in the nanoforest, prepared by self-assembly, can act as molecular wires to improve the electrical communication between the underlying electrode and redox proteins (covalently attached to the ends of the SWNT) [14, 15]. Willner's team [16] demonstrated that aligned reconstituted glucose oxidase (GOx) on the edge of SWCNT can be linked to an electrode surface (Figure 11.1). Such enzyme reconstitution at the end of CNTs represents an efficient approach for "plugging" an electrode into GOx. Electrons were, thus, transported along distances higher than 150 nm with the length of the SWCNT controlling the rate of electron transport. An interfacial electron transfer rate constant of 42 s^{-1} was estimated for the 50-nm-long SWCNT. The catalytic properties of metal nanoparticles can also facilitate the electrical contact between the redox centers of enzymes with electrode transducers. For example, gold nanoparticles were shown to be extremely useful as electron relays (electrical nanoplugs) for the alignment of GOx on conducting supports and wiring its redox center [17].

A wide range of enzyme electrodes based on oxidase or dehydrogenase enzymes rely on amperometric monitoring of the liberated hydrogen peroxide or NADH. The anodic detection of these species at conventional electrodes is often hampered by the large overvoltage encountered for their oxidation. The greatly enhanced redox activity of hydrogen peroxide [18] and NADH [19] at CNT-modified electrodes addresses these overvoltage limitations and makes these nanomaterials extremely attractive for numerous

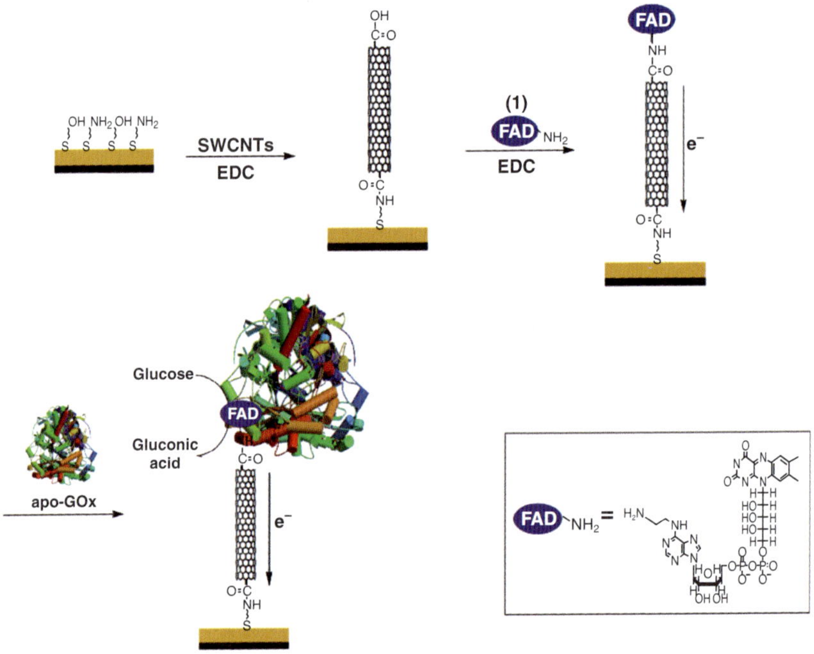

Figure 11.1 Assembly of SWCNT electrically contacted glucose oxidase electrode: linking the reconstituted enzyme, on the edge of the FAD-functionalized SWCNT, to the electrode surface (Reproduced with permission from [16]. Copyright © 2004, Wiley-VCH Verlag GmbH & Co, KGaA, Weinheim.)

oxidase- and dehydrogenase-based amperometric biosensors. The ability of CNTs to promote electron-transfer reactions is attributed to the presence of edge plane defects at their end caps. Carbon-nanotube-modified electrodes have also been shown to be extremely useful for circumventing surface fouling associated with the oxidation of the liberated NADH product [19]. The deposition of platinum nanoparticles onto the CNT has led to further improvements in the detection of the enzymatically liberated peroxide species [20]. In addition to CNT films, it is possible to use CNT-based inks [21] and pastes [22] for designing screen-printed and biocomposite, respectively, amperometric biosensors. The attractive electrocatalytic properties of metal nanoparticles (compared to bulk metal electrodes) can also benefit amperometric enzyme electrodes. For example, Niwa's team [23] dispersed 2 nm iridium nanoparticles in graphite-like carbon and used the resulting transducer for improved amperometric biosensing of glutamate.

Tao and coworkers [24] described a conducting-polymer nanosensor for detecting glucose based on a pair of nanoelectrodes, separated with a small (20–60 nm) gap connected by a polyaniline/glucose-oxidase film. The remarkably small dimensions of the new device, coupled with its very fast response and minimal oxygen consumption, makes it attractive for *in vivo* monitoring of glucose. Another promising approach for preparing conducting-polymer nanowire enzyme sensors involves electrodepolymerization within the channel between electrodes [25].

11.3.2 Nanomaterial-based Bioaffinity Sensors

The development of electrochemical DNA hybridization biosensors has received considerable research attention [26, 27]. Such DNA-sensing applications require high sensitivity through an amplified transduction of the oligonucleotide hybridization. Nanoparticle-based amplification protocols have led to the improved sensitivity of bioelectronic affinity assays by several orders of magnitudes [28]. In 2001, the groups of Wang [29] and Limoges [30] described the use of colloidal gold tags for the electronic detection of DNA hybridization. This strategy relies on capturing the nanoparticle tracer to the hybridized target, followed by a highly sensitive anodic-stripping voltammetric measurement of the metal tag. Analogous bioelectronic measurements of proteins based on sandwich immunoassays and gold nanoparticle tags have also been described [31]. Electronic DNA hybridization assays have been extended to other metal tracers, including silver [32] or iron [33]. Electrochemical measurements of silver nanoparticle tags can also be accomplished using potentiometric transduction in connection to highly sensitive silver ion-selective electrodes (Figure 11.2) [34]. Silver ions oxidatively released from silver enlarged gold nanoparticle labels in a sandwich immunoassay were, thus, detected by the potentiometric sensor. Quite often such nanoparticle-based electrochemical bioassays rely on the coupling biorecognition element to surfaces of magnetic beads, as it offers an effective minimization of nonspecific binding. The hybridization of probe-coated magnetic spheres with the metal-tagged targets results in three-dimensional network structures of magnetic spheres, crossed-linked together through the DNA and gold nanoparticles. The "magnetic" collection of such a magnetic-bead/DNA/metal-label assembly onto the electrode leads to the direct contact of the metal tag and the surface, and enables solid-state (chronopotentiometric) measurements without dissolving the metal tag [35]. This approach could facilitate the creation of magnetically addressable DNA arrays.

Several amplification processes can be used for dramatically enhancing the sensitivity of nanoparticle-based bioelectronic assays. For example, the metal nanoparticle tags can act as catalytic sites for the electroless deposition of other metals. Treatment of the gold-linked DNA-hybrid assembly with a silver ion in the presence of hydroquinone, thus, results in the catalytic deposition of silver on the gold tracer (acting as catalyst), leading to a dramatic (>100 fold) signal amplification [36]. Instead of enlarging spherical nanoparticle tracers, it is possible to enhance the sensitivity by using long nanowire tags [37].

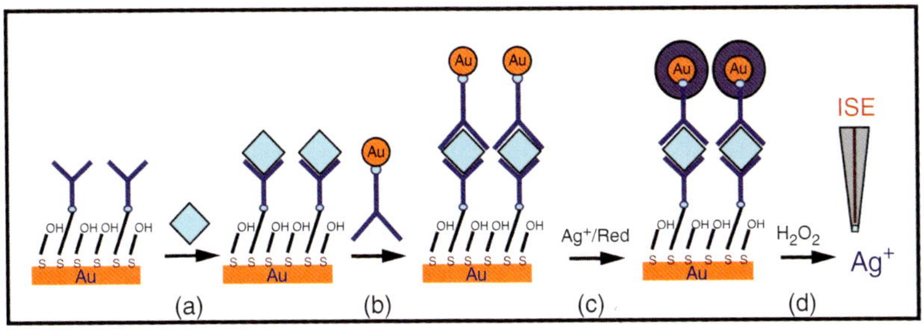

Figure 11.2 *Potentiometric biosensing of proteins with a ultrasensitive silver ion-selective microelectrode and silver nanoparticle label (Reproduced with permission from [34]. Copyright © 2006, American Chemical Society.)*

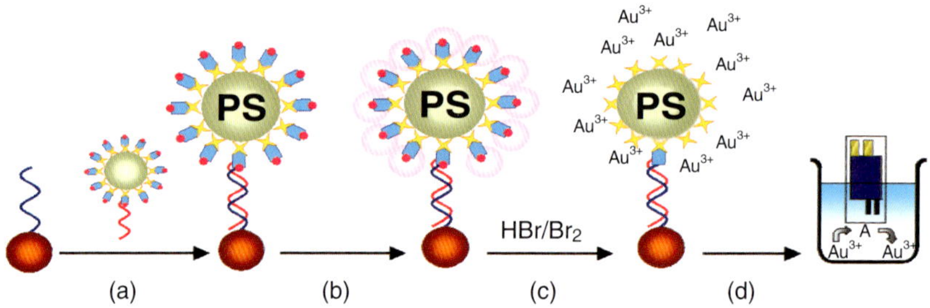

Figure 11.3 *Amplified bioelectronic detection of DNA hybridization, using polymeric microspheres carrying multiple gold nanoparticle tracers, catalytic enlargement of the gold particles, and a striping voltammetric signal transduction (Reproduced with permission from [38]. Copyright © 2004, Wiley-VCH Verlag GmbH & Co, KGaA, Weinheim.)*

Particularly useful for this task have been iridium nanowires. Wang's team described a triple-amplification bioassay, coupling the carrier-sphere amplifying units (loaded with numerous gold nanoparticles tracers) with the "built-in" preconcentration of the electrochemical stripping detection and a catalytic enlargement of the multiple gold-particle tags [38] (Figure 11.3). The success of these and other nanoparticle-based amplification protocols depends on our ability to maintain a low background response through proper attention to the surface-blocking chemistry and wash conditions.

Nanoparticle labels have also shown to be extremely useful for multiplexed biodetection. For example, inorganic (quantum dot) nanocrystals offer an electrodiverse population of electrical tracers as needed for designing electronic coding. Wang's team demonstrated the use of different inorganic-nanocrystal tags for a multitarget electronic detection of DNA [39] or proteins [40]. For example, three encoding nanocrystals (cadmium sulfide, zinc sulfide, and lead sulfide) were, thus, used to differentiate the signals of three target proteins in connection with a sandwich immunoassay and anodic stripping voltammetry of the corresponding metals (Figure 11.4). Each binding, thus, yields a distinct stripping voltammetric signal, whose position and size reflect the identity and level, respectively, of the corresponding protein target. This approach can be scaled up and multiplexed by using an automated parallel high-throughput microwell operation, with each microwell capable of carrying out multiple measurements. Libraries of electrical codes have been created by encapsulating different predetermined levels of multiple inorganic nanocrystals into polymeric carrier beads or depositing various metal tracers onto the pores of a host membrane [41]. The resulting electrochemical signatures reflect the predetermined proportions of the corresponding metals in such "identification" nanomaterials. Other attractive nanoparticle-based immunoelectrochemical bioassays of proteins were developed recently by the groups of Merkoçi [42, 43] and Rusling [44]. Various nanomaterial-based strategies for electrochemical immunoassays, reviewed recently by Rusling [44], are presented in Figure 11.5.

Nanoparticle-induced changes in the conductivity across a microelectrode gap can also be exploited for highly sensitive and selective electronic detection of DNA hybridization [45, 46]. The capture of the nanoparticle-tagged DNA targets by probes confined to the gap between the two microelectrodes, and a subsequent silver enlargement, results in a conductive metal layer across the gap, and leads to a measurable conductivity signal. Target DNA concentrations down to 500 fmol can, thus, be detected with remarkable

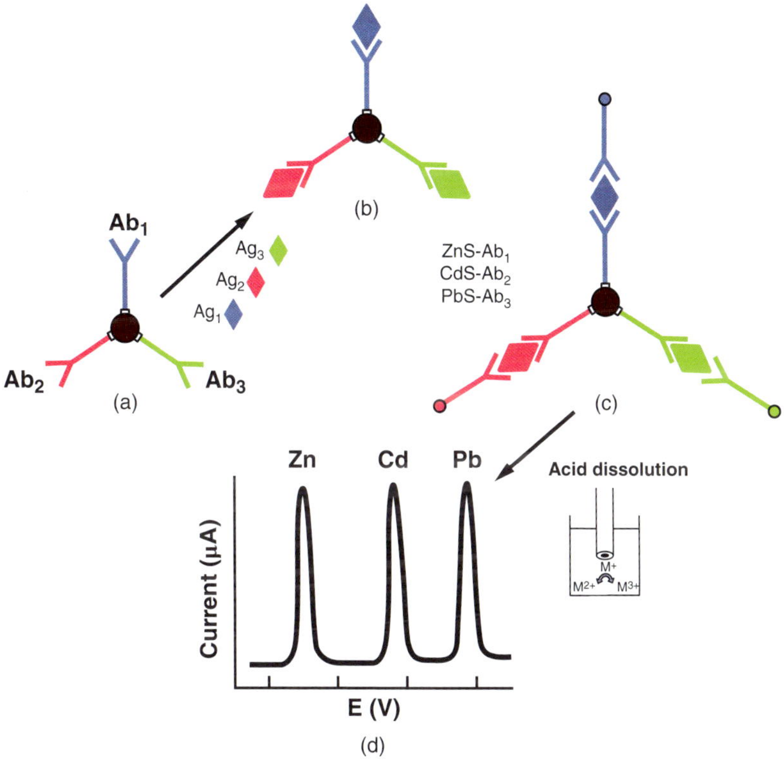

Figure 11.4 Multiantigen multiplexed sandwich immunoassay protocol based on different quantum dots nanocrystal tracers (Reproduced with permission from [40]. Copyright © 2004, American Chemical Society.)

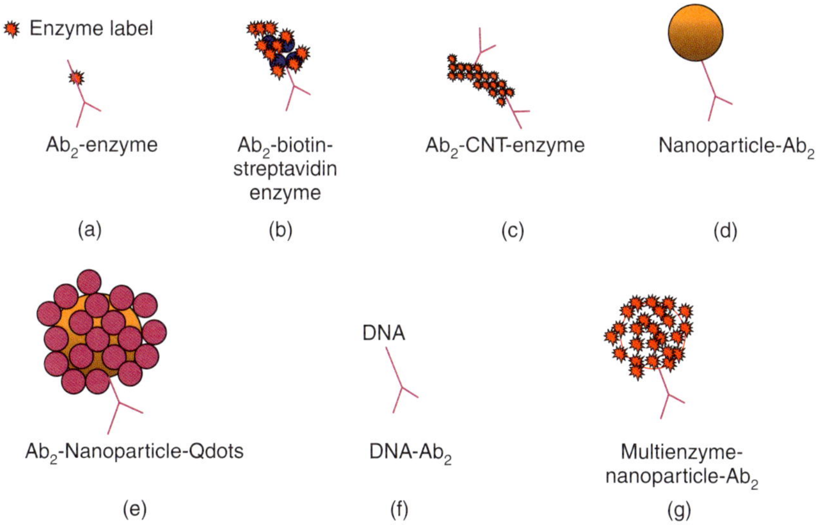

Figure 11.5 Amplification strategies for electrochemical immunosensors using nanoparticles or other moieties attached to secondary antibody (Ab2) (Reproduced with permission from [44]. Copyright © 2012, The Chemical Society of Japan and Wiley-VCH Verlag GmbH & Co, KGaA, Weinheim.)

point-mutation selectivity. This simple low-cost scheme offers the potential of the parallel readout of multiple electrode arrays. One-dimensional nanowires can also be used for bridging two closely spaced electrodes for label-free DNA detection. For example, p-type silicon nanowires – functionalized with PNA probes – have been shown extremely useful for real-time label-free conductometric monitoring of hybridization event [47]. This relies on the binding of the negatively charged DNA target that leads to an increase conductance, reflecting the increased surface charge.

Similar improvements were reported using nanowires and CNTs functionalized with other receptor molecules. For example, Lieber's team [48] reported on the use of nanowire devices for the direct real-time electronic detection of single virus molecules. Measurements made with nanowires modified with antibodies for influenza A displayed distinct and reversible conductivity changes upon binding and unbinding of single viruses. Conducting-polymer nanowire biosensors have also shown to be attractive for label-free bioaffinity sensing. For example, Mulchandani's group [49] reported on the real-time monitoring of nanomolar concentrations of biotin at an avidin-embedded polypyrrole nanowire. Similarly, noncovalent functionalization of CNTs was shown to be useful for label-free conductivity measurements of antibodies associated with human autoimmune diseases [50]. Nonspecific binding on the CNT was overcome by immobilizing polyethylene oxide chains.

CNTs can also lead to the ultrasensitive bioelectronic detection of DNA hybridization [51]. For example, multiwall CNT can be used as carriers for several thousands alkaline-phosphatase enzyme tracers and for accumulating the α-naphthol product of the enzymatic reaction (Figure 11.6). Such carriers relying on the alternate

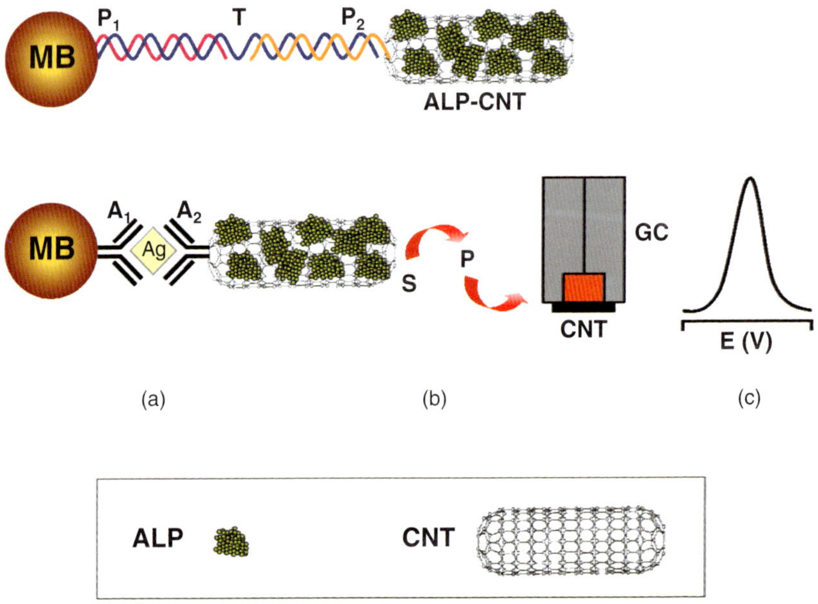

Figure 11.6 *Ultrasensitive bioassays of proteins and nucleic acids based on the amplification features of carbon-nanotube carriers and modified electrodes (Reproduced with permission from [51]. Copyright © 2004, American Chemical Society.)*

layer-by-layer electrostatic deposition of the enzyme on the CNT was used. Such CNT-based double amplification pathway (of both the recognition and transduction steps) allows the detection of DNA targets down to the 1.3 zmol level, hence indicating considerable promise for PCR-free DNA analysis.

The ability of CNTs to facilitate the adsorptive accumulation of the guanine nucleobase can lead to a dramatic amplification of label-free electrochemical detection procedures based on the intrinsic electroactivity of DNA [52]. The coupling of a CNT nanoelectrode array with the $Ru(bpy)_3{}^{+2}$-mediated guanine oxidation has facilitated the detection of subattomoles of DNA targets [53, 54]. Such CNT array was also applied for the label-free detection of DNA PCR amplicons, and offered the detection of less than 1000 target amplicons.

11.4 Future Prospects

The emergence of nanotechnology is opening new horizons for electrochemical biosensors. Recent years have witnessed the development of a variety of nanomaterial-based bioelectronic devices exhibiting novel functions. The use of nanomaterials in such sensing devices has taken off rapidly and will surely continue to expand. Nanoparticles, nanowires, and nanotubes have already made a major impact on the field of electrochemical biosensors, ranging from glucose enzyme electrodes to genoelectronic sensors. The resulting biosensors have shown to be an attractive alternative for a wide range of practical analytical applications.

What does the future hold for this technology? The unique properties of nanomaterials suggest that future interdisciplinary research could lead to a new generation of electrochemical biosensors and attractive new applications in the fields of environmental and food analyses. We are currently exploring nanoparticle-based protocols for the electronic detection of proteins. The use of nanoparticle tags for detecting and coding proteins is in its infancy, but the lessons learned from nucleic acid measurements should provide useful starting points. The monitoring of protein and protein interactions presents a greater challenge than that of nucleic acids, owing to the absence of (PCR-like) amplification technologies, the complexity of proteins, and their stronger nonspecific binding to solid supports. Nanoparticles comprising of mixed (recognition/shielding) monolayers are desired to fully utilize the potential of protein–nanoparticle hybrids. The addition of protein analysis to the arsenal of particle-based bioassays represents an important step in the direction of making particle bioelectronics a universal biodetection platform. Multiple electrode protein and DNA arrays based on nanoparticle-amplification platforms are, thus, expected in the near future.

One-dimensional nanotubes and nanowires are extremely attractive for a wide range of bioelectronic sensing applications. The ability to modify nanowires and nanotubes with biological recognition elements imparts high selectivity onto devices based on 1D nanostructures. While several novel sensing concepts based on 1D nanowires have been presented, incorporating these materials into routine functional devices remains a challenge. The successful bioelectronic utility of 1D nanostructures requires new nanofabrication capabilities and proper attention to the interconnection challenge, involving reproducible positioning of nanowires and nanotubes between closely spaced microelectrodes. Such attention to the nanotechnology/microtechnology interface is essential for assembling nanosensors into functional integrated devices. Proper attention should also be paid to the interface of these devices with the real world (i.e., to sample delivery issues). Such activity is expected to lead to powerful sensor arrays for parallel real-time monitoring of multiple analytes. The development of such biosensor arrays requires new methods for confining different biomolecules onto closely spaced 1D nanostructures.

A wide range of newly introduced nanoscale materials is expected to expand the realm of nanomaterial-based electrochemical sensors and biosensors. In addition, the judicious coupling of two different nanomaterials (i.e., inorganic/carbon nanohybrids) has been shown to offer further improvements in the analytical performance, superior to that observed when a single nanomaterial is used [55]. Such nanobioelectronic devices are expected to have a major impact upon ensuring our food safety or water quality, and upon other areas ranging from clinical diagnostics to security surveillance.

References

1. Hornyak, G.L., Moore, J.J., Tibbals, H.F., and Dutta, J. (2008) *Fundamentals of Nanotechnology*, CRC Press.
2. Fedlheim, D.L. and Foss, C.A. (2001) *Metal Nanoparticles: Synthesis, Characterization, and Applications*, CRC Press.
3. Niemeyer, C.M. (2001) Nanoparticles, proteins, and nucleic acids: biotechnology meets materials science. *Angew. Chem. Int. Ed.*, **40**, 4129.
4. Michalat, X., Pinaud, F.F., Tsay, J.M., Doose, S., Li, J.J., Wu, A.M., Gambhir, S.S., and Weiss, S. (2005) Quantum dots for live cells, in vivo imaging, and diagnostics. *Science*, **307**, 538–544.
5. Wang, J. (2005) Nanomaterial-based electrochemical biosensors. *Analyst*, **130**, 421–426.
6. Martin, C.R. (1996) Membrane-based synthesis of nanomaterials. *Chem. Mater.*, **8**, 1739–1746.
7. Rao, C.N., Satishkumar, C., Govindaraj, A., and Nath, M. (2001) Nanotubes. *ChemPhysChem*, **2**, 78–105.
8. Vairavapandian, D., Vichchulada, P., and Lay, M.D. (2008) Preparation and modification of carbon nanotubes: review of recent advances and applications in catalysis and sensing. *Anal. Chim. Acta.*, **626**, 119–129.
9. Wang, J. (2005) Carbon-nanotube based electrochemical biosensors: a review. *Electroanalysis*, **17**, 7–14.
10. Wang, J., Dai, J., and Yarlagadda, T. (2005) Carbon nanotube-conducting-polymer composite nanowires. *Langmuir*, **21**, 9–12.
11. Wang, J. (2002) Portable electrochemical systems. *Trends Anal. Chem.*, **21**, 226–232.
12. Wang, J. (2002) Real-time electrochemical monitoring: toward green analytical chemistry. *Acc. Chem. Res.*, **35**, 811–816.
13. Guindilis, A., Atanasov, P., and Wilkins, E. (1997) Enzyme-catalyzed direct electron transfer: fundamentals and analytical applications. *Electroanalysis*, **9**, 661–674.
14. Gooding, J.J., Wibowo, R., Liu, J.Q., Yang, W., Losic, D., Orbons, S., Mearns, F.J., Shapter, J.G., and Hibbert, D.B. (2003) Protein electrochemistry using aligned carbon nanotube arrays. *J. Am. Chem. Soc.*, **125**, 9006–9007.

15. Yu, X., Chattopadhyay, D., Galeska, I., Papadimitrakopoulos, F., and Rusling, J.F. (2003) Peroxidase activity of enzymes bound to the ends of single-wall carbon nanotube forest electrodes. *Electrochem. Commun.*, **5**, 408–411.
16. Patolsky, F., Weizmann, Y., and Willner, I. (2004) Long-range electrical contacting of Redox enzymes by SWCNT connectors. *Angew. Chem. Int. Ed.*, **43**, 2113–2117.
17. Xiao, Y., Patolsky, F., Katz, E., Hainfeld, J.F., and Willner, I. (2003) "Plugging into Enzymes": nanowiring of redox enzymes by a gold nanoparticle. *Science*, **299**, 1877.
18. Wang, J., Musameh, M., and Lin, Y. (2003) Solubilization of carbon nanotubes by Nafion toward the preparation of amperometric biosensors. *J. Am. Chem. Soc.*, **125**, 2408–2409.
19. Musameh, M., Wang, J., Merkoci, A., and Lin, Y. (2002) Low-potential stable NADH detection at carbon-nanotube-modified glassy carbon electrodes. *Electrochem. Commun.*, **4**, 743–746.
20. Hrapovic, S., Liu, Y., Male, K., and Luong, J.H.T. (2004) Electrochemical biosensing platforms using platinum nanoparticles and carbon nanotubes. *Anal. Chem.*, **76**, 1083–1088.
21. Wang, J. and Musameh, M. (2004) Carbon nanotube screen-printed electrochemical sensors. *Analyst*, **129**, 1–2.
22. Rubianes, M.D. and Rivas, G.A. (2003) Carbon nanotubes paste electrode. *Electrochem. Commun.*, **5**, 689–694.
23. You, T.Y., Niwa, O., Kurita, R., Iwasaki, Y., Hayashi, K., Suzuki, K., and Hirono, S. (2004) Reductive H_2O_2 detection at nanoparticle iridium/carbon film electrode and its application as L-glutamate enzyme sensor. *Electroanalysis*, **16**, 54–59.
24. Forzani, E.S., Zhang, H., Nagahara, L.A., Amlani, I., Tsui, R., and Tao, N.J. (2004) A conducting polymer nanojunction sensor for glucose detection. *Nano Lett.*, **4**, 1785–1788.
25. Ramanathan, K., Bangar, M.A., Yun, M., Chen, W., Mulchandani, A., and Myung, N.V. (2004) Individually addressable conducting polymer nanowires array. *Nano Lett.*, **4**, 1237–1239.
26. Palecek, E. and Fojta, M. (2001) Detecting DNA hybridization and damage. *Anal. Chem.*, **73**, 74A–83A.
27. Gooding, J.J. (2002) Electrochemical DNA hybridization biosensors. *Electroanalysis*, **14**, 1149–1156.
28. Wang, J. (2005) Nanomaterial-based amplified transduction of biomolecular interactions. *Small*, **1**, 1036–1043.
29. Wang, J., Xu, D., Kawde, A.N., and Polsky, R. (2001) Metal nanoparticle-based electrochemical stripping potentiometric detection of DNA hybridization. *Anal. Chem.*, **73**, 5576–5581.
30. Authier, L., Grossiord, C., Brossier, P., and Limoges, B. (2001) Gold nanoparticle-based quantitative electrochemical detection of amplified human cytomegalovirus DNA using disposable microband electrodes. *Anal. Chem.*, **73**, 4450–4456.
31. Dequaire, M., Degrand, C., and Limoges, B. (2000) An electrochemical metalloimmunoassay based on a colloidal gold label. *Anal. Chem.*, **72**, 5521–5528.
32. Cai, H., Xu, Y., Zhu, N., He, P., and Fang, Y. (2002) An electrochemical DNA hybridization detection assay based on a silver nanoparticle label. *Analyst*, **127**, 803–808.

33. Wang, J., Liu, G., and Merkoci, A. (2003) Particle-based detection of DNA hybridization using electrochemical stripping measurements of an iron tracer. *Anal. Chim. Acta.*, **482**, 149–155.
34. Chumbimuni-Torres, K.Y., Dai, Z., Rubinova, N., Xiang, Y., Pretsch, E., Wang, J., and Bakker, E. (2006) Potentiometric biosensing of proteins with ultrasensitive ion-selective microelectrodes. *J. Am. Chem. Soc.*, **128**, 13676–13677.
35. Wang, J., Xu, D., and Polsky, R. (2002) Magnetically-induced solid-state electrochemical detection of DNA hybridization. *J. Am. Chem. Soc.*, **124**, 4208–4209.
36. Wang, J., Polsky, R., and Danke, X. (2011) Silver-enhanced colloidal gold electrochemical stripping detection of DNA hybridization. *Langmuir*, **17**, 5739–5741.
37. Wang, J., Liu, G., and Zhou, J. (2003) Indium microrod tags for electrochemical detection of DNA hybridization. *Anal. Chem.*, **75**, 6218–6222.
38. Kawde, A. and Wang, J. (2004) Amplified electrical transduction of DNA hybridization based on polymeric beads loaded with multiple gold nanoparticle tags. *Electroanalysis*, **16**, 101–117.
39. Wang, J., Liu, G., and Merkoçi, A. (2003) Electrochemical coding technology for simultaneous detection of multiple DNA targets. *J. Am. Chem. Soc.*, **125**, 3214–3215.
40. Liu, G., Wang, J., Kim, J., and Jan, M.R. (2004) Electrochemical coding for multiplexed immunoassays of proteins. *Anal. Chem.*, **76**, 7126–7130.
41. Wang, J., Liu, G., and Rivas, G. (2003) Encoded beads for electrochemical identification. *Anal. Chem.*, **75**, 4661–4671.
42. Merkoçi, A. (2010) Nanoparticles-based strategies for DNA, protein and cell sensors. *Biosens. Bioelectron.*, **26**, 1164–1177.
43. Perfézou, M., Turner, A., and Merkoçi, A. (2012) Cancer detection using nanoparticle-based sensors. *Chem. Soc. Rev.*, **41**, 2606–2622.
44. Rusling, J.F. (2012) Nanomaterials-based electrochemical immunosensors for proteins. *Chem. Rec.*, **12**, 164–176.
45. Park, S., Taton, T.A., and Mirkin, C.A. (2002) Array-based electrical detection of DNA with nanoparticle probes. *Science*, **295**, 1503–1506.
46. Moller, R., Csaki, A., Kohler, J., and Fritzsche, W. (2001) Electrical classification of the concentration of bioconjugated metal colloids after surface adsorption and silver enhancement. *Langmuir*, **17**, 5426–5430.
47. Hahm, J. and Lieber, C.M. (2004) Direct ultrasensitive electrical detection of DNA and DNA sequence variations using nanowire nanosensors. *Nano Lett.*, **4**, 51–54.
48. Patolsky, F., Zheng, G., Hayeden, O., Lakadamyali, M., Zhuang, X., and Lieber, C. (2004) Electrical detection of single viruses. *Proc. Natl. Acad. Sci. U.S.A.*, **101**, 14017–14022.
49. Ramanathan, K., Bangar, M.A., Yun, M., Chen, W., Myung, N.V., and Mulchandani, A. (2005) Bioaffinity sensing using biologically functionalized conducting-polymer nanowire. *J. Am. Chem. Soc.*, **127**, 496–497.
50. Chen, R.J., Bangsaruntip, S., Drouvalakis, K.A., Shi Kam, N.W., Shim, M., Li, Y., Kim, W., Utz, P.J., and Dai, H. (2003) Noncovalent functionalization of carbon nanotubes for highly specific electronic biosensors. *Proc. Natl. Acad. Sci. U.S.A.*, **100**, 4984–4989.

51. Wang, J., Liu, G., and Jan, M. (2004) Ultrasensitive electrical biosensing of proteins and DNA: carbon-nanotube derived amplification of the recognition and transduction events. *J. Am. Chem. Soc.*, **126**, 3010–3011.
52. Wang, J., Kawde, A., and Mustafa, M. (2003) Carbon-nanotube-modified glassy carbon electrodes for amplified label-free electrochemical detection of DNA hybridization. *Analyst*, **128**, 912–916.
53. Li, J., Ng, H.T., Cassell, A., Fan, W., Chen, H., Ye, Q., Koehne, J., and Meyyappan, M. (2003) Carbon nanotube nanoelectrode array for ultrasensitive DNA detection. *Nano Lett.*, **3**, 597–602.
54. Koehne, J.E., Chen, H., Cassell, A.M., Yi, Q., Han, J., Meyyappan, M., and Li, J. (2004) Miniaturized multiplex label-free electronic chip for rapid nucleic acid analysis based on carbon nanotube nanoelectrode arrays. *Clin. Chem.*, **50**, 1886–1893.
55. Campuzano, S. and Wang, J. (2011) Nanobioelectroanalysis based on carbon/inorganic hybrid nanoarchitectures. *Electroanalysis*, **23**, 1289–1300.

12

Electrochemical Sensing on Microfluidic Chips

Alberto Escarpa, María Cristina González and Miguel A. López
Departamento de Química Analítica, Química Física e Ingeniería Química, Universidad de Alcalá, Ctra, Madrid-Barcelona km 33, 600, 28871, Alcalá de Henares, Madrid, Spain

12.1 Electrochemical Detection Implementation in Microfluidic Chips

The final objective of analytical miniaturized systems is represented by the concept of micro total analysis systems (micro-TASs) and lab-on-a-chip (LOC) technology [1, 2]. Micro-TAS concept was developed from the modification of the total analysis systems (TASs) by downsizing and integrating its analytical multiple steps (sample preparation, separation/reaction, and detection) onto single monolithic devices. The immediate consequence of this miniaturization in scale is that these devices allow fast analysis time and extremely low samples and reagents consumption.

However, one of the most relevant characteristics of these analytical microsystems is the possibility of handling fluidics on the nanoliter and even picoliter scale, which has widened the scope of micro-TAS to *microfluidics,* defined as the science and technology of systems that process or manipulate small amounts of fluidics ($10^{-9}-10^{-18}$ l), using channels measuring from tens to hundreds of micrometers [1].

An important "product" of microfluidic-LOC technology is the termed microfluidic chip (MC). An MC is an analytical system constituted at least by an injector (where a sample plug is loaded), a set of microchannels (where hydrodynamic or electrophoretic transport/separation of analytes is performed) interfaced suitably to reservoirs (where different solutions/samples are deposited) and a detector (where molecule detection is carried out). Microchannels and reservoirs are fabricated using microfabrication techniques and facilities (i.e., photolithography or micromolding) to form channels for sample injection,

Agricultural and Food Electroanalysis, First Edition.
Edited by Alberto Escarpa, María Cristina González and Miguel Ángel López.

separation, and analyte detection. Once all solutions, including those of the samples, are loaded, the samples are typically transferred into an injector region. Then their components are moved and/or separated and afterward detected.

Microchip platform (mostly into a monolithic piece) eliminates the necessity of most fluidic connections that otherwise link microfluidic components. Avoiding such connections greatly reduces sample dispersion, delay times, and dead volumes between different microchip compartments, and therefore, significantly increasing the separation power of such integrated miniaturized systems.

As relevant examples of MCs, microchip electrophoresis (ME) was one of the earliest examples of micro-TAS, and it constitutes one of the most representative examples of analytical microsystems [3]. The easy microfabrication of a network of channels using materials of well-known chemistry, and the possibility of using the electrokinetic phenomena to move fluids are among the most important factors to understand the relevance of ME to miniaturization. Since electrokinetic is easy to apply (only a pair of electrodes are needed), electroosmotic-driven flow (EOF) has been successfully implemented using different types of materials to manufacture channels, being glass the most commonly used. Microfabrication on polymers is faster and cheaper than on glass, so these materials have great potential for mass production. However, glass chips present the best EOF and the chemical modification on the surfaces of its microchannels is better known and easier as compared to polymers.

The typical layout of an ME (both a simple cross T-injector (Figure 12.1a), and a twin-T injector (Figure 12.1b)) is depicted in Figure 12.1. It has a network of channels with widths varying from 10 to 100 μm with typical straight separation channels between 3 and 10 cm. In a typical setup, buffer solutions are introduced on sample (S), running buffer (RB), and waste (W) reservoirs. The reservoir volume is often defined by their capacities (100–250 μl). Electric fields are applied to the reservoirs by high voltage power supplies (1–5 kV), and platinum electrodes are placed in the reservoirs for injection and separation.

Basically, two fluidic manipulations have to be used in ME: injection of a defined plug of sample and their analyte separation as it is illustrated in Figure 12.2. A high voltage is applied to the sample reservoir for a short time with the detection reservoir held at ground (*injection*). Sample is introduced directly into the separation channel by electrokinetic injection (the ability to dispense very well defined and small volumes of sample plugs with reproducibility is of paramount importance). After the injection is completed, the high voltage is switched back to the buffer reservoir and the separation is initiated (*run/separation*) [4].

Detection has been one of the main challenges for analytical microsystems, as very sensitive techniques are needed as a consequence of the ultrasmall sample volumes used in micron-sized environments. Electrochemical detection (ED) is a very suitable detection principle to be coupled in microchips because it presents the inherent ability for miniaturization without loss of performance and its high compatibility with microfabrication techniques. Similarly, it possesses high sensitivity, its responses are not dependent on the optical path length or sample turbidity, and it has low power supply requirements which are its additional advantages [5–9].

Both, amperometry and conductimetry have been successfully implemented as detection principles in microchip technology.

The amperometric detector design should ensure well-defined mass transport, minimal band broadening, and electrical isolation (decoupling) from the high separation voltage

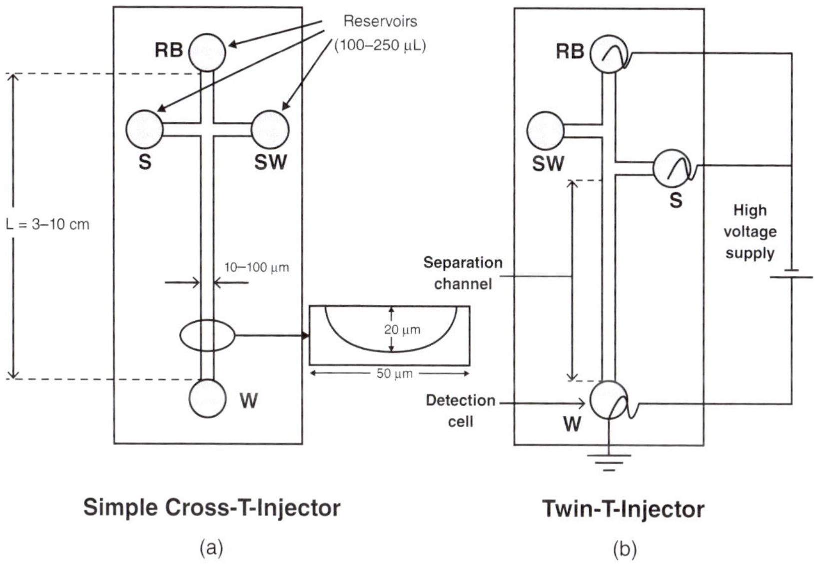

Figure 12.1 *Common microchip layouts. (a) Simple cross-T injector and (b) Twin-T injector*

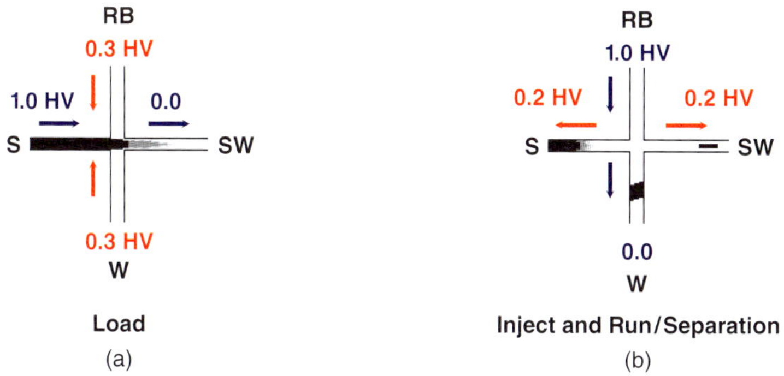

Figure 12.2 *Basic fluidic manipulations used in microchip electrophoresis: injection of sample and analytes separation. Arrows depict direction of flow of sample and separation buffer. (Adapted with permission from Ref. [4]. © 2004 WILEY-VCH Verlag GmbH & Co. KGaA, Weinheim.)*

(typically 1–5 kV). The latter is attributed to the fact that the current associated with the high separation voltage is usually several orders of magnitude larger than that measured at the electrochemical detector. High sensitivity, selectivity (via the applied potential and electrode material), simple handling, long-term stability, and rigidity are additional requirements.

Regarding amperometry implementation, the main approaches proposed have been taking into account the relative position between both working electrode-separation channel [5], where the configurations can be classified as: *end-channel, in-channel,* and *off-channel* detection as it is illustrated in Figure 12.3.

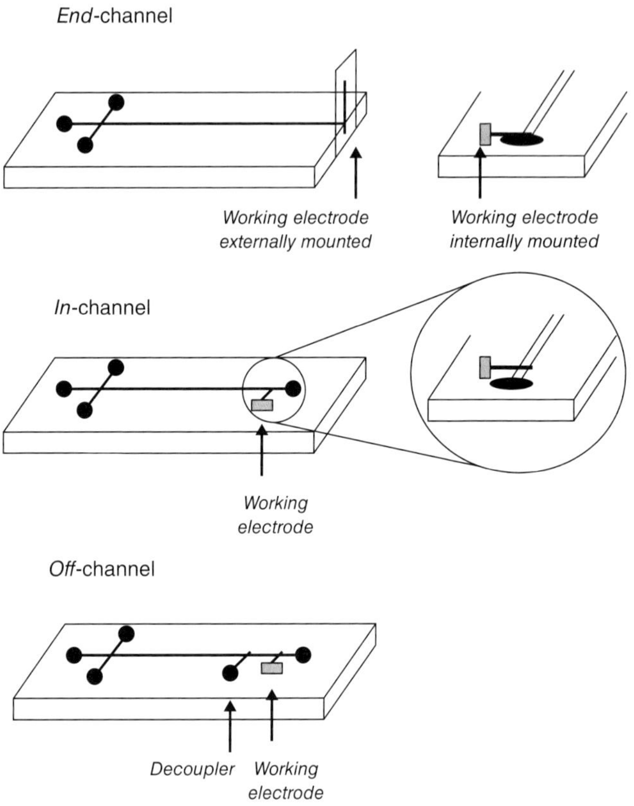

Figure 12.3 *Common configurations of electrochemical detection for microchip electrophoresis based on different capillary/working electrode arrangements:* end-*channel (external or internal),* in-*channel, and* off-*channel. (Adapted with permission from Ref. [5]. © 2002 WILEY-VCH Verlag GmbH & Co. KGaA, Weinheim.)*

End-channel detection involves the alignment of the electrode at the end of the separation channel (tens of micrometers) and it means that the electrode is outside the channel. Separation voltage has a minimal influence on the potential applied in an electrochemical detector because most of the voltage is dropped across the channel [5, 10]. The main advantage of this design is that no decoupler is required, and that the whole system is simpler and more rugged because the channel is all in one piece. However, the main disadvantage is a loss of separation efficiency, due to the relatively short length of the channel and the large distance between the end of the channel and the working electrode. This separation distance is also crucial for the noise signal obtained and it can lead to a complete loss of the analytical current.

In-channel strategies involve the placement of the working electrode directly within the separation channel. The analytes migrate over the electrode while still confined to the channel; thus eliminating the band broadening often observed with end-channel alignments [10].

Off-channel detection has also been employed to overcome the drawbacks of end-channel detection. Electrode placement in *off-channel* detection is similar to that in in-channel

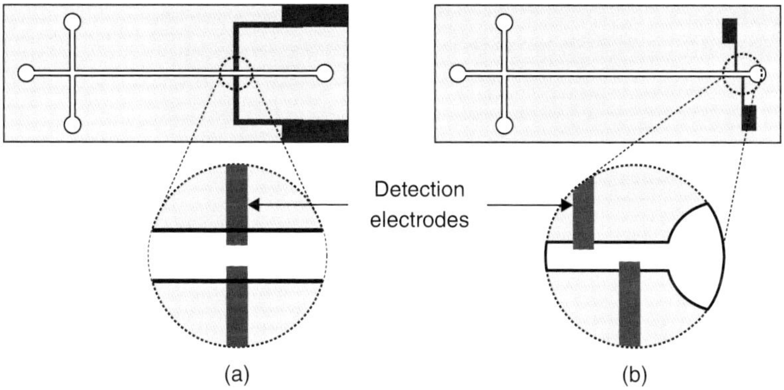

Figure 12.4 Configurations of conductivity detection for microchip electrophoresis: galvanic (a) and contactless (b)

detection, but the separation voltage is isolated from the amperometric current by using a decoupler. Conceptually speaking, the decoupler effectively shunts the separation voltage to ground, and a field-free region is created where analytes are pushed past the electrode by EOF generated prior to the decoupler [11].

Conductimetric detection is a less sensible, but universal detection technique that has been applied as a detection mode in ME, either in the galvanic (a pair of electrodes is placed in the separation channel for liquid impedance measurement) [12, 13] or the contactless mode (no contact between the pair of electrodes and separation channel solution) [14–16]. Both formats are illustrated in Figure 12.4. Contactless detection (CCD) is preferred for three reasons: (i) the electronic circuit is decoupled from the high-voltage applied for separation (no direct coupling between the electronics and the liquid in the channel), (ii) the formation of glass bubbles at the metal electrodes is prevented, and (iii) electrochemical modification or degradation of the electrode surface is prevented.

Food electro analysis on microfluidics and LOC technology has been carried out by following the two main approaches [17–22]: (i) improving selectivity toward separation systems with nonspecific detectors coupled after the separation of sample mixture in distinct zones of single analytes (format ME), or (ii) improving selectivity toward detection system using selective microfluidic (bio) sensors (format MC).

Selected examples involving ME for the separation of organic and inorganic food-related analytes, using amperometric and conductimetric detection, as well as involving MCs as biosensing platforms will be separately discussed in the following sections.

12.2 Microchip Electrophoresis with Electrochemical Detection for Food Analysis

12.2.1 Microchip Electrophoresis with Amperometric Detection for Organic Food Analytes

Detection of (natural) antioxidants, especially of polyphenolic compounds and vitamins, has been one of the most relevant and earliest studied examples on ME-AD in the field of food analysis because of its high significance. Carbon screen printed electrodes (CSPEs) coupled with ME have allowed a fast detection of food-related phenolic acids including chlorogenic, genistic, ferulic, and vanillic acids [23].The analytes were separated by capillary zone electrophoresis (CZE) because of the different extent of ionization of the phenolics under a borate buffer (pH 9.5). The use of 1 M nitric acid in the detector compartment dramatically improved the sensitivity/electroactivity of phenolics. The performed protocol used less than 300 s and an application to commercial wines without sample preparation was performed where only two of them (vanillic and chlorogenic) were actually detected.

Micellar electrokinetic chromatography (MEKC) with pulsed amperometric detection (PAD) using a polydimethylsiloxane (PDMS) microchip has been performed for the analysis of catechins and derivatives in green tea [24]. The detection of (+)-catechin, epigallocatechin gallate, (−)-epicatechin, and epicatechin gallate was achieved in less than 5 min improving the time employed using a high-performance liquid chromatography (HPLC) system (10 min).

Glassy carbon electrode (GCE) aligned to ME has also been explored for the fast and simultaneous detection of prominent natural antioxidants including examples of flavonoids and ascorbic acid [25]. Determination involved the electrophoretic separation of arbutin, (+)-catechin, and ascorbic acid in less than 4 min as well as the separation of different "couples" of natural antioxidants of food significance such as (+)-catechin and ascorbic acid, (+)-catechin and rutin, or arbutin and phlorizdin.

The sequential use of both ME reservoirs to integrate calibration and sample analysis on board the chip has been one of the important achievements in using this technology in the food analysis field [26–28]. In this strategy, a standard mixture is introduced in a reservoir (calibration reservoir) to obtain a calibration factor (defined as signal/concentration standard) and, sequentially, a food sample is introduced in the second one (sample-analysis reservoir). This useful strategy has been explored in determining water-soluble vitamins (pyridoxine, ascorbic, and folic acids) [27] as well as isoflavones [28]. Calibration and analysis protocols of vitamins were performed in less than 350 s, thereby considerably improving the overall analysis time and simplifying the analytical process being very useful for routine analysis in the agricultural-food sector.

ME-AD has also been creatively used to integrate and simplify on a microscale the traditional methods for complex natural antioxidant determination [29]. Basically, depending on the acid–base properties of the analytes, two approaches (*class-selective electrochemical index determination* – CSEID and *individual antioxidant determination* – IAD) were investigated for determining nine antioxidants ((+)-catechin, rutin, quercetin, chlorogenic, ferulic, caffeic, protocatechuic, vanillic, and gallic acids) in food samples. Firstly, the novel concept of CSEID is proposed, allowing a fast and reliable determination of the main antioxidant classes (flavonoids and phenolic acids) in less than 100 s at pH 5.0. At this target pH, all acids are negatively charged and all flavonoids remain neutral resulting in two zones exhibiting well-defined peaks. Also, an impressive separation of nine antioxidants at pH 9.0 is also offered in less than 260 s with the IAD approach showing the analytical potency of these microchips. A set of representative samples were analyzed including: apples and pears, red and white wines, and green tea tablets. In addition, in this work different integration strategies were further explored by using both reservoirs, looking for a simplification of the overall analytical process without losing the excellent analytical characteristics obtained in both approaches.

Figure 12.5 illustrates the schematics of the analytical simplification strategies for natural antioxidant determination with the corresponding electropherograms obtained. First, note that the numbers i–iv indicate the sequential buffer flow and sample/standard injection order of the analysis. The first strategy consisted of the integration of the two previously studied analytical approaches (sequential CSEID at pH 5.0 and IAD at pH 9.0) using one microchip reservoir for the application of each. This strategy has been demonstrated for standard solution (Figure 12.5a) and green tea extract samples (Figure 12.5b). In the first case, the standard (containing two flavonoids and two phenolic acids) was injected from one reservoir in 2-(*N*-morpholino)ethanesulfonic acid (MES) buffer at pH 5.0 resulting in two peaks (Figure 12.5a, electropherogram ii) corresponding to total flavonoids I ((+)-catechin and rutin) and total phenolic acids II (vanillic and chlorogenic acid). A consecutive sample injection from the other reservoir, but this time in borate buffer at pH 9.0, led to the complete separation of the four antioxidants (Figure 12.5a electropherogram iv). When applied to green tea extracts, this strategy gave rise to the electropherograms presented in Figure 12.5b. In this case, one peak was found at pH 5.0, corresponding to total flavonoids, while the injection in borate buffer at pH 9.0 resulted in the separation of (+)-catechin and rutin. The second strategy consisted of a simplified calibration and subsequent sample determination as it was described earlier. Figure 12.5c presents the electropherograms obtained for a calibration standard solution (containing (+)-catechin and gallic acid), injected by one of the microchip reservoirs (electropherogram ii), and the following total flavonoids CSEID in the sample, injected by the other reservoir (electropherogram iii).

CSEID approach has also been successfully transferred to the "easy-to-use" SU-8/Pyrex microchips. A disposable SU-8/Pyrex single-channel microchip integrating three 100 μm Pt electrodes (ME-AD) for CSEID of phenolic acids and flavonoids in complex natural extracts has been proposed [30]. Under strategic conditions, controlled by an MES buffer (25 mM, pH 5.0) and accordingly to the antioxidant acid–base properties, the simultaneous measurement of total acids and flavonoids indexes was achieved in less than 100 s with an excellent analytical performance. The reliability of ME-AD approach was demonstrated toward the high agreement between the total phenolics obtained using microchip approach with those obtained by the well-established HPLC-DAD (diode array detector) approach; revealing both identical order regarding to the total phenolic content in the target samples. These results suggested that the microchip approach is a reliable method for fast assessment of class-selective antioxidants constituting a very good alternative to the long analysis times and using toxic solvents required in HPLC. This excellent analytical performance is connected with the key features of the "ready-to-use" system employed in this work such as portability, full integration of AD (amperometric detection), easy-operation, ME-AD disposability, and a clear step to food analysis by the user.

The analysis of antioxidants in olive oil has also been explored using ME-AD. Three phenolic compounds (tyrosol, hydroxytyrosol, and oleuropein glucoside) were determined on real extra virgin olive oil samples adopting end-channel amperometric detection [31].

Amines have been also studied using ME-AD technology. A homemade miniaturized CE–AD system (Cu working electrode), has also been proposed for the fast determination of bioactive amines [32]. Because their formation and metabolism widely occur in living organisms, they are also present in a variety of different foods, primarily as a consequence of microbial amino-acid decarboxylation. The proposed method allowed three bioactive

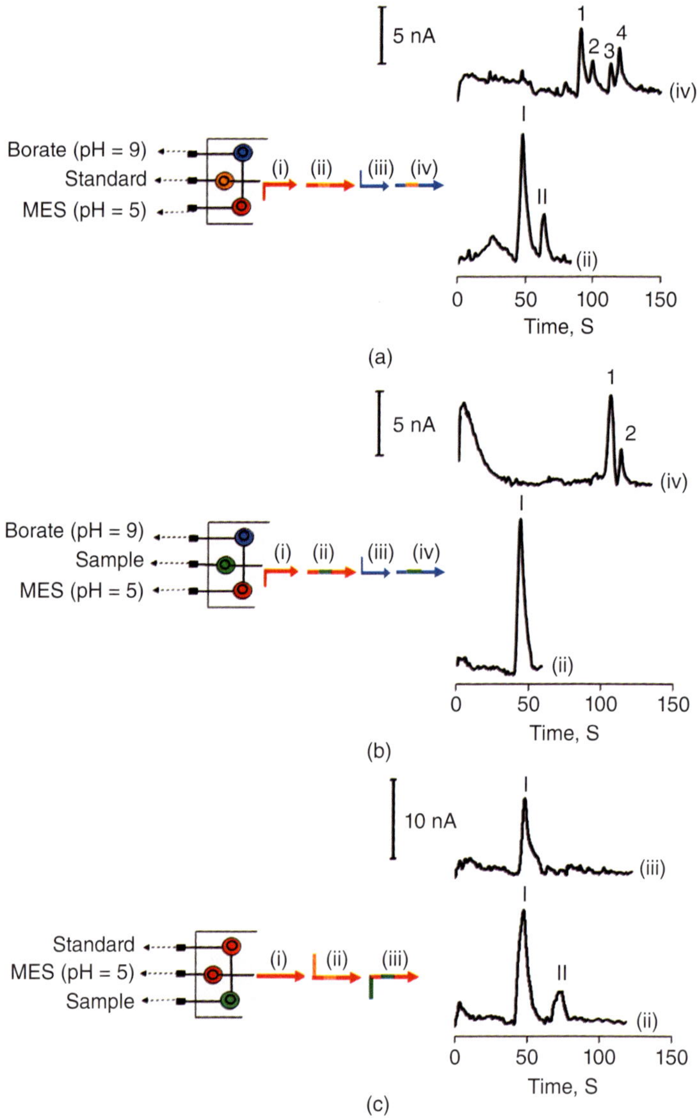

Figure 12.5 *Analytical integration and simplification strategies. In all cases: the numbers i–iv indicate the sequential buffer flow and sample/standard injection order of the analysis; peaks: I total flavonoids peak and II total phenolic acids peak; 1: (+)-catechin, 2: rutin, 3: chlorogenic acid, and 4: ferulic acid. (a) Integration of both CSEID (ii) and IAD (iv) in one microchip for a standard solution; (b) the same for green tea extract; and (c) integrated one point calibration CSEID, (ii) standard solution and (iii) green tea extract. (Adapted with permission from Ref. [29] Copyright © 2010, American Chemical Society.)*

amines (tryptamine, tryptophan, and tyramine) to be separated in a time of about 6 min with a successful application in wine and beer samples.

Other compounds of food significance that have recently been the focus of attention using ME-AD approach are dyes. Artificial dyes are commonly added to foodstuffs with the purpose of enhancing their attractiveness. Since almost all of these additives are prepared by chemical synthesis, their presence in foods is the cause of extensive consumer mistrust since they may lead to adverse health effects and because their consumption is regulated by law. A rapid analysis of azo-dyes in food by using ME-AD has been reported [33]. Five azo-dyes (yellow AB, Red 2G, Sunset Yellow, New Coccine, and Amaranth) were separated in less than 300 s and determined in soft drinks (liquid sample) and candies (solid sample), both with minimal sample preparation but still performed "off chip."

A new *in-channel* configuration for the amperometric detection of synthetic dyes in food has also been investigated [34]. The well-known advantages of *in-channel* configuration such as allowing better peak resolution and higher sensitivity due to the prevention of the dispersion of the analyte band were successfully demonstrated. Fast separation of the target dyes in the food samples (soft drinks, syrup, and candy) was performed in less than 200 s.

In an effort to build up a food LOC, preconcentration, separation, and AD, all of them integrated in a microchip, has also been proposed for the analysis of dyes in several food samples [35]. The microchip layout was comprised of three parallel channels. Two of them were for field-amplified sample stacking and the subsequent field-amplified sample injection steps, while the third one was reserved for the AD. The preconcentration method was used to analyze the dyes in several food samples (juice, traditional Korean alcohol, fish, noodles, and red pepper powders).

The fast and reliable detection of frauds in the quality control of flavors for food significance is also another potential possibility of the ME-AD approach. Vanilla flavor, obtained from *Vanilla planifolia* extracts, is a very unique fragrance and one of the most popular flavors used in the food industry. Vanillin (VAN) is a basic fragrance compound of the vanilla aroma, whereas ethyl vanillin (EVA) does not exist in nature. The basic vanilla flavor is VAN but *p*-hydroxybenzaldehyde (PHB), vanillic acid (VANA), and *p*-hydroxybenzoic acid (*p*-PHBA) are also constituents. A new microfluidic separation method (using simple CZE in 10 mM borate buffer pH = 9.5), involving EVA to provide unequivocal proof of frauds as well as all target marker flavors occurring in *Vanilla planifolia* extracts (VAN, PHB, VANA, and PHBA) has been proposed [36]. The elution order was strategically connected with sequential sample screening and analyte confirmation steps as it is shown in Figure 12.6. Thus, EVA is detected initially to distinguish between negative and positive samples. Then, if VAN is detected the sample is potentially positive, but final confirmation is performed by the simultaneous detection of PHB, VANA, and PHBA. To demonstrate the suitability of the approach, natural Vanilla extract (*Vanilla Planifolia*) and vanilla-related foodstuff samples were analyzed. During the analysis of Vanilla extract, the simultaneous absence of EVA and the presence of VAN, PHB, VANA, and PHBA was recorded in a very short time (less than 300 s); thus confirming their *expected* authenticity, and demonstrating the analytical potency of the microfluidic approach. Vanilla-related foodstuff samples were also analyzed by the microfluidic approach and two types of frauds were found: in black vanilla, the direct detection of EVA allowed the direct confirmation of the non-natural origin of the sample while the non-natural origin of white vanilla was confirmed toward the detection of VAN but the absence of PHB, VANA, and PHBA. The direct injection of

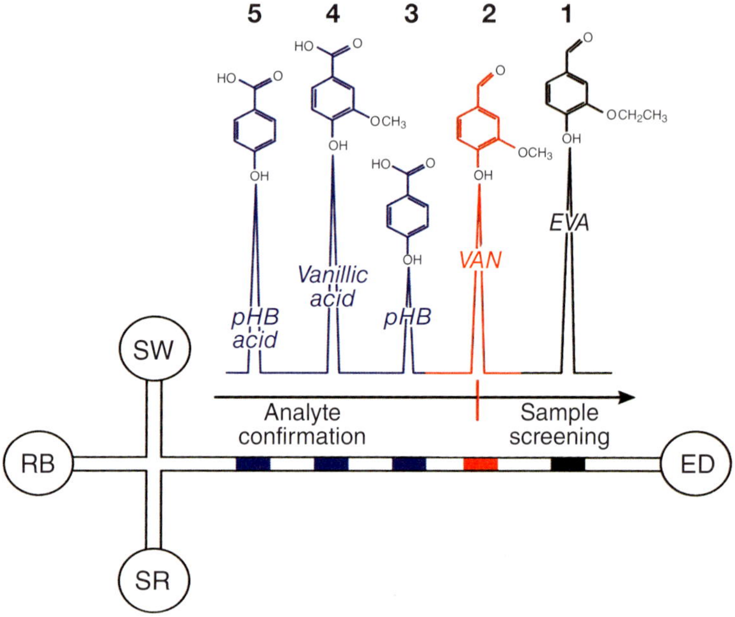

Figure 12.6 Scheme of the microfluidic chip following the sequential sample screening and analyte confirmation strategy proposed (RB, running buffer; SR, sample reservoir; SW, sample waste; and ED, electrochemical detector). (Adapted with permission from Ref. [36]. Copyright © 2009 WILEY-VCH Verlag GmbH & Co. KGaA.)

the extracts moved the migration time to higher values, but a good EOF stability was also achieved during real analysis (relative standard deviations (RSDs) for migration time less than 3% and ranged between 5% and 12% for peak heights; $n = 5$).

ME-AD approach for improving detection of some aldehydes with significance in foods using a suitable derivatization route has also been reported [37]. Several aldehydes such as formaldehyde (FA) and acetaldehyde (AA) are mostly minor compounds in food, and their occurrence can be an indication of quality deterioration, overheating, microbacterial fermentation, and/or off-flavor. A novel miniaturized capillary electrophoresis (CE) with the AD method has been developed for the fast determination of FA and AA in several food products without preconcentration. Because FA and AA usually exist as uncharged molecules with zero electrophoretic mobility, and the fact that these two compounds lack of chromophore for sensitive ultraviolet (UV) detection, effective, and sensitive determination of FA and AA in food samples is often a challenging task. In this work, an electroactive compound 2-thiobarbituric acid (TBA) was selected as the ideal derivatization reagent to facilitate the determination of FA and AA. Borate buffer (pH 9.0) allowed a good separation for all analytes within 13 min FA-TBA and AA-TBA adducts can be well separated with a low detection limit of 9.0×10^{-9} g/ml (13.2 fg) for FA. The proposed method was applied for the analysis of wines' samples and it was found that the average recovery of FA and AA was within the range of 96.1–106% with their RSDs at 2.8–10.1%, which indicated that the microchip approach was sufficiently accurate for the simultaneous determination of the two aldehydes.

12.2.2 Microchip Electrophoresis with Amperometric Detection for Inorganic Food Analytes

Some inorganic analytes of food significance has also been studied using ME-AD.

Related to the assessment of heavy metals which are alarming in foods, ME-AD approach has also been reported for their determination in juice samples [38]. Fast detection of prominent heavy metals (lead, cadmium, and copper) has been demonstrated on ME-AD platforms using CSPEs. Chemical separation (25 mM MES/L-histidine buffer solution of pH 7.0), electrokinetic conditions (+1200 V, 3 s injection), and detection potential (−0.8 V) were carefully optimized. Excellent recoveries in spiked juice samples (without oxygen-free condition) ranging from 92% to 108% ($n = 10$) were obtained, giving in addition an excellent reproducibility (RSDs < 7%).

The nitrite ion (NO_2^-) is one of the major pollutants in food and agricultural samples. Extracted samples (sonication and microwave) have been analyzed using an ME-AD coupled with a copper (3-mercaptopropyl) trimethoxysilane [Cu(II)-MPS] complex-modified carbon paste electrode (CPE) [39]. Moreover, it was demonstrated that the use of bare CPEs are not suitable for analyzing NO_2^- and even so NO_3^- [40] because of their low sensitivity and poor reproducibility. Indeed, MPS has attracted attention as an excellent electron transfer mediator due to some unique characteristics. It exhibits dual behavior both as thiols and as sol–gel functionalities. The thiol functional group in MPS is capable of coordinating with copper specifically; thus copper can be complexed with MPS and used as an electrocatalyst for the catalytic reduction of NO_2^- in an aqueous medium. The detection of NO_2^- was attributed to mediation or catalysis by the reduction of Cu(II)-MPS to Cu(I)-MPS at the modified electrode. The ME-AD conditions were a detection potential of −190 mV, a separation field strength of −150 V/cm, and an injection time of 5 s at −150 V/cm, with a separation buffer of 20 mM acetate with 3 mM cetyltrimethylammonium bromide (CTAB) (pH = 5.0) (cationic surfactants at low pH range permit NO_2^- not to be adsorbed into the inner wall of the glass channels, and to get a faster separation). An RSD of 4.3% ($n = 4$) and a limit of detection (LOD) (based on S/N = 3) of 0.35 ppm, with no apparent interference from other inorganic ions such as NO_3^-, Cl^-, PO_4^{3-}, organic ions, and other biological compounds commonly present in food samples has also been reported.

12.2.3 Microchip Electrophoresis with Conductometric Detection for Food Analysis

Conductivity has also successfully coupled to ME for food analysis.

Preservatives such as benzoate, sorbate, and vitamin C have been electrophoretically separated using poly(methyl methacrylate) (PMMA) microchip coupled to CCD [41]. A buffer composition of cyclodextrins HP-β-CD was needed to achieve suitable separation. An impressive application for the analysis of 12 inorganic and organic anions in less than 100 s, using the same detection route and employing different electrolyte solutions, has also been studied [42]. Ammonium, potassium, calcium, sodium, and magnesium as inorganic anions, as well as chloride, sulfate, oxalate, tartrate, malate, citrate, succinate, acetate, lactate, and inorganic phosphates, plus organic anions, were determined in several beverages including water, red wine, beer, and milk. This excellent work clearly reveals the relevant role of conductimetric detection in food routine analysis.

Monofluoroacetate (MFA) is an extremely toxin and CCD is potentially well suited for its detection. The rapid and quantitative determination of MFA in fruit juices and tap water has been proposed using ME with CCD [43]. The main feature of this interesting work was the direct determination (with very little sample preparation) of MFA in a complex matrix of fruit juices containing several interferences such as organic acids and small inorganic anions, as well as high sugar content. Direct injection without dilution was not possible because the channel walls of the microchip become coated with sugar molecules after several runs causing irreproducible results. This challenge was solved by dilution of sample since sensitivity was good. The interference of other related ions in the samples was avoided because they offered different conductivity according to their mass and charge. By working directly with diluted fruit juices, the total analysis time was only 5 min.

As we have stated earlier, the use of ME-based-polymers has several advantages with respect to glass such as cost and microfabrication simplicity. In this way, an interesting work dealing with a high-performance polymer polyether ether ketone (PEEK) microchip as a novel application in food environments has been reported in connection with a portable instrument [44]. In this work, CCD was also used as the detection principle.

Determination of tartaric, malic, succinic, acetic, and lactic acids, as well as sulfate in different wines (in 30 mM MES, 10 mM L-histidine at pH 5.6) as well as saccharides in different soft beverages (in 50 mM NaOH and 0.2 mM CTAB at pH 12.7) were proposed to demonstrate the analytical suitability of this novel microchip material. Of special interest was the suitability of PEEK for sugar determination since the strong background electrolyte (BGE) conditions required are not suitable for other conventional polymers commonly used in polymer microchip fabrication such as PMMA, revealing the role of PEEK in this kind of food applications.

12.3 Microfluidic Chips with Nanomaterial-Based Electrochemical Detection for Food Analysis

Nowadays, it is clear that nanotechnology offers valuable tools for building new architectonics for analytical microsystems. Apart from their high sensitivity and inherent miniaturization, another added functionality of ED is the opened opportunity to modify these surfaces suitably with nanomaterials to break frontiers in new food analysis applications. In food microfluidic analysis, two relevant nanomaterial examples have been explored as novel electrochemical detectors coupled to ME: carbon nanotubes (CNTs) [45, 46] and metallic nanowires (MNWs) [47].

However, CNTs are nanomaterials with unique geometrical, mechanical, electronic, and chemical properties which offer notably exciting possibilities involving the large active surface at electrodes of small dimensions, the enhancement of electronic transfer, and the strong sorption capacity. These properties have clear influence on their analytical sensitivity which is enhanced by the use of these nanomaterials. There are two main types of CNTs that can have high structural perfection: single-walled nanotubes (SWCNTs) which consist of a single graphite sheet seamlessly wrapped into a cylindrical tube, and multiwalled nanotubes (MWCNTs), which comprise an array of such nanotubes that are concentrically nested like rings of a tree trunk [48, 49].

The exploration of CNTs as electrochemical detectors of microfluidics has constituted a natural step toward the improvement of their analytical performance in terms of selectivity, sensitivity, and reproducibility because these nanomaterials inherently offer lowered detection potentials, higher surfaces, and better stability [50]. Indeed, the main analytical advantages expected and offered by using CNTs as electrochemical detectors in microfluidics [45, 46] are:

1. *Lower detection potentials:* The greater surface area of CNT-based electrodes leads to lower current densities and therefore lower "overpotentials." The electrocatalytic effect of CNT materials might have a strong effect on the electrocatalysis of analytes and therefore on lowering the detection potentials and in consequence, improving the overall selectivity of the analysis.
2. *Higher currents* because the greater surface area of CNT detectors enable larger scale redox conversion, therefore increasing the analytical sensitivity.
3. *Higher stability and resistance to passivation* originated from the greater surface area of the CNTs-based detectors. This characteristic implies better reproducibility because the resulting signal is prone to fouling.
4. In addition, one of the most unique perspectives derived by using CNTs as electrochemical detectors is the influence of CNTs on separation performance. In wall–jet ED (commonly used in end-channel configurations), nanomaterials can provide a higher heterogeneous electron transfer rate between analyte and electrode surface, which results in sharper and less tailing peaks and consequently resulting in a higher resolution power and higher peak capacity of the whole ME–ED system.

In summary, from an analytical point of view, CNTs inherently offer us an improved selectivity, sensitivity, and reproducibility as well as improved separation performance which can strategically been exploited in food analysis.

In food analysis using ME-ED technology, CNTs modification has been studied using both GCE and CSPEs for vitamin detection [51]. CSPE-CNTs modified have become the best approach since their improvement of sensitivity and lowered LODs with the advantages of easy surface modification, inherent miniaturization, and disposability. The performance of these CNT detectors for vitamins determination integrating calibration has also been evaluated with an excellent results. Vitamins calibration and analysis were carried out in less than 400 s with recoveries higher than 95%.

ME using different CSPE-CNTs (SWCNTs and MWCNTs) to analyze a wide group of analytes of food significance, such as dietary antioxidants, water-soluble vitamins, vanilla flavors, and isoflavones involved in representative food samples, has been deeply studied [52]. Ultrafast separations at lowered oxidation potentials resulted in well defined and resolved peaks with enhanced voltammetric current when compared to those obtained from unmodified screen-printed electrodes; thus making CNTs an ideal material for electrochemical sensing in food analysis. MWCNTs offered better performance as compared to SWCNTs.

The analytical suitability of the electrokinetic microfluidic platform with MWCNTs as detectors was further proposed based on its dual format/use as a flow and separation system, independently as it is depicted in Figure 12.7. Relevant applications of high significance, determination of total isoflavones, and fast detection of antioxidant profiles (arbutin,

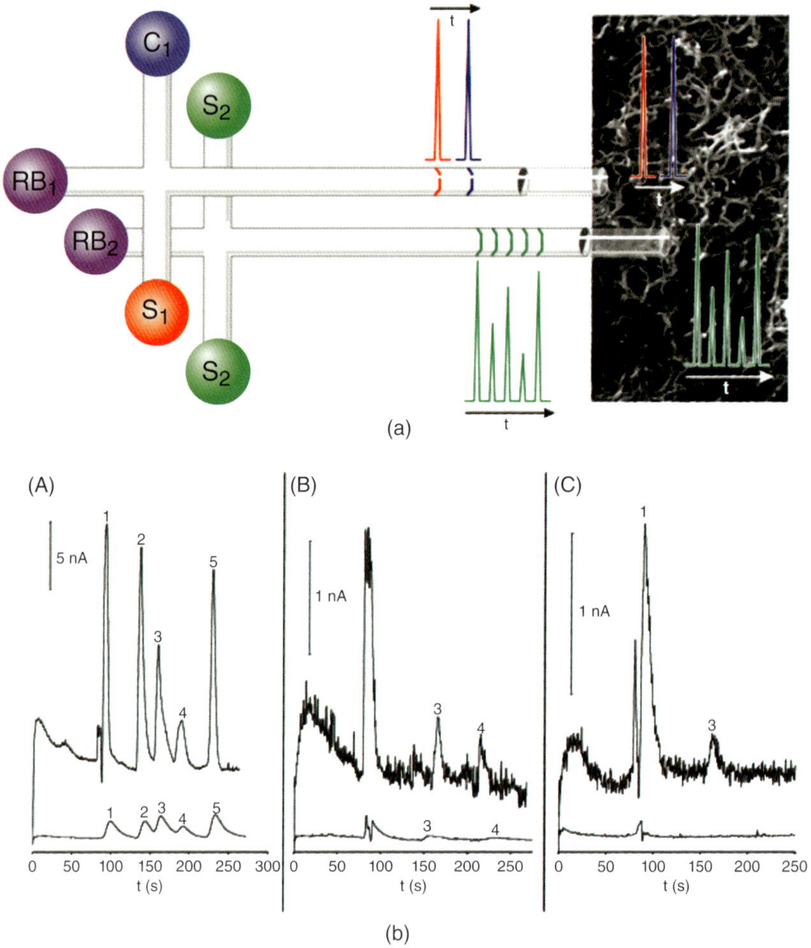

Figure 12.7 (a) Artwork illustrating dual format of microfluidics (FIA and ME separation system) with carbon nanotubes detectors for analytical domains (RB running buffer, C calibration reservoir, and S sample reservoir). (b) Microchip electropherograms using MWCNTs (upper) and CSPE (lower); (A) antioxidant standards, (B) apple, and (C) pear. Peaks: (1) arbutin, (2) phloridzin, (3) (+)-catechin, (4) rutin and (5) ascorbic acid. (Adapted with permission from Ref. [53]. Copyright © The Royal Society of Chemistry].)

phloridzin, (+)-catechin, rutin, and ascorbic acid were separated in less than 250 s) were chosen to demonstrate their analytical potential [53]. Figure 12.7a–c shows the sensitivity enhancement obtained for an antioxidant separation using the MWCNTs in comparison with bare CSPEs. The use of these unique materials in microfluidic-LOC technology has opened new expectations for food analysis.

Novel technologies using CNTs coupled to ME have also explored in food analysis. Single-walled carbon nanotube press-transferred electrodes (SWPTEs) are new disposable electrodes where CNTs act as exclusive electrochemical transducer [54, 55], being

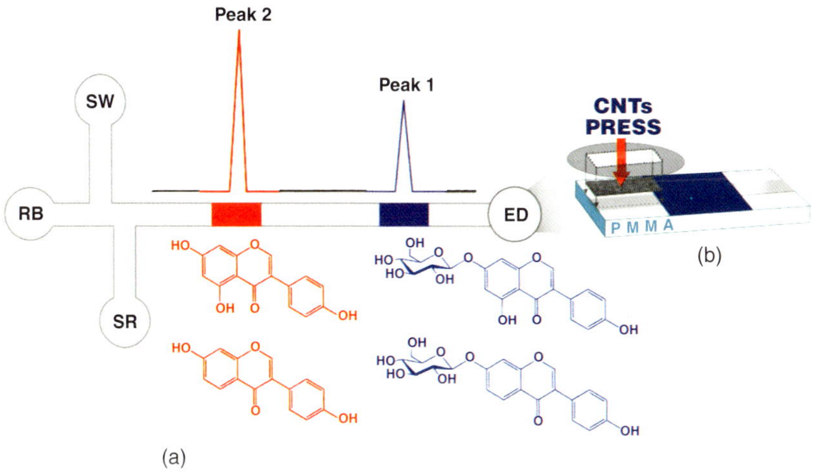

Figure 12.8 Strategy for the detection of class-selective isoflavone indexes on the microfluidic chip layout (a) and scheme of the lab-made SWPTE (b). RB: running buffer reservoir, SR: sample reservoir, SW: sample waste reservoir, and ED: electrochemical detection cell. (Adapted with permission from Ref. [55]. Copyright © The Royal Society of Chemistry)

an excellent alternative to common casting approaches. CNT from filtering of their homogeneous dispersions is transferred to PMMA substrate by pressure adopting the electrode geometry suitable for its use as detector in MCs. These pioneering SWPTEs coupled to microfluidics chips have been employed to develop their first application in food analysis [55]. A class-selective electrochemical isoflavone index determination has been proposed for the fast and reliable qualitative and quantitative assessment of class-isoflavone based on the co-migration of the total glycosides (TGs) and total aglycones (TAs) in less than 250 s with very good intra-SWPTEs repeatability (RSDs $\leq 8\%$, $n = 5$) and inter-SWPTEs reproducibility (RSDs $\leq 9\%$, $n = 3$). Figure 12.8 illustrates the strategy on the MC layout as well as detector used.

However, MNWs are 1D important blocks of nanotechnology and they can be formed by a variety of elements or oxides. MNWs are metallic or semiconducting nanostructures with cylindrical shape, completely filled without any hollows in the structure, because if the structure is hollow it would be a nanotube. They are one-dimensional anisotropic structures, small in diameter and large in surface-to-volume ratio. They can be found in different diameters from a few nanometers to 500 nm. The length of NWs (nanowires) is bigger, involving from a few micrometers (2–3) up to tens of micrometers. Because of their geometry, one of the most relevant features making them unique with respect to other nanostructures is that they can be easily designed and fabricated with different segments which add controlled functionality [47].

Using these nanomaterials, an interesting example in the field, is the coupling of copper nanowires (CuNWs) to ME. Indeed, CuNWs exhibit electrocatalysis toward carbohydrates becoming a "selective detector" with the expected enhanced sensitivity. This coupling has been explored for the fast and reliable analysis of monosaccharides in honey samples [56]. To this end, a representative group of nine honey samples were analyzed and the results were compared with those obtained by HPLC-RI (refractive index). ME-CuNWs approach allowed the separation of glucose and fructose in less than 250 s under optimized separation (20 mM NaOH + 10 mM H_3BO_3, pH = 12.0 separation voltage +1000 V) and detection (E = +0.70 V in 20 mM NaOH + 10 mM H_3BO_3, pH = 12.0) conditions. An excellent stability of EOF during sample analysis was achieved with RSDs for migration time <2% and for amperometric currents <9%. The quantitative contents for individual glucose and fructose obtained using ME-CuNWs in comparison with those obtained by HPLC-RI were highly in agreement with errors below 10%, indicating the reliability of the approach.

Nanomaterial "added the wished/wanted selectivity" and it is an illustrative example about how micro and nanotechnologies strategically driven shows an excellent potency in food analysis.

12.4 Microfluidic Electrochemical Biosensing Chips for Food Analysis

As previously stated, microfluidic platforms have gained a considerable attention as a network of integrated channels to perform the multiple operations associated with the sensing approach avoiding interconnections and dead volumes. Furthermore, this approach is clearly advantageous since it allows accurate, ultrasmall sample volumes (loop-size variable) to be introduced and it has accurate fluid-control and manipulation capabilities. However, detection continues to be one of the main challenges for microfluidic platforms because very sensitive techniques are needed due to the ultrasmall sample volumes used on micron-sized environment. In this sense, and following the common thread of this book, ED is a very well-suited detection principle for microfluidic platforms because of its inherent features of miniaturization without the loss of performance and its compatibility with the advanced micromachining and nanotechnologies. It can be advanced in the sense that the features derivate from the combination of microfluidic platforms, (bio)-sensing, and ED represent a powerful and exciting tool for biological and chemical analysis in the agricultural and food fields.

Stimulated by the current trend of micro-TAS and LOC technologies, biochemical sensor strategies have currently been incorporated in microfluidic systems for the realization of highly sophisticated systems. However, the miniaturization of microfluidic devices depends on micromachining technology and the integration of different components such as pumps and valves on a chip tends to be complicated.

The idea followed under this section is to give an overview of the potential of electrochemical microfluidic devices as (bio)-sensor platforms devoted to agricultural and food analysis. Although these types of microdevices have been reported in an important number of published works, those associated to food analysis are very limited [57, 58]. Related to this scope, most of the published works fall within electrochemical microfluidic immunosensors. Relevant examples in food analysis will be given in this section.

Combination of immunoassay and microfluidic platforms offers advantages related to the selectivity and sensitivity, associated to the antigen–antibody interaction with the remarkable features stated above for microfluidic platforms. Microfluidic immunoassays make use of a network of microchannels and/or immunoreactor chambers usually built in a monolithic platform (similar or even the same designs used in ME) from different materials as silicon, glass, or polymers with part of all the necessary components of an immunoassay procedure integrated. Furthermore, these microfluidic platforms are especially suitable for immunoassays from different point of view:

1. The long time associated to incubation stages in conventional immunoassay can be attributed to the inefficient mass transport for the immunoreagents to move them from the bulk solution to the wall surface where interaction takes place. In microchannels, the surface-area-to-volume ratio is higher which makes the diffusional distances dramatically reduced and produce lower analysis time.
2. Miniaturization reduces drastically the consumption of expensive reagents.
3. Fluid movement can be easily controlled, especially with electrokinetic fluidic motivation, through the adequate control of applied electric fields, or in a more complicated way by using of pumps, valves, mixers, and so on.

The advantages previously mentioned have converted the microfluidic immunoassay platforms into an emergent and powerful alternative in the field of food and agricultural analysis in which a complicated sample preparation step is highly required.

Two fundamental groups of analytes with highly concern in food analysis have been determined combining immunoassay and microfluidic platforms: toxins and food-borne pathogens. Today, one of the most important fields of food analysis is the accurate determination of extremely low levels of toxins. From this category, mycotoxins are natural and structurally diverse group of secondary metabolites produced by fungi on agricultural commodities in the field and during storage under a wide range of climatic conditions. Several hundred different mycotoxins have been discovered and they can be found as contaminants in cereals, related products used for feed, beverages as fruits juices and wines, foodstuffs, and their products worldwide. Mycotoxins are potent toxins and have a wide range of actions on animals and humans, (e.g., cyto-, nephron-, and neurotoxic, carcinogenic, mutagenic, immunosuppressive, and estrogenic effects) causing important health and economic problems. Thus, it is necessary to establish maximum permitted levels of these mycotoxins in different foods and feedstuffs, and develop fast, low cost, sensitive, and selective analytical methods.

Although different fluid-handling procedures have been reported in the literature, hydrodynamic pumping is the most used method to deliver fluids into microchannels. However, microfabrication of valves and pumps, besides being difficult and expensive to miniaturize, requires clean room facility technology. Associated to mycotoxin detection, both hydrodynamic and eletrokinetic flows have been addressed.

Related to electrochemical microfluidic immunosensors, a microfluidic cell of 7.50 μl has been used for performing a heterogeneous competitive electrochemical immunosensor for citrinin determination in rice samples [59]. The cell consists of two bodies: the top body, made of stainless, contains the counter electrode, internal holes that allow entry and exit of solution flow, as well as another channel that communicates with the reference electrode; the bottom body comprises the GCEs, one of them used to immobilize the citrinin–ovoalbumin conjugate (CIT-OVA) and the other used as detection unit. Different solutions flow in a stop and go mode for performing the heterogeneous competitive immunoassay. A limited amount of mouse anti-CIT IgG antibody is added to CIT from rice samples, competing immunologically with CIT labeled ovalbumin bonded at the GCE. This electrode surface was previously modified by electrodeposition of gold and subsequent modification with a cysteamine self-assembled monolayer to bind CIT-OVA. Formation of an immune complex with a secondary antibody rabbit anti-mouse labeled with the enzyme horseradish peroxidase (HRP) allowed the determination of the analyte. The HRP catalyses the oxidation of catechol to benzoquinone in the presence of hydrogen peroxide, which back electrochemical reduction to catechol that can be amperometrically detected on the GCE surface at −0.15 V vs. Ag/AgCl. The immunosensor presents a range of work between 0.5 and 50 ng ml^{-1}; an LOD of 0.1 ng ml^{-1} and a total assay time of 45 min, avoiding any sample pretreatment.

The same group has developed similar microfluidic electrochemical immunosensors for determining ochratoxina A, *Botrytis Cinereae* and zearalenone in fruits and feedstuffs, respectively [60–62]. The main body of these devices was made of polymeric materials as Plexiglas or PMMA, where a central channel (100 μm internal diameter) is designed as flow-through chamber containing the microfluidic immunosensor and the detector system. Other accessory channels are used to inject different solutions for performing the immunoassay. These microfluidic immunosensors make use of magnetical particles confined at the central channel as platform for specific antibody or analyte-derivatized immobilization. Following the principles of heterogeneous competitive immunoassay, the analyte from the sample and an HRP-analyte conjugated compete for binding to the specific antibody. After washing steps and using a working gold layer electrode deposited at central channel by sputtering technique, the enzymatic product is amperometrically measured and related to the analyte concentration in the sample.

The use of beads as support for antibody immobilization in microfluidic devices is broadly reported. Compared to microtiter wells used in ordinary immunoassays, the use of beads (mainly magnetic beads) present relevant advantages such as an increased surface-to-volume ratio, even compared to microchannels, reducing the diffusion distances. This fact outcomes assays with higher sensitivity due to high efficiency of interactions between samples and reagents. Easy manipulation and transport of the beads in a fluidic system, a variety of surface modifications available for an easy biomolecules attachment, are other relevant benefits.

Another interesting example of mycotoxin analysis has been developed using a total integrated electrokinetic magnetic bead-based electrochemical immunoassay for the control of permitted levels of zearalenone in infant foods [63, 64]. This work makes use of electrokinetic-EOF flow as pumping force to deliver samples and reagents into the microfluidic device. Electrokinetic-EOF-based fluidic motivation benefits these analytical platforms as the flow in multiple channels on a microchip can easily be controlled with a few electrodes. Electrokinetics is especially valuable in immunosensing devices, because it allows to modify the injected amounts of reagents and samples in different analytical steps by simply changing the electric field and time applied (acting as a variable well-controlled loop), as well as manipulating the fluidics in different directions including reversed and stopped flows with extreme simplicity and automation. In this work, following the mentioned principles, a strategy which implies the creative definition and use of simple double-T microchip layout to integrate all required immunoassay steps has been proposed as it is depicted in Figure 12.9. Different reagents and samples are automatically moved following a predefined potential and time program by electrokinetic injection and pumping. This strategy is based on physical separation of both zones (immunological complex formation and enzymatic reaction) in different channels of the microchip, avoiding one of the

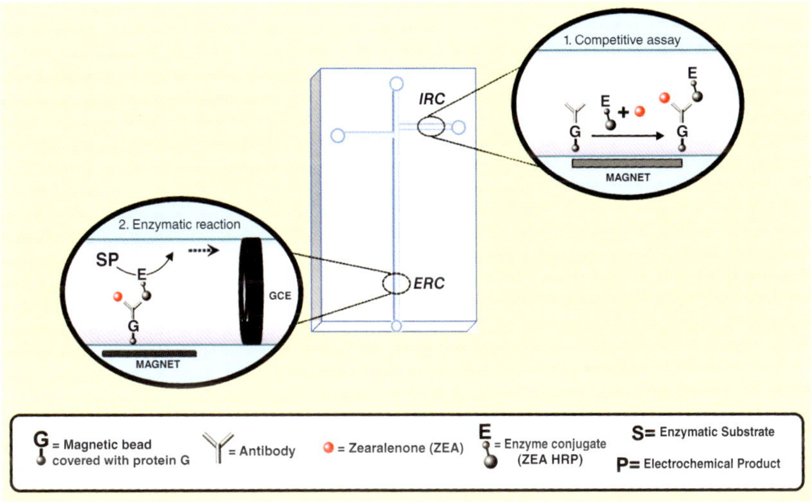

Figure 12.9 *Microfluidic chip layout and immunoassay principle (IRC, immunological reaction chamber and ERC, enzymatic reaction chamber). (Adapted with permission from Ref. [64]. Copyright © The Royal Society of Chemistry)*

greatest problems associated with microfluidic immunoassays as nonspecific adsorption to the channel walls surfaces. Immunoassay is carried out using a competitive scheme where the mycotoxin zearalenone and an HRP-labeled derivative compete for the binding sites of the specific antibody oriented immobilized onto magnetic beads through the use of protein G. Detection made by using an HRP enzyme, which in the presence of hydrogen peroxide and hydroquinone as electrochemical mediator, produces an electrical current proportional to the activity of the enzyme and, consequently, to the amount of zearalenone mycotoxin in the sample. Using this methodology, either solid or liquid food samples were analyzed, and those contaminated above the permitted levels of the mycotoxin could be easily discriminated. A suitable LOD of $0.4\,\mathrm{mg\,l^{-1}}$ well below the legislative requirements, and an extremely low systematic error of 2% from the analysis of a maize certified reference material, revealing additionally an excellent accuracy, have been obtained using this integrated microfluidic approach.

The second highly significant group of analytes is the food-borne pathogens. Tolerable levels of these agents are getting more stringent regulations due to the high concern of people for food safety. Bacteria such as *Escherichia coli* or *Staphylococcus aureus* are considered some of the most dangerous food-borne pathogens which require a rapid, sensitive, and reliable detection. Different works have been recently reported based on microfluidic immunosensors for food-borne pathogen detection. A PDMS-based microfluidic immunosensor chip integrated with nanoporous alumina membranes for *E. coli* and *S. aureus* has been reported [65]. Antibacterial antibodies were immobilized on alumina surface via self-assembled silane monolayer and introduced into a PDMS microfluidic device. Once targeted bacteria were captured by the complementary antibody on the nanoporous membrane, electrolyte current flow through the nanopores was blocked which led to impedance amplitude increase. Impedance spectrum is used to monitor impedance amplitude change before and after bacteria capturing by complimentary antibodies on the nanoporous alumina membrane. The higher the covered area ratio is, the larger the impedance amplitude will be monitored due to the increased blocking efficiency of electrolytic current flow through the nanoporous membrane. Cross-experiments between antibodies and nonspecific bacteria using impedance spectrum and sandwich-type immunoassay confirmed the specificity of this device for bacteria detection. The sensitivity of this technique was around 10^2 CFU/ml, which was higher than the traditional microelectrodes-based impedance sensor. Similar work was previously reported by the same group [66], where poly(ethylene glycol) hydrogel was fabricated on saline-modified nanoporous alumina surface for bacteria detection through the patterned antibody recognition mechanism. Low-frequency impedance spectrum is used to detect specific bacteria binding at low bacteria concentration of 10^2 CFU/ml.

A different biosensing strategy, describes a miniaturized micro flow injection biosensing system, containing a microbeads-based enzyme reactor for histamine detection for food analysis [67]. Histamine is a well-known factor to cause an allergy-like food poisoning, and it is produced in fish or fish product as a result of the bacterial decarboxylation of histidine which exists in high amounts in a lean fish of bluefish species like tuna, herring, markerel, and so on. It is important for fishery industries and food-processing companies to manage the freshness of the lean fish. The microfluidic device was fabricated using a build-up technique which is well known in printed-wiring assembly. An upper Pyrex wafer had an inlet port for the microbeads and solutions, and an outlet port for the waste. The bottom Pyrex wafer has the electrochemical detector where the electrodes were patterned using the photolithography technique. For performing the assay, the reactor contains histamine oxidase immobilized chitosan beads. This enzyme catalyses the oxidation of histamine to imidazole acetaldehyde, ammonia, and hydrogen peroxide. The amperometric detector combined with the nanoliter volume reactor by a miniaturized FIA (flow injection analysis) measures the current on oxidizing hydrogen peroxide generated from the enzyme reaction. Detection of histamine was carried out with high sensitivity and linearity from 1 μM to 1 mM in phosphate buffer solution and raw tuna samples.

12.5 Outlook

MC-ED in both, amperometry *end-channel* configuration and contactless conductivity has allowed fast separations and detection of prominent analytes (antioxidants, biogenic amines, vanilla flavors, dyes aldehydes, and metals) in food samples. Different strategies to improve the selectivity and sensitivity of the analysis by avoiding and/or making the sample preparation as simple as possible has also creatively addressed such as enhancing the peak capacity in order to perform direct injection, using the microchip platform to measure one target analyte/group of analytes (even with separation of related interferences), integrating some sample preparation steps on the microchip platform, and integrating new analytical tools from nanotechnology in the detection stage. However, due to its intrinsic complexity, the integration of sample treatment step on microchip technology is less developed than those related to separation and detection, being this issue one of the main challenges into the food analysis using LOC technology.

MC-ED has also become a novel biosensing platform where biological recognition material and suitable transducers can be cleverly integrated yielding a new sensor generation which could be termed as microfluidic (bio-)-sensors with high selectivity and sensitivity. These microfluidic biosensors have mainly been developed into immunoassay format focused on the detection of mycotoxin/toxins and food-borne pathogens as novel tools for food safety.

The synergies converged into the interfaces between LOC, nano, and biotechnologies using electrochemistry as detection principle represents a powerful and exciting tool for analysis in the food and agricultural field and one horizon plenty of promises.

Acknowledgments

Financial support from the AVANSENS program from the Community of Madrid (P2009/PPQ-1642) and CTQ2011-28135 from the Spanish Ministry of Economy and Competitiveness are gratefully acknowledged.

Acronyms

Micro-TAS	micro total analysis systems
LOC	lab-on-a-chip
MC	microfluidic chip
ME	microchip electrophoresis
EOF	electroosmotic-driven flow
S	sample
RB	running buffer
W	waste
ED	electrochemical detection
CCD	contactless detection
AD	amperometric detection
CSPEs	carbon screen printed electrodes
CZE	capillary zone electrophoresis
MEKC	micellar electrokinetic chromatography
PAD	pulse amperometric detection
PDMS	polydimethylsiloxane
HPLC	high-performance liquid chromatography
GCE	glassy carbon electrode
CSEID	class-selective electrochemical index determination
IAD	individual antioxidant determination
MES	2-(*N*-morpholino)ethanesulfonic acid
DAD	diode array detector
VAN	vanillin
EVA	ethyl vanillin
PHB	*p*-hydroxybenzaldehyde

VANA	vanillic acid
p-PHBA	*p*-hydroxybenzoic acid
RSD	relative standard deviation
FA	formaldehyde
AA	acetaldehyde
CE	capillary electrophoresis
UV	ultraviolet detection
TBA	2-thiobarbituric acid
CPE	carbon paste electrode
Cu(II)-MPS	copper (3-mercaptopropyl) trimethoxysilane
LOD	limit of detection
PMMA	poly(methyl methacrylate)
MFA	monofluoroacetate
PEEK	polymer polyether ether ketone
BGC	background electrolyte
CNTs	carbon nanotubes
MNWs	metallic nanowires
SWCNTs	single-walled carbon nanotubes
MWCNTs	multiwalled carbon nanotubes
SWPTEs	single-walled press-transferred electrodes
TG	total glycosides
TA	total aglycones
RI	refractive index
CIT	citrinin
OVA	ovoalbumin
HRP	horse radish peroxidase
CFU	colony formation units
FIA	flow injection analysis

References

1. Whitesides, G.W. (2006) The origins and the future of microfluidics. *Nature*, **442**, 368–373.
2. Janasek, D., Franzke, J., and Manz, A. (2006) Scaling and design of miniaturized chemical-analysis systems. *Nature*, **442**, 374–380.
3. van deer Berg, A. (2012) Riding the waves. *Lab Chip*, **12**, 2433–2434.
4. Geschke, O., Klank, H., and Tellesmann, P. (eds) (2004) *Mycrosystems Engineering of Lab-on-a-Chip Devices*, Wiley-VCH Verlag & Co. KgaA, Weinheim.
5. Vandaveer, W.R., Pasas, S.A., Martin, R.S., and Lunte, S.M. (2002) Recent developments in amperometric detection for microchip capillary electrophoresis. *Electrophoresis*, **23**, 3667–3677.
6. Wang, J. (2005) Electrochemical detection for capillary electrophoresis microchips: a review. *Electroanalysis*, **17**, 1133–1140.
7. Pumera, M., Merkoci, A., and Alegret, S. (2006) New materials for electrochemical sensing VII. Microfuidic chip platforms. *Trends Anal. Chem.*, **25**, 219–235.

8. Nyholm, L. (2005) Electrochemical techniques for lab-on-a-chip applications. *Analyst*, **130**, 599–605.
9. Sassa, F., Morimoto, K., Satoh, W., and Suzuki, H. (2008) Electrochemical techniques for microfluidic applications. *Electrophoresis*, **29**, 1787–1800.
10. Vandaveer, W.R., Pasas-Farmer, S.A., Fischer, D.J., Frankenfeld, C.N., and Lunte, S.M. (2004) Recent developments in electrochemical detection for microchip capillary electrophoresis. *Electrophoresis*, **25**, 3528–3549.
11. Chen, D.C., Hsu, F.L., Zhan, D.Z., and Chen, C.H. (2001) Palladium film decoupler for amperometric detection in electrophoresis chips. *Anal. Chem.*, **73**, 758–762.
12. Guijt, R.M., Baltussen, E., van der Steen, G., Schasfoort, R.B.M., Schlautmann, S., Billiet, H.A.H., Frank, J., van Dedem, G.W.K., and van der Berg, A. (2001) New approaches for fabrication of microfluidic capillary electrophoresis devices with on-chip conductivity detection. *Eletrophoresis*, **22**, 235–241.
13. Galloway, M., Stryjewski, W., Henry, A., Ford, S.M., Llopis, S., McCarley, R.L., and Soper, S.A. (2002) Contact conductivity detection in poly(methyl methacrylate)-based microfluidic devices for analysis of mono- and polyanionic molecules. *Anal. Chem.*, **74**, 2407–2415.
14. Pumera, M., Wang, J., Opekar, F., Jelinek, I., Feldman, J., Löwe, H., and Hardt, S. (2002) Contactless conductivity detector for microchip capillary electrophoresis. *Anal. Chem.*, **74**, 1968–1971.
15. Wang, J. and Pumera, M. (2002) Dual conductivity/amperometric detection system for microchip capillary electrophoresis. *Anal. Chem.*, **74**, 5919–5923.
16. Tanyanyiwa, J. and Hauser, P.C. (2002) High-voltage capacitively coupled contactless conductivity detection for microchip capillary electrophoresis. *Anal. Chem.*, **74**, 6378–6382.
17. Escarpa, A., González, M.C., González, A., and Blasco, A.J. (2007) Capillary-electrophoresis microchips: an opened gate to food analysis. *Electrophoresis*, **28**, 1002–1011.
18. Escarpa, A., González, M.C., López, M.A., González, A., Hervás, M., and García, M. (2008) Microchips for capillary electrophoresis: breakthroughs in real-world food analysis. *Electrophoresis*, **29**, 4852–4861.
19. Atalay, Y.T., Vermeir, S., Witters, D., Vergauwe, N., Nicolai, B., and Lammertyn, J. (2011) Microfluidic analytical systems for food analysis. *Trends Food Sci. Technol.*, **22**, 386–404.
20. Escarpa, A. (2012) Food electroanalysis: sense and simplicity. *Chem. Rec.*, **12**, 72–91.
21. Martín, A., Vilela, D., and Escarpa, A. (2012) Food analysis on microchip electrophoresis: an unpdated review. *Electrophoresis*, **33**, 2212–2227.
22. Escarpa, A. (2014) Lights and shadows on food microfluidics. *Lab Chip*, **14**, 3213–3224.
23. Scampicchio, M., Wang, J., Mannino, S., and Chatrathi, M.P. (2004) Microchip capillary electrophoresis with amperometric detection for rapid separation and detection of phenolic acids. *J. Chromatogr. A*, **1049**, 189–194.
24. Hompesch, R.W., García, C.D., Weiss, D.J., Vivanco, J.M., and Henry, C.S. (2005) Analysis of natural flavonoids by microchip-micellar electrokinetic chromatography with pulsed amperometric detection. *Analyst*, **5**, 694–700.

25. Blasco, A.J., Barrigas, I., González, M.C., and Escarpa, A. (2005) Fast and simultaneous detection of prominent natural antioxidants using analytical mycrosystems for capillary electrophoresis with glassy carbon electrode: a new gateway to food environments. *Electrophoresis*, **26**, 4664–4673.
26. Crevillén, A.G., Barrigas, I., González, M.C., and Escarpa, A. (2006) Microchip-electrochemistry route for rapid screening of hydroquinone and arbutin from miscellaneous samples: investigation of the robustness of a simple-cross injector. *Anal. Chim. Acta*, **562**, 137–144.
27. Crevillén, A.G., Blasco, A.J., González, M.C., and Escarpa, A. (2006) Fast and reliable route integrating calibration and analysis protocols for water-soluble vitamins determination on board microchip-electrochemistry platforms. *Electrophoresis*, **27**, 5110–5118.
28. Blasco, A.J., Crevillén, A.G., de la Fuente, P., González, M.C., and Escarpa, A. (2007) Electrochemical-valveless flow microsystems for ultra fast and accurate analysis of total isoflavones with integrated calibration. *Analyst*, **132**, 323–329.
29. Kovachev, N., Canals, A., and Escarpa, A. (2010) Fast and selective microfluidic chips for electrochemical antioxidant sensing in complex samples. *Anal. Chem.*, **82**, 2925–2931.
30. Castañeda, R., Vilela, D., González, M.C., Mendoza, S., and Escarpa, A. (2013) SU-8/Pyrex microchip electrophoresis with integrated electrochemical detection for class-selective electrochemical index determination of phenolic compounds in complex samples. *Electrophoresis*, **34**, 2129–2135.
31. Godoy-Caballero, M.P., Acedo-Valenzuela, M.I., Galeano-Díaz, T., Costa-García, A., and Fernández-Abedul, M.T. (2012) Microchip electrophoresis with amperometric detection for a novel determination of phenolic compounds in olive oil. *Analyst*, **137**, 5153–5160.
32. Chu, Q., Guan, Y., Geng, C., and Ye, J. (2006) Miniaturized capillary electrophoresis with amperometric detection: fast separation and detection of bioactive amines. *Anal. Lett.*, **39**, 729–740.
33. Dossi, N., Piccin, E., Bontempelli, G., Carrillo, E., and Wang, J. (2007) Rapid analysis of azo-dyes in food by microchip electrophoresis with electrochemical detection. *Electrophoresis*, **28**, 4240–4246.
34. Dossi, N., Toniolo, R., Pizzariello, A., Susmel, S., Perennes, F., and Bontempelli, G. (2007) A capillary electrophoresis microsystem for the rapid in-channel amperometric detection of synthetic dyes in food. *J. Eletroanal. Chem.*, **601**, 1–7.
35. Lee, K.S., Shiddiky, M.J.A., Park, S.H., Park, D.S., and Shim, Y.B. (2008) Electrophoretic analysis of food dyes using a miniaturized microfluidic system. *Electrophoresis*, **29**, 1910–1917.
36. Ávila, M., Zougagh, M., Escarpa, A., and Rios, A. (2009) Integrated microfluidic approach for simultaneous sample screening and analyte confirmation: application to the detection of frauds in vanilla samples. *Electrophoresis*, **30**, 3413–3418.
37. Zhang, D., Zhang, J., Li, M., Li, W., Aimaiti, G., Tuergsun, G., Ye, J., and Chu, Q. (2011) A novel miniaturised electrophoretic method for determining formaldehyde and acetaldehyde in food using 2-thiobarbituric acid derivatisation. *Food Chem.*, **129**, 206–212.

38. Chailapakul, O., Korsrisakul, S., Siangproh, W., and Grudpan, K. (2008) Fast and simultaneous detection of heavy metals using a simple and reliable microchip-electrochemistry route: an alternative approach to food analysis. *Talanta*, **15**, 683–689.
39. Shiddiky, M.J.A., Lee, K., Son, J., Park, D., and Shim, Y.J. (2009) Development of extraction and analytical methods of nitrite ion from food samples: microchip electrophoresis with a modified electrode. *Agric. Food Chem.*, **57**, 4051–4057.
40. Shiddiky, M.J.A., Won, M., and Shim, Y. (2006) Simultaneous analysis of nitrate and nitrite in a microfluidic device with a Cu-complex-modified electrode. *Electrophoresis*, **27**, 4545–4554.
41. Kuban, P. and Hauser, P.C. (2005) Application of an external contactless conductivity detector for the analysis of beverages by microchip capillary electrophoresis. *Electrophoresis*, **26**, 3169–3178.
42. Law, W.S., Kuban, P., Zhao, J.H., Li, F.Y., and Hauser, P.C. (2005) Determination of vitamin C and preservatives in beverages by conventional capillary electrophoresis and microchip electrophoresis with capacitively coupled contactless conductivity detection. *Electrophoresis*, **26**, 4648–4655.
43. Lu, Q., Wu, P., and Collins, G.E. (2007) Contactless conductivity detection of sodium monofluor acetate in fruit juices on a CE microchip. *Electrophoresis*, **28**, 3485–3491.
44. Muehlberger, H., Hwang, W., Guber, A.E., Saile, V., and Hoffmann, W. (2008) Polymer lab-on-a-chip system with electrochemical detection. *IEEE Sens. J.*, **8**, 572–579.
45. Pumera, M. and Escarpa, A. (2009) Nanomaterials as electrochemical detectors in capillary electrophoresis and microfluidics: fundamentals, designs and applications. *Electrophoresis*, **30**, 3315–3323.
46. Martín, A., González, M.C., López, M.A., and Escarpa, A. (2015) Multidimensional carbon allotropes as electrochemical detectors meet capillary and microchip electrophoresis. *Electrophoresis*, **36**, 179–194.
47. García, M., Batalla, P., and Escarpa, A. (2014) Metallic and polymeric nanowires for electochemical sensing and biosensing. *Trends Anal. Chem.*, **57**, 6–22.
48. Iijima, S. (1991) Helical microtubules of graphitic carbon. *Nature*, **354**, 56–58.
49. Ajayan, P.M. and Ijima, S. (1992) Smallest carbon nanotube. *Nature*, **358**, 23.
50. Wang, J., Tian, B., Chatrathi, M.P., Escarpa, A., and Pumera, M. (2009) Effect of heterogeneous electron-transfer rate on the resolution of electrophoretic separations based on microfluidics with end-column electrochemical detection. *Electrophoresis*, **30**, 3334–3338.
51. Crevillén, A.G., Pumera, M., González, M.C., and Escarpa, A. (2008) Carbon nanotubes disposable detectors in microchip capillary electrophoresis for water-soluble vitamins determination: analytical possibilities in pharmaceutical quality control. *Electrophoresis*, **29**, 2997–3004.
52. Crevillén, A.G., Ávila, M., Pumera, M., González, M.C., and Escarpa, A. (2007) Food analysis on microfluidic devices using ultrasensitive carbon nanotubes detectors. *Anal. Chem.*, **79**, 7408–7415.
53. Crevillén, A.G., Pumera, M., González, M.C., and Escarpa, A. (2009) Towards lab-on-a-chip approaches in real analytical domains base on microfluidic chips/electrochemical multi-walled carbon nanotube platforms. *Lab Chip*, **9**, 346–353.

54. Vilela, D., Garoz, J., Colina, A., González, M.C., and Escarpa, A. (2012) Carbon nanotubes press-transferred on PMMA substrates as exclusive transducers for electrochemical microfluidic sensing. *Anal. Chem.*, **84**, 10838–11084.
55. Vilela, D., Martín, A., González, M.C., and Escarpa, A. (2014) Fast and reliable class-selective isoflavone index determination on carbon nanotube press-transferred electrodes using microfluidic chips. *Analyst*, **139**, 2342–2347.
56. García, M. and Escarpa, A. (2014) ME-copper nanowires for fast and reliable determination of monossacharides in honey samples. *Electrophoresis*, **35**, 425–432.
57. Hervás, M., López, M.A., and Escarpa, A. (2012) Electrochemical immunosensing on board microfluidic chip platforms. *Trends Anal. Chem.*, **31**, 109–128.
58. Hervás, M., López, M.A., and Escarpa, A. (2012) Microfluidic chips as new platforms for electrochemical sensing, in *Chemical Sensors. Comprehensive Sensor Technologies*, Electrochemical and Optical Sensors, vol. **5** (ed G. Korotcenkov), Momentum Press, pp. 275–310.
59. Arévalo, F.J., Granero, A.M., Fernández, H., Raba, J., and Zón, M.A. (2011) Citrinin determination in rice samples using a micro fluidic electrochemical immunosensor. *Talanta*, **83**, 966–973.
60. Fernández-Baldo, M.A., Messina, G., Sanz, M.I., and Raba, J. (2010) Microfluidic immunosensor with micromagnetic beads coupled to carbon-based screen-printed electrodes for determination of *Botrytis cinerea* in tissue of fruits. *J. Agric. Food Chem.*, **58**, 11201–11206.
61. Fernández-Baldo, M.A., Gertolino, F.A., Fernández, G., Messina, G.A., Sanz, M.I., and Raba, J. (2011) Determitation of Ochratoxin A in apples contaminated with *Aspergillus ochraceus* by using a microfluidic competitive immunosensor with magnetic nanoparticles. *Analyst*, **136**, 2756–2762.
62. Panini, N.V., Salinas, E., Messina, G.A., and Raba, J. (2011) Modified paramagnetic beads in a microfluidic system for the determination of zearalenone in feedstuffs samples. *Food Chem.*, **125**, 791–796.
63. Hervás, M., López, M.A., and Escarpa, A. (2009) Electrochemical microfluidic chips coupled to magnetic bead-based ELISA to control allowable levels of zearalenone in bay foods using simplified calibration. *Analyst*, **134**, 2405–2411.
64. Hervás, M., López, M.A., and Escarpa, A. (2011) Integrated electrokinetic magnetic bead-based electrochemical immunoassay on microfluidic chips for reliable control of permitted levels of zearalenone in infant foods. *Analyst*, **136**, 2131–2138.
65. Tan, F., Leung, P.H.M., Liu, Z., Zhang, Y., Xiao, L., Ye, W., Zhang, X., Yi, L., and Yang, M. (2011) A PDMS microfluidic impedance immunosensor for *E. coli* 0157:H7 and *Staphylococcus aureus* detection via antibody-immobilized nanoporous membrane. *Sens. Actuators B*, **159**, 328–335.
66. Yu, J., Liu, Z., Liu, Q., Yuen, K.T., Mak, A.F.T., Yang, M., and Leung, P. (2009) A polyethylene glycol (PEG) microfluidic chip with nanostructures for bacteria rapid patterning and detection. *Sens. Actuators, A*, **154**, 288–294.
67. Ito, T., Hiroi, T., Amaya, T., Kaneko, S., Araki, M., Ohsawa, R., Yamamaura, A., and Matsumoto, K. (2009) Preliminary study of a microbeads based histamine detection for food analysis using thermostable recombinant histamine oxidase from *Arthrobacter crystallopoietes* KAIT-B-007. *Talanta*, **77**, 1185–1190.

13

Nanoelectrochemistry Applications Based on Electrospinning

Matteo Scampicchio[1*], *Maria Stella Cosio*[2], *Solomon Lemma Mengistu*[1] *and Saverio Mannino*[2]

[1]*Faculty of Science and Technology, Free University of Bolzano, Piazza Università 1, 39100, Bolzano, Italy*

[2]*Dipartimento di Scienze e Tecnologie Alimentari e Microbiologiche, Università degli Studi di Milano, Via Mangiagalli, 25, 20133, Milano, Italy*

13.1 A Note on Nanoelectrochemistry

Since the early talks of Richard Feynman in its *There's Plenty of Room at the Bottom* in 1959, there have been always a concern on what "nano" means and applies to. Recently, the Royal Society tried to make order between the definitions of nanoscience and nanotechnology, both referring to, respectively, the study or the preparation of materials from 100 nm down to the atomic level (approximately 0.2 nm) [1]. Later, the British Standard Institution defined nanomaterial as: *material having one or more external dimensions in the nanoscale or which is nanostructured.* [2] Accordingly, Murray makes a noticeable attempt to apply these concepts to nanoelectrochemistry of nanoparticles, nanoelectrodes, and nanopores, referring them as a dimensional scale of electrodes and electrochemical events [3].

A weakness of this interpretation is that most of the molecular events studied by electrochemistry already fit within the 1–1000 nm range of dimensions. Except to single atoms and molecules, which fall below this lower limit (which was introduced, indeed, to avoid single or small groups of atoms from being designed improperly as nano-objects), most of the matter studied by electrochemistry can be considered a part of nanoelectrochemistry,

*Corresponding author: matteo.scampicchio@unibz.it

Agricultural and Food Electroanalysis, First Edition.
Edited by Alberto Escarpa, María Cristina González and Miguel Ángel López.

including the study of micro- and nanoparticles, surfaces, colloids, emulsions, and aerosols and their mesoporous materials such as hydrogels and aerogels, together with their main applications in bioanalysis, catalysis, and electrocatalysis. Apparently, electrochemistry is already inherently "nano"!

To focus on the topic, in the next sections, the general term "nanoelectrochemistry" will be whittled down to the study of conductive surfaces coated with material having one or more external dimensions in the nanoscale and the applications will be limited to electrochemical sensors and biosensors modified with nonwoven nanofibrous membranes prepared by electrospinning.

13.2 Electrochemical Sensors Modified with Nanofibrous Membranes

Electrochemical sensors, modified with nanomaterial, have contributed to create great expectations in the last decade [4–8]. New materials such as carbon nanotubes [8–10], metal [3, 5, 11–16] and polymer nanoparticles [14, 17–20], carbon nanofibers [21–23], and boron-doped diamond nanograss [24] are ideal for electrochemical sensors due to their high surface area, high aspect ratio, and enhanced catalytic properties [25–30].

Regardless to the material used, electrochemical sensors suffer from several drawbacks, one of the most known is the passivation of its surface [31]. The oxidation of electroactive substances involves the formation of oxidized material at the electrode surface which blocks further reaction at the electrode [32]. For instance, during the electrochemical detection of biological analytes in matrices with a high content of phenolic compounds, the surface of the electrode is contaminated and the subsequent analysis is compromised [33].

Electrospun nanofibrous membranes can offer potential solutions to overcome the drawback. Electrodes coated with nanofibrous membranes can minimize or cancel passivation phenomena. As it will be shown later, such electrodes can be used for prolonged time in solution containing potential interfering substances without suffering from the problem of passivation. This is one of the potential applications described in the next section.

A second attractive feature of such membranes is their high surface available. In general, high surface available is desired to enhance reactivity of the materials, speed up adsorption or release mechanisms, and increase the amount of conjugated catalysts [34–36]. But, more interesting for electrochemical sensors, a material containing cavities, pores or, in general, empty volumes, implies a high permeability. These properties are fundamental because all the electrochemical detection mechanisms are based on the diffusion of one or more chemicals from the bulk solution to the electrode surface. For instance, a lower diffusion process, such as the one observed with electrodes coated with planar films, causes a lower current signals, lower sensitivity, higher detection potentials, and overall slower response. Instead, membranes prepared by electrospinning have high permeability and can offer a negligible resistance to the diffusion of chemicals.

Furthermore, nanofibrous membranes can be used to develop flexible and conductive nanofibrous electrodes, either with their modification with conducting polymers, or with the inclusion of activated multiwalled carbon nanotubes. The resulting membranes, made conductive by the incorporation of carbon nanotubes or conductive polymers, can be modeled in different forms, as desired, such as cylinders, balloons, spirals, and so on.

Finally, nanofibrous membranes will be used to develop biosensors by entrapping reactive molecules, catalysts, or other reactive materials in close proximity to the electrode surface without requiring any physical binding to the electrode. For instance, the distance between the active site of the enzyme entrapped in the membrane and the transduction element (the surface of the working electrode) can be minimized, increasing the intensity of the signal and the sensitivity of the sensor. Also, the large empty volume, typical of the nanofibrous membranes, is easily accessible and can be used for the massive loading of catalysts or other reactive materials in the development of biocatalytic solid-state membranes and biosensors. A detailed analysis on the main features and advantages offered by such nanofibrous biosensors will be finally described.

13.3 Introduction to Electrospinning

Electrospinning is a nanofiber production technique from a wide range of polymeric materials. In this section, an introduction to electrospinning process and starting materials, along with the applications of electrochemical sensors based on nanofibrous mats fabricated by the electrospinning process, will be discussed.

Electrospinning can be defined as a technique used *to produce nanofibers from a reservoir of reactive precursor species expelled through a nozzle, the tip of which is held at high voltage, the fibers being collected on a grounded plate.* [2]

The nanofibers are continuously produced with thickness from 10 to 1000 nm and potentially unlimited length. This distinguishes the fibers prepared by electrospinning with respect to those prepared by other techniques (such as artificial fiber spinning, thermally induced phase separation, molecular self-assembly, and the bio-fabrication alternatives) [37]. Fibers prepared by electrospinning are continuous filaments of potentially unlimited length, low cost, high production rate, and made with number of materials ranges, from natural, synthetics, and ceramics polymers [37–43]. A detailed survey of biopolymers used for electrospinning is reported by Kriegel *et al.* [44]. Among others, electrospinning of natural polymers is of great importance due to their implicit potential in foods and the medical sector [45]. Proteins and polysaccharides are two of the most favorable candidates to be electrospun [46].

The basic electrospinning setup needed to expel, eject, and collect the fibers is composed by a syringe, a high voltage power supply, and a collector. A schematic setup of a typical electrospinning unit is shown in Figure 13.1.

In this setup, a polymer dispersion is placed in the syringe equipped with a blunt stainless steel needle. A syringe pump controls the flow rate of the polymer solution. The system is generally horizontally aligned as in Figure 13.1, although vertical setups have also been described [47]. If the capillary is arranged in a vertical position, the solution may exit the capillary solely due to gravitational forces and a syringe pump may then not be needed. However, this setup allows for much less control than the use of a syringe pump, and consequently the horizontal setup or an arrangement including a syringe pump is preferred. The metallic capillary needle through which the polymeric solution is fed works as a positive electrode and is connected to a high voltage power supply operated in the positive DC mode at voltages between 1 and 50 kV. The setup also contains a grounded target collector on which the produced fibers are deposited and which works as the counter electrode. It completes the circuit and allows an electric field to be established between the capillary tip and the target. The involved amperages are typically very low and usually do not exceed a few microamperes [48].

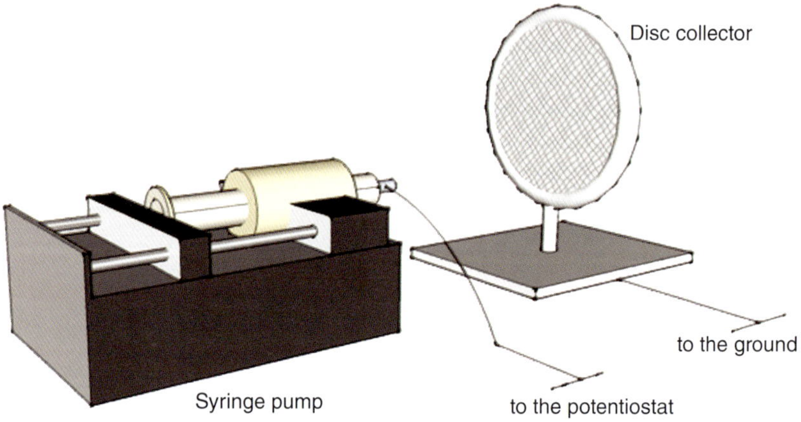

Figure 13.1 *Sketch of a simple electrospinning system. The basic setup comprises a syringe pump containing the polymer fluid. The high voltage power supply linked to the metal needle of the syringe pump. A grounded disk collector where the fibers ejected from the syringe is randomly collected*

Even though the theory behind electrospinning is quite complex, [49] the working principle of electrospinning is straightforward: A high voltage is applied to a drop of the polymer until the electrostatic repulsion between the polymer and the metallic needle reaches a critical potential and a filament is ejected and collected on a grounded collector. During the ejection, the solvent evaporates and the fluid jet stream continuously stretches, extends, and shrinks to very small diameter values until dried fibers are deposited onto the collector. The resulting fibers collected on an o-ring like target are shown in Figure 13.2.

Figure 13.2 *Picture of a homogeneous nylon-6 nanofibrous membranes still enclosed in the copper disk collector, just prepared by electrospinning. Note the disk collector over-coated by the fibers*

To control the final fiber properties, a number of variables must be taken into consideration. In addition to the choice of the polymer and solvent [50], it is of importance for other variables, such as (1) applied high voltage, flow rate, tip-to-collector distance, polymer concentration, conductivity, volatility of the solution, temperature, and relative humidity within the chamber [51–54]. All of them can vary the diameter [55], pore size or morphology [51, 56], and surface chemistry [57] of the resulting fibers [35, 44, 58, 59]. Operating diagrams that are able to predict the experimental parameters required for the process have been proposed [60].

13.4 Applications of Electrochemical Sensors Based on Electrospinning

13.4.1 Nanofibrous Membranes as Coating Material

Nanofibrous membranes have been primarily used as coating material to develop electrochemical sensors. In general, a piece of membrane (10–100 μm) can be simply placed over the electrode surface and kept tight thanks to an o-ring. Eventually, a drop of the solvent can be added to wet the membrane and improve its adhesion to the electrode surface. Figures 13.3 and 13.4 show a glassy carbon electrode coated with nanofibrous membranes.

With such configuration, the surface of the working electrode becomes protected by the functionalities of the membrane. Also, the electrode properties such as sensitivity and selectivity are changed due to this modification. But no chemical bond is formed between the electrode surface and the membrane. However, the electrode "*feels*" the membrane, in the sense that any chemical which diffuses toward the electrode surface, have first to pass through the membrane. The membrane functionalities can trigger different mechanisms or reactions, extending the potential applications of electrochemical sensors.

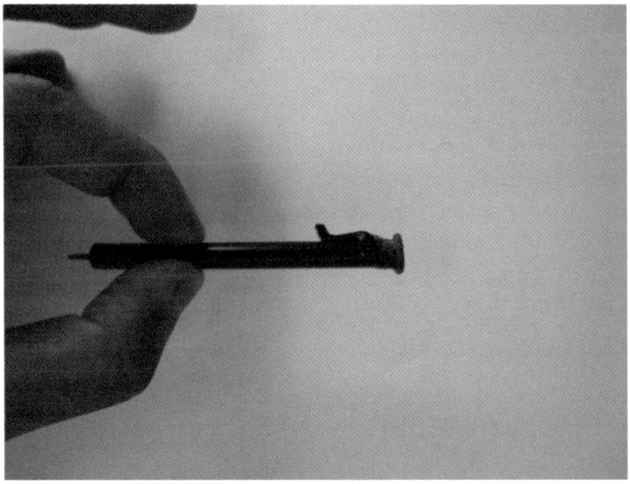

Figure 13.3 *Picture of the glassy carbon electrode coated with a nanofibrous membrane*

Figure 13.4 Front picture of the working electrode coated with a thin film nanofibrous membrane. Note the o-ring that keeps tight the membrane over the electrode surface

13.4.2 High Permeability

The main problem with electrodes coated with membranes is the higher resistance to the diffusion of the chemical species and the slower electron transfer mechanism. Chemical species that are blocked by the membrane cannot be detected. Species that decrease their diffusion can be oxidized or reduced but they show lower electrochemical currents and, for irreversible processes, a shifted peak potential, as expected from classical theory of electrochemical mechanism.

This problem can be overcome by using membranes prepared by electrospinning. The fibrous structure of membranes reduces the mass-transport resistance encountered by the chemical species passing through the channels of the membrane. As a result, electrodes coated with nanofibrous membranes offer, in principle, a negligible barrier to the analyte diffusion and high solvent permeability.

The high permeability can be explained by scanning electron microscopy images. Figure 13.5 shows random fibers of 100 nm thickness prepared by electrospinning.

There is not a general agreement if this morphology can be referred as "porous." IUPAC defines pores as cavities, channels, or interstices which are deeper than they are wide [61, 62]. Within this generous definition, the high permeability of the nanofibrous membranes can be attributed to their inherent high porosity.

13.4.3 Selective Barrier to Diffusion

The functionalities of the nanofibrous membranes together with their high surface available can be used to develop selective barriers toward the diffusion of specific molecules. For example, nylon-6 nanofibrous membranes exert such barrier to the diffusion of quercetin and others flavonoids such as myricetin and catechin. However, this property is not observed with ascorbic acid and other phenolics, such as caffeic, synaptic, syringic, vanillic, and gallic acid. The new selectivity exerted by the electrode coated with such membrane is exemplified in Figure 13.6.

Figure 13.5 SEM image of a polyamide nanofibrous membrane

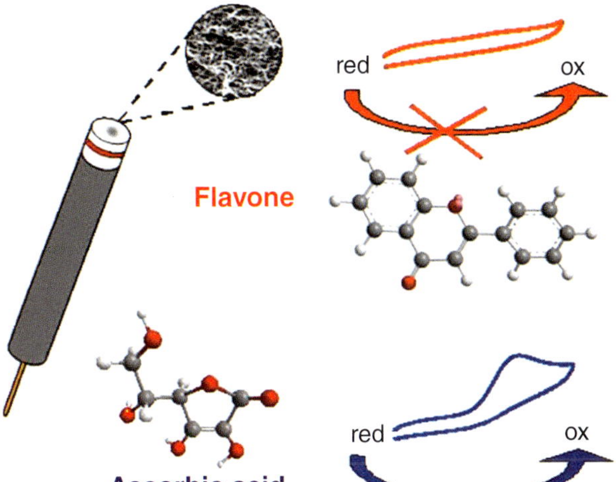

Figure 13.6 Sketch of the results achieved by the electrode coated with nanofibrous membranes with respect to quercetin and ascorbic acid

This selectivity can be explained by the different dissociation constant [63]. For instance, the dissociation constant of ascorbic acid ($pK_a = 4.1$) is of three orders of magnitude more acidic than the first dissociation constant of quercetin ($pK_a = 7.0$). At pH 6 or lower, quercetin is, in large extent, undissociated, while ascorbic acid is partially dissociated. In this condition of pH, quercetin is made hydrophobic and its adsorption on nylon is promoted; instead, ascorbic acid is made hydrophilic and its adsorption on nylon is prevented. Also, the electrostatic repulsion with the negative charge present on the nylon structure prevents its adsorption on the polymer. As a result, quercetin is adsorbed on the

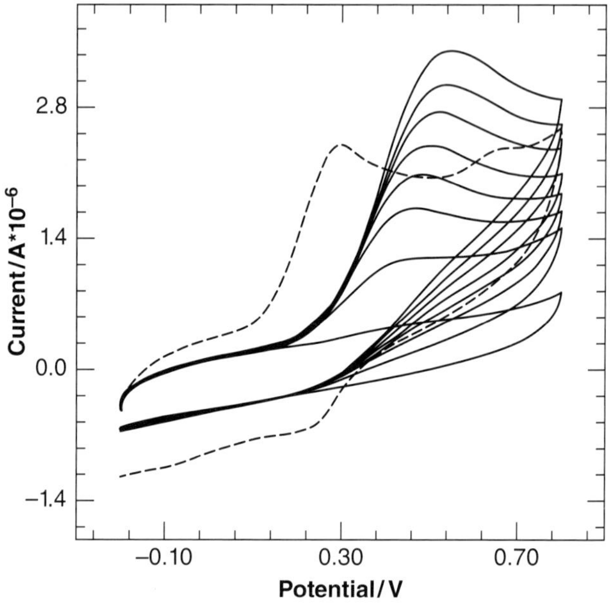

Figure 13.7 *Cyclic voltammetry of a glassy carbon working electrode coated with a polyamide nanofibrous membrane in a solution containing quercetin and ascorbic acid (solid lines). Also shown as dotted line, the signal of quercetin alone recorded at the bare electrode (without coating)*

membrane and it is not detected by the electrode. Ascorbic acid, instead, passes through the membrane, reaches the surface of the electrode and get oxidized (Figure 13.7).

The selective protection of the membrane can also be modulated by changing the pH of the solution. For example, the behavior of quercetin at pH higher than 7 is opposite. In this condition of pH, quercetin becomes dissociated and its hydrophobicity decreases. As a result, the quercetin adsorption on the membrane decreases linearly and above pH 10 is not present.

13.4.4 Conducting Nanofibrous Membranes

Most of the polymers used in electrospinning are nonconductive. In the next section, different procedures to prepare conductive nanofibrous membranes will be described and applied to the development of flat and flexible electrode sheet; thereby expanding the potential geometries used in electrochemistry.

A straightforward method to prepare conductive nanofibrous membrane (NFM) is from the electrospinning of conducting polymers (Figure 13.8), such as polypyrrole, polyaniline, polyethylene oxide, and polythiophene [64, 65]. Electrically conducting polymers are known to possess numerous features, which allow them to act as excellent materials for the development of sensing devices as well as for the immobilization of biomolecules in the design of biosensors [66].

Alternatively, it is also possible to develop conductive nanofibrous membranes from any kind of nonconductive polymers. At this purpose, conducting fibers made by polyamide

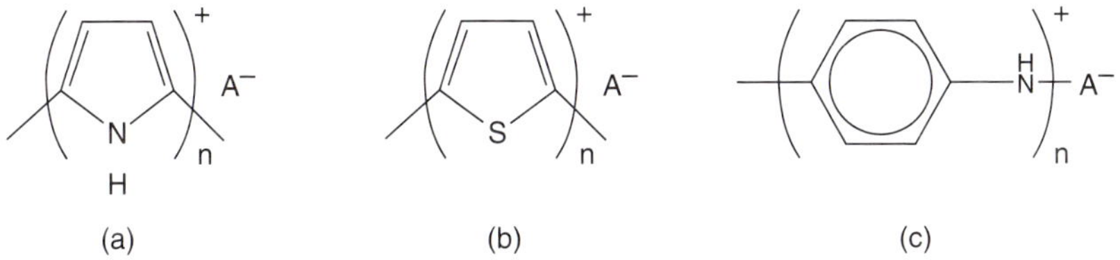

Figure 13.8 Families of conducting polymers: (a) polypyrrole; (b) polythiophene; (c) polyaniline

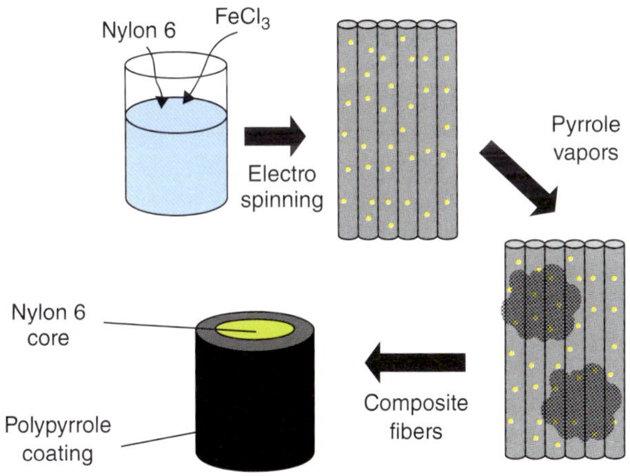

Figure 13.9 Procedure for the fabrication of composite conducting nanofibers

can be obtained by a two-step process: first, polyamide membranes are prepared by electrospinning starting with a polymer solution containing ferric chloride. Second, nonconductive membranes became conductive after their exposition to pyrrole vapors [67]. The basic steps of the method are shown in Figure 13.9. The resulting conductive nanofibrous membrane can be used as planar flexible sensors into an electrochemical cell (Figure 13.10). With such device, it was possible to detect anions such as phosphate and carbonate (Figure 13.11). Also shown in the figure, the corresponding calibration plot and the SEM images of the fiber mat before Figure 13.11(A) and after (B) its use on amperometric experiment [67].

Conductive nanofibrous membranes can also be prepared with carbon nanotubes [68]. Electrical properties of carbon nanotubes are well known [69]. For instance, Dai observed that electrochemical sensors functionalized with carbon nanotubes become not only sensitive to chemical doping by various molecules, but also they promote a change of conductance of up to two orders of magnitude [70]. This high electrical conductivity can be used for preparing conductive nanofibrous membranes made by unconductive polymer. There are several approaches to functionalize materials with carbon nanotubes [8, 9, 16, 71]. A simple one is reported by Fe *et al.* [72] and it is suitable to functionalize nonconductive polymer nanofibrous membranes and provide them enough conductivity to be used as electrochemical sensors. This is achieved by simply dip coating the membrane with a solution containing activated MWNT (0.5 g/l of MWNT and 0.3% wt/wt Triton X-100 in

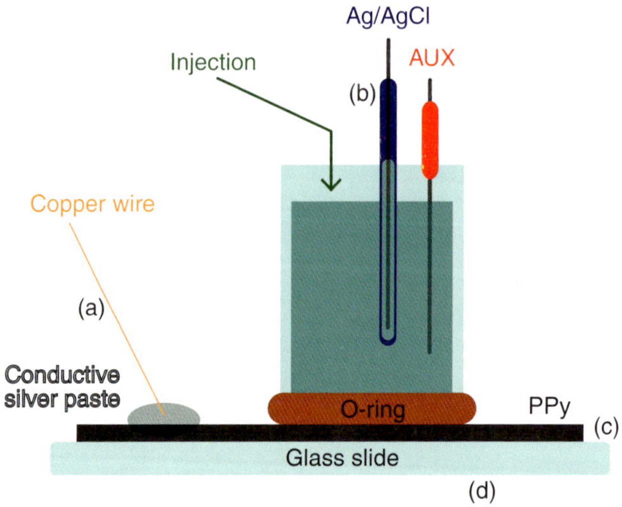

Figure 13.10 Electrochemical cell setup: (a) copper wire, (b) silver epoxy, (c) glass slide, and (d) conducting nanofiber mat; RE: reference electrode, AE: auxiliary electrode, WE: working electrode

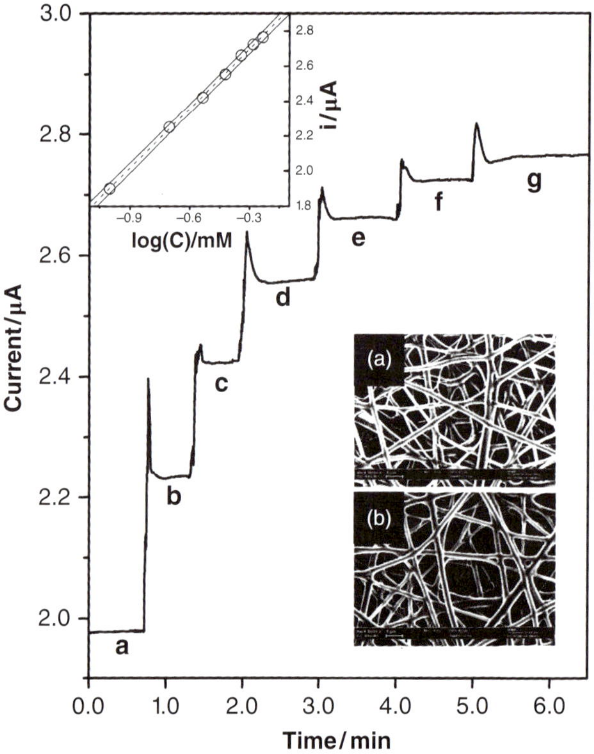

Figure 13.11 Amperometric i–t responses of increasing concentration of phosphate at the conductive screen-spun electrode (surface area 64 mm^2, glycine buffer 0.1 M, applied potential 0.7 V vs. Ag/AgCl 3 M). Also shown in the inset, the corresponding calibration plot and the SEM images of the fiber mat before (A) and after (B) its use on amperometric experiment

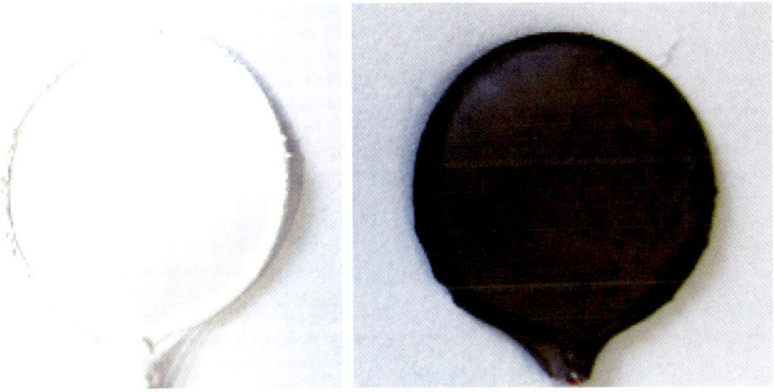

(a) (b)

Figure 13.12 *Picture of the electrospun nanofibrous membrane before (a) and after (b) the immobilization procedure with multiwalled carbon nanotubes*

water) for 5 h. The resulting membranes before and after the functionalization process are shown in Figure 13.12.

The morphology of the membranes functionalized with MWNT is shown in Figure 13.13. Fibers show their typical random orientation. Interestingly, also the MWNT appears randomly distributed, deployed around the surface of the fibers. This web-like morphology also provides a high textile strength. Sen *et al.* [73] reported that SWCNT-reinforced nanofibrous membranes made by polystyrene and polyurethane (PU) have a tensile

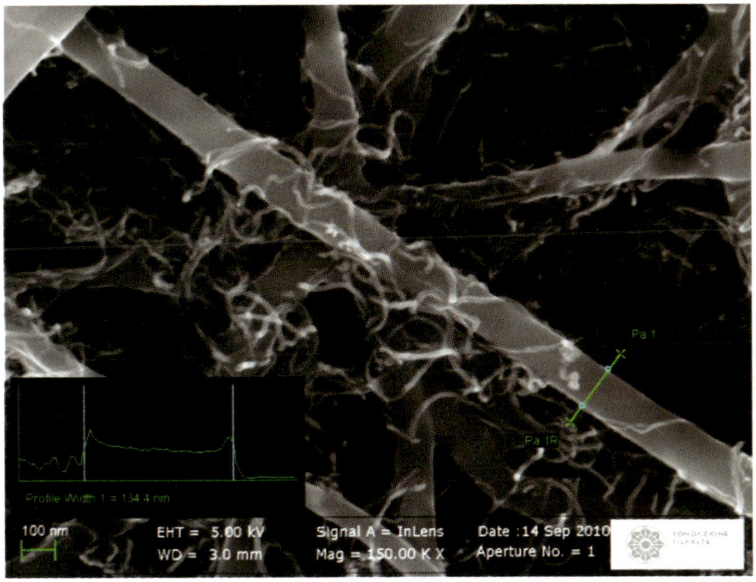

Figure 13.13 *SEM image of nanofibers after the immobilization procedure with carbon nanotubes*

strength and the modulus enhanced by 104% and 250% as compared to electrospun pure PU membranes.

Electrodes coated with such membranes have the great advantage of not suffering from the fouling of sulfhydryl compounds during their oxidation [72]. It is well known that the electrochemical detection of thiols is difficult because the oxidation of these reactive species is detrimental for the electrode performance [33, 74–79]. Instead, Fei *et al.* reported for the first time that the electrode coated with NFM/MWNT can be used to oxidize thiols functionalities without loss of the electrochemical performance of the sensor [72]. One reason is that NFM/MWNT sensor allows us to detect thiols at low potentials (+0.1 V vs. Ag/AgCl); reducing the number of potential interferents and increasing the rate of the desorption rate from the electrode surface. Furthermore, such sensor is quite sensitive and it is possible to dilute the sample, reducing the amount of passivating compounds.

The figures of merit of the sensors have been reported for the detection of cysteine. The dynamic range of the sensor was from 0.1 to 1.6 mM and the linearity ranged up to 0.4 mM ($R^2 = 0.997$). The sensitivity was 5.1 µA/mM and the detection limit was 15 µM (Figure 13.14). Analogous experiments using homocysteine and glutathione standard solutions, in the same range of concentrations, yielded similar responses.

A unique feature of this sensor is that at the end of each analysis, the MWNT membrane can be easily peeled-off, leaving the electrode surface back to its original electrochemical behavior. No trace of fouling was recorded on the electrode.

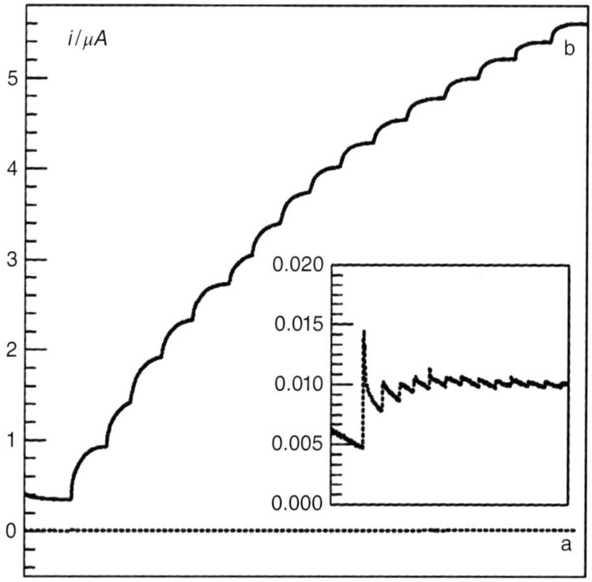

Figure 13.14 *Amperometric result of a working electrode coated with nanofibrous membranes modified with multiwalled carbon nanotubes after successive addition of cysteine (step of 0.1 mM). In the inset, it is shown the corresponding result obtained without the coating*

13.4.5 Biosensor Based on Nanofibrous Membranes

Electrochemical biosensors can be defined as integrated receptor–transducer device, which are capable of providing selective quantitative or semiquantitative analytical information using a biological recognition element [80, 81]. The biological recognition system (e.g., enzymes) translates information from the biochemical domain, usually an analyte concentration, into a chemical or physical output signal, which in the case of electrochemical biosensor is mostly an electrical signal [82, 83].

To design an electrochemical biosensor, different factors must be taken into account, including (i) the detected or measured parameter, (ii) the working principle of the transducer, (iii) the physical and chemical/biochemical model, (iv) the application, and (v) the technology and materials for sensor fabrication [84].

Materials used in electrochemical biosensors can greatly benefit from nanotechnology. Nanomaterials are especially used to immobilize the recognition elements and improve the transport of the electrical signal from the biorecognition element to the electrode surface (the transducer). A weak immobilization protocol affects the biosensor stability during use and storage and prejudices the overall functionality of the biosensor [85–89].

The immobilization of enzyme on nanofibrous membranes is an alternative and interesting solution with respect to the current state of the art [44, 46, 90–95]. First, nanofibrous membranes can load a larger amount of biorecognition elements [95, 96]. Thanks to the large surface area, a higher concentration of enzymes, antibodies, or cells can be bound or entrapped. Second, the thickness of the nanofibrous membranes, which can be as thin as tenth of microns, allows the positioning of the recognition element in close proximity to the electrode surface. This results in a very fast transduction signal [96]. The potential capability of the nanofibrous membranes to entrap a huge amount of enzymes in their three-dimensional structure can be observed by scanning electron microscopy (SEM). Figure 13.15 shows the SEM image of the fibers after their functionalization with glutaraldehyde. From Figure 13.5, it is possible to observe a more smooth surface of the fiber, indicating their coating with the cross linker.

Figure 13.16, instead, shows the fibers after their covalent modification with glutaraldehyde and glucose oxidase. In this case, the fibers appear surrounded by globular structures, which can be attributed to the enzymes. Also, the membranes preserve their high permeability which is still important to allow the diffusion of the product of the biocatalytic reaction or the mediator toward the electrode surface.

To exemplify the working principle of a mediated electrochemical biosensor, Figure 13.17 shows the typical response obtained with a glucose biosensor based on nanofibrous membranes. The cyclic voltammetry response of the mediator (dotted line) shows a simple reversible one electron signal. By adding glucose, an increase of the anodic current is observed, as expected from a catalytic process (Scheme 13.1) [97, 98]. No peaks are observed, but a large steady-state catalytic current.

The figures of merits achievable by this simple setup are promising (Figure 13.18). By using ferrocene methanol as mediator, and monitoring the signal generated in amperometric mode shows a linearity range of 1–15 mM of glucose, a sensitivity of 1.9 μA/mM (rsd = 2.9%). The limit of detection was 6×10^{-6} M. The decay of the electrode response was 0.5% per day (12% after 30 days with 1–2 h functioning per day) [99].

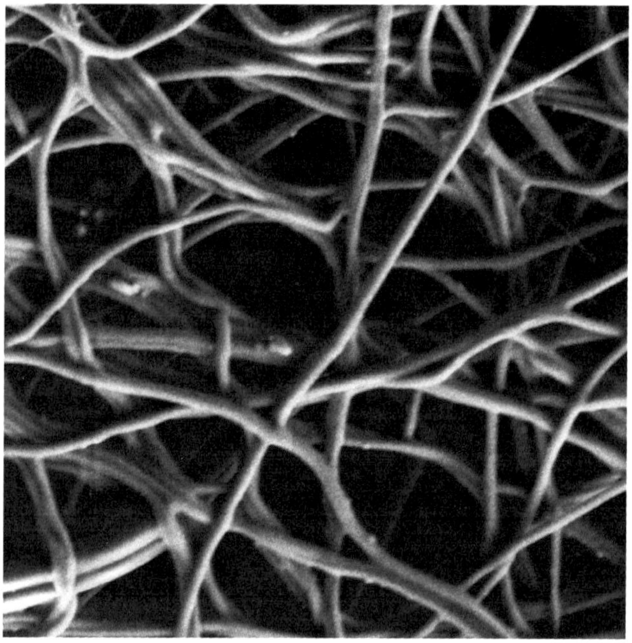

Figure 13.15 SEM image of the nanofibers prepared by electrospinning after their derivatization with glutaraldehyde

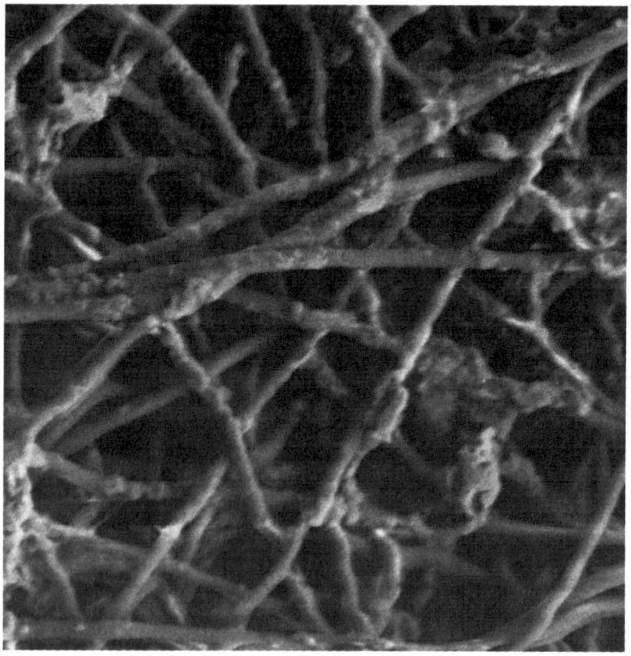

Figure 13.16 SEM image of the nanofibers prepared by electrospinning after their derivatization with glutaraldehyde and subsequent immobilization of glucose oxidase

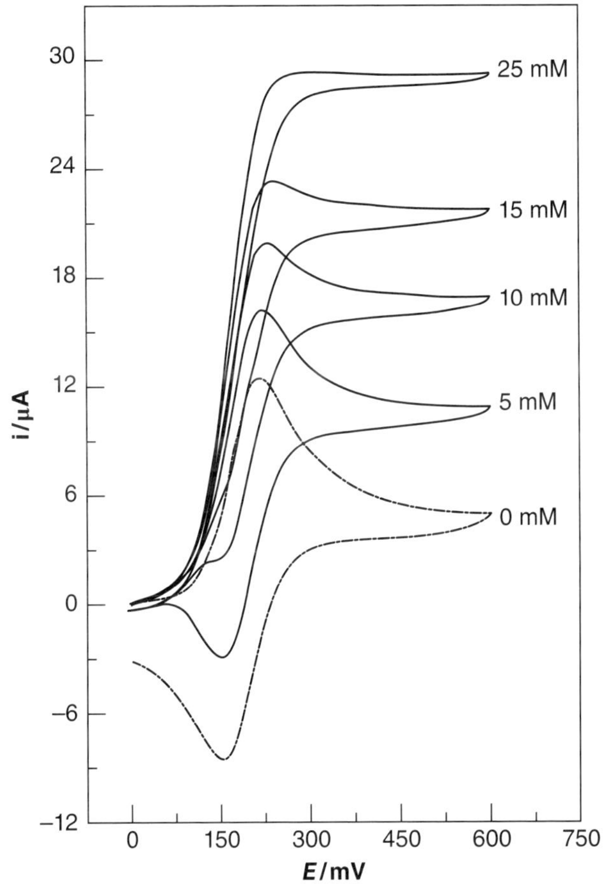

Figure 13.17 *Cyclic voltammetry for increasing addition of glucose (from 0 to 25 mM), obtained at a glassy carbon electrode coated with an electrospun nanofibrous membrane in which glucose oxidase was immobilized. The dotted line is the signal of the mediator without the presence of glucose*

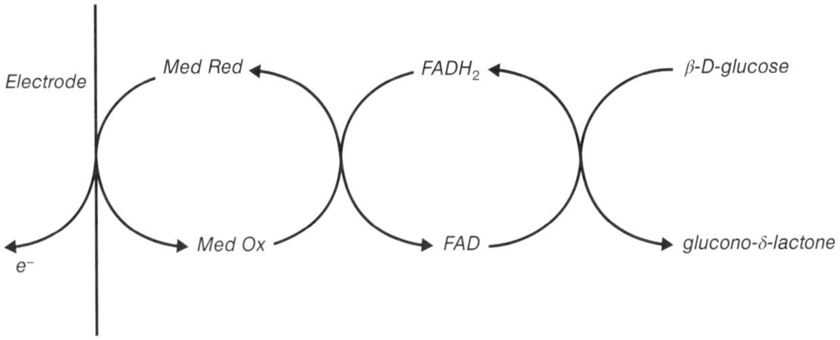

Scheme 13.1 *Catalytic cycle of a mediated electrochemical biosensor*

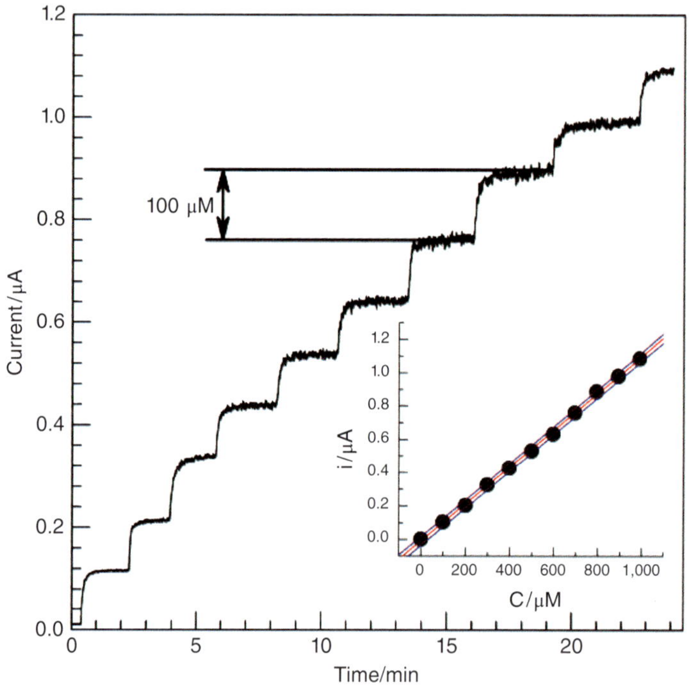

Figure 13.18 Amperometric response of the glucose biosensor after successive addition of glucose. Also shown as inset the corresponding calibration plot

Even with a simple first generation biosensor, working in amperometric mode at a fixed detection potential of +0.7 V (high enough to directly oxidize H_2O_2 at the electrode surface), it is possible to achieve a linearity between 1 and 9×10^{-3} M ($R^2 = 0.9997$) with a sensitivity of 1.11 μA mM^{-1} (rsd% = 0.97%) and a limit of detection between 2.5×10^{-6} M (calculated as S/N = 3). Short-term stability was excellent as shown in Figure 13.19. Long-term stability results in a decrease of 34% after one month of continuous use.

However, sensitivity of the biosensors based on nanofibrous membranes can be limited by the detrimental interaction of the enzyme with the nanostructure of the membrane. If the high surface area of the nanofibrous membranes is the cause of its high reactivity, it can also be the cause of altering the delicate three-dimensional structure of the enzyme. Preliminary experiments confirmed this hypothesis. This was observed with the tyrosinase biosensor [100]. Once the enzyme was immobilized on the nanofibrous membranes, it suffered from interference of several chemical species such as phenol-like molecules and sugars such as glucose and fructose. Moreover, severe changes in the catalytic behavior of the enzyme were observed. The change in enzyme activity was observed for glucose oxidase in relation to the pH. After immobilization, the optimal pH was shifted of about three units (Figure 13.20) [96].

The kinetic of the biocatalysis was determined before and after the immobilization of the enzyme on the nanofibrous membrane. The kinetic constants (K_m and V_{max}) were measured by studying the relationship between the steady-state current and the concentration

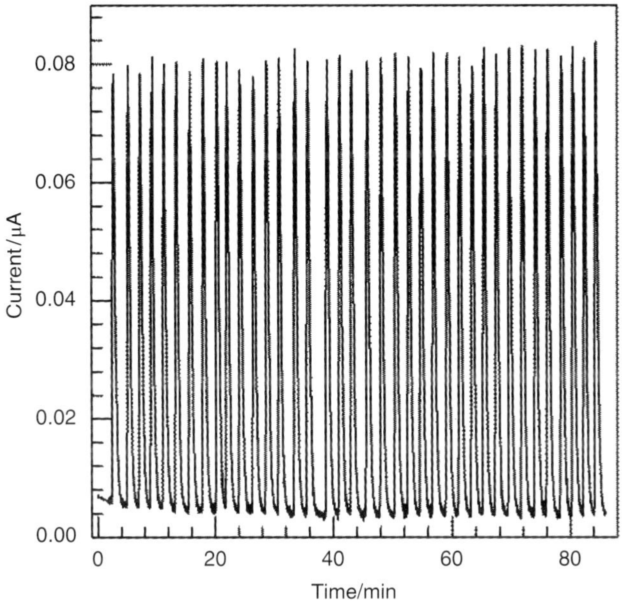

Figure 13.19 *Thirty-eight consecutive injections of glucose (0.5 mM) at a platinum electrode coated with a nanofibrous nylon-6 membrane where glucose oxidase was immobilized. Experimental conditions: phosphate buffer 0.10 M at pH 6.5 was used as carrier, at a flow rate of 1 ml min^{-1}. Detection potential was +0.70 V vs. Ag/AgCl (3 M)*

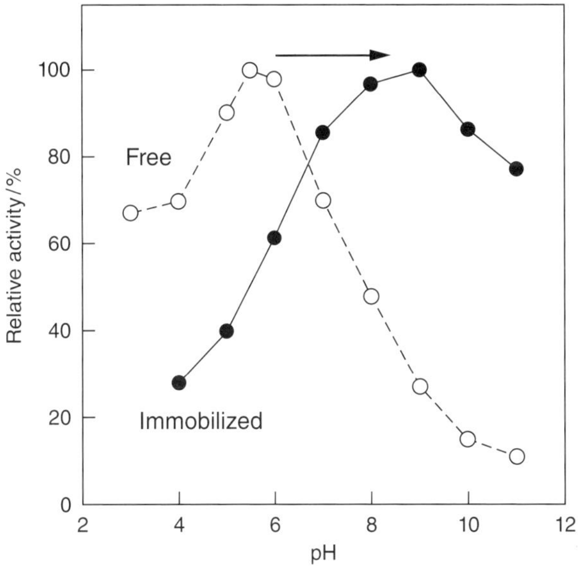

Figure 13.20 *Relative activity as a function of the pH of the bulk solution for glucose oxidase free (open circles) and immobilized onto NFM (full circles)*

of the substrate. A plot of l/iSS vs. 1/C yielded a "Lineweaver–Burk-like" relationship with the slope K_{mapp}/i_{Max} and intercept l/i_{Max}. As a result, the apparent Michaelis constant for the immobilized glucose oxidase was 1.7×10^{-2} M. This value is greater than those obtained in the case of homogeneous biocatalytic transduction, that is, when the enzymes are free in solution [101], but still comparable to the previous reports concerning glucose oxidase covalently bound to nylon [102]. Despite the fact that a low value for the apparent Michaelis–Menten constant implies more favorable conditions for glucose oxidase activity, it lowers the upper limit of the linear response range. Therefore, the increase in the magnitude of K_m is still desirable since it allows the extension of the upper limit of the linear response range.

These experiences indicate a strong relationship between NFM and the protein. The high concentration of functional groups in the surface area of the NFM can be responsible for stretching of the immobilized proteins, and changes on their original configuration.

References

1. *Nanoscience and Nanotechnologies* The Royal Society, 2004, ISBN 0 85403 604 0, http://www.nanotec.org.uk/finalReport.htm.
2. *BSI British Standards*, Terminology for Nanomaterials, PAS 136 (2007) BSI British Standards.
3. Murray, R.W. (2008) Nanoelectrochemistry: metal nanoparticles, nanoelectrodes, and nanopores. *Chem. Rev.*, **108**, 2688–2720.
4. Arico, A., Bruce, P., Scrosati, B., Tarascon, J., and Van Schalkwijk, W. (2005) Nanostructured materials for advanced energy conversion and storage devices. *Nat. Mater.*, **4**, 366–377.
5. Katz, E. and Willner, I. (2004) Integrated nanoparticle–biomolecule hybrid systems: synthesis, properties, and applications. *Angew. Chem. Int. Ed.*, **43**, 6042–6108.
6. Kimmel, D.W., LeBlanc, G., Meschievitz, M.E., and Cliffel, D.E. (2012) Electrochemical sensors and biosensors. *Anal. Chem.*, **84**, 685–707.
7. Privett, B.J., Shin, J.H., and Schoenfisch, M.H. (2010) Electrochemical sensors. *Anal. Chem.*, **82**, 4723–4741.
8. Wang, J. and Musameh, M. (2003) Carbon nanotube/teflon composite electrochemical sensors and biosensors. *Anal. Chem.*, **75**, 2075–2079.
9. Wang, J. (2005) Carbon-nanotube based electrochemical biosensors: a review. *Electroanalysis*, **17**, 7–14.
10. Lawrence, N., Deo, R., and Wang, J. (2005) Comparison of the electrochemical reactivity of electrodes modified with carbon nanotubes from different sources. *Electroanalysis*, **17**, 65–72.
11. Pumera, M., Sanchez, S., Ichinose, I., and Tang, J. (2007) Electrochemical nanobiosensors. *Sens. Actuators, B*, **123**, 1195–1205.
12. Scampicchio, M., Wang, J., Blasco, A.J., Arribas, A.S., Mannino, S., and Escarpa, A. (2006) Nanoparticle-based assays of antioxidant activity. *Anal. Chem.*, **78**, 2060–2063.
13. Hrapovic, S., Liu, Y., Male, K., and Luong, J. (2004) Electrochemical biosensing platforms using platinum nanoparticles and carbon nanotubes. *Anal. Chem.*, **76**, 1083–1088.

14. Katz, E., Willner, I., and Wang, J. (2004) Electroanalytical and bioelectroanalytical systems based on metal and semiconductive nanoparticles. *Electroanalysis*, **16**, 19–44.
15. Luo, X., Morrin, A., Killard, A., and Smyth, M. (2006) Application of nanoparticles in electrochemical sensors and biosensors. *Electroanalysis*, **18**, 319–326.
16. Wang, J. (2005) Nanomaterial-based electrochemical biosensors. *Analyst*, **130**, 421–426.
17. Fu, L., Liu, H., Zhang, H., Li, C., Zhang, T., Wu, Y., Holze, R., and Wu, H. (2006) Synthesis and electrochemical performance of novel core/shell structured nanocomposites. *Electrochem. Commun.*, **8**, 1–4.
18. Gangopadhyay, R. and De, A. (2000) Conducting polymer nanocomposites: a brief overview. *Chem. Mater.*, **12**, 608–622.
19. Jang, J. (2006) Conducting polymer nanomaterials and their applications. *Emissive Mater.: Nanomater.*, **199**, 189–259.
20. Shipway, A. and Willner, I. (2001) Nanoparticles as structural and functional units in surface-confined architectures. *Chem. Commun.*, **20**, 2035–2045.
21. Lu, X., Zhou, J., Lu, W., Liu, Q., and Li, J. (2008) Carbon nanofiber-based composites for the construction of mediator-free biosensors. *Biosens. Bioelectron.*, **23**, 1236–1243.
22. Marken, F., Gerrard, M., Mellor, I., Mortimer, R., Madden, C., Fletcher, S., Holt, K., Foord, J., Dahm, R., and Page, F. (2001) Voltammetry at carbon nanofiber electrodes. *Electrochem. Commun.*, **3**, 177–180.
23. Ambrosi, A., Sasaki, T., and Pumera, M. (2010) Platelet graphite nanofibers for electrochemical sensing and biosensing: the influence of graphene sheet orientation. *Chem. Asian J.*, **5**, 266–271.
24. Wei, M., Terashima, C., Lv, M., Fujishima, A., and Gu, Z. (2009) Boron-doped diamond nanograss array for electrochemical sensors. *Chem. Commun.*, **24**, 3624–3626.
25. Campuzano, S. and Wang, J. (2011) Nanobioelectroanalysis based on carbon/inorganic hybrid nanoarchitectures. *Electroanalysis*, **23**, 1289–1300.
26. Chen, J., Miao, Y., He, N., Wu, X., and Li, S. (2004) Nanotechnology and biosensors. *Biotechnol. Adv.*, **22**, 505–518.
27. Hu, X. and Dong, S. (2008) Metal nanomaterials and carbon nanotubes-synthesis, functionalization and potential applications towards electrochemistry. *J. Mater. Chem.*, **18**, 1279–1295.
28. Lahiff, E., Lynam, C., Gilmartin, N., O'Kennedy, R., and Diamond, D. (2010) The increasing importance of carbon nanotubes and nanostructured conducting polymers in biosensors. *Anal. Bioanal.Chem.*, **398**, 1575–1589.
29. Li, H., Liu, S., Dai, Z., Bao, J., and Yang, X. (2009) Applications of nanomaterials in electrochemical enzyme biosensors. *Sensors*, **9**, 8547–8561.
30. Sarma, A.K., Vatsyayan, P., Goswami, P., and Minteer, S.D. (2009) Recent advances in material science for developing enzyme electrodes. *Biosens. Bioelectron.*, **24**, 2313–2322.
31. Gilroy, D. and Conway, B. (1965) Kinetic theory of inhibition and passivation in electrochemical reactions. *J. Phys. Chem.*, **69**, 1259–1267.
32. Gattrell, M. and Kirk, D. (1993) A study of electrode passivation during aqueous phenol electrolysis. *J. Electrochem. Soc.*, **140**, 903–911.

33. Lawrence, N., Beckett, E., Davis, J., and Compton, R. (2002) Advances in the voltammetric analysis of small biologically related compounds. *Anal. Biochem.*, **303**, 1–16.
34. Li, W., Laurencin, C., Caterson, E., Tuan, R., and Ko, F. (2002) Electrospun nanofibrous structure: a novel scaffold for tissue engineering. *J. Biomed. Mater. Res.*, **60**, 613–621.
35. Greiner, A. and Wendorff, J.H. (2007) Electrospinning: a fascinating method for the preparation of ultrathin fibres. *Angew. Chem. Int. Ed.*, **46**, 5670–5703.
36. Bognitzki, M., Czado, W., Frese, T., Schaper, A., Hellwig, M., Steinhart, M., Greiner, A., and Wendorff, J. (2001) Nanostructured fibers via electrospinning. *Adv. Mater.*, **13**, 70–72.
37. Luo, C.J., Stoyanov, S.D., Stride, E., Pelan, E., and Edirisinghe, M. (2012) Electrospinning versus fibre production methods: from specifics to technological convergence. *Chem. Soc. Rev.*, **41**, 4708–4735.
38. Teo, W.E. and Ramakrishna, S. (2006) A review on electrospinning design and nanofibre assemblies. *Nanotechnology*, **17**, R89–R106.
39. Jin, L., Wang, T., Zhu, M., Leach, M.K., Naim, Y.I., Corey, J.M., Feng, Z., and Jiang, Q. (2012) Electrospun fibers and tissue engineering. *J. Biomed. Nanotechnol.*, **8**, 1–9.
40. Long, Y., Yu, M., Sun, B., Gu, C., and Fan, Z. (2012) Recent advances in large-scale assembly of semiconducting inorganic nanowires and nanofibers for electronics, sensors and photovoltaics. *Chem. Soc. Rev.*, **41**, 4560–4580.
41. Aleshin, A. (2006) Polymer nanofibers and nanotubes: charge transport and device applications. *Adv. Mater.*, **18**, 17–27.
42. Huang, Z., Zhang, Y., Kotaki, M., and Ramakrishna, S. (2003) A review on polymer nanofibers by electrospinning and their applications in nanocomposites. *Compos. Sci. Technol.*, **63**, 2223–2253.
43. Li, D. and Xia, Y. (2004) Electrospinning of nanofibers: reinventing the wheel? *Adv. Mater.*, **16**, 1151–1170.
44. Kriegel, C., Arrechi, A., Kit, K., McClements, D.J., and Weiss, J. (2008) Fabrication, functionalization, and application of electrospun biopolymer nanofibers. *Crit. Rev. Food Sci. Nutr.*, **48**, 775–797.
45. John, M.J. and Thomas, S. (2008) Biofibres and biocomposites. *Carbohydr. Polym.*, **71**, 343–364.
46. Supaphol, P., Suwantong, O., Sangsanoh, P., Srinivasan, S., Jayakumar, R., and Nair, S.V. (2012) Electrospinning of biocompatible polymers and their potentials in biomedical applications. *Biomed. Appl. Polym. Nanofibers*, **246**, 213–239.
47. Duan, B., Dong, C., Yuan, X., and Yao, K. (2004) Electrospinning of chitosan solutions in acetic acid with poly(ethylene oxide). *J. Biomater. Sci., Polym. Ed.*, **15**, 797–811.
48. Park, S., Park, K., Yoon, H., Son, J., Min, T., and Kim, G. (2007) Review. Apparatus for preparing electrospun nanofibers: designing an electrospinning. *Polym. Int.*, **56**, 1361–1366.
49. Hohman, M., Shin, M., Rutledge, G., and Brenner, M. (2001) Electrospinning and electrically forced jets. I. Stability theory. *Phys. Fluids*, **13**, 2201–2220.
50. Son, W., Youk, J., Lee, T., and Park, W. (2004) The effects of solution properties and polyelectrolyte on electrospinning of ultrafine poly(ethylene oxide) fibers. *Polymer*, **45**, 2959–2966.

51. Deitzel, J., Kleinmeyer, J., Harris, D., and Tan, N. (2001) The effect of processing variables on the morphology of electrospun nanofibers and textiles. *Polymer*, **42**, 261–272.
52. Zong, X., Kim, K., Fang, D., Ran, S., Hsiao, B., and Chu, B. (2002) Structure and process relationship of electrospun bioabsorbable nanofiber membranes. *Polymer*, **43**, 4403–4412.
53. Shin, Y., Hohman, M., Brenner, M., and Rutledge, G. (2001) Experimental characterization of electrospinning: the electrically forced jet and instabilities. *Polymer*, **42**, 9955–9967.
54. Reneker, D.H. and Yarin, A.L. (2008) Electrospinning jets and polymer nanofibers. *Polymer*, **49**, 2387–2425.
55. Fridrikh, S., Yu, J., Brenner, M., and Rutledge, G. (2003) Controlling the fiber diameter during electrospinning. *Phys. Rev. Lett.*, **90**, 144502.
56. Rebollar, E., Cordero, D., Martins, A., Chiussi, S., Reis, R.L., Neves, N.M., and Leon, B. (2011) Improvement of electrospun polymer fiber meshes pore size by femtosecond laser irradiation. *Appl. Surf. Sci.*, **257**, 4091–4095.
57. Deitzel, J., Kosik, W., McKnight, S., Tan, N., DeSimone, J., and Crette, S. (2002) Electrospinning of polymer nanofibers with specific surface chemistry. *Polymer*, **43**, 1025–1029.
58. Greiner, A., Wendorff, J.H., Yarin, A.L., and Zussman, E. (2006) Biohybrid nanosystems with polymer nanofibers and nanotubes. *Appl. Microbiol. Biotechnol.*, **71**, 387–393.
59. Agarwal, S., Wendorff, J.H., and Greiner, A. (2010) Chemistry on electrospun polymeric nanofibers: merely routine chemistry or a real challenge? *Macromol. Rapid Commun.*, **31**, 1317–1331.
60. Hohman, M., Shin, M., Rutledge, G., and Brenner, M. (2001) Electrospinning and electrically forced jets: II. Application. *Phys. Fluids.*, **13**, 2221–2236.
61. Rouquerol, J., Avnir, D., and Everett, D., *et al.* (1994) Guidelines for the characterization of porous solids. *Stud. Surf. Sci. Catal.*, 1994, **87**: 1–9.
62. Rouquerol, J., Avnir, D., Fairbridge, C., Everett, D., Haynes, J., Pernicone, N., Ramsay, J., Sing, K., and Unger, K. (1994) Physical and biophysical chemistry division commission on colloid and surface chemistry including catalysis. *Pure Appl. Chem.*, **66**, 1739–1758.
63. Scampicchio, M., Bulbarello, A., Arecchi, A., and Mannino, S. (2008) Electrospun nanofibers as selective barrier to the electrochemical polyphenol oxidation. *Electrochem. Commun.*, **10**, 991–994.
64. Reneker, D.H., Yarin, A.L., Zussman, E., and Xu, H. (2007) Electrospinning of nanofibers from polymer solutions and melts. *Adv. Appl. Mech.*, **41**, 43–195.
65. Long, Y., Li, M., Gu, C., Wan, M., Duvail, J., Liu, Z., and Fan, Z. (2011) Recent advances in synthesis, physical properties and applications of conducting polymer nanotubes and nanofibers. *Prog. Polym. Sci.*, **36**, 1415–1442.
66. Gerard, M., Chaubey, A., and Malhotra, B. (2002) Application of conducting polymers to biosensors. *Biosens. Bioelectron.*, **17**, 345–359.
67. Granato, F., Scampicchio, M., Bianco, A., Mannino, S., Bertarelli, C., and Zerbi, G. (2008) Disposable electrospun electrodes based on conducting nanofibers. *Electroanalysis*, **20**, 1374–1377.

68. Xie, X., Mai, Y., and Zhou, X. (2005) Dispersion and alignment of carbon nanotubes in polymer matrix: a review. *Mater. Sci. Eng., R*, **49**, 89–112.
69. Trojanowicz, M. (2006) Analytical applications of carbon nanotubes: a review. *TrAC, Trends Anal. Chem.*, **25**, 480–489.
70. Dai, G., Liu, C., Liu, M., Wang, M., and Cheng, H. (2002) Electrochemical hydrogen storage behavior of ropes of aligned single-walled carbon nanotubes. *Nano Lett.*, **2**, 503–506.
71. Merkoci, A., Pumera, M., Llopis, X., Perez, B., del Valle, M., and Alegret, S. (2005) New materials for electrochemical sensing VI: carbon nanotubes. *TrAC, Trends Anal. Chem.*, **24**, 826–838.
72. Li, F., Scampicchio, M., and Mannino, S. (2011) Carbon nanotube-adsorbed electrospun nanofibrous membranes as coating for electrochemical sensors for sulfhydryl compounds. *Electroanalysis*, **23**, 1773–1775.
73. Sen, R., Zhao, B., Perea, D., Itkis, M., Hu, H., Love, J., Bekyarova, E., and Haddon, R. (2004) Preparation of single-walled carbon nanotube reinforced polystyrene and polyurethane nanofibers and membranes by electrospinning. *Nano Lett.*, **4**, 459–464.
74. Nekrassova, O., Lawrence, N., and Compton, R. (2003) Analytical determination of homocysteine: a review. *Talanta*, **60**, 1085–1095.
75. Lawrence, N., Davis, J., and Compton, R. (2000) Analytical strategies for the detection of sulfide: a review. *Talanta*, **52**, 771–784.
76. White, P., Lawrence, N., Davis, J., and Compton, R. (2002) Electrochemical determination of thiols: a perspective. *Electroanalysis*, **14**, 89–98.
77. Scampicchio, M., Lawrence, N.S., Arecchi, A., and Mannino, S. (2007) Electrochemical reduction of Ellman's reagent: a novel selective detection protocol for thiol compounds. *Electroanalysis*, **19**, 2437–2443.
78. Lawrence, N., Davis, J., Jiang, L., Jones, T., Davis, S., and Compton, R. (2000) Selective determination of thiols: a novel electroanalytical approach. *Analyst*, **125**, 661–663.
79. Scampicchio, M., Lawrence, N.S., Arecchi, A., Cosio, S., and Mannino, S. (2007) Voltammetric detection of free sulfhydryl compounds in food flours. *Electroanalysis*, **19**, 85–90.
80. Thevenot, D.R., Toth, K., Durst, R.A., and Wilson, G.S. (2001) Electrochemical biosensors: recommended definitions and classification. *Biosens. Bioelectron.*, **16**, 121–131.
81. Thevenot, D., Toth, K., Durst, R., and Wilson, G. (1999) Electrochemical biosensors: recommended definitions and classification. *Pure Appl. Chem.*, **71**, 2333–2348.
82. Wilson, R. and Turner, A. (1992) Glucose oxidase: an ideal enzyme. *Biosens. Bioelectron.*, **7**, 165–185.
83. Wang, J. (2005) Nanomaterial-based amplified transduction of biomolecular interactions. *Small*, **1**, 1036–1043.
84. Zhang, S., Wright, G., and Yang, Y. (2000) Materials and techniques for electrochemical biosensor design and construction. *Biosens. Bioelectron.*, **15**, 273–282.
85. Kandimalla, V.B., Tripathi, V.S., and Ju, H. (2006) Immobilization of biomolecules in sol–gels: biological and analytical applications. *Crit. Rev. Anal. Chem.*, **36**, 73–106.
86. Salimi, A., Sharifi, E., Noorbakhsh, A., and Soltanian, S. (2007) Electrochemical glucose biosensors. *Biosens. Bioelectron.*, **22**, 3146–3153.

87. Scouten, W., Luong, J., and Brown, R. (1995) Enzyme or protein immobilization techniques for applications in biosensor design. *Trends Biotechnol.*, **13**, 178–185.
88. Williams, R. and Blanch, H. (1994) Covalent immobilization of protein monolayers for biosensor applications. *Biosens. Bioelectron.*, **9**, 159–167.
89. Zhu, Z., Garcia-Gancedo, L., Flewitt, A.J., Xie, H., Moussy, F., and Milne, W.I. (2012) A critical review of glucose biosensors based on carbon nanomaterials: carbon nanotubes and graphene. *Sensors*, **12**, 5996–6022.
90. Agarwal, S., Wendorff, J.H., and Greiner, A. (2008) Use of electrospinning technique for biomedical applications. *Polymer*, **49**, 5603–5621.
91. Manesh, K.M., Santhosh, P., Gopalan, A., and Lee, K. (2007) Electrospun poly(vinylidene fluoride)/poly(aminophenylboronic acid) composite nanofibrous membrane as a novel glucose sensor. *Anal. Biochem.*, **360**, 189–195.
92. Bhardwaj, N. and Kundu, S.C. (2010) Electrospinning: a fascinating fiber fabrication technique. *Biotechnol. Adv.*, **28**, 325–347.
93. Kim, J., Jia, H., and Wang, P. (2006) Challenges in biocatalysis for enzyme-based biofuel cells. *Biotechnol. Adv.*, **24**, 296–308.
94. Liang, D., Hsiao, B.S., and Chu, B. (2007) Functional electrospun nanofibrous scaffolds for biomedical applications. *Adv. Drug Delivery Rev.*, **59**, 1392–1412.
95. Wang, Z., Wan, L., Liu, Z., Huang, X., and Xu, Z. (2009) Enzyme immobilization on electrospun polymer nanofibers: an overview. *J. Mol. Catal. B: Enzym.*, **56**, 189–195.
96. Arecchi, A., Scampicchio, M., Brenna, O.V., and Mannino, S. (2010) Biocatalytic nylon nanofibrous membranes. *Anal. Bioanal.Chem.*, **398**, 3097–3103.
97. Bourdillon, C., Demaille, C., Moiroux, J., and Saveant, J. (1993) New insights into the enzymatic catalysis of the oxidation of glucose by native and recombinant glucose oxidase mediated by electrochemically generated one-electron redox co-substrates. *J. Am. Chem. Soc.*, **115**, 2–10.
98. Wang, J. (2008) Electrochemical glucose biosensors. *Chem. Rev.*, **108**, 814–825.
99. Scampicchio, M., Arecchi, A., Lawrence, N.S., and Mannino, S. (2010) Nylon nanofibrous membrane for mediated glucose biosensing. *Sens. Actuators, B*, **145**, 394–397.
100. Arecchi, A., Scampicchio, M., Drusch, S., and Mannino, S. (2010) Nanofibrous membrane based tyrosinase-biosensor for the detection of phenolic compounds. *Anal. Chim. Acta*, **659**, 133–136.
101. Bourdillon, C., Bourgeois, J.P., and Thomas, D. (1980) Covalent linkage of glucose oxidase on modified glassy carbon electrodes. Kinetic phenomena. *J. Am. Chem. Soc.*, **102**, 4231–4235.
102. Portaccio, M., El-Masry, M., Diano, N.R., De Maio, A., Grano, V., Lepore, M., Travascio, P., Bencivenga, U., Pagliuca, N., and Mita, D.G. (2002) An amperometric sensor employing glucose oxidase immobilized on nylon membranes with different pore diameter and grafted with different monomers. *J. Mol. Catal. B: Enzym.*, **18**, 49–67.

14

Electrochemical Impedance Spectroscopy

Araceli González-Cortés
Departamento de Química Analítica, Facultad de Ciencias Química, Universidad Complutense de Madrid, Complutense de Madrid, Avda. Complutense, 28040, Madrid, Spain

14.1 Introduction

The wholesomeness of food is the real proviso for healthy life. Food freed from microbial and chemical cross-contaminations adds on to its hygienic and nutritive value. Infectious diseases spreading every day through food have become a life-threatening problem for millions of people around the world. In fact, food or food products are the potent transmitting agent of more than 250 known diseases [1].

Electrochemical and electrical impedance has been shown as powerful techniques to be applied in the food industry and have been used for monitoring and detecting pathogens and drug residues in foodstuffs [2]. Actually, in recent years, impedimetric biosensors which integrate biological recognition technology and impedance have gained widespread application in the field of bacteria detection [3]. Moreover, the presence of corrosion processes occurring on the metallic materials caused by adsorption of proteins surfaces represents a big problem in the food industry. Electrochemical impedance spectroscopy (EIS) has demonstrated its usefulness to investigate the fouling of these materials [4]. This technique is also suitable to differentiate between fresh and frozen-thawed salmon, permitting the detection of fraud, such as the sale of frozen-thawed products offered as fresh fish [5], and to assess salt content in food products such as cured ham or pork loin using impedance spectroscopy [6].

EIS is a powerful electrochemical technique capable of detecting small changes occurring at the solution–electrode interface. Accordingly, EIS has been extensively exploited for

Agricultural and Food Electroanalysis, First Edition.
Edited by Alberto Escarpa, María Cristina González and Miguel Ángel López.

the characterization of materials and surface modification procedures, as well as for the monitoring of binding events. During the last decade, impedance spectroscopy has been widely used for probing various types of biomolecular interactions (immunosensors, DNA hybridization, rapid biomolecular screening, cell culture monitoring) [7].

The fact that impedance measurements can be performed in a reagentless assay format makes it very attractive for biosensor development. The equipment required is compact, does not contain mobile parts, and is easy to miniaturize, suggesting that EIS biosensors could be easily used in-field with minimal requirements. Furthermore, impedance combines rapid response, low detection limits, cost-effectiveness, and the possibility of performing real-time monitoring of the samples, in contrast to established strategies for pathogen detection such as culture, polymerase chain reaction (PCR), enzyme linked immunosorbent assay (ELISA), or those sensors based on sandwich assay formats [8]. A number of authors have exploited EIS for pathogen detection, by monitoring either the changes in the medium conductivity caused by bacterial growth/metabolism, or the changes in the solution–electrode interface due to microorganism nonspecific adsorption or specific capture onto the sensor surface.

14.2 Impedance Spectroscopy – Theoretical Background

For impedance measurements, a small sinusoidal AC voltage of small amplitude (typically 5–10 mV) is applied, and the current response, that differs in amplitude and phase (phase difference, ϕ) with the applied voltage, is determined. The impedance is calculated as the ratio between them [9].

The in-phase current response determines the real (resistive) component of the impedance, while the out-of-phase current response determines the imaginary (capacitive) component. The AC probe voltage should be small enough so that the system response is linear, allowing simple equivalent circuit analysis. A second reason is to avoid disturbing the system. Impedance methods are quite powerful, in that they are capable of characterizing physicochemical processes of widely differing time constants, sampling electron transfer at high frequency and mass transfer at low frequency.

Impedance results are commonly fitted to equivalent circuits of resistors and capacitors, such as the Randles circuit shown in Figure 14.1, which is often used to interpret simple electrochemical systems. It describes an electrochemical cell where a single-step Faradaic process in the presence of diffusion may occur. The use of a large-area electrode was particularly important for the impedance measurements since the impedance of a counter electrode must be negligible compared to the impedance of the working electrode [4].

In Figure 14.1, R_{ct} is the charge-transfer resistance, which is inversely proportional to the rate of electron transfer; C_{dl} is the double-layer capacitance; R_s is the electrolyte resistance between the working and reference electrodes, the double-layer capacitance C_{dl} and the Faradaic impedance due to the charge-transfer process Z_f at the working electrode–electrolyte interface. Z_f is normally subdivided into a charge-transfer resistance R_{ct} and the so-called Warburg impedance Z_w, which reflects the influence of the mass transport of the electroactive species on the total impedance of the electrochemical cell. Thus, for those diffusion-limited processes, Z_w becomes dominant, whereas for those charge-transfer-controlled processes, Z_f is just R_{ct}.

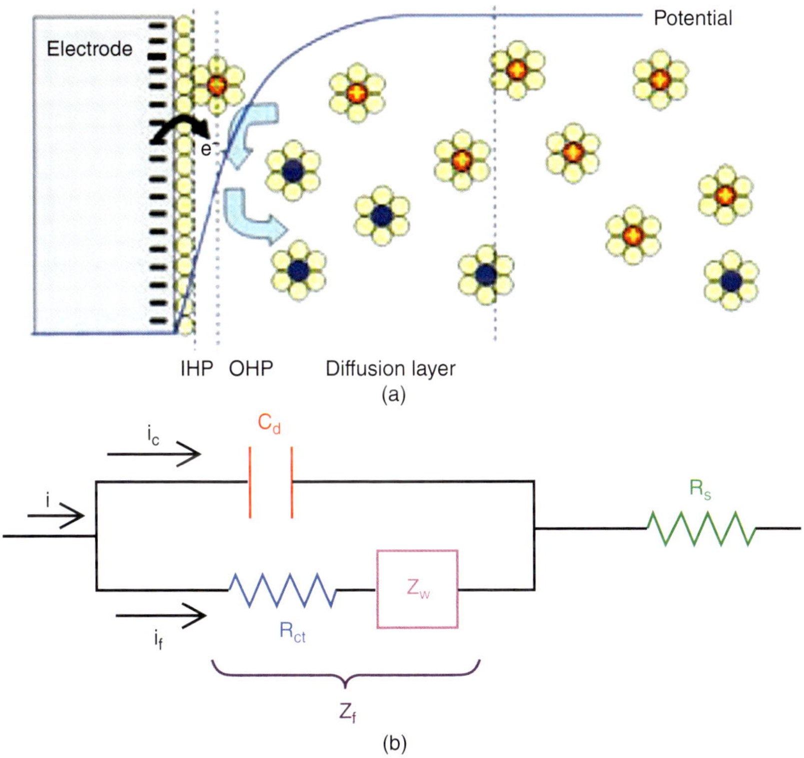

Figure 14.1 *A simple electrified interface, in which the vertical dotted lines in (a) are represented by the electronic components in (b). (a) The oxidants (red) with a positive charge diffuse toward the negatively charged electrode, accept electrons from the electrode at the interface, become the reductants (blue), and diffuse to the bulk of the solution. The oxidant is also a counterion to the electrode. No specific adsorption is considered at the interface. IHPs and OHPs are the inner and outer Helmholtz planes, respectively. (b) An equivalent circuit representing each component at the interface and in the solution during an electrochemical reaction is shown for comparison with the physical components.* C_{dl}, *double-layer capacitor;* R_{ct}, *polarization resistor; W, Warburg resistor; and* R_s, *solution resistor. From Ref. [10] with permission. Copyright © 2003, American Chemical Society*

All these components can be identified by studying the variation in impedance of an electrochemical cell over a wide frequency range (normally 100 kHz to 0.1 Hz). A common way of showing the resulting data is the complex plane or Nyquist plot (Figure 14.2a), in which the real Z' vs. the imaginary Z'' components of the impedance are plotted. In this plot, two separate processes are very well differentiated, that is, a semicircle, observed at higher frequencies, relating to a charge-transfer-controlled process, the intercept of which with the X axis gives R_s and R_{ct} values, and a straight line with a slope of 1, characteristic of the lower frequency range, represents the diffusionally limited electrochemical process, Z_w, whose extrapolation to the X axis allows calculation of the Warburg coefficient (σ), from which the diffusion coefficients of the electroactive species can be estimated.

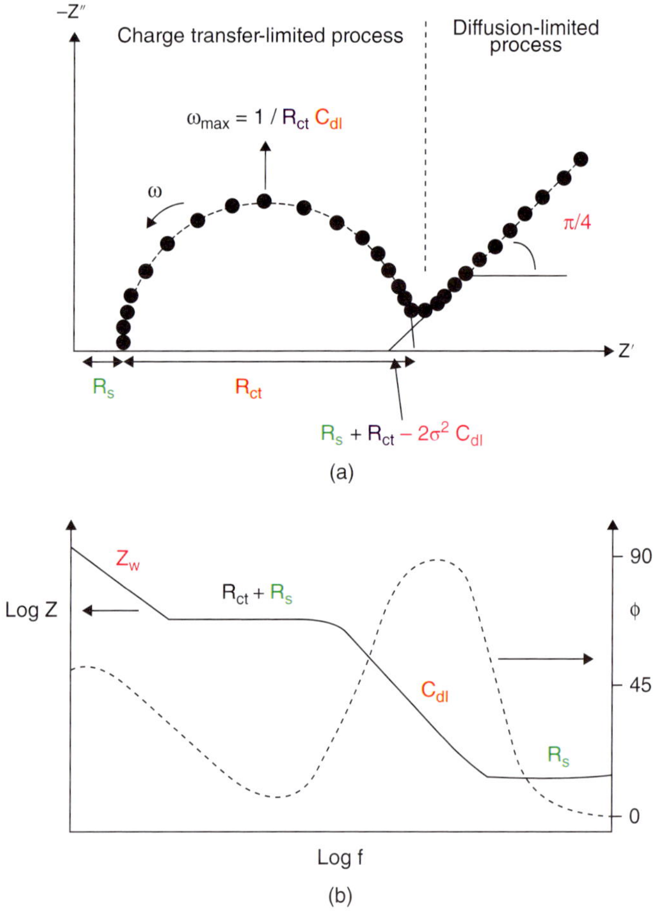

Figure 14.2 *Nyquist (a) and Bode (b) plots arising from the Randles circuit shown in Figure 14.1. (Each dot of the Nyquist plot represents the impedance at a given frequency.) Adapted from Ref. [11], with permission from Elsevier*

From the frequency at the top of the semicircle, ω_o, where Z'' is maximum and equal to 0.5 R_{ct}, the time-relaxation constant (τ) for the Faradaic process can also be calculated:

$$\omega = \frac{1}{C_{dl}\,R_{ct}}; \quad \tau = C_{dl}R_{ct}$$

Another way of presenting impedance data is the Bode plot, in which the logarithm of the absolute value of Z and the phase (ϕ) are plotted against the logarithm of frequency (f). Figure 14.2b shows the Bode plots for the Randles circuit. Unlike the Nyquist plots, these data presentations give direct information about f and ϕ that help ascertain the different constituent phases of the system more easily. Thus, in those frequency regions where a resistive behavior is dominant, a horizontal line is observed for the log Z vs. log f representation and a ϕ close to 0° is measured. Also, capacitive behavior within a frequency region is described by a straight line with a slope of 1 in the log Z vs. log f plots and a ϕ around 90°, whereas diffusion-controlled phenomena (Warburg impedance) would give a straight line with a slope of 1/2 and a ϕ of 45°.

If an analyte affects one or more of these equivalent circuit parameters and these parameters are not affected by interfering species, then impedance methods can be used for analyte detection. R_s arises primarily from the electrolyte resistance and is analytically useful mainly in conductivity sensors. The Warburg impedance, which can be used to measure effective diffusion coefficients, is seldom useful for analytical applications. The equivalent circuit elements in Figure 14.1 that are most often useful for analyte detection are R_{ct} and C_{dl}.

If EIS measurements were carried out with a constant electrolyte composition, C_{dl} would be mainly dependent on the electrode area. So, any electrode modifier of insulating features decreases the double-layer capacitance as compared to the pure metal electrode. Thus, the double-layer capacitance usually arises from the series combination of several elements, such as analyte binding (C_{anal}) to a sensing layer (C_{sens}) on an Au electrode (C_{Au}).

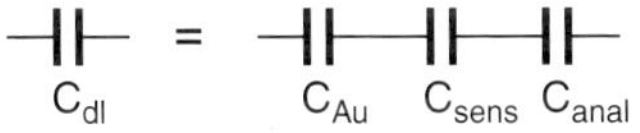

In this case, the measured capacitance is

$$\frac{1}{C_{dl}} = \frac{1}{C_{Au}} + \frac{1}{C_{sens}} + \frac{1}{C_{anal}}$$

for a sensing layer and analyte layer that are continuous. In many cases, the capacitance at the Au electrode–sensing layer interface is large and can be neglected. The sensitivity is then determined by the relative capacitance of the analyte layer and the sensing layer.

For each dielectric layer, the capacitance per unit area depends on the layer thickness (δ) according to

$$\frac{C}{A} = \frac{\varepsilon_{dl}}{\delta} \quad \text{where} \quad \varepsilon_{dl} = \varepsilon_o \varepsilon_\rho$$

where A is the electrode area, ε_{dl} is the dielectric permittivity introduced into the double-charged-layer molecules, ε_o is the dielectric constant of the vacuum, and ε_ρ is the effective dielectric constant of the layer separating the ionic charges and the electrode surface, so capacitance is most sensitive to binding of large analytes, such as proteins or cells.

One difficulty with capacitive sensors is that their sensitivity depends on obtaining the proper thickness of the original sensing layer. If the original sensing layer is too thin, then the underlying electrode surface may be partially exposed, allowing for nonspecific interactions from interfering species. However, if the original sensing layer is too thick, then the detected AC impedance current is dramatically reduced, as well as the change in capacitance upon analyte binding.

Sometimes, particularly when the electrode surface is rough and/or the bulk properties of the electrode material are inhomogeneous, the electronic properties of the interface cannot be described sufficiently well with a capacitive element, and a constant phase element (CPE) should be introduced instead of C_{dl}. CPEs are identified in impedance spectra due to the appearance of depressed semicircles in Nyquist plots (with their center below the real axis) and phase-angle values lower than 90° in Bode plots [9, 11, 12].

It is important to distinguish between non-Faradaic and Faradaic sensors [13]. In *Faradaic impedance sensors,* a redox species is alternately oxidized and reduced by the transfer of an electron to and from the metal electrode. Thus, Faradaic EIS requires the addition of a redox-active species and DC bias conditions such that it is not depleted. The interaction between a charged redox couple and the electrode surface is reflected by the charge-transfer resistance (R_{ct}). In most impedance biosensor studies, a redox probe is added to the analyte to decrease the measured impedance value. Otherwise, the AC current response that is measured becomes quite small, which may decrease the experimental sensitivity [14]. The choice of the redox probe depends on various parameters such as the charge, hydrophobicity/hydrophilicity, size of the redox couple, and the chemical and physical properties of the modified electrodes. The $Fe(CN)_6^{3-/4-}$ has been the most frequently used redox probe for the development of impedance biosensors. The electron-transfer resistance, R_{ct}, controls the electron-transfer kinetics of the solubilized redox probe at the electrode interface. Thus, the insulating modifier on the electrode is expected to retard the interfacial electron-transfer kinetics and to increase the electron-transfer resistance. So, R_{ct} can also be quite sensitive to analyte binding, particularly for detection of large species, such as proteins or cells, which significantly impede electron transfer. For analyte binding (R_{anal}) to a sensing layer (R_{sens}) on an Au electrode (R_{Au}), the measured resistance is

$$R_{ct} = R_{Au} + R_{sens} + R_{anal}$$

These resistances are also connected as series elements in the equivalent electronic circuit,

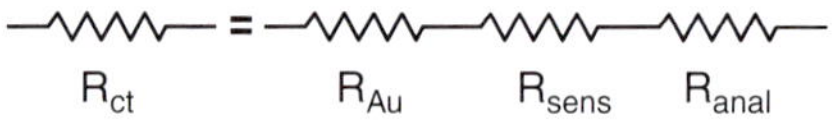

The resistance at the interface between the Au electrode and the sensing layer is typically negligible.

Impedance sensing is most useful for large species that significantly perturb the sensing interface. Generally, decreases in the double-layer capacitance and increases in charge-transfer resistance (R_{ct}) consistent with the addition of insulating layers of proteins on the electrode surface were observed. Many of the examples of impedance sensors that we will discuss later in this chapter monitor R_{ct} as a measure of analyte concentration.

Non-Faradaic impedance sensors employ no redox probes in the electrolyte solution, and only non-Faradaic impedance is operative. The electron transfer characterizing parameters, R_{ct} and Z_W, become infinite, and the equivalent circuit can be simplified:

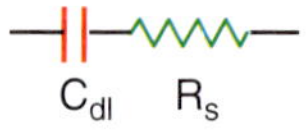

The variable component in the circuit is presented by C_{mod} and C_{anal}, and it affects the imaginary part of the impedance, Z'', considering only the dense part of the double-charged layer:

$$Z'' = \frac{1}{\omega\ (C_{Au} + C_{mod} + C_{anal})}$$

and impedance changes arise primarily from displacement of water and ions by molecules binding to the sensor surface.

One of the most important parameters governing the technique is the applied frequency. At low frequencies ($f < 1$ mHz) the impedance value is basically determined by the DC conductivity of the electrolyte solution. At very high frequencies ($f > 100$ kHz), inductance of the electrochemical cell and connecting wires could contribute to the impedance spectra. Thus, the analytically meaningful impedance spectra are usually recorded at frequencies where they are mainly controlled by the interfacial properties of the modified electrodes (10 mHz $< f <$ 100 kHz).

The term capacitive biosensor usually designates a sensor based on a non-Faradaic scheme, usually measured at a single frequency.

Generally, the impedance detection techniques can be classified into two types depending on the presence or absence of specific biorecognition elements. The first type works by measuring the impedance change caused by binding of targets to bioreceptors (antibodies and nucleic acids) immobilized onto the electrode surface, while the detection principle of the second type is based on metabolites produced by bacterial cells as a result of growth and no biorecognition elements are used on the surface of electrodes for the capture of bacterial cells [3].

14.3 Chemical Sensors

EIS has become a very useful tool in the field of metal packaging with the base materials steel and aluminum, aiming to enhance the corrosion stability and to minimize metal migration into food. Accordingly, this technique has been extensively applied in corrosion studies where variations in surface properties of materials occur as a result of changes in surface chemical composition.

The heating of biological fluids in the food industry often causes deposition (fouling) on the processing equipment. Ommanovic *et al.* used EIS to investigate the interfacial behavior of β-lactoglobulin, one of the main constituents of milk deposits on heat exchangers, at an austenitic stainless steel surface over the temperature range 299–343 K at an open circuit potential [4]. When a stainless steel electrode is immersed in an electrolyte solution, it spontaneously starts to passivate by forming and aging the protective Cr/Fe-rich oxide film on its surface. This significantly inhibits further dissolution of the alloy and, thus, enhances its corrosion stability. However, the addition of β-lactoglobulin to the phosphate buffer solution produced a decrease in charge-transfer resistance as a result of the increased dissolution rate of the passive film. This value represents a direct measure of the metal dissolution rate, that is, corrosion rate.

The isoelectric point for β-lactoglobulin is 5.2, and therefore it carries a negative charge at pH 7.0; however, at the same pH, a hydrophilic stainless steel surface at the open circuit potential has a net positive charge. This will facilitate the involvement of negatively charged carboxylate groups of the protein as anchoring sites in the contact region between the proteins and the stainless steel surface. The observed decrease in the charge-transfer resistance values was a result of the increased dissolution rate of the passive film, which was shown to be very sensitive to the amount of adsorbed protein.

Figure 14.3a shows an example of an EIS spectrum recorded on a stainless steel electrode in a phosphate buffer solution. The presentation of the data in the form of a Bode impedance plot clearly reveals the presence of three distinguishable frequency-dependent segments.

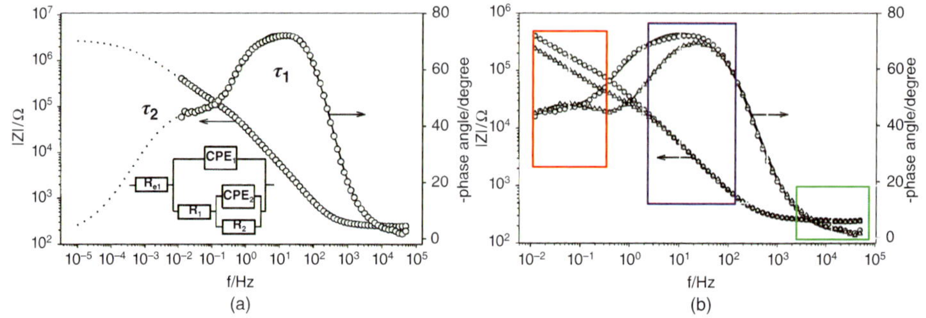

Figure 14.3 (a) Bode plot of the stainless steel electrode recorded at the open circuit potential and 299 K in the protein-free solution, showing two time constants τ_1 and τ_2. The symbols are experimental values, the solid line is the best fit, and the dashed lines are the extrapolated curves obtained using the same fitting parameters. Inset: electrical equivalent circuit used to model the EIS data. (b) Bode plot of the stainless steel electrode recorded at the open circuit potential (OCP) and 299 K in the (O) protein-free solution and (Δ) solution containing 0.25 g/l β-lactoglobulin. From Ref. [4] with permission from Elsevier

In the high-frequency region, the absolute impedance curve is almost independent of the frequency, with the phase angle value approaching zero degree. This is a typical response for a resistive behavior and corresponds to a resistance of the phosphate solution between the working and the reference electrode.

In the medium-frequency region, a linear relationship can be observed between the absolute impedance and the frequency with a slope of −0.82 and phase angle of approximately −72°. This frequency region corresponds to a capacitive behavior of the electrode/electrolyte interface. In the lower frequency region, around 200 mHz, the phase angle curve approaches −50° and then further decreases to about −42°, although at a lower rate. At the same time, a change in the slope of the impedance (|Z|) curve toward lower values can be clearly noticed, indicating an increase in a resistive behavior of the electrode, relative to a capacitive behavior. However, the region where the absolute impedance is independent of the frequency (DC limit) is not attained in the investigated frequency range. Experimental data are fitted using the electrical equivalent circuit composed of two RC time constants in parallel (inset to Figure 14.3).

In the equivalent circuit of the figure, it is obvious that the outer subcircuit with the lower time constant (CPE_1-R_1, τ_1), detected at high frequencies, is related to the fast charging/discharging processes at the electrode/electrolyte interface, where CPE_1 stands for the double-layer capacitance and R_1 for the charge-transfer resistance. Moreover, the simulated DC limit is extended to the very low frequencies (dotted Z-line), thus indicating the presence of a process with a very high time constant. Therefore, the inner subcircuit (CPE_2-R_2, τ_2) represents a slower process, probably associated with the mass transport of electroactive species through the formed Cr/Fe-rich passive oxide film.

The EIS spectrum recorded in a protein-containing solution (Figure 14.3b, triangles) shows that the addition of the protein did not produce any change in the capacitive behavior of the electrode, and the impedance curves are overlaid in the medium-frequency region. However, as the resistive behavior predominates, a significant difference between the spectra can be noticed.

The rapid assessment of the quality and useful life of coatings package, employed to protect from corrosion in the case of canned foodstuffs is an important issue for both packaging manufacturers and food industries. EIS has been used for direct quality control of lacquered tinplated cans, one of the most common packaging materials [15]. This method allows the influence of type of lacquer and tinplate support to be evaluated when coming in contact with cooked octopus in brine. The principle behind this study is that a perfectly protected (lacquered) tin plate is expected to behave as a pure capacitor, due to the dielectric (insulating) properties of the lacquer. According to the AC theory [16], a metal covered with an undamaged coating exhibits a high impedance and the Bode blot of this sample (metal/dielectric coating) is a straight line with a slope of -1 (log $|Z|$ vs. log f plot), whereas the phase angle (phase vs. log f plot) reaches 90° over a wide frequency range. In the case of a damaged coating, water penetrates into the coating and forms a new liquid/metal interface under the damaged area with corrosion phenomena being initiated at this new interface. In this case, the behavior of the system is not purely capacitive as a resistive element (R), parallel to the capacitor (C), is thus introduced.

In summary, EIS succeeded in identifying the insufficient protection provided by the lacquer at stressed areas of samples that showed no evident defect after visual and microscopic examination.

EIS has also been used successfully as a nondestructive transduction method for electrochemical recognition of Zr^{4+} on gold electrodes modified with hydroxamated self-assembled monolayers (SAMs). The sensor was successfully examined by recovery tests in synthetic, tap, and river water samples [17]. Zr(IV) was selectively accumulated on the topside of hydroxamated surface, and the surface blocked the charge-transfer process between metallic base and a redox probe, *p*-benzoquinone (PBQ)/H_2Q, in solution (Figure 14.4). This redox probe was selected according to two criteria: (i) stability against any possible chemical reaction, either in the solution or on the modified surface, and (ii) maximum sensitivity of the recorded analytical signal for constant conditions. In this case, no electrostatic interaction between the surface and the probe occurred. A low detection limit (7.8×10^{-10} M Zr(IV)) and wide linear range calibration plot (1.0×10^{-9}–5.0×10^{-5} M Zr(IV)) were obtained. The method avoided ZrO_2 microcrystalline formation and allows easy regeneration.

Pathogenic bacteria secrete various virulence factors that can directly interact with the outer lipid bilayer membrane of eukaryotic cells, inducing cell death by apoptosis or necrosis. Such virulence factors account for much of the toxic action associated with bacterial infection; therefore the detection of such proteins could provide a methodology for sensing/detection of pathogenic bacteria in, for example, food or human tissue.

Tun *et al.* used a tethered bilayer lipid membrane (TBLM) as a sensing platform for the detection of the clinically relevant pathogenic bacteria *Staphylococcus aureus* MSSA 476 and *Pseudomonas aeruginosa* PAO1 via their secreted toxins in the growth supernatant broth by EIS. TBLMs were used to detect and discriminate between the supernatant growth media of pathogenic and nonpathogenic bacteria. The damaging effect of bacterial toxins on the TBLMs was persistent and was indicated by a continuous decrease in resistance and an increase in capacitance with time. The supernatant solution of nonpathogenic *E. coli* DH5α bacteria caused no significant change in TBLM impedance [18].

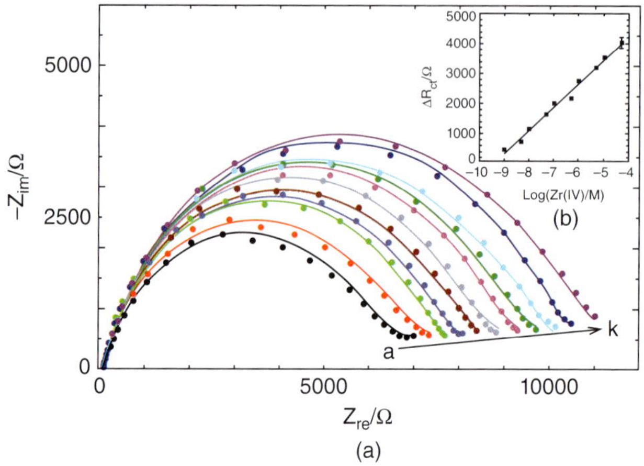

Figure 14.4 (A) EIS complex plane plots obtained in the presence of 5 mM PBQ/H_2Q redox probe on the gold cysteamine (Au-CA)-nitrilodihydroxamic acid (NDHA) self-assembled monolayer (SAM) electrode in 0.1 M acetate buffer solution (ABS), pH 3.5, and 0.1 M $NaClO_4$, after 20 min of accumulation of Zr(IV) from 0.23 M $HClO_4$ solutions containing: (a) 0.0 M, (b) 1.0×10^{-9} M, (c) 5.0×10^{-9} M, (d) 1.0×10^{-8} M, (e) 5.0×10^{-8} M, (f) 1.0×10^{-7} M, (g) 5.0×10^{-7} M, (h) 1.0×10^{-6} M, (i) 5.0×10^{-6} M, (j) 1.0×10^{-5} M, and (k) 5.0×10^{-5} M Zr(IV). (B) The inset shows the calibration curve obtained from complex plane plots ($\Delta R_{ct} = R_{ct} - R_{ct,b}$). From Ref. [17] with permission from Elsevier

14.4 Electrochemical Biosensors Based on Impedance Spectroscopy

An impedance biosensor consists of a biologically sensing element integrated with a physical transducer that transforms the signal resulting from the interaction of the analyte with the biological element into an impedance signal. The EIS sensitivity is affected further by pinhole effects, which occur when the employed immobilization procedure does not result in the total coverage of the electrode surface. As a result, redox molecules defuse through immobilized biomolecules and reach bare spots on the electrode surface. Such events, together with nonspecific adsorption of targeted biomolecules onto the bare spots, contribute a great deal to insensitive EIS responses. Though pinhole effects occur in almost all biosensors, nonspecific adsorption is more prevalent in immunosensors and microbial biosensors. On the other hand, EIS signals resulting from DNA biosensors are grossly affected by repulsions between the negatively charged phosphate backbone and redox anions such as $Fe(CN)_6^{3-/4-}$ that make the quantitation of DNA hybridization reactions rather difficult. Thus, several blocking agents including bovine serum albumin (BSA) and 1-mercaptohexanol (MCH) have been used to eliminate both nonspecific adsorption and pinhole effects. In addition, electrostatic interactions between DNA phosphate backbone and anions have been minimized using positively charged DNA binders such as $Ru(NH_3)_6^{3+}$. Nevertheless, the success of EIS-based biosensors depends on a careful selection of electrode materials that are compatible with the preferred immobilization chemistry. In fact, the immobilization method is the key process in the construction of impedimetric biosensors, since the efficiency of biomolecule immobilization on the electrode surface can profoundly affect the analytical performance of impedance biosensors [3].

14.4.1 Enzymatic Biosensors

Impedance spectroscopy has been extensively used to follow changes of the interfacial properties of electrodes upon immobilization of enzymes and to characterize biocatalytic processes at enzyme-modified electrodes. Faradaic impedance spectroscopy can be used to study the kinetics of the electron transfer originating from bioelectrocatalytic reactions. It should be noted, that for characterizing redox-active biomolecules by impedance spectroscopy no additional redox probe is added to the electrolyte solution, and the measured electron-transfer process corresponds to the entire bioelectrocatalytic reaction provided by the biocatalyst. Under the condition that the enzyme is not saturated by the substrate, the electron-transfer resistance of the electrode is also controlled by the substrate concentration. Thus, the substrate concentration can be analyzed by the impedance spectroscopy following R_{ct} values [9].

Bourigua *et al.* described an enzyme biosensor for detecting trimethylamine (TMA) which is a parameter for estimating fish freshness. The biosensor implies covalent attachment of the flavin containing monooxygenase 3 (FMO_3) enzyme and ferrocenyl groups to an electropolymerized polypyrrole film. It was demonstrated that the ferrocenyl groups are able to act as redox probe and monitor the oxidation of TMA catalyzed by the FMO_3 enzyme. The amperometric biosensor of TMA provided a sensitivity of $6\,(\mu A/cm^2)/\mu g\,ml^{-1}$. Monitoring the TMA by EIS showed the same sensitivity and improved the dynamic range for TMA detection. In fact, a low detection limit of $0.4\,\mu g\,ml^{-1}$ TMA and a dynamic range from 0.4 to $80\,\mu g\,ml^{-1}$ were obtained by EIS. This biosensor for TMA analysis seems to be a very promising analytical tool for the evaluation of fish freshness in real samples, such as horse-mackerel fish [19].

In order to prevent bacteria growth at moderate temperature and increase raw milk preservation during transport, many producers add hydrogen peroxide at different quantities. However, a concentration higher than 4 mM is forbidden by several food hygiene centers all over the world since such H_2O_2 level could disturb biological life and let milk not be suitable for subsequent transformation. Esseghaier *et al.* developed an impedimetric biosensor able to detect as low as 50 μM hydrogen peroxide in raw milk samples which is 80 times lower than the maximum admissible concentration in raw milk supplies [20]. The biosensor employed a gold electrode modified with a mixed self-assembled pyrroloquinoline quinine monolayer formed by 1,2-dipalmitoyl-*sn*-glycero-3-phosphoethanolamine-*N*-(biotinyl) (biotinyl-PE) and 16-mercaptohexadecanoic acid. Horseradish peroxidase conjugated streptavidin was attached on the SAM through biotin–streptavidin interaction and EIS was applied in the presence of $Fe(CN)_6^{3-}/Fe(CN)_6^{4-}$ as a redox probe. The Nyquist plot diameter showed a gradual decrease with H_2O_2 concentration as a consequence of the change in organic layer electroconductivity after peroxide reduction. The impedimetric biosensor was able to detect as low as 10 μM of hydrogen peroxide and the variation of log(Zo/Z) at 100 mHz vs. H_2O_2 concentration plot exhibited a linear shape with signal saturation next to 100 mM (*Zo*/*Z* is the ratio of the total impedance of HRP-strep layer to the system impedance after H_2O_2 injection). The constructed biosensor was successfully tested in milk samples.

Shamsipur *et al.* prepared a robust and effective catalase (CAT)-nanocomposite film modified-glassy carbon electrode using multiwalled carbon nanotubes (MWCNTs) and a room temperature ionic liquid, 1-butyl-3-methylimidazolium hexafluorophosphate ([bmim][PF6]), by a layer-by-layer self-assembly method [21]. The ionic liquid assembled on the GC/MWCNTs was not only responsible for a better immobilization of the enzyme, but was also found to largely accelerate the electron transfer between CAT and the electrode. The variation of $1/R_{ct}$ vs. H_2O_2 concentration produced a calibration curve over a wide linear range of 5–1700 nM and a limit of detection (LOD) of 0.25 nM. The applicability of the GC/MWCNTs/[bmim][PF_6]/CAT modified electrode was evaluated by the determination of hydrogen peroxide in milk.

14.4.2 Immunosensors

Impedance spectroscopy (including Faradaic impedance in the presence of a redox probe and non-Faradaic capacitance measurements) represents an important electronic transduction means of antigen–antibody binding interactions on electronic elements [22]. The development of electrochemical immunosensors suggests the use of alternative detection schemes based on impedance or capacitance because antibody–antigen binding does not involve electrochemical reaction. The use of EIS reports detection limits that vary from the nanomolar to picomolar range, which are sufficient for many analytical applications. However, the application of impedance immunosensors has been limited by their sensitivity to nonspecific adsorption, which often limits the selectivity of biosensors applied to complex matrices. Some interesting recent examples are commented below.

Liu *et al.* compared Faradaic and non-Faradaic impedance strategies with amperometric measurements for sensitive detection of antibodies to a peptide sequence from the major peanut allergen, Arachis hypogaea 2 (Ara h2). They use a glutathione-decorated gold nanoparticles (GSH-AuNP) sensor platform to attach a film containing a peptide featuring the amino acid sequence from Ara h2 having specific affinity for an antipeanut antibody IgE. An Ara h2-2 peptide layer was successfully attached onto GSH-AuNP-coated sensors by amidation between amine groups on the peptide and carboxylic acid groups of GSH. As mentioned above, one potential disadvantage of impedance sensing is the influence of nonspecific binding. The measurement itself is nonspecific, since any biomolecule in the sample that binds to the sensor will contribute to the increase in R_{ct}. The relative influence of nonspecific binding sites is decreased when enzyme-catalyzed precipitation is adopted, since the impedance change is generated mostly from precipitation on the sensor surface produced by the horseradish peroxidase (HRP) label reaction. In this case, nonspecific binding sites on gold nanoparticle (AuNP)-peptide electrodes were blocked by incubating the sensors with saturated casein. After IgY (chicken antipeanut antibodies) was bound from the serum, HRP-conjugated goat antichicken IgY binds to IgY on the sensor surface to fabricate a sandwich immunoassay in which TMB (3,3′,5,5′-tetramethylbenzidine) is oxidized by peroxide-activated HRP to produce an insoluble oxidized product at pH 3.5 to amplify the impedance of Ab2-HRP linked immunosensors. This sensor configuration, using non-Faradaic impedance by applying 0 V vs. saturated calomel electrode (SCE) with 1 ± 0.1 Hz frequency, had the best LOD of 5 pg/ml and the best linear range of over five orders of magnitude (from 5 pg/ml to 1 μg/ml) for the IgY antibody in undiluted calf serum. This LOD was 100-fold lower than label-free impedimetric immunosensors (0.5 ng/ml) and 60-fold lower than when using HRP-Ab2 in amperometric immunosensors (0.3 ng/ml) [23].

Another recent example reports a label-free electrochemical impedimetric immunosensor for the sensitive detection of metabolite of furaltadone, 5-morpholino-3-amino-2-oxazolidone (AMOZ), one of the most common nitrofuran antibacterial agents

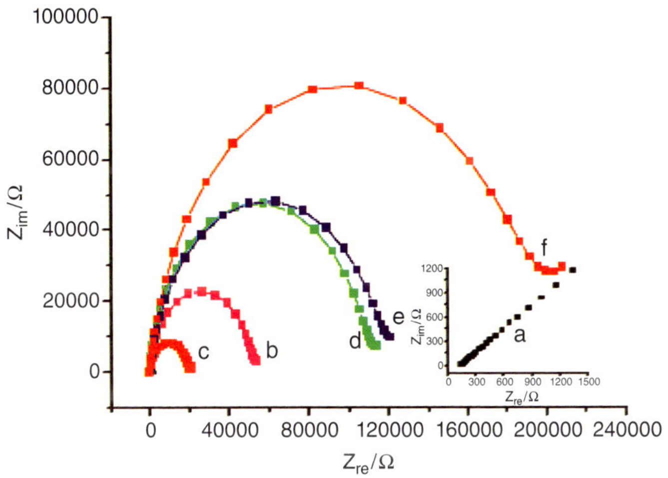

Figure 14.5 *Nyquist plots (Z_{im} vs. Z_{re}) for Faradaic impedance measurements in the presence of 0.1 mol/l phosphate buffer solution containing 1.0 mmol/l $K_3[Fe(CN)_6]/K_4[Fe(CN)_6]$ (1:1 mixture) and 0.1 mol/l KNO_3 at: (a) bare Au electrode, (b) dithiol/Au, (c) nano-Au/dithiol/Au, (d) AMOZ-Ab/nano-Au/dithiol/Au, (e) BSA/AMOZ-Ab/nano-Au/dithiol/Au, and (f) AMOZ/BSA/AMOZ-Ab/nano-Au/dithiol/Au. The concentration of AMOZ was 1.0×10^3 ng/ml. The frequency range is from 10^{-1} to 10^6 Hz with a signal amplitude of 10 mV. From Ref. [24] with permission from Elsevier*

(antibiotics) widely used in food-producing animals. The immunosensor design involved a monoclonal antibody against AMOZ (AMOZ-Ab) immobilized on a gold electrode modified with 1,4-benzenedithiol and AuNP self-assembled layers. Nyquist plots for Faradaic impedance measurements in the presence of 0.1 mol/l phosphate buffer solution containing 1.0 mmol/l $K_3[Fe(CN)_6]/K_4[Fe(CN)_6]$ (1:1 mixture) and 0.1 mol/l KNO_3 are shown in Figure 14.5. At the bare gold electrode, a straight line was observed (curve a). After self-assembling adsorption of 1,4-benzenedithiol molecules on the electrode surface, the diameter of the semicircle increased, implying that the monolayer of 1,4-benzenedithiol obstructed the electron transfer of the electrochemical probe (curve b). Then, the self-assembled AuNP monolayer on the surface of dithiol/Au (curve c) suggested a low electron-transfer resistance of the redox probe indicating the successful adsorption of AuNPs through the -SH group of the 1,4-benzenedithiol monolayer. When nano-Au/dithiol/Au was immersed in AMOZ-Ab solution, the diameter of the semicircle increased instead (curve d), demonstrating that the positively charged AMOZ-Ab had been successfully immobilized on the surface of the modified electrode via electrostatic interactions with the negatively charged AuNPs and retarded the interfacial electron transfer. After AMOZ-Ab/nano-Au/dithiol/Au was blocked by BSA, R_{ct} increased slightly (curve e). When the immunosensor was used to detect AMOZ, R_{ct} increased again (curve f) because of the interaction between AMOZ-Ab and AMOZ, which acted as the inert electron- and mass-transfer blocking layer, thereby significantly hindering the diffusion of ferricyanide toward the electrode surface.

Under the optimized conditions, the relative change in impedance was proportional to the logarithmic value of AMOZ concentrations in the range of 1.0 to 1.0×10^6 ng/ml (r = 0.9991) with the detection limit of 1.0 ng/ml. The impedimetric immunosensor was successfully applied to the determination of AMOZ in animal derived food samples, such as chicken muscle, pork, swine casing, shrimp, egg, and honey, with good agreement with the results obtained by HPLC-MS/MS [24].

An electrochemical impedance immunosensor has been reported recently for the detection of Sudan I, a red azo dye used in food, by covalently conjugating the Sudan I monoclonal antibodies with *o*-mercaptobenzoic acid on a gold electrode. Co-addition of 1-ethyl-3-(3-dimethylaminopropyl) carbodiimide (EDC) and N-hydrosuccinimide (NHS) could condense antibodies reproducibly and densely on the SAM. The Sudan I concentration was measured through the increase of impedance values in the corresponding specific binding of the Sudan I and Sudan I antibody. A linear relationship between the increased electron-transfer resistance (R_{ct}) and the logarithmic value of Sudan I concentrations was found in the range of 0.05–50 ng/ml with the detection limit of 0.03 ng/ml. Using hot chili as a model sample, an acceptable recovery of 96.5–107.3% was obtained. The results were validated by conventional HPLC method with good correlation [25].

Prodromidis's group proposed impedimetric immunosensors for the direct detection of *Salmonella typhimurium* in pure cultures of type strains and inoculated milk samples using Faradic impedimetric measurements. Polyclonal anti-*Salmonella* was cross-linked, in the presence of glutaraldehyde, on gold electrodes modified with a mixed SAM of 11-amino-1-undecanethiol and 6-mercapto-1-hexanol at a constant 1 + 3 proportion, respectively. The mixed SAM was also deposited in the presence of triethylamine, which was used to prevent the formation of interplane hydrogen bonds among amine terminated thiols. The alteration of the interfacial features of the electrodes due to different modification or recognition steps, was measured by Faradic EIS in the presence of a hexacyanoferrate(II)/(III) redox couple. To achieve the working simplicity expected by a biosensor, immunoreaction was performed directly in cultures. This resulted in the elimination of various centrifugation and washing steps, which are used for the isolation of bacteria cells from the culture, thus making the proposed immunosensors promising candidates for on-site applications. Moreover, a substantial increase in the signal occurs when it is compared with that obtained for the immunoreaction carried out in a buffer solution, where the cell population remains unchanged during the immunoreaction. It is also important to point out that the proposed method is insensitive to dead cells as, in contrast to viable cells, they cannot proliferate, and thus their presence has almost no contribution to the measuring signal. A meticulous study regarding the type and the concentration of different post-blocking agents, which were used for blocking the nonspecific binding sites of the sensors, was conducted, and optimized immunosensors were successfully used for the direct detection of *S. typhimurium* in milk samples. For a detection time of 2 h, BSA-blocked immunosensors can provide reliable analytical signals for a concentration of three orders of magnitude lower to the infectious dosage of *S. typhimurium.* It is important to point out that none of the analytical trials we run gave false negative (*S. typhimurium*) or positive (*E. coli*) results, thus constituting strong evidence for the validity of the proposed immunosensors for the detection of *S. typhimurium* in milk samples [26].

An example of the selective detection of *E. coli* involved an impedimetric immunosensor developed using SAM-modified gold screen-printed electrodes (SPEs) [27]. In this work, two different immunosensor configurations were tested and compared. The first one implied the covalent immobilization of the specific antibodies on AuSPEs using the homobifunctional cross-linker 3,3′-dithiobis[sulfosuccinimidylpropionate] (DTSSP) SAM. The other one was based on selective thiolation of the antibody using sulfo-LC-SPDP (sulfosuccinimidyl 6-[3′-(2-pyridyldithio) propionamide] hexanoate), which is a heterobifunctional linker with a cleavable thiol–thiol bond which reacts selectively with primary amines and thiol groups present in the antibody to generate a selected number of specific chemisorption sites through which the antibody can chemisorb to form a SAM. The performance of both designs upon binding of *E. coli* cells to the electrode was evaluated by measuring the electron-transfer resistance by EIS in the presence of $[Fe(CN)_6{}^{3-}]/[Fe(CN)_6{}^{4-}]$ as a redox probe. Figure 14.6 schematizes the two impedimetric immunosensor configurations developed in this work.

The stepwise assembly with both immunosensor designs was monitored by using EIS and CV. Figure 14.7a shows the obtained Nyquist plots for a bare AuSPE (curve 1), a DTSSP-AuSPE (curve 2), an anti-*E. coli*-DTSSP-AuSPE (curve 3), the latter modified electrode after blocking with ethanolamine (curve 4), and after formation of the antibody–*E. coli* complex (curve 5). The formation of the DTSSP SAM resulted in the expected increase in the R_{ct} value. When antibodies were immobilized on the DTSSP-AuSPE, the penetration of the redox probe was reduced and, accordingly, R_{ct} increased notably. Blocking of the residual succinimidyl groups in the SAM with ethanolamine promoted the redox probe transfer to the electrode surface, and a decrease in R_{ct} was observed. This fact could be due either to electrostatic attractions or to the ability of ethanolamine to replace weakly immobilized antibodies. The further attachment of bacterial cells to the modified electrode surface gave rise to an additional barrier toward the access of the redox probe to the electrode, resulting in a further increase in the electron-transfer resistance. R_{ct} changed from 761 to 998 Ω, that is, increased 31% after the formation of the antibody–*E. coli* complex.

A similar set of EIS/CV experiments to those commented above were carried out using 2-pyridyl-disulfide activated antibodies. This immunosensor configuration exhibited a much higher increase in the R_{ct} value (162%) (see Figure 14.7b) and, accordingly, the possibility of achieving a much higher sensitivity for the assay. These results are likely due to a larger amount of immobilized antibody using this approach, as a consequence of the higher packing density of the SAM achieved with the thiolated antibodies. Furthermore, the 2-pyridyl-disulfide exhibits a higher reactivity against antibodies when compared with DTSSP, thus giving rise to a larger amount of immobilized antibodies.

Electron-transfer resistance (R_{ct}), as well as the increase in electron-transfer resistance (ΔR_{ct}) values were taken as analytical signals, according to the equation:

$$\Delta R_{ct} = R_{ct\ (antibody-bacteria)} - R_{ct\ (antibody)}$$

where $R_{ct\ (antibody-bacteria)}$ is the electron-transfer resistance value measured after antibody immobilization on the electrode, and $R_{ct\ (antibody)}$ is the electron-transfer resistance value upon the antibody–*E. coli* conjugate formation. In the case of using anti-*E. coli*-DTSSP-AuSPE immunosensors, the ΔR_{ct} values were measured as:

$$\Delta R_{ct} = R_{ct\ (antibody-bacteria)} - R_{ct\ (ethanolamine)}$$

where $R_{ct\ (ethanolamine)}$ is the resistance value after the blocking treatment.

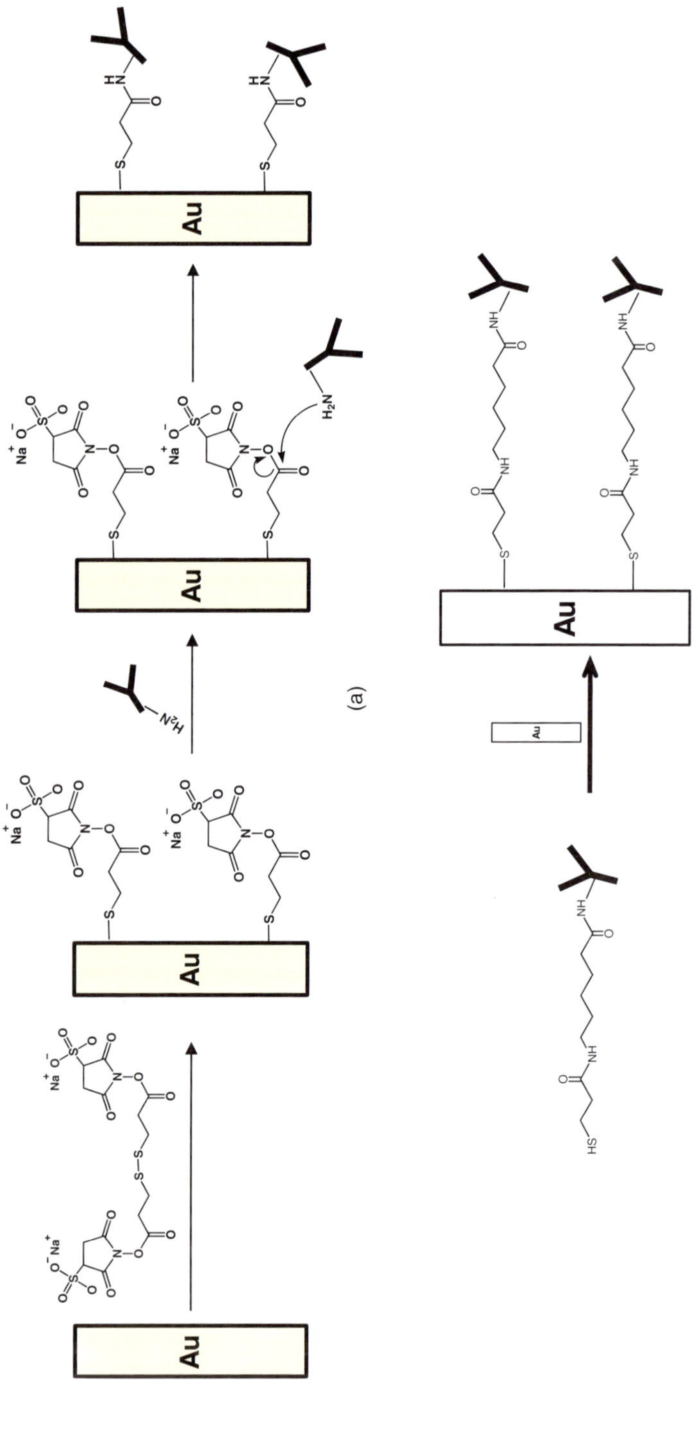

Figure 14.6 *Schemes of the immunosensor configurations: (a) covalent immobilization of anti-E.coli antibodies on the DTSSP-modified AuSPE and (b) immobilization of thiolated antibodies onto the AuSPE. Adapted from Ref. [27] with permission from Elsevier*

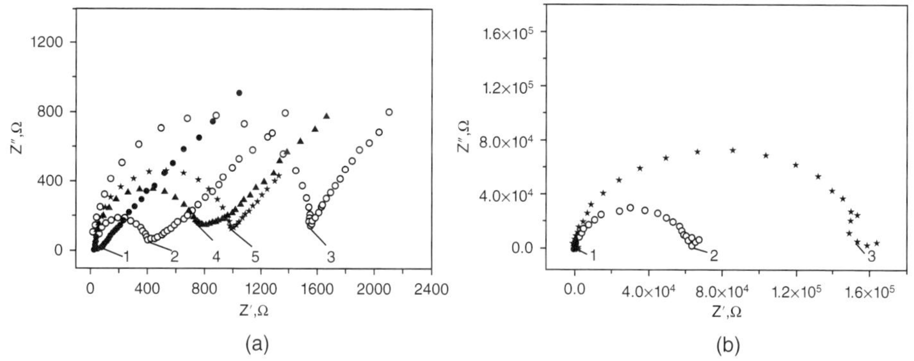

Figure 14.7 *Complex impedance plots obtained with a bare AuSPE (curves 1), a DTSSP-modified AuSPE (curve 2a), an anti-*E. coli*-DTSSP-AuSPE (curve 3a), an anti-*E. coli*-DTSSP-AuSPE after blocking with ethanolamine (curve 4a), an* E. coli*-anti-*E. coli*-DTSSP-AuSPE (curve 5a), a thiolated-anti-*E. coli*-AuSPE (curve 2b), and an* E. coli*-thiolated-anti-*E. coli*-AuSPE (curve 3b). Operating conditions: EIS, 5 mM $[Fe(CN)_6]^{3-/4-}$ (1:1) in 0.1 M KCl, 0.01–10 000 Hz frequency range with a 0.01 V r.m.s. signal. Experimental conditions (a) [anti - E. coli] = 10 µg/ml, $t_{anti\text{-}E.\ coli}$ = 60 min, $t_{E.\ coli}$ = 60 min, [E. coli] = 1.0 × 10^5 cfu/ml. (b) [sulfo - LC - SPDP] = 40 mM, $t_{sulfo\text{-}LC\text{-}SPDP}$ = 60 min, [anti - E. coli] = 200µg ml^{-1}, [DTT] = 150 mM, t_{DTT} = 45 min, $t_{thiolated\text{-}anti\text{-}E.\ coli}$ = 15 h, [E. coli] = 1.0 × 10^3 cfu/ml. From Ref. [27] with permission from Elsevier*

Regarding the anti-*E.coli*-DTSSP-immunosensor, ΔR_{ct} varied linearly with the logarithmic value of the *E. coli* cells concentration over three orders of magnitude, 1.0×10^4 and 1.0×10^7 cfu/ml, and the LOD was 1.0×10^4 cfu/ml. The configuration using the thiolated antibodies gave rise to a better analytical performance, exhibiting a linear relationship in the 5 to 1.0×10^8 cfu/ml range and a LOD, with no need for preconcentration or pre-enrichment steps, of 3.3 cfu/ml. The immunosensors showed a high selectivity against *Staphylococcus aureus* (*S. aureus*) and *Salmonella choleraesuis* (*S. choleraesuis*). The usefulness of the thiolated antibodies-based immunosensor for the rapid analysis (1 h) of 10 cfu/ml *E. coli* inoculated river and tap water samples was demonstrated.

Zhiley *et al.* described a sensitive electrochemical immunosensor for the detection of deoxynivalenol (DON), one of the most common mycotoxin contaminants of cereal-based food products. It makes use of a glassy carbon electrode modified with a composite constructed from fullerene (C_{60}), ferrocene, and an ionic liquid. The components were immobilized on the surface of the electrode using chitosan cross-linked with epichlorohydrin. Then, the antibody to DON was covalently conjugated to the surface which then was blocked with serum albumin. The synergistic electrocatalytic activity of fullerene and ferrocene remarkably enhanced the electron-transfer speed between the modified layer on the electrode and $Fe(CN)_6^{3-/4-}$ in the solution and resulted in increasing sensitivity of the immunosensor. The ionic liquid, 1,3-dibutylimidazolium bis (trifluoromethylsulfonyl)amine, provided a biocompatible microenvironment for the antibody. Under optimal conditions, a linear relationship between the electron-transfer resistance and DON concentration was found in the DON concentration range of 0.001–0.3 ng/ml. solution. The detection limit of the label-free impedance detection assay was 0.0003 ng/ml. Also, it was found that the electron-transfer resistance change can retain 98% for the immunosensor (from

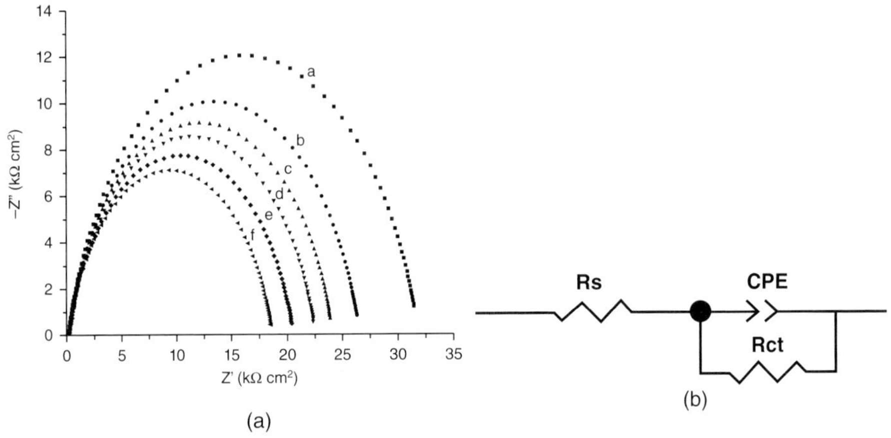

Figure 14.8 *(A) EIS responses of a Pt/PANI-PV-SO*$^{3-}$ *anti-OTA immunosensor to standard OTA solutions in PBS (a, b, c, d, e, and f are 0, 2, 4, 6, 8, and 10 ng/ml OTA antigen, respectively). (E = 0 V). (B) equivalent circuit of Pt/PANI-PV-SO*$^{3-}$ *(*R_s *= solution resistance, CPE = constant phase element;* R_{ct} *= charge-transfer resistance). From Ref. [29] with permission. Copyright © 2011 Wiley-VCH Verlag GmbH & Co. KGaA*

6936 to 6797 Ω) over a 180-day storage period at 4 °C, thus indicating that the immunosensor had excellent stability and could be applied to determine DON in routine analysis. The method was satisfactorily applied to the determination of DON in Chinese food [28].

A common way to immobilize antibodies involving conducting polymers is entrapment within the polymeric chains. Muchindu *et al.* fabricated an impedimetric immunosensor for ochratoxin A (OTA) composed of polyaniline-polyvinyl sulfonate and anti-OTA antibody on a Pt disk electrode. Aniline doped with PV-SO_3^- was polymerized on a Pt electrode in 1 M HCl, because an acidic medium is required to produce polyaniline emeraldine salt which is the polymer's most conducting state. The surface of the polymer was reduced in 2 ml phosphate buffered saline (PBS) at −500 mV vs. Ag/AgCl for 600 s. After reduction was complete, PBS was removed from the cell and replaced with the negatively charged OTA antibody solution. Oxidation was performed at 700 mV vs. Ag/AgCl for 600 s. During oxidation, the antibody was electro-statically attached to the polymer surface. The antigen exists as monoanion (OTA^-) and dianion (OTA^{2-}) at neutral pH and binding of this charged antigen to the OTA immunosensor reduced the charge-transfer resistance as shown in Figure 14.8a by the decrease in both imaginary and real impedance. The equivalent circuit in Figure 14.8b was used to fit the responses of the immunosensor to standard OTA solutions. The R_{ct} values were subtracted from those of the blank PBS (i.e., response at 0 ng/ml OTA) to obtain the normalized values that are plotted in Figure 14.9.

The detection limit of the sensor was calculated as 10 pg/kg and it was applied to the analysis of certified reference materials of roasted coffee, wheat, and corn with OTA concentrations of 2.5, 8.6, and 21.1 mg/kg, respectively. The result for corn is comparable to that obtained with the ELISA technique as well as the vendor specifications [29].

Yang *et al.* reported a capacitive immunosensor for the detection of *Salmonella spp.* based on grafted ethylene diamine and self-assembled AuNP monolayers on a glassy

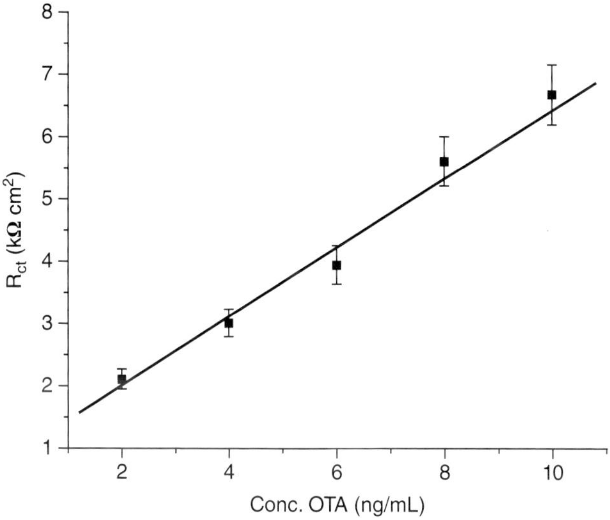

Figure 14.9 *Calibration plot for Pt/PANI-PV-SO^{3-}/anti-OTA immunosensors. From Ref. [29] with permission. Copyright © 2011 Wiley-VCH Verlag GmbH & Co. KGaA*

carbon electrode and immobilization of *Salmonella* monoclonal antibodies through physical adsorption. Based on a Bode plot of phase angle and logarithm of impedance magnitude vs. the logarithm of frequency (Figure 14.10), a frequency of 193 Hz was chosen to measure capacitance.

The relative change in capacitance ($\%\Delta C$) was taken as the measured parameter:

$$\%\Delta C = (C_{\text{Ag-Ab}} - C_{\text{BSA}}/\ C_{\text{BSA}}) \times 100$$

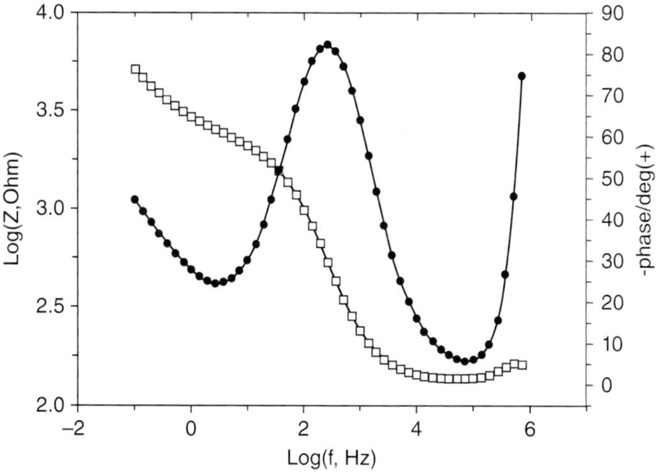

Figure 14.10 *Bode plot for the determination of the frequency dependence on the impedance and phase angle for the immunosensor. From Ref. [30] with permission from Elsevier*

where C_{Ag-Ab} is the value of the capacitance after *Salmonella* spp. coupling to the antibody (Ab) on the BSA/Ab/AuNP/ED/GCE and C_{BSA} is the capacitance after blocking the remaining adsorption-reactive sites by BSA. $\%\Delta C$ values were proportional to the logarithmic values of *Salmonella* spp. concentration in the range of 1.0×10^2 to 1.0×10^5 cfu/ml ($r = 0.991$), with the detection limit of 1.0×10^2 cfu/ml. The immunosensor could be regenerated repeatedly by immersing in glycine–HCl buffer solution (pH 2.8) and was shown to be stable after two months of storage. The capacitive sensor was successfully used for the detection of *Salmonella* spp. in lab-processed commercial pork samples [30].

Li *et al.* developed a label-free capacitive immunosensor for rapid and sensitive detection of *E. coli* O157:H7. The immunosensor was fabricated by immobilizing affinity-purified anti-*E. coli* O157:H7 antibodies onto SAMs of 3-mercaptopropionic acid (MPA) on the surface of a quartz crystal Au electrode activated by immersion in EDC/NHS solution to convert the terminal carboxylic group of MPA to an active NHS ester. Bacteria suspended in solution became attached to the immobilized antibodies when the immunosensor was tested in liquid samples. The change in capacitance caused by the bacteria was directly measured and an equivalent circuit was introduced to simulate the capacitive immunosensor (Figure 14.11a). This consisted of ohmic resistance (R_s) of the electrolyte; double-layer capacitance (C_{dl}) including C_{ins}, C_{rec}, and C_{GC}; electron-transfer resistance (R_{ct}); and Warburg impedance (Z_w). The first capacitor, C_{ins}, closest to the metal surface describes the insulating layer of MPA; the second capacitor, C_{rec}, constitutes the grafting molecules, the bound recognition element, binding of analyte, and any contribution from the Stern layer; and the third capacitor, C_{GC}, represents the concentration-dependent diffuse layer.

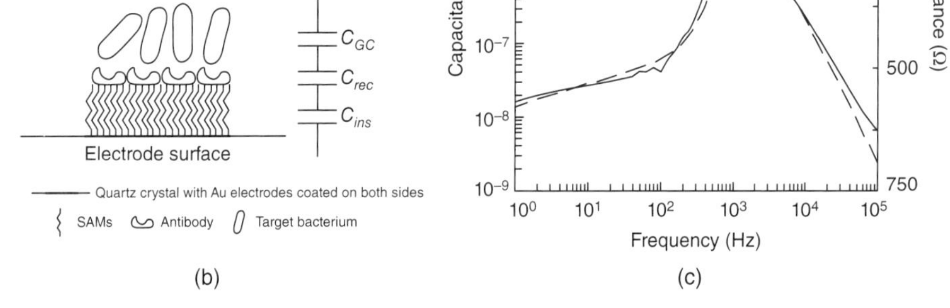

Figure 14.11 *(a) Equivalent circuit of the capacitance measurement system. (b) Schematic representation of the double-layer capacitance construction. (c) Measured and fitting data of capacitance and resistance of the immunosensor. The capacitance was measured in 0.01 M PBS (0.138 M NaCl, 0.0027 M KCl) containing 10 mM* $[Fe(CN)_6]^{3-/4-}$. *Amplitude voltage, 5 mV; DC potential, equilibrium potential of* $[Fe(CN)_6]^{3-/4-}$; *frequency range, 1 Hz to 100 kHz;* E. coli *O157:H7,* 2.2×10^5 *cells. Solid line, measured data; dot line, fitting data. From Ref. [31] with permission from Elsevier*

Moreover, i, i_c, and i_f represent total current, double-layer current, and Faradaic current, respectively. A schematic representation of the double-layer capacitance construction is depicted in Figure 14.11b. Figure 14.11c (solid line) shows the capacitance and resistance of cells (2.2×10^5 cfu ml^{-1}) on the electrode. To validate the equivalent circuit, 50 data points on the capacitance were automatically selected by the software and used as inputs to the equivalent circuit; the simulated results are shown by the dotted line in Figure 14.11c.

A linear relationship between the C_{dl} shift and logarithmic value of the cell concentration was found in a range of bacterial cell concentrations from 2.2×10^2 to 2.2×10^5 cfu ml^{-1}. A detection limit of 2.2×10^2 cfu ml^{-1} was obtained in 1 h and the immunosensor was applied for the detection on pathogens in food samples such us ground beef, milk, and spinach samples [31].

One way to enhance the response of a label-free immunosensor is to increase the amount of an immobilized antibody on the electrode so that a larger immunological complex can occur, and therefore produce a larger signal. In this context, multilayer AuNPs networks have been employed to allow for further increases of the electrode surface area [32]. Instead of using a multilayer of AuNPs, only one layer of nanoparticles on a self-assembled thiourea monolayer (SATUM) was employed by Chullasat *et al.* [33]. Then a layer of mercaptosuccinic acid (MSA), a thiol compound, was adsorbed on the AuNP layer and activated with EDC/NHS to immobilize the antibody. Chloramphenicol (CAP) was chosen as a model analyte because it is a synthetic antibiotic with a low molecular weight that is widely used as an anti-microbial agent. The schematic illustration of the modified electrode is shown in Figure 14.12.

The CAP–anti-CAP interaction was monitored via the change of the imaginary part of the impedance (Z'') as a function of time at a single frequency using a flow injection system. The optimum frequency to monitor the immunoreaction was determined from a Bode plot in the region where the system shows nearly ideal capacitor behavior, that is, the plot of logarithm of impedance against logarithm of frequency is a straight line with a slope of -1 and the phase angle closest to $-90°$ (Figure 14.13, inset). The carrier buffer was first passed though the system and the Z'' base-line was recorded (Figure 14.13). When the CAP was injected, it bounded to the immobilized anti-CAP on the modified electrode which is like adding another capacitor in series thus leading to the increase of impedance ($\Delta Z''$) which corresponded with the concentration of CAP. A regeneration solution (NaOH 50 mM) was then injected to remove CAP from the immobilized anti-CAP, and when the signal base-line was recovered a new analysis cycle was applied ($\Delta Z''_2$). The calibration plot of the SATUM/AuNPs/MSA modified electrode provided a wide linear range of $(0.50–10) \times 10^{-16}$ M. The sensor electrode provided good selectivity for CAP detection and is well-suited for the detection of ultra trace levels of CAP in shrimp samples. The results agreed well with those obtained by the high-performance liquid chromatography coupled with a photo diode array detector.

Another label-free impedimetric flow-injection immunosensor was developed for the direct detection of penicillin G [34]. Antipenicillin G was immobilized on a gold electrode modified with a SAM of thioctic acid. The real-time monitoring of impedance was carried out at the optimum frequency of 160 Hz. The detection limit of penicillin G in milk was reported to be 3.0×10^{-15} M.

Chai *et al.* described the direct detection of the biological toxin Ricin in acidic environment without pH adjustment by electrochemical impedance on nano-porous aluminum

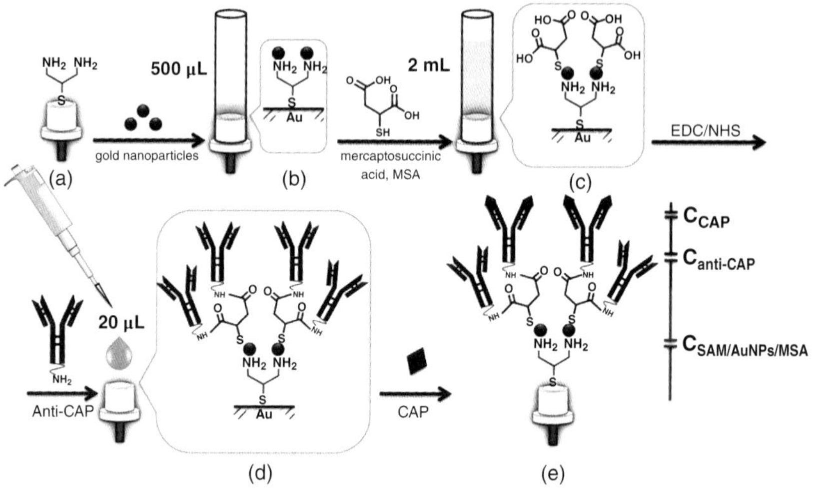

__Figure 14.12__ Schematic illustration of the modified electrode: the gold electrode with a self-assembled thiourea monolayer (SATUM) (a) a small tube containing the gold nanoparticle solution was fitted on the electrode, gold nanoparticles was adsorbed on the electrode via the SATUM (SATUM/AuNPs) (b), mercaptosuccinic acid adsorbed on the AuNP layer (SATUM/AuNPs/MSA) (c), and activated with EDC/NHS to immobilize the antibody (anti–CAP) (d). The layers on the modified gold electrode surface are represented by capacitors in series (e) where CSATUM/AuNPs/MSA is the capacitance related to the SATUM/AuNPs/MSA layer, $C_{anti\text{-}CAP}$ is the capacitance related to the chloramphenicol antibody layer and C_{CAP} is the capacitance related to the binding between C_{AP} and immobilized antibody on the modified electrode. From Ref. [33] with permission from Elsevier

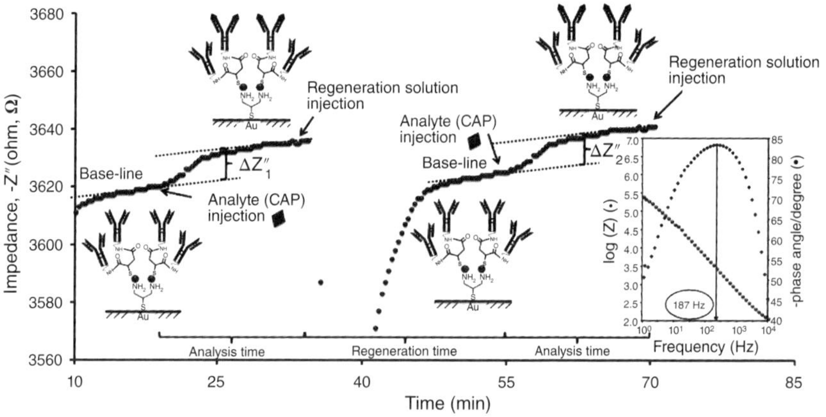

__Figure 14.13__ Impedance response of the modified electrode monitor at a single frequency. Inset: Bode plot. From Ref. [33] with permission from Elsevier

substrate hydrophobically modified via SAM of 3-aminopropyltriethoxysilane (APTES) and covalent cross-linking of Ab. It is a challenge to develop the immunosensor for the direct detection of biological toxins in the acidic diagnostic samples since acids interfere with immunoreactions. Generally, immunoreaction occurs around neutral pH, and the rate of immunoreaction is extremely low in acidic environment. Accordingly, traditional immunoassays such as ELISA, radioimmunoassay (RIA), and electrophoretic immunoassay frequently include the step of pH adjustment prior to the analysis of target toxins in order to avoid the interference by acids on immunoreaction.

It was suggested that the adsorption of hydroxides, negatively charged, on the hydrophobic body of APTES-SAMs by the polarization during EIS analysis offsets the effect of acids on the immunological activity of the immobilized Ab and helps to keep the immunological activities of the immobilized Ab on an electrochemical impedance immunosensor in an acidic environment. This immunosensor could detect the presence of the biological toxin in 20 min at 10 kHz and was potentially applicable to the detection of the biological toxin in acidic foods at different pH: milk (pH 6.5), vegetable soup (pH 4.8), and tomato juice (pH 3.9), which can be categorized to neutral, low acid, and acid foods [35].

14.4.2.1 Impedimetric Immunosensors Using Interdigitated Array Microelectrodes

Microelectrodes have been used in the fabrication of impedimetric biosensors due to their ability to contribute to the miniaturization of the sensor and improvement of the sensitivity. In this context, interdigitated array microelectrodes (IDμEs), which show some advantages over macrosized electrodes including detecting small amounts of generated electrode products, eliminating the need for a reference electrode, providing simple means for obtaining a steady-state current response and low response times, have been widely employed to fabricate impedimetric sensors [3, 36]. The IDμEs are capable of monitoring the changes of the electrical properties in the immediate neighborhood of the electrode surface. Every IDμE has two electrodes that were used to connect them to the impedance analyzer surfaces. Figures 14.14 and 14.15 show the electric field distribution of an interdigital sensor. It is reported that 95% of the current between the two finger electrodes in an IDμE flows above the electrode surface within a distance that equals the sum of electrode width and space and the impedance is almost completely determined by processes close to the surface [39].

Therefore, the size of the IDμE electrode must be chosen for different biological entities in consideration of their sizes (could be nanometer to micrometer) to achieve a sensitive biosensor for that type of biological entity. For example, micrometer IDμE electrodes are suitable for sensing of biological cells, while nanometer IDμEs are more proper for sensing DNA, up to 200 nm for a probe–DNA binding, and in the order of 10–100 nm for an

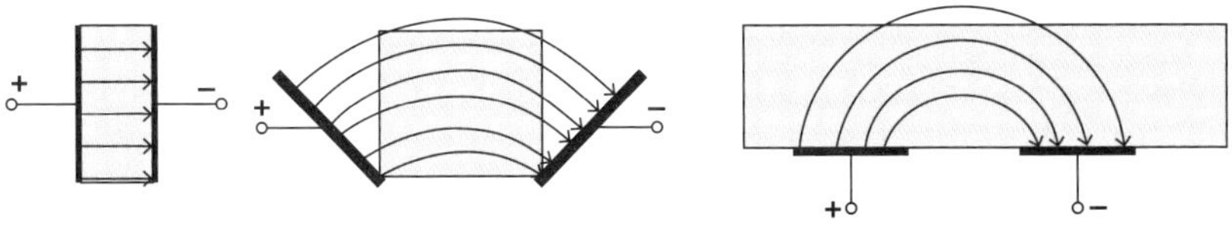

Figure 14.14 An interdigital sensor behaving as a parallel plate capacitor. Electric field distribution of a parallel plate capacitor. Adapted from Ref. [37] with permission from Elsevier

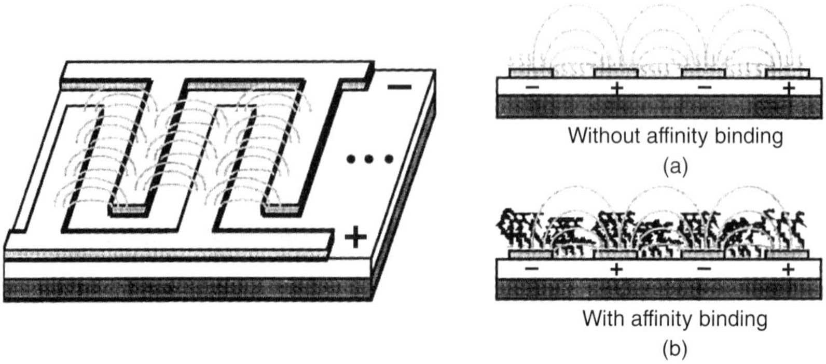

Figure 14.15 *Principle of an interdigitated impedimetric affinity biosensor: (a) without and (b) with molecules to be detected. From Ref. [38] with permission from Elsevier*

antigen–antibody layer [40]. Another advantage is that large active areas are structured at once and many sensors can be put in an array.

Valera *et al.* developed a label-free electrochemical impedance immunosensor by passive immobilization of antigen, atrazine-haptenized protein (atrazine-BSA), on IDμEs using a silicon gasket to form wells on the wafer. For the assay development, atrazine standards were added to the silicon gasket wells (50 μl/well), followed by the antibody [41]. The changes of the IDμE surface showed evident modifications in the impedance. Adding atrazine during the competition step implies that a fraction of this reagent will not be available and hence, the captured amount will be lower than in the reference sample. This change in the antibody concentration is a measure of the atrazine concentration used. The scheme of the immunosensor reaction is shown in Figure 14.16.

The impedance measurements were carried out in the absence of any redox probe, and thereby, the absence of redox mediators, and the subsequent non-Faradaic behavior is an advantage for measures in real matrixes, where the presence of multiples substances hinders the Faradic response. The measurement of the impedance as a function of the frequency was interpreted using the electronic equivalent circuit depicted in Figure 14.17; the Nyquist plots are shown in Figure 14.18. In addition to the interfacial parameters, the effects of the interdigitated μ-electrodes and that of the contacts were considered including a parallel capacitance to account for the natural capacitance of the IDμE ($C_{\text{ID}\mu\text{E}}$) and a series resistance to account for the contact resistance (R_c).

The possibility of working at a single frequency, in the range of middle frequencies, is an important advantage in order to reduce the complexity and cost of the control circuit. This immunosensor demonstrated high sensibility to atrazine, detecting concentrations below the maximum residue level (MRL) (100 ppb) required by the European Union for atrazine in wine grapes which can represent an interesting solution for the agro-food field.

The same authors developed a much more robust label-free impedimetric immunosensor, based on the same IDμE arrays conveniently derivatized through covalent procedures to ensure reliability and to prevent nonspecific adsorption [42]. Atrazine-BSA was covalently immobilized on the surface of the IDμEs previously activated with (3-glycidoxypropyl)trimethoxysilane via the side chain amino groups of lysines or arginines with the epoxy groups on the device surface. Before, the gold electrodes were blocked

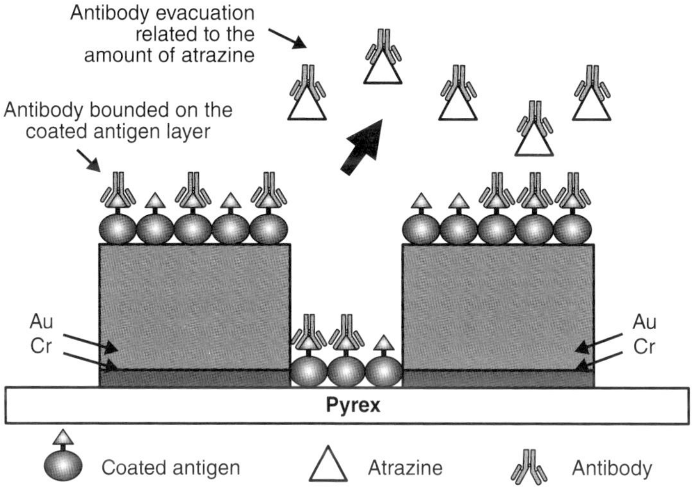

Figure 14.16 *Scheme of the immunosensor reaction for the development of an impedimetric sensor for atrazine detection. An amount of the specific antibody is bounded on the coated antigen layer. The rest is evacuated of the IDμEs which is related to the atrazine concentration. From Ref. [41] with permission from Elsevier*

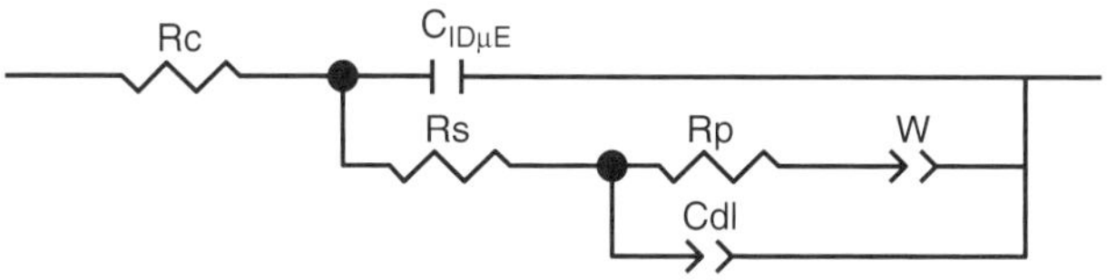

Figure 14.17 *Equivalent circuit used to model impedance data in the impedimetric immunosensor for atrazine detection*

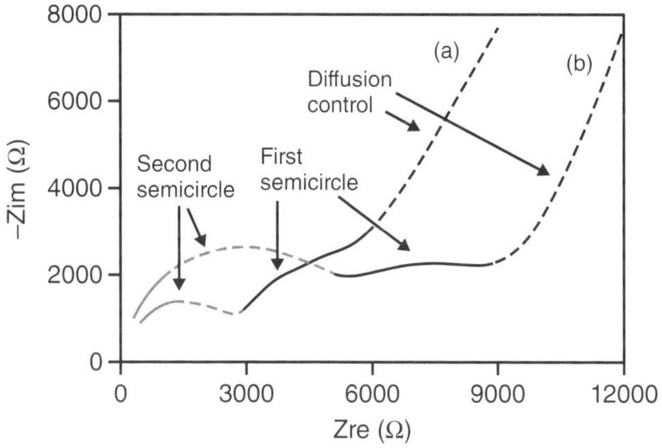

Figure 14.18 *Nyquist plot of the impedance spectrum corresponding to the development of an impedimetric immunosensor for atrazine detection: (a) Step I: antigen immobilization and (b) Step II: incubation step. The impedance spectra include three regions: (i) diffusion control, (ii) first semicircle, and (iii) second semicircle. From Ref. [41] with permission from Elsevier*

using *N*-acetylcysteamine to prevent nonspecific adsorptions. A variation in the response was observed when the same device was consecutively measured which was attributed to the reorientation of the diverse domains and lateral chains of the proteins due to the electrical field applied. However, the response became stable after the third measurement. Although impedance measurements were recorded on the whole frequency spectrum, the performance of repetitive measurements demonstrated that for routine purposes the recording the variation of R_s was sufficient. The R_s value increased significantly with the antibody binding. A clear decrease in the ΔR_s value was observed when the pesticide concentration increased as expected in a competitive assay. The covalent attachment of the biomolecules allowed increasing the detectability of the immunosensor (LOD of 0.04 μg/l) in comparison with the previous value of 5.76 μg/l reported by the same authors. This detectability was sufficient to perform analysis of water samples in compliance with the EC directives (0.1 μg/l). However, for more complex samples such as red wine, it was necessary to introduce sample treatment methods to avoid nonspecific signals derived from the matrix. By passing the sample through a C18-SPE cartridge, the immunosensor was able to detect atrazine in wine at concentrations far below the MRL (50 μg/l) required by EC.

14.4.3 Genosensors and Aptasensors

DNA-based biosensors have gained much importance for detection of target genes responsible for diseases, in food industry, environment, and in agriculture. Commonly, the development of biochemical DNA sensors requires the immobilization of nucleic acids, acting as the sensing interface, on the electronic transducers and the electronic transduction of the hybridization occurring with the analyte DNA on the transducer. It is worth mentioning that since the nucleic acids and the resulting nucleic acid/DNA complexes are oligoanionic polymers, their immobilization on surfaces generates a negatively charged interface that repels negatively charged redox labels, and particularly multi-charged negative ions such as $[Fe(CN)_6]^{3-/4-}$. The repulsion of the redox label is anticipated to inhibit the interfacial electron transfer and thus leads to the enhanced electron-transfer resistance, R_{ct} [9].

In the case of nucleic acid as a bioreceptor for pathogen detection, the identification of a target analyte's nucleic acid is achieved by matching the complementary base pairs that are often the genetic components of an organism. For example, Dinckaya *et al.* developed nucleic-acid-based impedimetric biosensors for rapid and selective detection of aflatoxin M1. In order to immobilize a thiol-modified single stranded DNA (HSDNA) probe that specifically bound to aflatoxin M1, a SAM of cysteamine and AuNPs on the SAM were prepared on gold electrodes, layer-by-layer. A layer self-assembly of colloidal Au resulted in a larger electrode surface area and easier attachment of sDNA. In this work, the R_{ct} value was first recorded as approximately 83 Ω for the bare gold electrode. The impedance spectrum of the cysteamine-modified gold electrode was an almost straight line. It was influenced by an effect stabilizing the electrostatic interaction between the electrode surface and the redox probe. Moreover, it caused a small increase in the thickness of the surface. The AuNPs did not alter the R_{ct} value. Due to the nonconductivity of the HS-DNA probe, after the application of the DNA probe, the electron-transfer of $[Fe(CN)_6]^{4-/3-}$ was blocked and R_{ct} increased remarkably. The temperature considerably affected the biosensor performance. Typically, when the temperature was below 35 °C, R_{ct} values decreased by at least 65%

for high aflatoxin M1 concentrations. For relatively low concentrations the decrease rate was about 40%. It is likely that the temperature increased the movement of the molecules and of course the kinetic energy to transport them onto the electrode surfaces effectively. Consequently, working temperature was chosen as 35 °C. The biosensor provided a linear response to aflatoxin M1 over the concentration range of 1–14 ng/ml and was applied successfully for the determination of aflatoxin M1 in real milk samples [43].

One of the most recent and requested DNA-sensor applications is the detection of genetically modified organisms (GMOs) in foods. A GMO is referred to a living organism whose genome has been modified by the introduction of an exogenous gene able to express an additional protein that confers new characteristics. At present, analytical methods for GMOs can detect both the new protein expressed by the genetically modified DNA sequence or the sequence itself. The detection of the former is generally based on enzymatic immunoassay analysis. However, since food processing often leads to protein denaturation, the analysis of DNA, which is much more stable than proteins, is preferred for screening of both raw ingredients and processed products.

Bonanni *et al.* report the impedimetric detection of transgenic insect-resistant Bt maize DNA, using carbon nanotubes as sensing platform. Carbon nanotubes were modified with carboxylic groups which allowed the covalent immobilization of the oligonucleotide sequence under study (Figure 14.19). The impedimetric technique allowed either the detection of complementary DNA sequences or their quantification.

The use of streptavidin modified nanoparticles permitted rapid formation of an easily detected conjugate with biotinylated DNA through a streptavidin–biotin interaction, thus avoiding the need of synthesizing DNA modified with nanoparticles, a more time-consuming procedure. The protocol employing a competitive binding assay allowed the detection of unlabeled DNA both completely complementary or with one or two mismatches. In order to visualize the presence and distribution of AuNPs, a silver enhancement

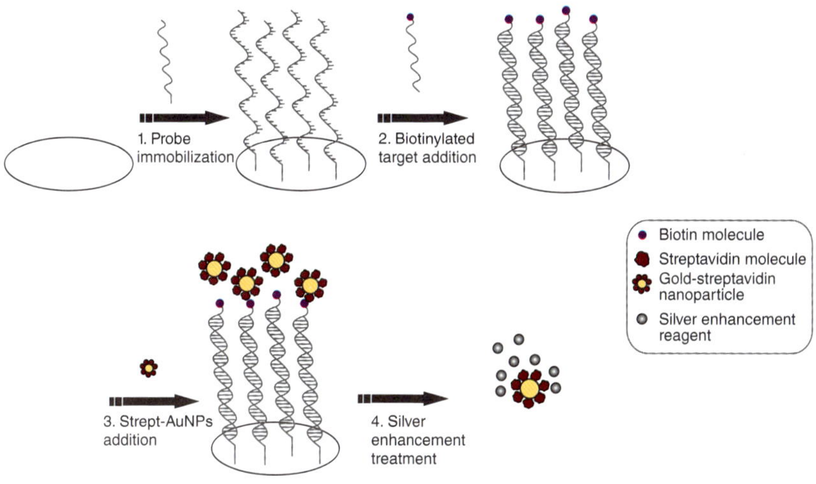

Figure 14.19 *Scheme of the experimental procedure for impedimetric genosensing of oligonucleotide sequences specific for transgenic insect resistant Bt maize. From Ref. [44] with permission from Elsevier*

treatment was applied to electrodes already modified with DNA–nanoparticles conjugate, allowing direct observation by scanning electron microscopy [44].

Commercial production processes involve heat treatment, which often denatures food proteins, thereby altering protein tertiary structure and thus interfering with protein detection. DNA remains are intact longer under heat and pressure processing and can therefore provide the basis of a robust assay for the detection of allergen residues in foods. Sun *et al.* developed a novel electrochemical DNA sensor by using a stem–loop probe for peanut allergen Ara h 1 detection [45]. The stem–loop structured DNA probes are superior to linear probes in several aspects for the detection of nucleic acid. The greatest advantage is the superior mismatched discrimination ability. They also have great potential for real-time monitoring analysis. The probe was modified with a thiol at its 5′ end and a biotin at its 3′ end. The biotin-tagged "molecular beacon"-like probe was attached to the surface of a gold electrode to form a stem–loop structure by self-assembly through facile gold–thiol affinity. 6-MCH was used to cover the remnant bare region. The stem-loop probe was "closed" when the target was absent, and then the hybridization of the target induced the conformational change to "open," state in favor of the formation of the thermodynamically more stable, rigid target-probe duplex. Then, the biotin at its 3′ end moved away from the electrode surface, which changed the electron-transfer efficiency (see Figure 14.20).

Compared with the bare gold electrode, the stem-loop probe modified electrode occupied a large area on the electrode surface, the electrode could not efficiently exchange electrons with the solution, and the electron-transfer efficiency reduced. After immobilization of MCH, the R_{ct} increased. After hybridization with complementary target DNA, the stem–loop structure opened, and the target-probe duplex formed, moving the attached biotin group away from the electrode surface. The addition of target increased the R_{ct} value. The electrochemical impedance response increased with the increase of the target concentration. A linear relationship between ΔR_{ct} and the logarithm of the complementary target concentration was found in the range of 1 fM to 0.1 nM, with a detection limit of 0.35 fM (Figure 14.21). Moreover, a complementary target could be discriminated from one-base mismatch and noncomplementarity. The proposed strategy was successfully applied to detect Ara h 1 in the peanut DNA extracts of peanut milk beverage, and the concentration of it was 3.2×10^{-13} mol/l.

Electrochemical biosensors based on DNA aptamers (aptasensors) are of growing interest due to their high sensitivity and selectivity. This is particularly due to unique properties of

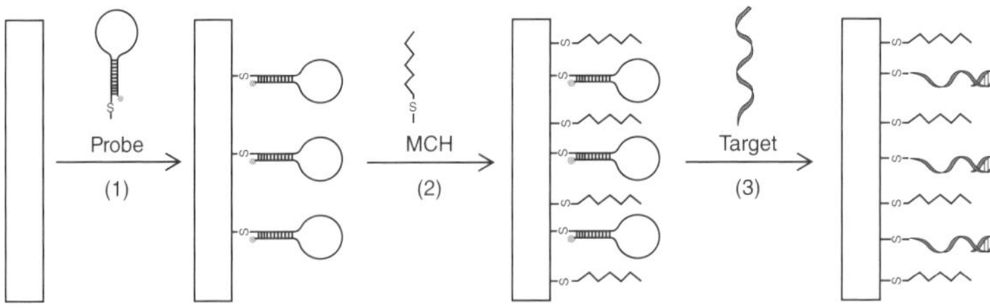

Figure 14.20 Schematic illustration of the fabrication of the DNA sensor for peanut allergen Ara h 1 detection. (1) Immobilization of stem – loop probe to the gold electrode; (2) immobilization of 6-mercaptohexanol (MCH) to block the remaining bare region; and (3) hybridization with complementary target DNA. Adapted from Ref. [45] with permission. Copyright © 2012, American Chemical Society

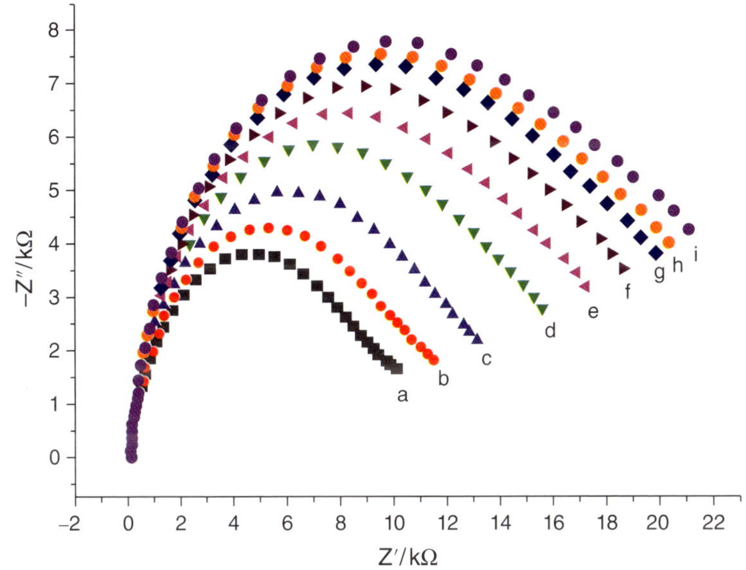

Figure 14.21 *Nyquist plots, −Z″ vs. Z′ for an Au electrode in PBS buffer solution containing 2.5 mM* $Fe(CN)_6^{3-}/Fe(CN)_6^{4-}$*: after immobilization of the hairpin probe and MCH, then hybridization at 37 °C with (a)* 10^{-15} *M, (b)* 10^{-14} *M, (c)* 10^{-13} *M, (d)* 10^{-12} *M, (e)* 10^{-11} *M, (f)* 10^{-10} *M, (g)* 10^{-9} *M, (h)* 10^{-8} *M, and (i)* 10^{-7} *M of complementary target DNA. From Ref. [45] with permission. Copyright © 2012, American Chemical Society*

DNA or RNA aptamers – the single stranded nucleic acids with high affinity to proteins or to other low and macromolecular compounds, which is comparable with that of antibodies. These special characteristics open new routes in biosensor assembling for further practical applications, such as detection of toxicants in food or environment.

Using the redox probe $[Fe(CN)_6]^{-3/-4}$ it is possible to amplify the detection of affinity interactions at surfaces, such as proteins using DNA aptamers as receptors by electrochemical impedance. Because a DNA aptamer is negatively charged the anionic redox probe will repeal from the sensor surface causing an increase of the charge-transfer resistance (R_{ct}). Binding of protein resulted in a decrease (lysozyme) or increase (thrombin) of the negative charge at the surface, causing a decrease or increase of R_{ct}, respectively. This approach has been used so far mostly for the detection of proteins. However, it has been shown that this method can also be applied for the determination of small molecules such as OTA. In the aptasensor developed by Castillo *et al.*, thiolated DNA aptamers specific to OTA have been immobilized by chemisorption to the surface of a gold electrode [46]. Prior to immobilization, aptamers solution was incubated in a 90 °C water bath and then exposed to ice cooling for providing proper folding of the aptamers. Under the experimental conditions used (pH 7), OTA is negatively charged; thus, binding of OTA to aptamers should increase the negative charge of the surface and induce repealing of the redox probe. As a result the R_{ct} value should increase. So, the charge-transfer resistance, R_{ct}, increased with increasing OTA concentration in the range of 0.1–100 nM. The LOD (0.12–0.40 nM) depended on the configuration of the aptamers. The sensor discriminated between OTA and ochratoxin B

(OTB) and N-acetyl-l-phenylalanine (NAP). The aptasensor was regenerable in 1 mM HCl and it was validated in coffee, flour, and wine spiked with OTA with satisfactory recovery.

14.5 Nonelectrochemical Interfacial Impedance

Impedance was applied more than 10 years ago in the field of microbiology as a means of detecting and quantifying pathogenic bacteria in food. This method is based on the measurement of changes in the electrical impedance of a culture medium or a reaction solution resulting from the bacterial growth. Usually due to microbial metabolic processes, conductance, and capacitance increase while impedance decreases. In this category, no biorecognition elements are used on the surface of electrodes for the capture of bacterial cells. The impedance change in the medium is mainly produced by the release of ionic metabolites from the live cells, and thereby decreases the impedance of the media; thus, it could distinguish between viable and dead cells. Such a method has been developed as a rapid method that can detect bacteria in 3–12 h (vs. 24–72 h for standard plate count) [3].

In the food industry, microbial tests are performed to screen the product for dangerous pathogens as well as to guarantee that the total microbial concentration is below the allowed threshold values. The sampling inspection can be conducted at the site and if the results are suspicious, therefore further analysis of contaminated sample can be done in the laboratory using expensive techniques [47]. The impedance method was approved by the Association of Official Analytical Chemists International (AOAC) as a first action method for screening Salmonella in food samples [48].

Several commercial analytical instruments are based on this principle. These systems include the Bactometer (BioMerieux, Nuertingen, Germany), the Malthus system (Malthus Instruments Ltd., Crawley, UK), the Rapid Automated Bacterial Impedance Technique (RABIT; Don Whitley Scientific Ltd., Shipley, UK), and the Bac-Trac (Sy-Lab, Purkersdorf, Austria). However, these measurement systems are not suitable for on-the-spot applications, so many efforts have been made to minimize the instruments [3].

Grossi *et al.* [49] developed an embedded portable biosensor system for the determination of bacterial concentration. This system is composed of an incubation chamber, containing the sample under test, and two electronic boards: one dedicated to measuring the sample electrical characteristics, and the other to controlling the sample temperature, fixed at a value suitable to enhance bacterial growth. The system composed of the sample and a couple of electrodes stimulated with a low frequency (200 Hz) sinusoidal voltage can be modeled by means of the series of a resistance (accounting for the conductivity of the medium and of the electrode/electrolyte interface) and a capacitance (due to the double layer present at the electrode/electrolyte interface). The material of the electrodes is very important, since metal ions released into the sample can be toxic to bacteria and inhibit their growth. From this point of view, platinum and stainless steel electrodes have been found to have no negative effect at the low applied field, while silver and gold electrodes exhibit bacteriostatic properties. After the time needed for stabilization, both the Z resistive and reactive components remain essentially constant (at a "baseline" value) until the bacterial concentration in the sample under test reaches a threshold value (in the order of 10^7 cfu/ml) sufficient to induced appreciable changes in the sample electrical characteristics. When such a condition is reached, the impedance begins to decrease. The time at which the electrical parameters begin to deviate from their baseline value is called detect time (DT) and, since DT has a linear relationship with the logarithm of bacterial concentration, the latter can be easily worked out. The developed system was validated by means of a series of experiments performed on ice-cream samples (where no additional media for dilution were used) and fresh water samples (where yeast extract 3% medium was added to monitor total bacterial concentration and Mac Conkey broth for selective growth of coliform strains). In both cases, reliable measurements were obtained, in good agreement with data obtained by standard plate count, normally used as a reference. As for competitiveness, this system provides much faster results compared with the standard plate count technique.

The use of impedance measurements to analyze the coagulation process provides a basis for the control of the soymilk coagulation process. Li *et al.* developed a method to characterize and analyze the soymilk coagulation process and to evaluate the influence of coagulation temperature on this process by monitoring the electrical impedance.

An electrical equivalent circuit model, composed of a resistor and a capacitor which were connected in series, was adopted to represent the soymilk–electrode system. High frequency impedance measurements presented an increase in resistance and constant value in reactance for different coagulation times. This resistance is attributed to the bulk resistance of soymilk mixture and changed minimally when the coagulation process is finished [50].

Kim *et al.* proposed a plug-type, disposable electrode using a gold-coated silicon wafer, polydimethylsiloxane polymer to fix two electrodes, and a borosilicate glass tube to construct an impedimetric sensor instrument. The developed sensor could be used for *in situ* real-time monitoring of bacterial growth in a lab-scale fermentor by measuring impedance signals without the risk of introducing contamination. The impedance changes of an *E. coli* solution at 0.01 MHz frequency of applied AC for 13 h could be correlated with the bacteria concentration as estimated by a hemocytometer. In conclusion, impedance increased during the measurement period, as the concentration of *E. coli* increased. The correlation between the *E. coli* concentration and the impedimetric data was linear [51].

Meat aging affects meat tenderness and flavor and, for this reason, the evaluation of the degree of aging is important to satisfy the quality requirements of consumers. Moreover, studying the aging process allows optimizing storage times of meat pieces and therefore improving meat eating quality. The complex processes which occur during post-mortem time change the electric properties of the samples; therefore, the study of meat electric properties during the postmortem period could be a useful tool for evaluating the meat aging process in a rapid and non expensive way. During aging, there is a slow decrease in impedance due to progressive deterioration of the cell membrane. Furthermore, the electrical anisotropy, which represents the variation in impedance according to the direction of the electrical field in the meat sample, decreases linearly with the mechanical resistance of muscle fibers during aging. Dame *et al.* developed a ratio of anisotropy based on the bovine muscle electric properties during the aging process along the longitudinal and transverse directions [52].

A system to carry out a nondestructive punctual measurement of salt content in meat samples using impedance spectroscopy was designed by Masot *et al.* The choice of the electrodes and their configuration in EIS experiments is one of the most critical points. The designed system includes a concentric needle electrode that is introduced into the food sample. Collected impedance data were then statistically processed. Partial least squares (PLS) was applied to a set of samples in order to obtain a model. Studies were carried out in NaCl aqueous solutions using minced meat samples. In both cases, the model was applied to a second set of samples and a very remarkable correlation between observed and predicted NaCl content was obtained. These results are promising for the development of easy-to-use and low-cost instrumentation for salt content analysis [6]. More recently, Fernández-Segovia *et al.* developed a nondestructive, easy-to-use system, based on impedance spectroscopy to differentiate between fresh and frozen-thawed salmon. The sensor employed was an electrode designed using thick-film technology. This technology includes high-resolution screen-printing methods. It dispenses pastes or inks with different electrical characteristics (conductive, resistive, and dielectric) on an insulating substrate, in order to form an electronic circuit. The design of this electrode, with a pointed end, facilitates its penetration into the sample. The impedance measurements were carried out by inserting the sensor into the sample perpendicularly to the muscular fibers of the fish. The penetration depth of the electrode into the sample was 10 mm in the direction perpendicular to the fiber direction.

The cell membrane acts as an insulator at low frequencies, behaving like a capacitor. The membranes are intact in fresh fish, but in frozen-thawed samples, an important number of cells membranes are destroyed. This produces on one hand the reduction of the capacitive component of the biological tissues, which decreases the value of the impedance phase. On the other hand, the destruction of the cells membranes increases the amount of free electrolytes in the tissue, increasing its conductivity and producing a decrease of the impedance module [5].

As has been shown above, a number of samples can be analyzed at a single time using impedance microbial technology but the sensitivity of electrical impedimetric sensors is less as compared to other sensors. Some studies have demonstrated the promise of interdigitated microelectrodes in impedance measurements for monitoring the growth of bacteria to improve analytical performance [1].

Varshney *et al.* designed a simple double IDμEs based flow cell to detect viable *E. coli* O157:H7 cells from the samples after an enrichment growth in a low conductivity growth medium [53]. The impedance change was due to the growth of *E. coli* O157:H7 cells that were analyzed as a function of frequency. An equivalent circuit was designed to curve fit experimental results and to determine parameters of the electrical circuit responsible for impedance change caused by the growth of *E. coli* O157:H7 cells. The equivalent circuit consisted of two double-layer capacitors (C_{dl}) (one for each set of electrodes) connected in series with bulk medium resistor (R_s). CPE was used to model the double-layer capacitance of the electrodes. This is commonly favored over the use of a simple capacitor. C_{dl} accounted for the effect of ionic species on the capacitance near the surface of an electrode. R_s represented the bulk resistance of the solution, accounting for the change in conductivity and charge transport across the bulk solution. The equivalent circuit analysis indicated that the change in impedance values due to bacterial growth was caused by both double-layer capacitance and bulk medium resistance, and these parameters in turn were the function of the amount of charged ions present in the medium.

A double IDμEs based flow cell was successfully used in an impedance biosensor for the detection of *E. coli* O157:H7 in a range from 8.0 to 8.2×10^8 cfu/ml after an enrichment growth of 14.7 and 0.8 h, respectively. A logarithmic linear relationship between detection time (T_D) in h and initial cell concentration (N_0) in cfu/ml was $T_D = -1.73 \log N_0 + 14.62$, with $R^2 = 0.93$. Specificity of this impedance biosensor could be ensured by the use of a selective growth medium. Impedance measurement in a flow cell can be useful to design a portable instrument for monitoring bacterial cell growth and their metabolites.

The same authors used magnetic beads coated with a bio-recognition element for the specific separation and concentration of cells prior to impedance measurement. After separation and resuspension of bacterial cells in low conductivity 0.1 M mannitol solution, the concentrated sample was uniformly spread on the surface of IDμE. The impedance sensor detected a minimum of 7.4×10^4 and 8.0×10^5 cfu/ml of *E. coli* O157:H7 in pure culture and ground beef samples, respectively. The concentration of bacterial cells attached to nanoparticle–antibody conjugates in the active layer of IDμE above the surface of electrodes was achieved with the help of a magnet placed underneath the electrode chip. This improved the sensitivity of the biosensor by 35% as compared to no use of the magnet. These experiments were performed on an open chip [54].

The same technique was used in a flow cell embedded with gold IDμE [55]. The flow cell was used to detect bacterial cells in a detection volume of 60 nl. This impedance biosensor was able to detect as low as 1.6×10^2 and 1.2×10^3 cells of *E. coli* O157:H7 cells present in pure culture and ground beef samples, respectively. The total detection time from sampling to measurement was 35 min. In both of these experiments, immunomagnetic separation (IMS) based on nanoparticles was used to separate the target bacterial cell from the food matrix and avoid the tedious processes of filtration and centrifugation for sample preparation [56]. This technique has a major advantage of additional separation and concentration of bacterial cells from crude samples before impedance measurement. In this way, the background noise caused by the nontarget components in the samples could be reduced, improving the signal-to-noise ratio. Additionally, the surface of electrodes can be used multiple times as there is no bio-recognition element present directly on the surface of electrodes.

Yang *et al.* present a simple and rapid impedance method to detect Salmonella cells by making use of the impedance properties of bacterial cell suspensions using IDμE. It was found that bacterial cell suspensions in deionized water (DI) with different cell concentrations could generate different electrical impedance spectral responses, whereas cell suspensions in PBS solution could not produce any significant differences in impedance spectra in response to different cell concentrations. The characteristics of the spectrum along with the equivalent circuit in Figure 14.22a suggest that there are electrochemical reactions occurring on the IDμE in the cell suspension in DI water. The electron-transfer resistance (R_{ct}) and Warburg impedance (Z_w) are the parameters indicating the electrochemical reactions and the diffusion of the electrochemical active species in the suspension. This implies that cells may release some electrochemical active composites to the DI water. However, in the cell suspension in PBS, the impedance spectrum does not show any characteristics related to electrochemical active parameters, which implies that cells may not release active electrochemical species into PBS [57].

A linear relationship between the impedance and the logarithmic value of the cell concentration was found in the 10^6–10^{10} cfu/ml range. The detection limit (the time at which the impedance starts to decrease and crosses a certain threshold value) was calculated to be 3.45×10^6 cfu/ml. To achieve the selectivity of this method, these authors also demonstrated the feasibility of integrating magnetic separation to this impedance method.

A novel planar interdigitated sensor-based sensing system was developed for detection of dangerous marine biotoxins in seafood [47]. All sensors have the same effective area

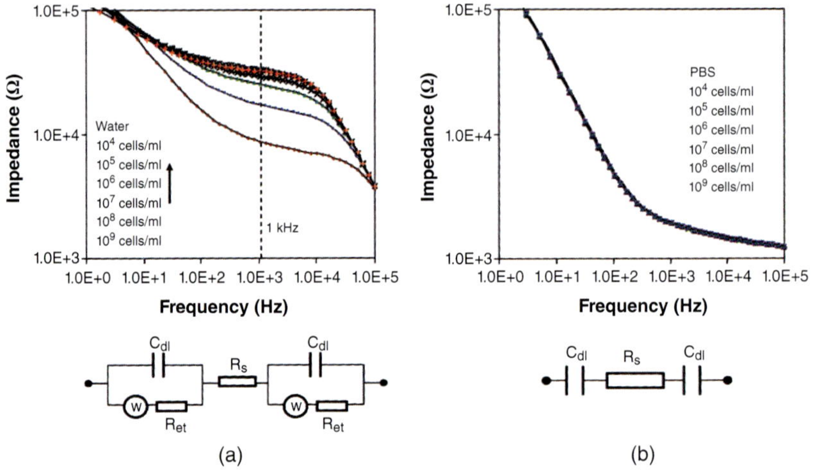

Figure 14.22 Electrical impedance spectroscopy for detection of Salmonella suspensions using interdigitated microelectrodes in (a) DI water and (b) PBS with the cell concentrations in the range of 10^4 to 10^9 cfu/ml, along with water and PBS as controls, together with their fitting spectra and the equivalent circuits. Frequency range: 1 Hz to 100 kHz. Amplitude: ±50 mV. From Ref. [57] with permission from Elsevier

but different configurations. The sensor, designed to have 2 positive electrodes at each end separated by 11 negative electrodes, has better sensitivity compared to other configurations. The changes in this sensor sensitivity were analyzed with mussels before and after adding proline (which is structurally close to domoic acid chemical structure). The presence of proline on the mussel surface and also injected proline to the mussels was very clearly detected by the sensor. The authors showed that the interdigitated sensor could be used to detect small amount of domoic acid in mussels (approximately 12.6 μg/g of domoic acid in mussel meat).

Rahman *et al.* fabricated new types of planar interdigital sensors with different coating materials to assess their response to endotoxins by impedance spectroscopy. A carboxyl-functional polymer, APTES, and thionine were chosen to be coated onto FR4 interdigital sensors. The chosen coating materials have carboxylic or amine functional groups, which were optimized to be stable in water. All coated sensors were immobilized with PmB (Polymyxin B) which has specific binding properties to lipopolysaccharide. Data were analyzed using principal component analysis correlating the impedance parameters of the test sample. The developed interdigital sensing approach is able to detect contaminated food samples, including milk, drinking water, cucumber, beef, and chicken meat, with endotoxin at 0.05 unit/ml [58].

14.6 Conclusions and Perspectives

The examples commented in this chapter show fairly well that impedance sensors and biosensors constitute an excellent tool for food analysis due to their inherent characteristics of low cost, ease of miniaturization, and label-free operation. This label-free nature is a very important advantage of impedimetric detection as the labeling processes make the experiments relatively complex and might affect the affinity between the probes and their targets especially in the development of biosensors. However, despite the many advantages of impedimetric sensing devices, after two decades of research efforts and hundreds of publications, no product based on impedance-based biosensors has enjoyed widespread commercial success. To date, instrumentation is lacking for sensor arrays and for hand-held on-site applications. All published affinity impedance sensors are based on bulky single-channel EIS instrumentation. Only limited efforts have been expended in miniaturization of the electronics for on-site applications. Therefore, further efforts should be devoted to developing commercial products which will require improved stability, reduced volume, increased sensitivity, and lowered costs. The integration of nanomaterials may be used to address the issue of inadequate sensitivity of dielectric spectroscopy in routine biochemical analysis and cellular assays. The continued use of nanomaterials could allow the fabrication of cheap and portable impedance components. Future research could involve the development of improved immobilization chemistry that is compatible with dielectric spectroscopy. This may significantly reduce the noise signals associated with reduced sensitivity in EIS. Future progress in this field requires the collaboration between chemists, engineers, physicists, and clinicians, and could lead to automated diagnostic equipment for clinical assays.

References

1. Arora, P., Sindhu, A., Dilbaghi, N., and Chaudhury, A. (2011) Biosensors as innovative tools for the detection of food borne pathogens. *Biosens. Bioelectron.*, **28**, 1–12.
2. Dweik, M., Stringer, R.C., Dastider, S.G., Wu, Y., Almasri, M., and Barizuddin, S. (2012) Specific and targeted detection of viable *Escherichia coli* O157:H7 using a sensitive and reusable impedance biosensor with dose and time response studies. *Talanta*, **94**, 84–89.
3. Wang, Y., Ye, Z., and Ying, Y. (2012) New trends in impedimetric biosensors for the detection of foodborne pathogenic bacteria. *Sensors*, **12**, 3449–3471.
4. Omanovic, S. and Roscoe, S.G. (2000) Interfacial behavior of β-lactoglobulin at a stainless steel surface: an electrochemical impedance spectroscopy study. *J. Colloid Interface Sci.*, **227**, 452–460.
5. Fernández-Segovia, I., Fuentes, A., Aliño, M., Masot, R., Alcañiz, M., and Barat, J.M. (2012) Detection of frozen-thawed salmon (Salmo salar) by a rapid low-cost method. *J. Food Eng.*, **113**, 210–216.
6. Masot, R., Alcañiz, M., Fuentes, A., Schmidt, F.C., Barat, J.M., Gil, L., Baigts, D., Martínez-Máñez, R., and Soto, J. (2010) Design of a low-cost non-destructive system for punctual measurements of salt levels in food products using impedance spectroscopy. *Sens. Actuators, A*, **158**, 217–223.
7. Prodromidis, M.I. (2010) Impedimetric immunosensors – A review. *Electrochim. Acta*, **55**, 4227–4233.
8. Mejri, M.B., Baccar, H., Baldrich, E., del Campo, F.J., Helali, S., Ktari, T., Simonian, A., Aouni, M., and Abdelghani, A. (2010) Impedance biosensing using phages for bacteria detection: generation of dual signals as the clue for in-chip assay confirmation. *Biosens. Bioelectron.*, **26**, 1261–1267.

9. Katz, E. and Willner, I. (2003) Probing biomolecular interactions at conductive and semiconductive surfaces by impedance spectroscopy: routes to impedimetric immunosensors, DNA-sensors, and enzyme biosensors. *Electroanalysis*, **15**, 913–947.
10. Park, S.-M. and Yoo, J.-S. (2003) Electrochemical impedance spectroscopy for better electrochemical measurements. *Anal. Chem.*, **75**, 455A–461A.
11. Fernández-Sánchez, C., McNeil, C.J., and Rawson, K. (2005) Electrochemical impedance spectroscopy studies of polymer degradation: application to biosensor development. *TrACs Trends Anal. Chem.*, **24**, 37–48.
12. Suni, I.I. (2008) Impedance methods for electrochemical sensors using nanomaterials. *TrACs Trends Anal. Chem.*, **27**, 604–611.
13. Daniels, J.S. and Pourmand, N. (2007) Label-free impedance biosensors: opportunities and challenges. *Electroanalysis*, **19**, 1239–1257.
14. Wang, J., Carmon, K.S., Luck, L.A., and Suni, I.I. (2005) Electrochemical impedance biosensor for glucose detection utilizing a periplasmic *E. coli* receptor protein. *Electrochem. Solid-State Lett.*, **8** (8), H61–H64.
15. Pournaras, A.V., Prodromidis, M.I., Katsoulidis, A.P., Badeka, A.V., Georgantelis, D., and Kontominas, M.G. (2008) Evaluation of lacquered tinplated cans containing octopus in brine by employing X-ray microanalysis and electrochemical impedance spectroscopy. *J. Food Eng.*, **86**, 460–464.
16. Macdonald, J.R. (1987) *Impedance Spectroscopy*, John Wiley & Sons, Inc., New York.
17. Shervedani, R.K. and Bagherzadeh, M. (2009) Electrochemical impedance spectroscopy as a transduction method for electrochemical recognition of zirconium on gold electrode modified with hydroxamated self-assembled monolayer. *Sens. Actuators B*, **139**, 657–664.
18. Tun, T.N. and Jenkins, A.T.A. (2010) An electrochemical impedance study of the effect of pathogenic bacterial toxins on tethered bilayer lipid membrane. *Electrochem. Commun.*, **12**, 1411–1415.
19. Bourigua, S., Ichi, S.E., Korri-Youssoufi, H., Maaref, A., Dzyadevych, S., and Renault, N.J. (2011) Electrochemical sensing of trimethylamine based on polypyrrole–flavin-containing monooxygenase (FMO3) and ferrocene as redox probe for evaluation of fish freshness. *Biosens. Bioelectron.*, **28**, 105–111.
20. Esseghaier, C., Bergaoui, Y., ben Fredj, H., Tlili, A., Helali, S., Ameur, S., and Abdelghani, A. (2008) Impedance spectroscopy on immobilized streptavidin horseradish peroxidase layer for biosensing. *Sens. Actuators B*, **134**, 112–116.
21. Shamsipur, M., Asgari, M., Maragheh, M.G., and Moosavi-Movahedi, A.A. (2012) A novel impedimetric nanobiosensor for low level determination of hydrogen peroxide based on biocatalysis of catalase. *Bioelectrochemistry*, **83**, 31–37.
22. Owino, I.O.K. and Sadik, O.A. (2005) Impedance spectroscopy: a powerful tool for rapid biomolecular screening and cell culture monitoring. *Electroanalysis*, **17** (23), 2101–2113.
23. Liu, H., Malhotra, R., Peczuh, M.W., and Rusling, J.F. (2010) Electrochemical immunosensors for antibodies to Peanut allergen ara h2 using gold nanoparticle-peptide films. *Anal. Chem.*, **82**, 5865–5871.
24. Jin, W., Yang, G., Wu, L., Wang, Q., Shao, H., Qin, A., Yu, B., Li, D., and Cai, B. (2011) Detecting 5-morpholino-3-amino-2-oxazolidone residue in food with label-free electrochemical impedimetric immunosensor. *Food Control*, **22**, 1609–1616.

25. Xiao, F., Zhang, N., Gu, H., Qian, M., Bai, J., Zhang, W., and Jin, L. (2011) A monoclonal antibody-based immunosensor for detection of Sudan I using electrochemical impedance spectroscopy. *Talanta*, **84**, 204–211.
26. Mantzila, A.G., Maipa, V., and Prodromidis, M.I. (2008) Development of a faradic impedimetric immunosensor for the detection of salmonella typhimurium in milk. *Anal. Chem.*, **80**, 1169–1175.
27. Escamilla-Gomez, V., Campuzano, S., Pedrero, M., and Pingarron, J.M. (2009) Gold screen-printed-based impedimetric immunobiosensors for direct and sensitive *Escherichia coli* quantisation. *Biosens. Bioelectron.*, **24**, 3365–3371.
28. Zhilei, W., Xiulan, S., Zaijun, L., Yinjun, F., Guoxiao, R., Yaru, H., and Junkang, L. (2011) Highly sensitive deoxynivalenol immunosensor based on a glassy carbon electrode modified with a fullerene/ferrocene/ionic liquid composite. *Microchim. Acta*, **172**, 365–371.
29. Muchindu, M., Iwuoha, E., Pool, E., West, N., Jahed, N., Baker, P., Waryo, T., and Williams, A. (2011) Electrochemical ochratoxin A immunosensor system developed on sulfonated polyaniline. *Electroanalysis*, **23**, 122–128.
30. Yang, G.-J., Huang, J.-L., Meng, W.-J., Shen, M., and Jiao, X.-A. (2009) A reusable capacitive immunosensor for detection of *Salmonella* spp. based on grafted ethylene diamine and self-assembled gold nanoparticle monolayers. *Anal. Chim. Acta*, **647**, 159–166.
31. Li, D., Feng, Y., Zhou, L., Ye, Z., Wang, J., Ying, Y., Ruan, C., Wang, R., and Li, Y. (2011) Label-free capacitive immunosensor based on quartz crystal Au electrode for rapid and sensitive detection of *Escherichia coli* O157:H7. *Anal. Chim. Acta*, **687**, 89–96.
32. Wang, M., Wang, L., Yuan, H., Ji, X., Sun, C., Ma, L., Bai, Y., Li, T., and Li, J. (2004) Immunosensors based on layer-by-layer self-assembled Au colloidal electrode for the electrochemical detection of antigen. *Electroanalysis*, **16**, 757–764.
33. Chullasat, K., Kanatharana, P., Limbut, W., Numnuam, A., and Thavarungkul, P. (2011) Ultra trace analysis of small molecule by label-free impedimetric immunosensor using multilayer modified electrode. *Biosens. Bioelectron.*, **26**, 4571–4578.
34. Thavarungkul, P., Dawan, S., Kanatharana, P., and Asawatreratanakul, P. (2007) Detecting penicillin G in milk with impedimetric label-free immunosensor. *Biosens. Bioelectron.*, **23**, 688–694.
35. Chai, C., Lee, J., and Takhistov, P. (2010) Direct detection of the biological toxin in acidic environment by electrochemical impedimetric immunosensor. *Sensors*, **10**, 11414–11427.
36. Berggren, C., Bjarnason, B., and Johansson, G. (2001) Capacitive biosensors. *Electroanalysis*, **13**, 173–180.
37. Mukhopadhyay, S.C., Deb Choudhury, S., Allsop, T., Kasturi, V., and Norris, G.E. (2008) Assessment of pelt quality in leather making using a novel non-invasive sensing approach. *J. Biochem. Biophys. Methods*, **70**, 809–815.
38. Laureyn, W., Nelis, D., Van Gerwen, P., Baert, K., Hermans, L., Magnée, R., Pireaux, J.-J., and Maes, G. (2000) Nanoscaled interdigitated titanium electrodes for impedimetric biosensing. *Sens. Actuators B*, **68**, 360–370.
39. Laschi, S. and Mascini, M. (2006) Planar electrochemical sensors for biomedical applications. *Med. Eng. Phys.*, **28**, 934–943.

40. Yang, L. and Bashir, R. (2008) Electrical/electrochemical impedance for rapid detection of foodborne pathogenic bacteria. *Biotechnol. Adv.*, **26**, 135–150.
41. Rodríguez, A., Valera, E., Ramón-Azcón, J., Sánchez, F.-J., Marco, M.-P., and Castañer, L.M. (2008) Single frequency impedimetric immunosensor for atrazine detection. *Sens. Actuators B*, **129**, 921–928.
42. Ramón-Azcón, J., Valera, E., Rodríguez, Á., Barranco, A., Alfaro, B., Sánchez-Baeza, F., and Marco, M.P. (2008) An impedimetric immunosensor based on interdigitated microelectrodes (IDμE) for the determination of atrazine residues in food samples. *Biosens. Bioelectron.*, **23**, 1367–1373.
43. Dinçkaya, E., Kınık, Ö., Sezgintürk, M.K., Altuğ, Ç., and Akkoca, A. (2011) Development of an impedimetric aflatoxin M1 biosensor based on a DNA probe and gold nanoparticles. *Biosens. Bioelectron.*, **26**, 3806–3811.
44. Bonanni, A., Esplandiu, M.J., and del Valle, M. (2009) Impedimetric genosensors employing COOH-modified carbon nanotube screen-printed electrodes. *Biosens. Bioelectron.*, **24**, 2885–2891.
45. Sun, X., Guan, L., Shan, X., Zhang, Y., and Li, Z. (2012) Electrochemical detection of peanut allergen ara h 1 using a sensitive DNA biosensor based on stem–loop probe. *J. Agric. Food Chem.*, **60**, 10979–10984.
46. Castillo, G., Lamberti, I., Mosiello, L., and Hianik, T. (2012) Impedimetric DNA aptasensor for sensitive detection of ochratoxin A in food. *Electroanalysis*, **24**, 512–520.
47. Syaifudin, A.R.M., Jayasundera, K.P., and Mukhopadhyay, S.C. (2009) A low cost novel sensing system for detection of dangerous marine biotoxins in seafood. *Sens. Actuators B*, **137**, 67–75.
48. Association of Official Analytical Chemists (AOAC) (1996) Salmonella in food, automated conductance methods: AOAC official method 991.38, in *Official Methods of Analysis of AOAC International*, 16th edn, AOAC International, Gaithersburg, MD.
49. Grossi, M., Lanzoni, M., Pompei, A., Lazzarini, R., Matteuzzi, D., and Ricco, B. (2010) An embedded portable biosensor system for bacterial concentration detection. *Biosens. Bioelectron.*, **26**, 983–990.
50. Li, X., Toyoda, K., and Ihara, I. (2011) Coagulation process of soymilk characterized by electrical impedance spectroscopy. *J. Food Eng.*, **105**, 563–568.
51. Kim, Y.-H., Park, J.-S., and Jung, H.-I. (2009) An impedimetric biosensor for real-time monitoring of bacterial growth in a microbial fermentor. *Sens. Actuators B*, **138**, 270–277.
52. Damez, J.L., Clerjon, S., Abouelkaram, S., and Lepetit, J. (2008) Beef meat electrical impedance spectroscopy and anisotropy sensing for non-invasive early assessment of meat ageing. *J. Food Eng.*, **85**, 116–122.
53. Varshney, M. and Li, Y. (2008) Double interdigitated array microelectrode-based impedance biosensor for detection of viable *Escherichia coli* O157:H7 in growth medium. *Talanta*, **74**, 518–525.
54. Varshney, M. and Li, Y. (2007) Interdigitated array microelectrode based impedance biosensor coupled with magnetic nanoparticle–antibody conjugates for detection of *Escherichia coli* O157:H7 in food samples. *Biosens. Bioelectron.*, **22**, 2408–2414.
55. Varshney, M., Li, Y., Srinivasan, B., and Tung, S. (2007) A label-free, microfluidics and interdigitated array microelectrode-based impedance biosensor in combination with

nanoparticles immunoseparation for detection of *Escherichia coli* O157:H7 in food samples. *Sens. Actuators B*, **128**, 99–107.

56. Varsheney, M., Yang, L., Su, X.-L., and Li, Y. (2005) Magnetic nanoparticle–antibody conjugates for the separation of *Escherichia coli* 0157:H7 in ground beef. *J. Food Prot.*, **68**, 1804–1811.
57. Yang, L. (2008) Electrical impedance spectroscopy for detection of bacterial cells in suspensions using interdigitated microelectrodes. *Talanta*, **74**, 1621–1629.
58. Rahman, M.S.A., Mukhopadhyay, S.C., Yu, P.-L., Goicoechea, J., Matias, I.R., Gooneratne, C.P., and Kosel, J. (2013) Detection of bacterial endotoxin in food: new planar interdigital sensors based approach. *J. Food Eng.*, **114**, 346–360.

Part III

Industrial Implications

15

Electroanalysis in Food Process Control

Maria Stella Cosio[1], Simona Benedetti[1], Matteo Scampicchio[2] and Saverio Mannino[1]

[1]Dipartimento di Scienze per gli Alimenti, la nutrizione e l'ambiente, Università degli Studi di Milano, Via Celoria 2, 20133, Milano, Italy

[2]Faculty of Science and Technology, Free University of Bolzano, Piazza Università 1, 39100, Bolzano, Italy

Competitiveness, productivity, quality, and waste reduction are frequently heard in the industry circles. The pressure is grooving to make high-quality products at lower cost and with less waste. In response to these needs, a new subdiscipline of analytical chemistry has emerged: process analytical chemistry (PAC) [1]. Unlike traditional chemical analysis, which is performed in well-equipped laboratories with the aim to identify and quantify small amounts of analytes, the goal of PAC is to supply quantitative and qualitative information about a chemical process that can be used not only to monitor and control the process, but also to optimize its efficient use of energy, time, and raw materials. In addition, it is possible to simultaneously minimize plant effluent release and improve product quality.

The PAC approach is quite different from the traditional way of food process control, where samples are taken at different point of the process and then transported to the central laboratory, stored, and analyzed with sophisticated instrumentation. In comparison, PAC analyzers are located in or right next to process and are considered physically and operationally part of the process. Besides the location of the instrumentation, time frame is another characteristic that distinguish PAC from the traditional analytical philosophy. In fact, a centralized laboratory usually can take hours or day before returning the results while PAC data are immediately available for process control and optimization. This in turn lowers the "quality cost" of production, that is, the loss suffered when a batch of product

Agricultural and Food Electroanalysis, First Edition.
Edited by Alberto Escarpa, María Cristina González and Miguel Ángel López.

fails to meet specifications and has to be reprocessed, sold at reduced cost or discarded. In practice, we can distinguish four ways of process analysis:

- Off-line
- At-line
- On-line
- In-line

The first two ways can be considered the traditional operational way since both require taking the sample from the process and subsequent analysis in a central laboratory in the first way (off-line) or in proximity of the process line in the second case (at line); both using more or less sophisticated instrumentation.

In the on-line way to operate, an automated sampling system is used to extract the sample and present it to an analytical instrument for measurement. This makes this way distinct from the first two. Further, it is possible to subdivide the on-line way into two categories: intermittent methods that require injection of a portion of the sample stream into the instrument and continuous methods that permit the sample to flow continuously through the instrument.

The classic examples of an intermittent on-line process analyzer are gas chromatography and liquid chromatography. However, an interesting intermittent technique to consider is flow injection analysis (FIA); a technique that may be thought of as an on-line alternative to manual wet chemical analysis [2]. The major advantages of FIA are high throughput, reliability, and precision. However, an even greater range for FIA will be forthcoming as new technologies are added. First is the capability to perform on-stream liquid–liquid extractions. Second is the incorporation of a chromatographic step. Third is the inclusion of sophisticated multichannel, rapid-scan spectroscopic/amperometric detectors to provide a more robust single-component analysis, or simultaneous multicomponent analysis. As these capabilities are added to FIA, it will become more apparent that this technology is an appealing paradigm for developing an automated universal microlaboratory that integrates the functions of sample acquisition, clean-up, concentration, separation, sensing, and identification. Process engineers will find the approach particularly attractive because it closely resembles the flow-stream concept used in automated chemical-processing lines.

However, the major disadvantage of on-line process analysis is the need to construct a separate analytical line that properly samples the main stream and presents it to the instrument at a suitable temperature and pressure. This has led to the fourth way of acting: in-line PAC. Here, chemical analysis is done *in situ*, directly inside the process line, using a probe that is chemically sensitive. In its ideal form, the device might resemble a typical industrial temperature probe. Such an implementation obviously would be attractive for electrochemists since the detection system can be based on micro and nanosensors based on ion-sensitive field-effect transistor (ISFET) or chemical modified field-effect transistor (CHEMFET) technology for measuring, for instance, pH or ions. However, these types of devices require a careful design since they will especially require in the food processing industry as some type of encapsulation to avoid the direct contact with the food matrix.

Process control is now recognized as one of the most critical functions in any overall program of quality assurance. This has resulted from a new approach to manufacturing in which the emphasis has shifted from the detection to the prevention of nonconformances and defects. The key elements of process control must include a well-conceived plan of controls to be exercised in relation to the specific product and process, along with an explicitly well-documented set of instructions for personnel carrying out the procedures. The use of the hazard analysis critical control points (HACCPs) techniques and statistical procedures are tools for process control in well-established food-quality assurance. This approach to quality manufacturing using process control as a key operational activity can provide assurances that quality targets can be achieved consistently in a cost-effective manner.

An effective program of process control must aim to manufacture or formulate product which meets the specifications consistently, and ensure that only acceptable products are released from each step of the overall process. This makes process control a key operational element of quality assurance strategy. Recently, the food industry has been influenced by the pharmaceutical industry where process analytical technology (PAT) is becoming compulsory. The main principle is that quality should be designed or built-in to a process or product.

The Food and Drug Administration (FDA) has recently released guidelines for its PAT initiative. This guidance represents the FDA's current thinking on the topic. According to FDA website (http://www.fda.gov), PAT is "a system for designing, analyzing, and controlling manufacturing through timely measurements (i.e., during processing) of critical quality and performance attributes of raw and in-process materials and processes with the goal of ensuring final product quality." The goal of PAT is "to understand and control the manufacturing process, which is consistent with our current drug quality system; quality cannot be tested into products, it should be built-in or should by design."

The main PAT objective is to increase process knowledge and therefore ensure a consistent predefined product using data from on-line and real-time analyses.

Quality control and efficiency using PAT is based on in-process data rather than from laboratory sampling and is representative of the manufacturing process in its entirety. Using this sort of approach, it is expected that the following improvements will be achievable:

- Reducing production cycle times by using on-, in-, and/or at-line measurements and controls.
- Preventing rejects, scrap, and re-processing.
- Considering the possibility of real-time release.
- Increasing automation to improve operator safety and reduce human error.
- Facilitating continuous processing to improve efficiency and manage variability.
  - Using small-scale equipment (to eliminate certain scale-up issues) and dedicated manufacturing facilities.
  - Improving energy and material use and increasing capacity.

PAT will allow quality to be built into the manufacturing process and subsequently enable continuous process improvements to be made. It includes scientifically based process design that identifies key measures of output quality and the critical process parameters that affect them, appropriate measurement techniques, statistical tools and feedback process control strategies that combine to ensure the desired quality of the final product.

There are a number of tools outlined in the FDA guidance expected to be utilized by the PAT initiative that will support process control. These tools can be categorized as follows:

- Modern process analyzers or PAC tools
- Multivariate data acquisition and analysis tools

- At-, on- or in-line tools
- Closed-loop feedback
- Process and endpoint monitoring and control tools
- Continuous improvement and knowledge management tools.

Some, or all, of these tools may be used in combination to provide monitoring in the formation applicable to the quality assurance of the manufacturing process.

It is important to note that the term *analytical* in PAT is viewed broadly to include chemical, physical, microbiological, mathematical, and risk analysis conducted in an integrated manner.

15.1 Sensors in Food Process

Technological advancements in process monitoring, control, and industrial automation over the past decades have contributed greatly to improve the productivity of virtually all manufacturing industries throughout the world.

Nowadays, in the era of digital control systems (DCS), one can plug in a laptop or use a wireless hand tool to instantly establish access to all the data, displays, and intelligence that reside anywhere on the DCS network. This capability, in combination with the self-tuning, self-diagnosing, and optimizing features of modern process control, makes both start-up activity and operational routines much easier and more efficient.

For what concern the food production, quality control is an important aspect from the point of view of providing foods of acceptable nutritional value, and for providing safety of products. Several characteristics such as size, shape, density, maturity, moisture content, oil content, flavor, firmness, tenderness, color, defects, blemishes, and so on, are routinely used in the quality control of agricultural and biological food products. Until recently, most analytical techniques, used in quality control, required isolation of the food component of interest and subsequent analysis. The original properties of the product are, therefore, destroyed during sample preparation and analysis. Oftentimes, such analyses are expensive, time-consuming, and require sophisticated instrumentation, and hence are not suited for on-line quality control of food products.

The development of sensors in the food science and industry faces many difficulties and there is still a lack of "real-time" sensors for controlling the sterilization and the sanitary quality of the products. In fact, the food-processing environment demands that process sensors must satisfy the following requirements:

- No hygiene risk
- Physically robust and stable
- Simple to operate
- Real time
- Sensitive
- Nondestructive.

However, the present development techniques in the fields of physics of materials and information processing should lead to new progress in the knowledge of food processes and the design of new sensors. The technology of metal–oxide–semiconductor

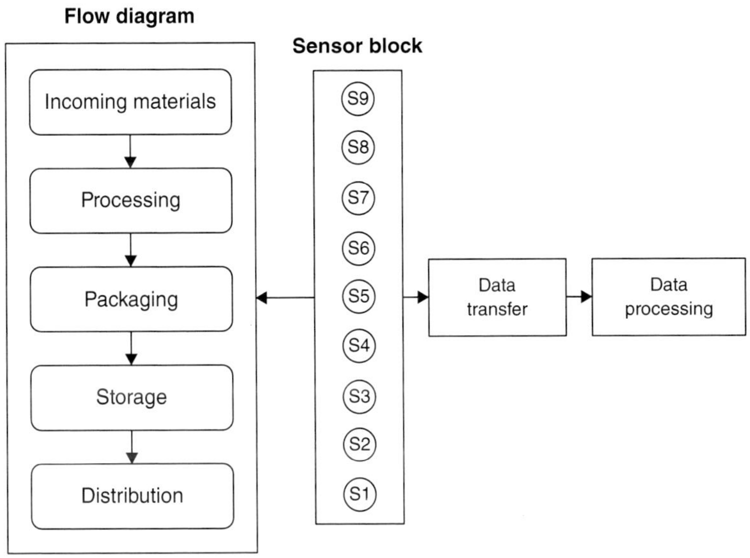

Figure 15.1 *Schematic representation of on-line sensors in a food-processing line*

field-effect transistor (MOSFET) and field-effect transistor (FET) constituted of metal oxides semiconductors can be used to give to the sensors selectivity with respect to some ions (ISFET), such as H^+, K^+ Na^+, Ca^{++}, and so on by modifying the type of sensitive surface. The FETs can also be used to detect enzyme (ENFET(enzyme field effect transistor)) and gases (GASFET (gas field effect transistor)) or to develop immune-sensitive field effect transistor (IMFET).

These sensors can be used as simple single sensors or they can form a sensor array for multiple analytes detection. The sensor array can also be formed by unselective sensors that can generate a response pattern that can be analyzed by powerful methods of multivariate statistics. In principle, appropriate sensors can be placed along all the food chain as shown in Figure 15.1.

At present, in the food sector remote sensing is greatly expanding since many hygiene problems can arise with contact sensors. The noninvasive era represents the ultimate in desirability. Because the probe does not physically contact the sample, the sampling problem is greatly alleviated. This era obviously has a great deal in common with remote sensing and nondestructive evaluation.

Recent progress in the development of instrumentation utilizing the optical properties of food products has provided several nondestructive techniques for quality evaluation. Most optical methods of nondestructive nature make use of the characteristic absorption spectra of components of interest. Such methods are highly sensitive, rapid, and reproducible, and have been successively used in routine on-line quality control of a large number of samples.

Spectroscopic sensors, in combination with multivariate data analysis, will often win the competition for most favorable process analyzer as compared to other during-process measurements. The most commonly used spectroscopic techniques for process monitoring are ultraviolet–visual absorption (UV–vis), near-infrared (NIR) and infrared (IR) absorption,

Raman scatter, and excitation emission fluorescence spectroscopy (EEFS). They use interaction with radiation from different regions of the electromagnetic spectrum to identify the chemical nature of molecules. For example, absorption of UV and VIS radiation causes valence electron transitions in molecules which can be used to measure species down to parts per million concentrations; for fluorophores (i.e., EEFS) determination can even go down to parts per billion levels. Whereas UV, VIS, and EEFS are limited to a smaller, select group of molecules, the NIR, IR, and Raman scatter spectroscopy techniques are probing molecular vibrations present in almost any species; their quantification limits are somewhat higher but can still be impressive. The reader is referred to textbooks for further details on basic principles of these spectroscopic techniques [3].

The development and utilization of electroanalytical methods in on-line process control is closely connected with the development of new instrumentation. New challenges lie ahead in improvement of data acquisition speed, level of automation, improvement in data analysis, as well as in implementation of large arrays of micro- and nanoelectrodes often in combination with complementary spectroscopic techniques. This places new demands on electrochemical instrumentation with respect to sensitivity, accuracy, and simultaneous acquisition of signals of different kinds. However, electrochemical sensors can be applied during food processing, product development, and food safety and biosecurity (nanosensors and nanotracers), only if a suitable sampling apparatus and appropriate cell development is designed for a particular application, that is, a sampling probe capable of withdrawing sterile, cell-free samples from fermenters and bioreactors.

In addition, electrochemical sensors are difficult to place in-line mainly for hygienic problems; the best way is to adapt them for on-line monitoring with the aid of a sampling probe (Figure 15.2) durable in design and construction to withstand the temperatures, pressures, viscosities, shear forces, and chemicals typical of the process under analysis.

At present spectroscopy, in all its variations, appears the most widely accepted, among analytical techniques for PAT approach in the food industry. However, the main advantage of the electrochemical sensor technology over spectroscopy is that electronic-nose (e-nose) and electronic-tongue (e-tongue) can be tailored to the monitoring process in a way that complementary information nonobtainable with spectroscopy can be acquired.

In the following sections, two systems that have a chance to be used during food processing will be discussed: the so called e-nose and e-tongue.

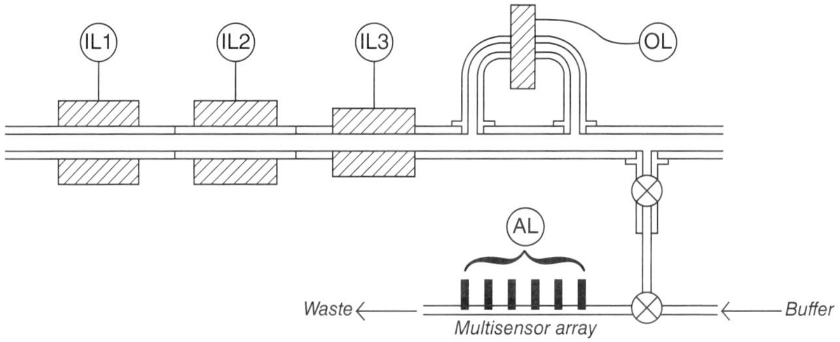

Figure 15.2 *Sampling probe for on-line application*

15.2 Electronic Nose

At the beginning of the 1990s, the term "artificial" or "e-nose" appeared and was defined by Gardner and Bartlett in 1993 as an "instrument which comprises an array of electronic chemical sensors with partial specificity and an appropriate pattern recognition system, capable of recognizing simple or complex odors."

A simple odor is usually represented by only or few types of odor molecules, whereas a complex odor is caused by a mixture of hundreds of different types of molecules. Odors (or, scientifically, odorant molecules) are generally light (molecular mass up to 300 Da), small, polar, and often hydrophobic [4].

The two main components of an e-nose are the sensing system and the automated pattern recognition system. The sensing system can be an array of several different sensing elements (e.g., chemical sensors), where each element measures a different property of the sensed chemical, or it can be a single sensing device (e.g., spectrometer) that produces an array of measurements for each chemical, or it can be a combination. Each chemical vapor presented to the sensor array produces a signature or pattern characteristic of the vapor.

Operationally, an e-nose consists of a sensor array, a gas pump as sampling system, and a serial interface for data transfer to a remote PC where an appropriate pattern-classification system is used as signal processing (Figure 15.3).

Modern instruments can be easily adapted to particular requirement in process-control applications. By adopting a particular sampling strategy, sensor array can be operated in the laboratory as well as on-line for process-control or environmental-monitoring applications. In the laboratory, the detector can be used with a headspace autosampler. Adsorbent trapping units are also available for on-line measurement with a gas chromatographymass spectrometry(GCMS) instrument. The different sensors are built into a very small sensor chamber and usually can be ordered for a particular application. In other words, the gas detection devices can be easily adapted to particular requirement in process-control applications.

The key principle involved in the e-nose concept is the transfer of the total headspace of a sample to a sensor array that detects the presence of volatile compounds in the headspace and a pattern of signals provided that are dependent on the sensors selectivity and sensitivity and the characteristics of the volatile compounds in the headspace [5].

The e-nose in practice acts as the human nose. In fact, the human receptors convert the chemical responses to electronic nerve impulses that are propagated by neurons through a complex network before reaching the higher brain for interpretation. Similarly, the chemical sensors in the e-nose react with the sample and produce electrical signals. A computer reads the unique pattern of signals and interprets them with some form of "intelligent" pattern classification algorithm.

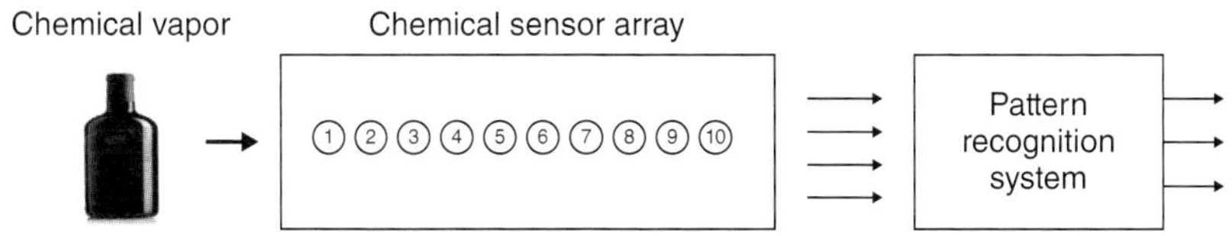

Figure 15.3 *Basic elements of an electronic-nose*

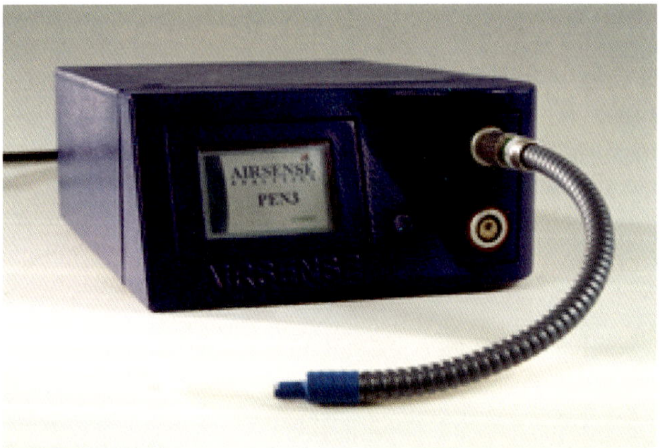

Figure 15.4 A commercial electronic-nose from Airsense

Several commercial intelligent gas sensor array instruments are now available in the market covering a variety of chemical sensor principles, system design, and data analysis techniques (Figure 15.4).

The ideal sensors to be integrated in an e-nose should fulfill the following criteria [6]: high sensitivity toward chemical compounds, that is, similar to that of the human nose (down to 10^{-12} g/ml); low sensitivity toward humidity and temperature; medium selectivity, they must respond to different compounds present in the headspace of the sample; high stability; high reproducibility and reliability; short reaction and recovery time; robust and durable; easy calibration; easily processable data output; small dimensions.

By chemical interaction between odor compounds and the gas sensors, the state of the sensors is altered giving rise to electrical signals which are registered by the instrument. In this way, the signals from the individual sensors represent a pattern which is unique for the gas mixture measured and is interpreted by multivariate pattern recognition techniques. Samples with similar odors generally give rise to similar sensor response patterns and samples with different odorous (both pleasant and unpleasant) show differences in their patterns. The sensors of an e-nose can respond to both odorous and odorless volatile compounds. Sensors for the e-nose must be sensitive to various hydrophobic compounds with a mass range from 18 to 300 mass units.

15.3 Electronic Nose Technologies

Different gas sensor technologies are available, but only four technologies are currently used in commercialized e-noses: metal oxide semiconductors (MOSs); MOSFETs; conducting organic polymers (CPs); and piezoelectric crystals (bulk acoustic wave – BAW). Others, such as fiber optic [7], electrochemical [8], and bi-metal sensors, are still in the developmental stage and may be integrated in the next generation of the e-nose. The choice of the sensors used is important; the sensors used should be able to detect gases that are important for the measured sample. Currently, major limitations of e-noses to be overcome include high-application specificity, dependence on extensive calibration, and sensitivity of gas sensors to humidity. Humidity is present in large quantities in almost every type of foods, and especially limits the performance of e-nose based on conducting polymers. Thus, a device designed with the hope of solving almost every problem relating to odor, evaluating samples in real time under conditions much more severe than in a controlled laboratory, is still to be achieved. In all cases, the goal is to create an array of differentially sensitive sensing elements.

15.3.0.1 Metal Oxide Semiconductors

MOS gas sensors are high gap ionic semiconductors. A change in the surrounding environment modifies electrical transport properties of MOS. These sensors rely on changes of resistance induced by the adsorption of gases and subsequent surface reactions [9]. They consist of a ceramic substrate (round or flat) heated by wire and coated by a metal oxide semiconducting film. Theoretically, there are no limitations for using any materials for solid-state gas sensors design independently of their physical, chemical, structural, or electrical properties. Receptor function provides the ability of the oxide surface to interact with the target gas, and transducer function provides the ability to convert the signal caused by chemical interaction of the oxide surface into electrical signal. Surface of metal oxides is responsible for receptor function of solid-state gas sensors.

Due to the high operating temperature (200–650 °C), the organic volatiles transferred to the surface of the sensors are totally combusted to carbon dioxide and water, leading to the change in the resistance. MOS sensors are the most used one in commercial e-nose as they have small dimension, low cost, low power consumption, on-line operation, and high compatibility with microelectronic processing.

15.3.0.2 Metal Oxide Semiconductor Field-Effect Transistors

The MOSFET is a transistor used for amplifying or switching electronic signal.

When a voltage is applied between the gate and body terminals, the electric field generated penetrates through the oxide and creates at the semiconductor–insulator interface an "inversion layer" or "channel" through which current can pass. Varying the voltage between the gate and body modulates the conductivity of this layer and thereby controls the current flow between drain and source.

When polar compounds interact with this metal gate, the electric field, and thus the current flowing through the sensor, is modified. The recorded response corresponds to the change in voltage necessary to keep a constant preset drain current [10]. The selectivity and sensitivity of MOSFET sensors may be influenced by the operating temperature (50–200 °C), the composition of the metal gate, and the microstructure of the catalytic metal. MOSFET sensors are thought to be very robust. The major advantage of the MOSFET transistor is that they have a relatively low sensitivity to moisture and use low power for accomplishing its purpose and the dissipation of power in terms of loss is very little.

15.3.0.3 Conducting Organic Polymers

CP sensors like MOS sensors rely on changes of resistance by adsorption of gas. These sensors comprise a substrate (such as a fiber glass or silicon), a pair of gold-plated electrodes, and a conducting organic polymer such as polypyrrole, polyaniline, or polythiophene as a sensing element. Such conducting polymer elements are simply prepared and readily modified chemically to respond to a broad range of analytes. In addition, these sensors yield a fairly rapid low-power DC electrical signal in response to the vapor of interest, and their signals are readily integrated with software- or hardware-based neural networks for the purposes of analyte identification.

In general, CP sensors show a good sensitivity especially for polar compounds. However, their low operating temperature (<50 °C) makes them extremely sensitive to moisture. Although such sensors are resistant to poisoning, they have a lifetime of only about 9–18 months.

A variety of polymer films that have distinctly different electrical resistance responses to various vapors can be prepared to obtain diagnostic array signal patterns in response to a series of odorants. However, their use in PAC is very limited.

15.3.0.4 Piezoelectric Crystal Sensors

Piezoelectric crystal sensors are passive solid-state electronic devices, which can respond to changes in temperature, pressure, and most importantly, to changes in physical properties at the interface between the device surface and a foreign fluid or solid. Such changes in physical properties include variations in interfacial mass density, elasticity, viscosity, and layer thickness.

In recent years, the incorporation of various chemically sensitive layers has enabled the transition from the microbalance to the mass sensor and resulted in the explosive growth of piezoelectric sensors.

This type of sensor operates by observing the propagation of an acoustic wave through the solid-state device. Sensing is achieved by correlating acoustic wave-propagation variations to the amount of analyte captured at the surface and then to the amount or concentration of analyte present in the sample exposed to the sensor, or to the changes in physical properties of interfacial thin films.

Measurement of the frequency shift and resonant resistance change of acoustic wave is one of the most accurate types of physical measurement. The frequency and resonant resistance of piezoelectric crystal is influenced by a large number of parameters mentioned above. A selective sensor is obtained when a sensor surface is coated with selectively interacting thin film.

Changes in physical properties are induced within a thin layer of material at one sensor surface in virtually all piezoelectric acoustic sensor applications. This thin surface layer often consists of a different material than the sensor bulk and is used to induce selective adsorption or absorption to the thin layer of an analyte of interest.

The sensitivity of piezoelectric detectors is directly proportional to the square of resonant frequency, and inversely proportional to the surface area. Therefore, a SAW device is considered to be an excellent transducer of chemical vapor sensor. Besides sensitivity, the SAW device could be made at relatively low cost, and miniaturized with precise and reproducible characteristics using photolithographic techniques. In addition, the lithographic fabrication capability easily permits a complex circuit to be present on the same crystal surface.

Piezoelectric crystal sensors are made of tiny disks, usually quartz, lithium niobate ($LiNbO_3$) or lithium tantalite ($LiTaO_3$), coated with materials such as chromatographic stationary phase, lipids, or any nonvolatile compounds that are chemically and thermally stable [11]. Since piezoelectric sensors may be coated with an unlimited number of materials, they present the best selectivity. However, the coating technology is not yet well controlled, which induces poor batch-to-batch reproducibility.

15.4 Electronic Noses for the Food Industry

Currently, the biggest market for e-noses is the food industry but mainly for off-line applications [12]. These include quality assessment in food production, inspection of food quality by odor, control of food cooking processes [13] inspection of fish, monitoring the fermentation process, checking rancidity of mayonnaise, verifying if orange juice is natural, monitoring food and beverage odors [14], grading whiskey, inspection of beverage containers, checking plastic wrap for containment of onion odor, and automated flavor control [15] to name a few. In some instances, e-noses can be used to augment or replace panels of human experts. In other cases, e-noses can be used to reduce the amount of analytical chemistry that is performed in food production especially when qualitative results will do. The monitoring of flavors/aroma is probably the most investigated area of e-nose. Evaluation of flavor in foods can be performed in different steps of production to monitor sensory properties prediction, quality and freshness, harvest time, geographic origin, and processing methods.

During recent years there has been increasing interest in the certification of the geographical origin of food products, since authenticity and quality issues can be often associated with a given geographical origin. Authenticity of oils and wines has been broadly studied, since these products are easily adulterated [16], and e-nose (Pen 2, Airsense) has been used, in combination with multivariate analysis, to verify the geographical origin and uniqueness of Garda extra virgin olive oils. E-nose technologies successfully used to characterize and classify wines from different geographical origins [17, 18]. E-noses can be applied, with the appropriate sampling techniques, to solve problems such as origin, variety, and aging, rather than detecting individual compounds [19, 20].

However, recently it has been shown that an e-nose can be easily employed to monitor the product from the raw material, through the process, to final product, including shelf-life, by analyzing volatile compounds released in each stage of processing. Commercial instruments are also in the market but the most important applications concerning post-processing application is monitoring the shelf-life of the product. In fact, flavor and aroma quality are important sensory attributes for consumers and should be measured during the shelf-life. By measuring the release of volatiles during storage, a time-dependent model can be made for the prediction of shelf-life or storage time.

Many examples can be found in different fields of food industry from fruits and vegetables to cheese and fish, and so on. Fruits and vegetables are one of the major growing segments in food retail establishments; fresh cut fruit are still under study because of the difficulties in preserving their fresh-like quality over prolonged periods [21]. Environmental applications of e-noses include analysis of fuel mixtures [22], detection of oil leaks [23], testing groundwater for odors, and identification of household odors [24]. Potential applications also include identification of toxic wastes, air quality monitoring, and monitoring factory emissions.

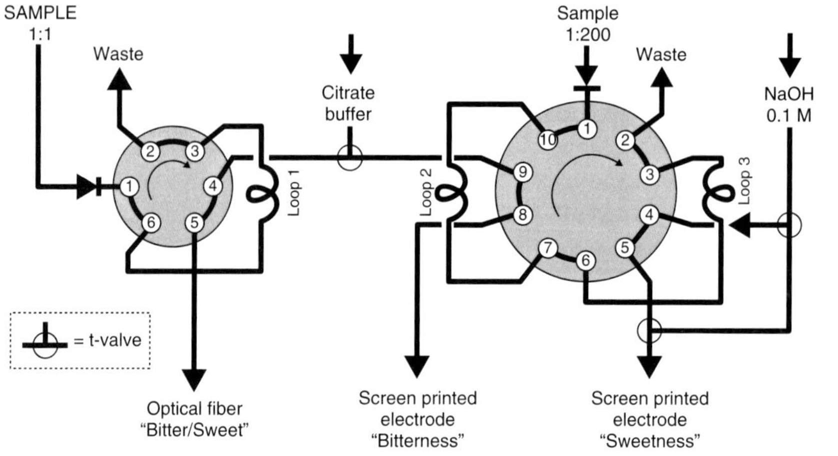

Figure 15.5 *A schematic FIA system with in-line amperometric/spectroscopic sensors*

15.5 Electronic Tongue

The "e-tongue"can be defined as an analytical instrument composed of an array of nonselective chemical sensors with partial specificity to different solution components and an appropriate pattern recognition system, capable to recognize quantitative, and qualitative composition of simple and complex solutions.

There are two major sensors classes that use liquid electrolytes: amperometric and potentiometric sensors. The earliest example of an amperometric gas sensor, the Clark oxygen sensor used for the measurement of oxygen in the blood is more than 40 years old. The amperometric sensor produces current signal, which is related to the concentration of the analyte by Faraday's law and the laws of mass transport. It is operated in a region where mass transport is limiting and therefore has a linear response with concentration of the analyte. This type of sensor has now been developed in many different geometries with different types of material coated or uncoated to enhance sensitivity and selectivity (Figure 15.5).

E-tongues can be found in different versions according to the detectors employed: electrochemical (potentiometric, voltammetric, amperometric, impedimetric, or conductimetric), optical, mass and enzymatic sensors (biosensors). One example is a taste sensor made by potentiometric electrodes coated with a lipid layer, another one is a taste analyzer made by ChemFET.

E-tongues have reportedly been used to obtain data for sourness, bitterness, and astringency for foodstuffs such as beers, wines, and teas. This involved detecting sensory attributes of bitter, sweet, sour, fruity, caramel, artificial, burnt, intensity, and body using potentiometric/amperometric chemical sensors along with the same pattern recognition techniques described above for the e-nose technology.

Taste sensation is the result of physicochemical interactions of food molecules with a complex system of hundreds of cell buds located randomly all over the tongue.

For amperometric sensors, there are essentially four classes including metal, conducting polymer, phtalocyanine film, and biosensors. Nanocomposite materials are still at the development stage.

Three different detection modes can be utilized including fixed potential preferred in flow systems and for biosensors, step pulse potential and sweeping potential of which the last two are preferred in batch systems. Up and coming are microchip capillary electrophoresis coupled with amperometric screen-printed electrodes [25]. Metal sensors are also used, but lack selectivity. They are more useful for classification applications than for evaluation of taste, like predicting sensorial descriptors of Italian red dry wines of different origins [26].

Conducting polymer sensors such as polypyrrole and polyaniline show a variation in conductivity with adsorption of different analytes. E-tongues based on conducting polymers have been used to evaluate bitterness, as well as sweet, bitter, acid, salty, and astringent tastes. The advantage of conducting polymers is the rapid adsorption/desorption and partial selectivity through modifying the dopant of the polymer. Phthalocyanine film electrodes can have different chemical properties. These are coordination compounds where a transition metal is coordinated with a phthalocyanine ring.

Biosensors for e-tongue are systems with a biochemical transducer, an enzyme, and solid electrode in intimate proximity. Enzymes are oxidases that consume oxygen and produce hydrogen peroxide or the reduced form of β-nicotinamide adenine dinucleotide (phosphate) NAD(P)H as a dehydrogenase. Improvement in the performance of metal sensors, conductive polymers or biosensors is linked to scaling down of size to nanodimensions which increases the surface-to-volume ratio of the sensors, lowering detection limits.

New techniques are being applied to develop miniaturized sensor arrays such as screen printing for thick film and electron beam evaporation, thermal vacuum deposition, and pulsed laser deposition for thin-film technique. Microfabrication techniques were used to prepare a sensor array for use in a voltammetric e-tongue by depositing gold (Au), platinum (Pt), iridium (Ir), and rhodium (Rh) on a silicon wafer.

E-tongue sensors often use a lipid membrane as a recognition element that changes taste-relevant substances into electric potential charge across thin membranes and the membrane potential is independent of thickness of the lipid membrane, which must have durability and reproducibility even when repeatedly rinsed. The concentration of the lipid in the membrane can be optimized and affects the detection limit of the sensor to adsorptive taste substances [27].

Unfortunately, when applied in-line in food processing, all these systems show certain weaknesses. Therefore, the development of sensors with diverse sensitivity, selectivity, and overall robustness is needed.

Probably, nanostructured electrode materials and sophisticated electrode architectures are the key factors for the successful development of state-of-the-art devices with optimized performance for on-/in-line applications.

In particular, nanofibrous membranes may be one of the essential components of a sensing or biosensing unit (chemical, electric, electrochemical, optical, or thermal). In this case, they can be used as a detection aid preventing fouling phenomena, occurring onto detector surface (e.g., electrodes), or eliminating interferences and controlling the operating regime of the sensor [28].

Furthermore, nanotechnologies might enable the incorporation of sensors in the packaging which may detect deterioration in food quality resulting in more accurate sell-by dates for perishable foods which would, in turn, improve food safety and reduce wastage.

When available, sensors can be embedded in a packaging film to detect food-spoilage organisms and trigger a color change to alert the consumer that the shelf-life is ending or has ended. For instance, an array of sensors (e-tongue) has been included in a packaging material.

The development of innovative nanostructured sensors may enhance traceability and security of food products. Nanotechnology may provide materials with recognition groups, systems, detectors of pathogens, contaminants, or devices to track food products from manufacturing and processing to shipping. There are nanobar codesgin in the market that contain metal stripes varying in width, length, and amount, such that billions of combinations can be created to tag individual products [29].

The barcodes have been primarily used to assure brand and authenticity in pharmaceuticals, but applications could be forthcoming in tracing food batches. Combined with pathogen sensors, the barcodes that must be read by modified microscopes could trace sources of outbreak. There are antibacterial surfaces for the machines involved in food processing or production. With the conventional methods of detection it takes hours to days. With molecular electronic based methods, biosensors, e-nose, microarrays, nanobiosensors (based on microfluidic, nanomaterials, DNA, etc.) it takes seconds to minutes. A biosensor developed by an Agricultural Research Service scientist, Athens, uses fluorescent dye particles attached to bacteria antibodies. If Salmonella bacteria are present in the food being tested, the nanoculturing results with these two examples of instantaneous sensors, sized dye particles become visible. No need to send out to the lab and wait days for results.

15.6 Pattern Recognition Models

We have completed an outline of some of the possibilities for chemical sensing in process analysis. How does one analyze the data from the array or instruments and sensors that have been placed in/on-line? Unlike classical methods, e-tongue and e-nose sensors do not provide information on the nature of the compounds under investigation, but only give a digital fingerprint of the status of the food product at the point and time of measurement. Usually a large amount of data is generated depending on the type of sensing units placed in the process line. For this reason, an increase in the use of multivariate statistical analysis for the elaboration of such amounts of data and for the evaluation and interpretation of useful information should be developed. The potential advantages of such an approach include the reduction in complexity of the sensor coating selection, the ability to characterize complex mixtures without the need to identify and quantify individual components and that it can be exploited to search for a structure and correlation in the data, or to make a model from a calibration set of data which then is used to predict new data. In practice, the sensors fingerprint is analyzed by multivariate statistical models. These models evaluate the data matrix constituted by the analyzed fingerprints, recognize potential patterns for a number of species, and calculate decision rules to identify these patterns. Then, this can be used to classify unknown patterns.

The main purposes of multivariate analysis are data reduction (unsupervised analysis) and data modeling like regression and/or classification models (supervised analysis).

In this discussion, we will consider PCA and artificial neural network (ANN), both extensively applied to e-noses and e-tongues results both for discrimination and monitoring purposes.

PCA is an unsupervised technique whose main idea is to explore data analysis and to reduce dimensionality with a minimum loss of information. This is achieved by projecting the data onto fewer dimensions that are chosen to exploit the relationships between the variables. It also measures qualitative associations between variables. To analyze the results, a line of best fit through a system of points in space is obtained. This technique allows the similarities and differences between objects and samples to be better assessed [30].

ANNs are applied for classification purposes. The basic idea of ANN is to simulate the function of the human brain. Similar to a biological neuron, an artificial neuron receives a series of input information connected to a weight factor, which is adjustable during network training. Neurons form layers with intra- or inter-layer connections, resulting in feedback or feedforward networks. Layers between the input and the output layers are called hidden layers. The learning process for developing a neural network can be either supervised or unsupervised if it needs or not target outputs as teacher, respectively.

Several different types of ANN are available and the most popular is the backpropagation approach. In this procedure, input patterns presented to the input layer, for example, signals from an array of chemical sensors, generate a flow of activation to the output layer. Errors in the output are then fed back to the input layer to modify the weights of the interconnections. It should be emphasized that backpropagation does not describe a network but represents a learning algorithm. In this way, the network can be trained with known parameters, such as sensor array responses to sets of known chemicals.

Since a neural network can arrive at different solutions for the same data, if different values of the initial network weights are provided, the network should be trained several times. The goal is to try and find a neural network model for which multiple trainings approach the same final mean squared error (MSE).

In a practical approach, the computer program is provided with a training set consisting of the fingerprint of each samples and the property measured by an independent reference method. It is important that the training set consists of representative samples with variations spanning that of the set of all samples that might reasonably be encountered during processing. Once the dataset is acquired, the computer searches for correlations between the spectra and the sought-for property. Eventually a transformation vector is produced that, when multiplied by the spectral vector, yields an estimate for the property. Once the derived transformation vector has been validated with a double-blind study of a second set of samples for which spectra and property have been measured, the method can be ready for use.

15.7 Sampling

The norm ISO 17025 "General requirements for the competence of testing and calibration laboratories" at the point 5.7 states that "The laboratory shall have a sampling plan and procedures for sampling when it carries out sampling of substances, materials, or products for subsequent testing or calibration. The sampling plan as well as the sampling procedure shall be available at the location where sampling is undertaken. Sampling plans shall, whenever reasonable, be based on appropriate statistical methods. The sampling process shall address the factors to be controlled to ensure the validity of the test and calibration results." Furthermore, it defines sampling as a "procedure whereby a part of a substance, material or product is taken to provide for testing or calibration of a representative sample of the whole."

These general definitions are also valid for process analysis into food manufacturing but the process should be considered in terms of process dynamics. Process sampling and careful consideration of the process dynamics is thus of paramount importance when implementing PAT.

A sampling plan is a detailed outline of which measurements will be taken at what times, on which material, in what manner, and by whom. Sampling plans should be designed in such a way that the resulting data from the parameters of interest will contain information from a representative sample.

The objectives and sampling purposes to be achieved should be clear when developing the sampling procedures to be adopted.

Furthermore, in food systems sampling should be done in a well-defined area to avoid difficulties in the execution of procedures, reduce the risk of contamination and cross-contamination, enable the proper execution of laboratory analysis and include all necessary safety and health precautions to the sampler and environment. It is then of paramount importance that personnel responsible for the sampling activities should be trained on the applicable procedures.

All tools and auxiliary materials should be inert, and in a clean condition before and after their use.

The key criterion to which a sampling system must be designed is that the sample must be taken from a well-mixed and representative point in the flow. This requirement is also true for on-line analyzers and this requires that for instance sampling loops are so designed that loop velocity is sufficiently high to maintain homogeneity at all times and mounted inside a receiver housing, allowing them to be easily isolated and serviced without needing to shut down the main pipeline. The receiver housing can be cooled if required and may contain multiple receivers with manual or automatic changeover.

Besides precision, specificity, sensitivity, and practicality, the following parameters should be considered when choosing the most appropriate sampling method:

1. *Measurement frequency.* The frequency of the measurements determine how detailed the true signal can be approximated. Depending on the process dynamics rapid on-line sensors can be of a tremendous advantage compared to slow off/at-line laboratory equipment.
2. *Time of measurement.* The time each individual measurement takes (e.g., integration of detector response or time for manual at-line grab sampling) also determines the usefulness of the response. Every variation inside the measurement cycle will be averaged, and the reconstructed signal is thus a smoothed representation. In practice, for sampling process systems, this parameter will represent a compromise between process dynamics and the measurement uncertainty.
3. *Carry-over effect between measurements.* Some process sampling designs and process sensors can have memory effects which give carry-over signal and leads to a smoothed (or more correctly convoluted) reconstruction curve deviating from the true process signal. A typical example is the delayed response of process temperature sensors due to the protective shielding required for harsh conditions which acts as (unwanted) isolator for the measurement element. Fiber-based, noninvasive spectroscopic sensors in combination with proper process sampling design and placement have a strong advantage in this respect.

4. *Dead-time between sampling and measurement.* The so-called dead-time between sampling and measurement is a parameter of paramount importance to consider in at-/on-/in-line measurements, especially when the results of the measurement are used to modify in real time other parameters influencing the process.
5. *Uncertainty of measurements.* The last and often most important parameter to consider is the measurement uncertainty of the process analyzer. In combination with the statistical model used for prediction, it will determine if the sensor is robust and sensitive enough for a given application. All analytical equipment are prone to errors and often more so when implemented in a harsh and disordered process environment. The ratio of measurement uncertainty and the natural variation in the process determines if a sensor is of value for that application.

By considering these five parameters it is possible to establish if a given process analyzer fulfills the demands for a specific monitoring and control task.

15.8 Conclusions

A desired goal of the PAT framework is to design and develop processes that can consistently ensure a predefined quality at the end of the manufacturing process. Such procedures would be consistent with the basic tenet of quality by design and could reduce risks to quality and regulatory concerns while improving efficiency.

The savings to be obtained from installation of the present generation of process analyzers can be enormous. The new generation of on-line spectroscopic/amperometric instruments and the accompanying multivariate analysis promise even more capabilities, with the possibility of developing new processing methods that previously could not be considered because of the lack of adequate control.

However, there is a need of a major integration among process engineers and analytical chemists. In other words, the time is right for analytical chemists to collaborate with process engineers with a goal of optimizing processes, using analytical instruments that are able to give real-time information from the raw material to the end of the food product including shelf-life.

Perhaps the most exciting promise of PAC is to make process lines so efficient and so well-controlled that zero-defect products are released together with a decrease in effluent. Imagine the good will such efficient plants might generate in the community.

References

1. Koch, K.H. (1999) *Process Analytical Chemistry*, Springer.
2. Ruzicka, J. and Hansen, E. (1988) in *Flow Injection Analysis*, 2nd edn (ed J.D. Winefordner), Wiley. ISBN: 978-0-471-81355-2
3. Mirschel, G., Savchuk, O., Scherzer, T., and Genest, B. (2013) Process control of printing processes with in-line NIR spectroscopy and elimination of the influence of the substrate on the prediction of the coating weight. *Prog. Org. Coat.*, **76**, 83–93.
4. Craven, M.A., Gardner, J.W., and Bartlett, P.N. (1996) Electronic noses-development and future prospects. *TrAC, Trends Anal. Chem.*, **15** (9), 486–493.

5. Gardner, J.W. and Bartlett, P.N. (1993) A brief history of electronic noses. *Sens. Actuators*, **B18**, 211–220.
6. Bartlett, P.N., Blair, N., Gardner, J.W.1993. *Electronic Nose. Principles, Applications and Outlook*. ASIC, 15 Colloque, Montpellier, 478–486.
7. Dickinson, T.A., White, J., Kauer, J.S., and Walt, D.R. (1996) A chemical detecting system based on a cross-reactive optical sensor array. *Nature*, **382**, 697–700.
8. Baltruschat, H., Kamphauser, I., Oelgeklaus, R., Rose, J., and Wahlkamp, M. (1997) Detection of volatile solvents using potentiodymanic gas sensors. *Anal. Chem.*, **69**, 743–748.
9. Kohl, D. (1992) Oxidic semiconductor gas sensors, in *Gas Sensors* (ed G. Sberveglieri), Kluwer Academic Publishers, Dordrecht, pp. 43–88.
10. Lundstrom, I., Spetz, A., Winquist, F., Ackelid, U., and Sundgren, H. (1990) Catalytic metals and field-effect devices useful combination. *Sens. Actuators*, **B1**, 15–20.
11. Guilbault, G. and Jordan, J.M. (1988) Analytical uses of piezoelectric crystal: a review. *Crit. Rev. Anal. Chem.*, **19**, 1–28.
12. Pisanelli, A.M., QutobA.A., TraversP., SzyskoS., PersaudK.C.1994. Application of multi array polymer sensors to food industries. *Life Chem. Rep.*, **11**, 303–308.
13. Almeida, A. (1994) Neural Network in Oven Prevents Overcooking. The Institute, 3.
14. Gardner, J.W., Hines, E.L., and Wilkinson, M. (1990) Application of artificial neural network to an electronic olfactory system. *Meas. Sci. Technol.*, **1**, 446–451.
15. Moriizumi, T., T. Nakamoto, & Y. Sakuraba (1992). *Pattern Recognition in Electronic Noses by Artificial Neural Network Models*. Vol. **212, 217–236**, Springer, Netherlands (doi: 10.1007/978-94-015-7985-8_14).
16. Cosio, M.S., Ballabio, D., Benedetti, S., and Gigliotti, C. (2006) Geographical origin and authentication of extra virgin olive oils by an electronic nose in combination with artificial neural networks. *Anal. Chim. Acta*, **567** (2), 202–210.
17. Buratti, S., Benedetti S., Scampicchio, M., Pangerod, E.C. Characterization and classification of Italian Barbera wines by using an electronic nose and an amperometric electronic tongue. *Anal. Chim. Acta*2004**525**, 133–139.
18. Reid, L.M., O'Donnell, C.P., and Downey, G. (2006) Recent technological advances for the determination of food authenticity. *Trends Food Sci. Technol.*, **17** (7), 344–353.
19. Berna, A.Z., Lammertyn, J., Saevels, S.,Di Natale, C., Nicolaı, B.M. 2004. Electronic nose systems to study shelf life and cultivar effect on tomato aroma profile. *Sens. Actuators,* **B97**, 324–333.
20. Cynkar, W., Damberg, R., Smith, P., and Cozzolino, D. (2010) Classification of tempranillo wines according to geographic origin: combination of mass spectrometry based electronic nose and chemometrics. *Anal. Chim. Acta*, **660**, 227–231.
21. Soliva-Fortuny, R.C. and Martín-Belloso, O. (2003) New advances in extending the shelflife of fresh-cut fruits: a review. *Trends Food Sci. Technol.*, **14**, 341–353.
22. Lauf, R.J. and Hoffheins, B.S. (1991) Analysis of liquid fuels using a gas sensor array. *Fuel*, **70**, 935–940.
23. Shurmur, H.V. (1990) The fifth sense: on the scent of the electronic nose. *IEE Rev.*, **1**, 95–98.
24. Keller, P.E., Kouzes, R.T., and Kangas, L.J. (1994) Three neural network based sensor systems for environmental monitoring. IEEE Electro 94 Conference Proceeding, Boston, MA, pp. 377–382.

25. Wang, J., Mannino, S., Camera, C., Chatrathi, M.P., Scampicchio, M., and Zima, J., (2005) Microchip capillary electrophoresis with amperometric detection for rapid separation and detection of seleno amino acids. *J. Chromatogr*, **A1091** (1-2), 177–188.
26. Legin, A., Rudniskaya, A., Lvova, L., Vlasov, Y., Di Natale, C., and D'amico, A. (2003) Evaluation of Italian wine by the electronic tongue: recognition quantitative analysis and correlation with human sensory perception. *Anal. Chim. Acta*, **484**, 33.
27. Toko, K. (2000) *Biomimetic Sensor Technology*, Cambridge University Press, Cambridge.
28. Scampicchio, M., Bulbarello, A., Arecchi, A., Cosio, M.S., Benedetti, S., Mannino, S.2012. Electrospun nonwoven nanofibrous membranes for sensors and biosensors*Electroanalysis***24**,4, 719–725.
29. Bulbarello, A., S. Sattayasamitsathit , A. Crevillen , J. Burdick , S. Mannino , P. Kanatharana , P. Thavarungkul , A. Escarpa , J. Wang 2008.Striped alloy nanowire optical reflectance barcodes prepared from a single plating solution*Small*, **4**,5, 597.
30. Adams, M.J. (1995) *Chemometrics in Analytical Spectroscopy*, Royal Society of Chemistry, Cambridge, pp. 70–79.

16

Instrumental Aspects of Food Analysis by Electrochemical Methods

Wendell K. T. Coltro[1], *Maria F. Mora*[2] *and Carlos D. Garcia*[3]
[1]*Institute of Chemistry, Federal University of Goias, Goiania, GO, 74001-970, Brazil*
[2]*Jet Propulsion Laboratory, California Institute of Technology, 4800 Oak Grove Drive, Pasadena, CA, 91109, USA*
[3]*Department of Chemistry, Clemson University, Clemson, SC 29634, USA*

16.1 Introduction

The food industry and governmental agencies must guarantee the quality and safety of the food supply. This task requires determination of food composition and characteristics during the development process, through production, and after a product is in the market. Foods are complex mixtures that require proper and specific methods of analysis to guarantee quality, meet government regulations, and ensure public health [1]. Thus, the choice of appropriate analytical methods and associated instrumentation is a critical step in the analytical process. Depending on the characteristics of the sample and the type of analysis required there are a variety of techniques available, including flow injection analysis [2], sequential injection [3, 4], chemiluminescence [5], ultrasound [6], electronic tongues [7, 8], electrophoresis [9, 10], optical spectroscopy [11, 12], nuclear magnetic resonance [13, 14], surface plasmon resonance [15], mass spectrometry [16, 17], and chromatography [18–20]. It is not uncommon either to couple these techniques to sample clean-up [21] and preparation procedures such as microextraction [21] or microwave digestion [22]. In addition, a large body of food-related literature has focused on the development of methodologies targeting groups of compounds such as proteins [23], carbohydrates [24], metabolites [25], tanins [26, 27], contaminants [28, 29], endocrine-disruptors [30],

Agricultural and Food Electroanalysis, First Edition.
Edited by Alberto Escarpa, María Cristina González and Miguel Ángel López.

phenols [31, 32], dyes [33, 34], carotenoids [35], metals [36], and chiral molecules [37]. In this ocean of possibilities, electrochemical detection (ECD) clearly stands out as one of the most versatile techniques, offering great advantages for food analysis. ECD can be used to analyze a wide range of analytes, ranging from small inorganic ions [38], organic molecules, and polymers, to full microorganisms [39]. Furthermore, ECD features remarkable mass sensitivity (approaching that of fluorescence) while offering portability, inherent miniaturization of both the detector and control instrumentation, independence from sample turbidity, optical path length, or substrate clarity, low cost, minimal power consumption, and high compatibility with advanced micromachining and microfabrication technologies. In addition, ECD does not scale linearly with decreasing electrode size, a limitation common to most optical methods. Therefore, it is easier to miniaturize the detection elements and still maintain (or even improve) the signal level. Furthermore, with the use of microfabrication, it is possible to integrate multiple electrodes into a single device thus improving the ability to perform more selective and/or more sensitive detection modes.

One of the most important aspects related to the recent advancement of electroanalytical techniques for food analysis is probably the development of instrumentation. Although great progress has been achieved in the last few years in terms of performance, integration, and applicability of ECD, the basic architecture of electrochemical systems has not changed much since initially conceived [40–43]. Recent developments, however, have yielded smaller systems, greater precision and accuracy in the control parameters, and lower noise. The latter is a critical point because it has pushed the limit of detection (LOD) from micromolar to nanomolar or (for specific applications) to picomolar. Despite the clear advantages associated with electrochemical detectors, the training of personnel and somehow limited availability of commercial instruments [44] have traditionally limited the development of electrochemical methods applied to agricultural and food-related samples. In lieu of such considerations, this chapter aims to provide a brief overview of the instrumental aspects linked to ECD in food samples. It is important to mention that although many designs have been described, in some cases even avoiding the use of a formal potentiostat [45, 46], this chapter intents to highlight representative examples of the instruments and approaches recently presented. It is also important to note that due to minimal customization requirements, almost any electrochemical detector can be used for most of the samples pertinent to this book.

16.2 Principles

All ECD methods are based on the measurement of electronic data (current, potential, or resistance) produced in a single step from a group of electrodes. The signal (i.e., current) can be then amplified, filtered, and processed within the same domain, avoiding the use of optical components and transducers, simplifying the instrumental setup, and lowering the overall cost of the system. In the most typical setup, a potentiostat controls the voltage difference between the working electrode (WE) and a reference electrode (RE) and measures the current circulating between the working and the auxiliary electrode (AE). A simplified scheme of the main electronic components included in a potentiostat is shown in Figure 16.1, where E_{P1} and E_{P2} are two input signals (that can be combined before reaching the WE). The circuit on the left is used to control the potential applied to the

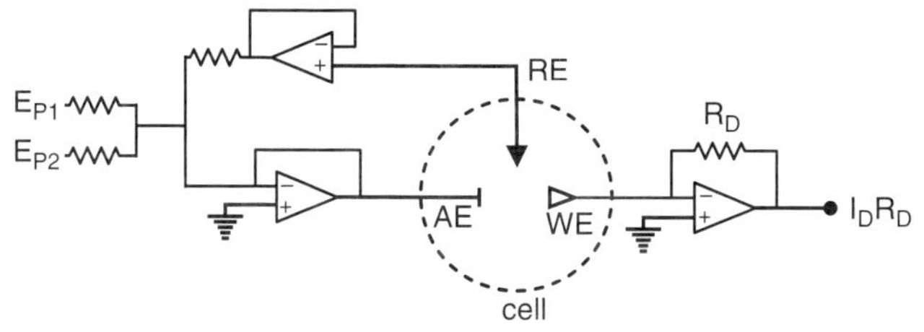

Figure 16.1 *Basic potentiostat circuit to control the potential applied to the working electrode. Adapted from Ref. [47] with permission from Oxford University Press*

cell (via the AE) and the circuit on the right is used to measure the current (via the WE and the variable resistor, R_D) circulating during the experiment. Although many modes of ECD have been developed, probably most food-related applications are based on amperometry, cyclic voltammetry, potentiometry, or conductimetry [48, 49]. As the instruments required to perform each of these techniques are slightly different, they will be discussed separately. Although pertinent details will not be discussed in this chapter, most modern electroanalytical systems are controlled by a personal computer and, as recently pointed by Gulaboski and Pereira [50], operate in a digital fashion, producing a series of potential steps (instead of a truly lineal change) and yielding a better discrimination of the Faradaic over the capacitive currents. It is also important to mention that in most reports, the electrodes are arranged (or even entirely integrated [51, 52]) in a cell, which is filled with an electrolyte. This solution should be selected in such a way that its electrical resistance does not limit the current circulating through the system. This phenomenon is known as *iR* drop and could result in significant errors during electrochemical measurements.

16.3 Instrumentation for Electrochemical Detection

16.3.1 Instruments for Voltammetric Techniques

Voltammetry comprises a series of electroanalytical techniques in which information about the analyte is collected by measuring the current as a function of the potential applied to the WE [53]. The specific name of the technique is typically associated with the waveform applied to the WE.

16.3.1.1 Amperometry

In the simplest case of voltammetry, a constant potential is applied to the detection electrode and the current is recorded as function of time. Being the most straightforward electrochemical technique that can be applied for detection of readily electroactive compounds, it requires simple instrumentation and has the advantages of being inexpensive and easily integrated with a variety of instruments including HPLC [54] and capillary electrophoresis [10, 55–57].

Several companies now offer standard bench-top instrumentation that can perform amperometry with background currents that yield to a LOD in the micromolar to nanomolar range. As expected, this range depends on the analyte, technique, and experimental conditions selected for the analysis and can be further improved by the implementation of filters (e.g., LOD of 100 amol for dopamine) [58]. Among the commercial instruments that have been used for food-related samples it is worth mentioning Metrohm (http://www.metrohm.com/) [34, 59], ECO Chemie (http://www.ecochemie.nl/, now part of Methrohm) [60], CH Instruments (http://www.chinstruments.com/) [61, 62], Bioanalytical Systems (http://www.basinc.com/) [63], Palm Instruments BV (http://www.palmsens.com/) [64], Pinnacle Tech. (http://www.pinnaclet.com/) [65, 66], Arvin Instruments (http://www.arbin.com/products/materials-research), ESA Biosciences (http://www.esainc.com/, now part of Thermo Scientific) [67], Dionex (http://www.dionex.com/, now also part of Thermo Scientific), Biodevice Technology Ltd (http://www.biodevicetech.com/) [68], DropSens (http://www.dropsens.com/) [69], Micrux Technologies (http://www.micruxfluidic.com/), Pine Instruments (http://www.pineinst.com/) [70], and Antec (http://www.myantec.com/). Although some of these instruments have been specifically designed to be coupled with separation-based systems, they can be customized and used for virtually all electrochemical applications. Metrohm, for example, has developed the *Autolab* potentiostats, which can be coupled with additional modules including electrochemical quartz crystal microbalance, impedance analysis, multiple cells, as well as pH [71] and temperature probes. At a lower cost, CH Instruments also offers a wide array of instruments with interfacing capabilities. Bioanalytical Systems Inc. (BASi) offers two electrical detector packages (LC-4C and liquid chromatography (LC) Epsilon electrochemical detectors) that can be either used alone or coupled with flow injection and separations systems. Dionex and Antec have developed electrochemical detectors with noticeable sensitivity to be coupled with high-performance anion-exchange chromatography (HPAE) or LC.

Amperometry is probably one of the most common electroanalytical techniques used in food analysis and there are numerous examples in the literature. Among others, it is worth mentioning analysis of cholesterol [72], vitamins [73, 74], carbohydrates [75, 76], antioxidants [4, 77–79], pesticides [80, 81], and toxins [82]. It is also important to point out that, although not discussed in this chapter, the same instrumental configuration used in amperometry can be used for the development of amperometric biosensors [83–86], electrochemical ELISA assays [87–89], and electrochemical tongues [90, 91].

16.3.1.2 Pulsed Amperometric Detection

Although amperometric detection is widely applied and can reach LODs in the low nanomolar levels, it is limited by the low number of naturally electrochemically active compounds and the loss of signal due to electrode fouling. This phenomenon, which is associated with the adsorption of either matrix components or analyte byproducts to the electrode surface, has been thoroughly described for the ECD of phenolic compounds [92–95]. To overcome these problems, a potential pulse can be applied to the detection electrode (a noble metal such as Au or Pt) to remove adsorbed materials. There are two main forms of pulsed ECD: pulsed amperometric detection (PAD) and integrated pulsed amperometric detection (iPAD). In PAD, a sequence of a high positive potential (to clean the electrode surface), a negative potential (to reactivate the electrode surface) and a third potential (to detect the target analyte) is used. In iPAD, a similar approach is taken, but the detection step is replaced by a triangular waveform. Although slight differences in the electrochemical mechanism [96] impact the selectivity and sensitivity of each technique, both approaches (often used in conjunction with HPLC or CE) can provide significant advantages when the analyte is not naturally electrochemically active or when rapid electrode fouling occurs (e.g., with carbohydrates, amines, and alcohols) [97–100]. Several metallic electrodes (Au, Pt, Ag, C) can be used with this technique leading to LOD in the pico- or fentomol range. Examples of the use of this detector include the analysis of iodine [101], carbohydrates [102], and peptides [103]. Because PAD requires the hardware to process information at a higher rate than amperometry (typical pulse times 1–500 ms), this option is not always available in commercial instruments and has been included in only a handful of custom systems. In the former category, Dionex offers a series of detectors that can be coupled to high-affinity anion exchange chromatography for the analysis of underivatized analytes (e.g., sugars, alcohols, and amines). Recently, Coutinho *et al.* [104] developed a system that utilizes a PCI-6014B multifunction data acquisition board, a SH6850 shielded cable, a SC-2050 cable adapter, and a CB-50LP terminal block (all from National Instruments, Austin, TX). This instrument is capable of performing amperometry, PAD, and coulometry. The PCI-6014B multifunction board features 16 single-ended (or 8 differential) gain programmable analog inputs multiplexed to a 16-bit analog-to-digital converter (ADC), two 16-bit digital-to-analog converters (DACs), 8 digital I/O lines, two 24-bit counters/timers, digital trigger, and one direct access memory (DMA) channel shared by all resources on the device. The board is capable of acquiring an analog signal in a frequency up to 200 kHz and generating an analog signal up to 10 kHz. The analog input can be configured to four input ranges: ±50 mV, ±500 mV, ±5 V, and ±10 V. This board was interfaced to a 1.4 GHz Pentium 4 personal computer running LabVIEW 7.1 (National Instruments; Austin, TX) under Windows XP.

16.3.1.3 Voltammetry

As it can provide some of the most basic electrochemical information related to the reactivity of the selected analyte (peak potential and peak current) most instruments that perform amperometry can also perform some of the most basic voltammetric techniques. These techniques determine the current as a function of the potential applied to the WE (in a conventional three-electrode cell) and can be performed with relatively simple instrumentation [105, 106]. As different signals can be combined in the input ports of the instrument, multiple variations of the technique have been developed including cyclic voltammetry, linear sweep voltammetry, linear sweep stripping voltammetry, stripping voltammetry [107, 108], fast-scan cyclic voltammetry [109], square-wave voltammetry [110],and sinusoidal voltammetry [111].

Applications of voltammetric techniques to agriculture and food science abound in the literature and most developments have been geared toward surface modifications of the WE to improve its reactivity or selectivity. For example, Gholivand *et al.* [112] designed a molecularly imprinted polymer to improve the recognition ability of the WE and reached a dynamic linear range and LOD of 5.0–1000 and 3.2 nM of cyanazine, respectively. In combination with cathodic stripping, the same group obtained similar performance for propazine in food samples [113]. The technique has been also used for the detection of metallic cations [114–117], vainillin [118], pesticides [119, 120], antioxidants [121], and dyes [122, 123].

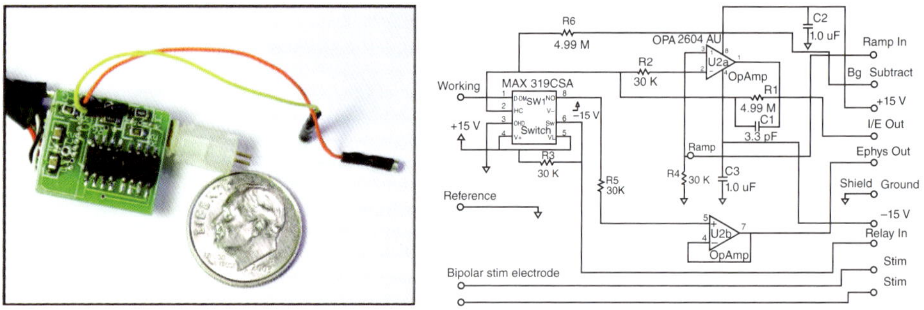

Figure 16.2 *Photography and associated circuits of the miniaturized headstage used for combined Echem/Ephys freely moving experiments adjacent to a U.S. dime. The orange and yellow connecting wires are for reference and working electrodes. Adapted from Ref. [128] with permission from American Institute of Physics*

In terms of instrumentation, many research groups have developed custom instruments for voltammetry during the last two decades [124–126]. More recently, a universal electrochemistry instrument (UEI) for fast-scan voltammetry (two-electrode setup [127]) was developed by Mark Wightman's group [128]. In this system, current flows from the WE to the RE (performing the duties of an AE). The authors also point out that as the EI-400 (ESA Biosciences, no longer offered) most two-electrode instruments use operational amplifiers as current-to-voltage converters with the noninverting input at ground potential and the inverting input connected to the WE (held at ground potential). In contrast, the UEI consists of an operational amplifier in I–E mode with the ramp signal applied to the noninverting input while the WE is connected to the inverting input. In this case, the voltage at the electrode follows the voltage applied to the noninverting input of the amplifier. At the same time, the current that flows between the WE and RE (as a result of the applied voltage) is summed at the inverting input and adds to the voltage seen at the output of the amplifier (Figure 16.2).

16.3.2 Instrumentation for Potentiometric Techniques

Because it is routinely used to measure pH, potentiometry is probably one of the best-known and most frequently used electrochemical methods. In potentiometry, the voltage difference between two electrodes is measured at near-zero current using a voltmeter. As a consequence of the simplicity of this instrumental setup, most of the research in this area has been directed to the improvement of the indicator electrodes, most often of the ion-selective kind. In general, the instruments required to perform potentiometry range from simple high-impedance noninverting amplifiers [124, 129–131] to research-grade pH/mV meters [132]. The latter allow reaching LOD in the micromolar to nanomolar range for routine applications, either in direct measurements [133, 134] or when the electrodes are integrated in a flowing system [135]. As many other instruments, the precision and range offered by these systems are typically inversely proportional to their price and span from a few hundred dollars to several thousand. As commercial sources of these instruments abound, they will not be discussed in this chapter. Circuits to read ion-selective electrodes can also be assembled using standard Arduino boards (less than \$50). As an

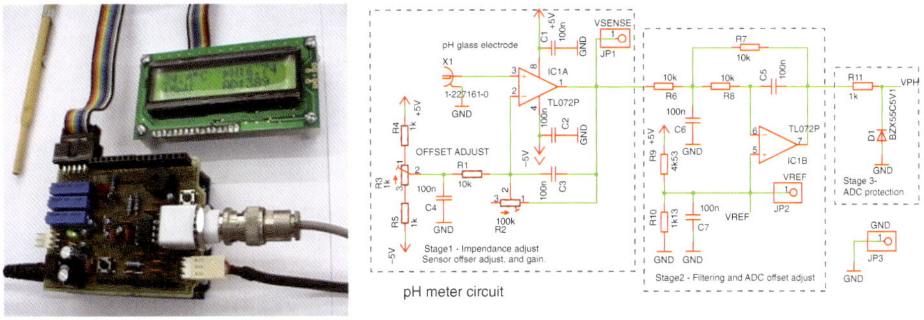

Figure 16.3 *Picture of the complete* phduino *and schematics of the associated electronic circuit. The project's page includes a full list of supplies and calibration instructions. Images courtesy of Carlos A. Neves*

example, Figure 16.3a shows a picture of an open-source system presented as part of the *phduino project* (http://code.google.com/p/phduino/) by Dr. Carlos A. Neves at the University of Sao Paulo, Brazil. Due to the modularity offered by the main board, this instrument can be interfaced, programmed, and expanded beyond the pH and temperature probes already included. In addition, the operational amplifier (op-amp) selected for this application (OPA129, see stage 1 in Figure 16.3b) leads to low noise and a resolution of 0.02 units in pH.

In combination with stripping analysis, the LOD found by potentiometry could be further improved to concentrations ranging between parts per million [136, 137] and parts per billion [138] (depending on the analyte and food matrix). The basic experimental setup needed for a potentiometric stripping experiment simply comprises a potentiostatic circuitry, a three-electrode electrochemical cell, and an impedance recorder [139]. Although the analytical procedure involved in potentiometric stripping analyses can be manually performed using a standard potentiostat, specific instruments (PSA ION 3, http://www.steroglass.it/ or TraceLab, http://www.radiometer-analytical.com/) are currently available.

16.3.3 Instruments for Conductometric Techniques

Conductimetric detection is a universal detection mode that can be suited for food analysis when used in combination with a separation technique such as ion chromatography or CE. The detection can be performed with the electrodes in direct (contact) or indirect (contactless) contact with the solution present in the cell. In direct conductivity, a small amplitude alternate signal is applied to the excitation electrode aiming to minimize the electrolysis effects on the conductivity measurements. Since the capacitance of the electrical double layer is relatively high, low-frequency signal is used in this mode [140]. Due to practical difficulties and limited stability of the electrodes, the contact mode is generally not used and has been replaced by contactless mode. Capacitively coupled contactless conductivity detection (C^4D) is a conductimetric detection mode where the electrodes are positioned outside the cell [141]. In 1998, Zemann *et al.* [142] and da Silva and do Lago [141] implemented an axial arrangement where the cell is formed by two tubular electrodes with a gap (typically between 0.5 and 2 mm). Figure 16.4a shows the diagram of the electronic circuit initially proposed by do Lago. In this case, a function generator FG-2002C (Goldstar,

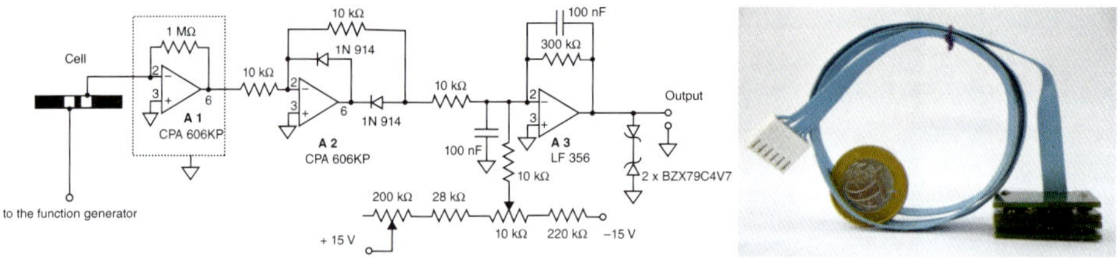

(a) (b)

Figure 16.4 *(a) Electronic circuit of a C⁴D detector. The current-to-voltage converter A1 is kept in a grounded metallic case placed as near as possible to the cell. A2 and related components compose a precision rectifier. The adjustable resistors compensate the baseline before a 30× gain is applied by A3 to the dc signal. The back-to-back Zener diodes protect the A/D converter against overvoltage. (b) Photograph of the* openC^{4}D Lite *the detector next to a coin. Reproduced from permission from Ref. [141]. Copyright © 1998, American Chemical Society*

Seoul, Korea) was used to generate a 600-kHz 20-V_{pp} sinusoidal signal, which was applied to one of the electrodes. The receiver electrode was connected to the input of a current-to-voltage converter. The current-dependent voltage was rectified and amplified in the last stage (A3), which also compensates the baseline. As later described in greater detail [143, 144], the best frequency of operation can be optimized considering the dimensions of the capillary and electrodes, the gap between the electrodes, and the quality of the components used. The use of a contactless mode offers noticeable advantages for electrophoresis systems, where the application of an electric field is required to promote the separation. Some of the difficulties avoided by C^{4}D include electrode fouling, bubble formation (electrolysis), and electrical interference [57, 140, 145–149]. In the last 10 years, the use of C^{4}D on microchip-CE has gained much popularity due to the easiness of implementation and compatibility with microfabrication technologies, which enables easy integration of separation channels and detection electrodes. In this contactless mode on both capillary- and chip-based systems, the capacitance of the wall of the capillary/channel is very small and, for this reason, a high-frequency signal must be applied to the electrodes. The current recorded by the receiver electrode reflects the conductivity changes inside the cell which is directly proportional to the concentration of ionic species in the gap. Even with the electrodes positioned outside the capillary, conductivity measurements can be taken with LODs in the micromolar level for inorganic ions [140, 145].

A few years ago, Francisco and do Lago [150] also introduced a compact and high-resolution version of a C^{4}D system. This version does not comprise moving parts (potentiometers and connectors) or a bench-top waveform generator. The detector has about 6.5 cm^3 and a 1.1-MHz oscillator, optimized to minimize noise levels when 20–100 μm ID capillaries are used. In addition, high resolution was achieved by using a 21-bit ADC. Based on the design shown in Figure 16.4a, the same group also developed an open-source detector (*openC^{4}D Lite*, https://sites.google.com/site/openc4d/home). The resulting device (shown in Figure 16.4b) can be integrated to an *MCU Lite* (also described), which has four ports to install openC^{4}D Lite detectors, two channels to control thermal markers, one external trigger, and enable communication to a PC (open-source software is also provided) using a universal asynchronous receiver/transmitter (UART) to USB cable developed by

FTDI. The authors remarked that the instrumental advances and the associated open-source project recently reported can spread out the use of this universal detector for food analysis. Furthermore, the inherent portability of the C^4D system could make it an ideal instrument for on-site food testing. One example that combines creativity and the instrumental simplicity of this home-made conductivity detection is the analysis of monoalkyl carbonates in carbonated alcoholic beverages (sparkling wine, beer, and mixed drinks) [151]. By using two detectors, the authors not only demonstrated the presence of monoethyl carbonate (MEC) in the selected drinks but also proposed that the low pH values were responsible for the larger concentrations of MEC observed in lager beer and mixtures of rum and cola. In addition, the possibility to control not only a C^4D detector but also a series of valves using open-source software (Arduino) has been recently demonstrated [152].

An array of commercial instruments is also available to perform conductivity detection. Among others, it is worth mentioning the ICS-5000+ (Dionex), model 432 (Waters), model 650 (Chomtech), and a series of detectors and modules from eDAQ (http://www.edaq.com). Most of these systems feature small cell volumes (<0.1 μl) and large dynamic ranges (up to 15 000 μS/cm).

Examples of the use of conductivity detectors (coupled to capillary electrophoresis or ion chromatography) for the analysis of food samples include the determination of phytic acid [153], perchlorate [154], carnitine [155], biogenic amines [156, 157], caffeine [158], and carbohydrates [159]. Conductivity detectors can also be used in combination with microchips. However, most pertinent reports using microchip-C^4D are linked to the analysis of beverages [145]. Although the main disadvantage of C^4D is its poor sensitivity (LODs in the high micromolar range for most ions), it can be seamlessly coupled to a preconcentration step. As an example of such approach, Mark and coworkers [160] combined C^4D with single-drop microextraction and determined volatile aliphatic amines linked to the degradation of seafood samples. The LODs reported by this method were below 400 ng/ml.

16.3.4 Instruments Developed for Portability

In addition to the bench-top instruments, different companies have also commercially available portable/handheld and battery- or USB-powered potentiostats, which are lightweight, relatively inexpensive, and used-friendly. These portable systems are particularly useful for on-site assessments of food quality and safety. Metrohm, CH Instruments, BASi, DropSens, PalmSens, Pine Instruments, MicruX Technologies, microfluidic ChipShop, and Pinnacle are some of the outstanding companies with electrochemical instrumentation dedicated for on-site applications. As examples, the instruments from the last three companies will be described and discussed independently due to their specificity or integration with other instrumental components. The model 910 PSTAT mini potentiostat, from Metrohm, with dimensions of $80 \times 54 \times 23\ mm^3$ ($l \times w \times h$) and 94 g provides working potentials of ±2.048 V and current ranges from 200 μA down to 10 nA. CH Instruments offer several instrument models to meet different requirements. The model 1200B is a handheld system, which allows ±7.5 V compliance voltage ensuring a working potential range of ±2.4 V. These models are able to measure currents down to 100 pA. The company BASi markets three versions of portable potentiostats (called of Petit Ampère) with dimensions of $180 \times 100 \times 65\ mm^3$ ($l \times w \times h$) and 0.9 kg that are designed for basic ECD in biosensor applications. The potential and current ranges provided by the Petit Ampère instruments are ±1.5 V and 200 pA to 200 μA, respectively. DropSens and PalmSens also offer instruments with similar capabilities including potential (±4.0 V) and current (1 nA–10 mA or 1 nA–100 mA) ranges. The model μSTAT 400 is the more recent portable potentiostat from DropSens. It is Li-ion battery-powered, with dimensions of 125 × 95 × 40 mm (*l* × *w* × *h*) and 150 g. The data can be transmitted to a computer through USB, RS232, or Bluetooth connections. The EmStat instrument series are the ultimate version supplied by PalmSens and one of the smallest electrochemical interfaces available on the market weighing only 85 g. It is powered through USB connection and housed into an anodized aluminum box with dimensions of 67 × 50 × 28 mm (*l* × *w* × *h*). Pine Instruments supplies both bench-top and portable potentiostats. The latter series comprises currently three models (WaveNow, WaveNano, and WaveNowXV) of instruments housed in compact boxes (30 mm thick and 175 mm long) weighing less than 200 g each. The WaveNano potentiostat is a low current version from Pine Research Developments capable of measuring currents from 1 mA down to 100 pA and applying potential of ±4.0 V. On the other hand, the potentiostat model WaveNowXV is an extended voltage version which allows the application of ±10 V. These instruments may be optionally battery-powered enabling their use for on-site applications. Likewise Pine Instruments, BioDevice Technology Ltd. (BDT) has also developed and commercialized a compact potentiostat called BDT miniSTAT100. It provides applied voltage and current ranges of ±4.0 V and 1 nA to 100 μA, respectively.

In addition to commercial instruments, considerable attention has been dedicated to the development of custom compact systems for chemical analysis based on ECD. Because these systems address specific shortcomings of commercial instruments (e.g., portability, connectivity, cost, etc.), remarkable advances have been reported in literature during the last years [161]. In order to provide a portable device, an electrochemical sensor based on an SOC-based chip was developed [162]. The potentiostat was constructed using a C8051F020 mixed-signal microprocessor, a current-to-voltage converter, three inverter amplifiers, two low-pass filters, and an RS232 serial data transfer interface. The C8051F020 chip is an 8051-based CPU with 12-bit DACs, 12-bit ADCs, and digital peripherals. A personal computer is used to control the experimental process and collect the experimental data of potentiostat. The proposed potentiostat design specifications are described as follows: the programmable counter voltage ranges between −3 and 0 V under the resolution 1 mV, the measured reference voltage ranges between 0 and 1 V under the resolution 1 mV, and the measured current ranges between 0 and 100 μA under the resolution 0.1 μA. Jiang *et al.* developed a miniaturized electrochemical detector using a USB to UART BRIDGE CP2101 a 8-bit microcontroller, an operational amplifier, and a DC–DC converter [163]. To improve the signal-to-noise ratio (<1 pA), low-pass filters were placed at both the input and output ends in the OPA129 circuit and the OPA129 input pad was designed as short as possible and shielded with ground pads. Although the potentiostat has moderate accuracy, it features small size, low cost, and high portability. Another design of a potentiostat based on a CMOS sensing circuit was presented and used for the detection of dopamine at the micromolar level [164]. The microelectrode array and the CMOS die are shown in Figure 16.5a and a detail of the circuit is shown in Figure 16.5b, and c.

The anode (generator) electrode is connected to the circuit and its bias is provided through the virtual ground of the operational amplifier shown in Figure 16.5b. The gate of the M1 transistor provides isolation between the op-amp and the sensing node. The potential at the sensing node is regulated by the use of a large external capacitor C1, which is typically

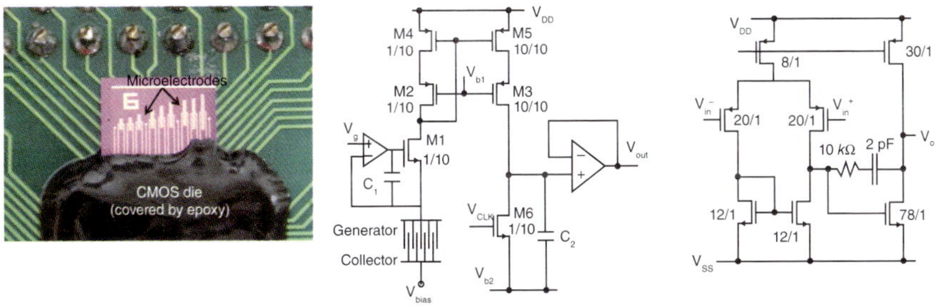

(a) (b) (c)

Figure 16.5 (a) Micrograph of the microelectrode die and the CMOS die on a small printed circuit board. (b) Schematic of the CMOS circuit for dopamine detection. (c) Schematic of the op-amp design as shown in (b). The labeled transistor width and length are in micrometers. Reproduced with permission from [164]. © IOP Publishing. Reproduced with permission. All rights reserved

in the order of nanofaraday. The amplifier is a conventional two-stage op-amp design with p-type input transistors. It has an open-loop gain of 67 dB, a unity-gain frequency of 32 MHz and a phase margin of 70°. The final output voltage is produced through a buffer circuit, also made of the same operational amplifier. The oxidation current produced at the anode is amplified by the wide-swing cascade current mirror (M2–M5) with a current gain of 10. The amplified current is then converted to a voltage by the use of the integrating capacitor C2 with a value of 4 pF. An external clock signal V_{CLK} constantly resets the output and controls the integration time through the M6 transistor. The buffer input is reset to the potential provided by the voltage V_{b2}. Simulation results show that the circuit can provide a constant current gain in the 0.2 nA–0.4 μA range. By removing the input current source of the current mirror in simulation, an input-referred leakage current of 6.4 pA produces a 0.8 V output in a 50 ms span. A similar approach was then used by Kim *et al.* [165] to develop a CMOS-based electrochemical biosensor array with high throughput capability (100 electrodes). Although the potentiostat was used to measure neurotransmitters, they demonstrated the operation of an electrode array with integrated high-gain potentiostats and output time-division multiplexing with minimum dead time for readout.

One of the difficulties in the design of potentiostats is the accuracy of the current measurements. In most traditional systems, the current-to-voltage conversion is performed either across a large resistance or a capacitor, limiting the resolution to the nanoampere range. To overcome this difficulty, a potentiostat for glucose measurement in an industrial setting has been developed [166]. The instrument uses standard industrial communication protocols, profibus, and a 4–20 mA current loop, for remote control and monitoring. Experimental results showed 0.01% accuracy with 1 mV resolution for voltage applications and 0.01% accuracy with 1 pA resolution for current measurements. More recently, the architecture and design of a compact, multichannel, hybrid-multiplexed potentiostat for performing electrochemical measurements on continuously biased electrode arrays was also presented [167]. The proposed architecture featured a combination of sequential and parallel measurements, to enable high performance while keeping the system low-cost and compact. The accuracy of the signal readout is maintained by following a special multiplexing approach, which ensures the continuous biasing of all the WEs of the array. After sampling the results, a digital calibration technique factors out errors from component inaccuracies. A prototype printed circuit board (PCB) was designed and built using off-the-shelf components for the real-time measurement of the amperometric signal of 48 electrodes, exhibiting high linearity (R^2>0.999) and a resolution of 400 pA.

Taking advantage of the increased availability and computing capabilities of mobile phones, several groups have integrated them with analytical devices. Recently, Lillehoj presented a system controlled by a cell phone (Nexus One running Android 2.3.4 operating system) and capable of performing electropolymerization, electric-field pulsed mixing, and amperometric detection (see Figure 16.6) [168]. A USB interface enabled the communication between the phone and a Teensy microcontroller that controls the potentiostat through a UART serial communication protocol.

Very recently, the capacity of cellular phones to play music was applied to achieve potentiostatic control (Figure 16.7). In this report [169], Delaney *et al.* demonstrated that by making an electrical connection to the sensor via the phone's audio socket and playing an appropriate sound file, a WE can be suitably polarized to initiate the electrochemical reaction leading to electrochemiluminescence emission. They showed how both the electrochemical excitation and photonic detection processes can be software-controlled using

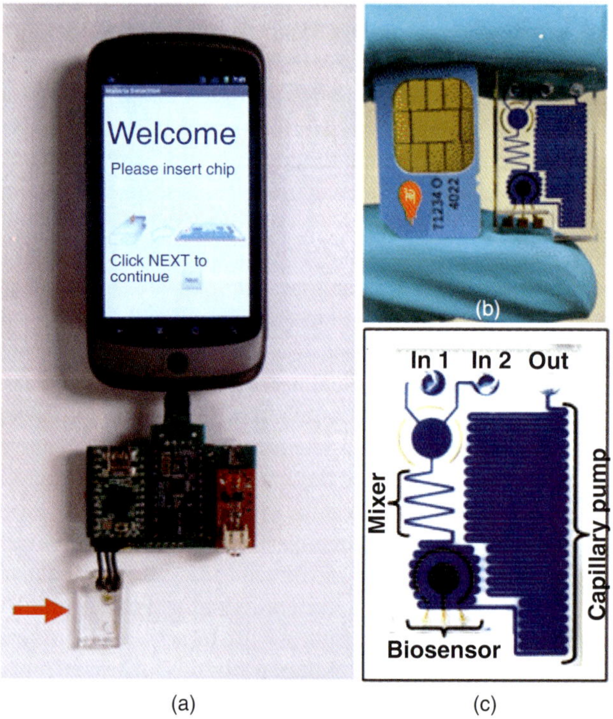

Figure 16.6 *Photograph of the assembled prototype including a microfluidic chip and a mobile phone SIM card for comparison. Photograph of the assembled prototype (a). Arrow indicates the microfluidic device. For comparison, a mobile phone SIM card (b) along the microfluidic chip (c). Reproduced with permission from Ref. [168]. © The Royal Society of Chemistry*

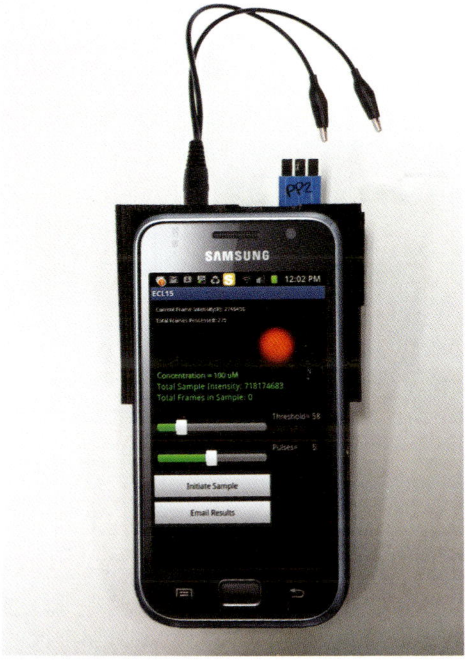

Figure 16.7 *Arrangement used for mobile-phone-based ECL sensing. The audio jack supplies the potential to the paper microfluidic sensor, while the resultant emission is detected by the camera in video mode. Reproduced from Ref. [169] with permission from Elsevier*

a custom-written mobile phone application and demonstrated the possibility of sensing 2-(dibutylamino)ethanol with an LOD of 100 μM. G. Whitesides' group has also recently presented a cell phone-based electrochemical detector capable of performing a wide variety of important electroanalytical techniques (chronoamperometry, cyclic voltammetry, differential pulse voltammetry, square wave voltammetry, and potentiometry) [170].

Based on the customized examples presented above, it is possible to envision the development and acceptance of complete systems equipped with other instrumental components and wireless control and data transmission. This arrangement would be useful for on-site food and agriculture analysis, where the monitoring of preservatives, hazardous, or environmental compounds may be required for routine or quality control.

16.3.5 Low-Cost Potentiostats

As mentioned by many authors, an important factor to consider when selecting an analytical methodology is the overall cost (e.g., instrumentation, supplies, etc.). Striking a balance between instrumentation cost and functionality, a series of inexpensive potentiostats have been recently presented. As an example of a system that can be used by students in advanced analytical and integrated laboratory courses, an inexpensive potentiostat (less than $50) was developed using a 3 cm × 13 cm PCB [171]. For this potentiostat (shown in Figure 16.8), a PIC18F452 microcontroller was selected because it includes a

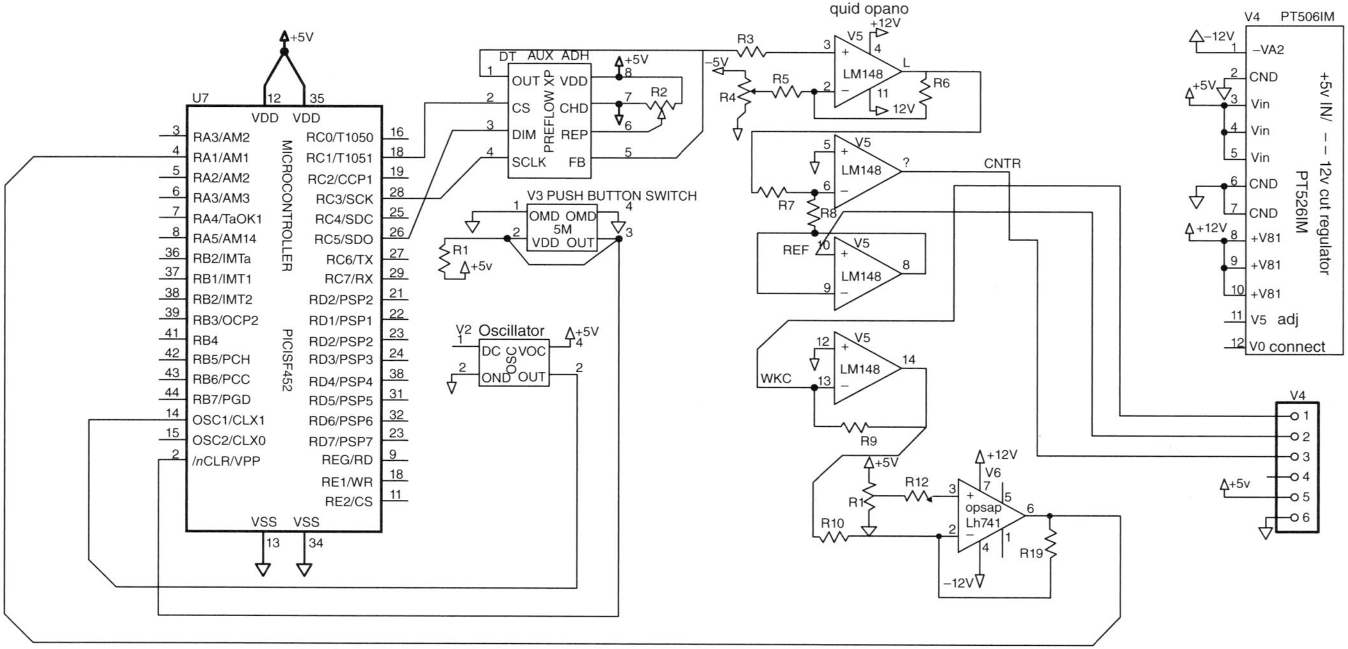

Figure 16.8 *Schematic of potentiostat system designed using PIC18F452. Reproduced with permission The Chemical Educator, http://chemeducator.org [171]*

built-in ADC, a serial peripheral interface (SPI), and a timer-capture module. The ADC was used for measuring the electrochemical current equivalent voltage. The SPI was used for communication between the microcontroller and DAC and the timer-capture module is used for updating the DAC and monitoring the output of the ADC at precise time intervals. The output voltage swing of the serial DAC is variable and depends on the voltage on the reference pin (V_{REF} pin of the MAX5354). A potentiometer was connected to the V_{REF} pin and the output voltage swing of the DAC was controlled by changing the potentiometer settings. The output of the serial D/A is unipolar; to change the output voltage to bipolar an op-amp level shifter circuit is used. A buffer is inserted between the CE and the level-shifted voltage; the buffer prevents loading of earlier op-amp stages from the current drawn by the electrochemical cell. The RE is connected to the noninverting terminal of a voltage-follower circuit. The voltage-follower circuit helps in maintaining a constant potential at the RE without drawing much current. The WE is connected to the inverting terminal of a current-to-voltage converter; this, in effect, keeps the WE tied to ground and also produces a voltage equivalent of the electrochemical current at the output of the voltage converter. The electrochemical current, now converted to a voltage, is level-shifted and attenuated/amplified to bring the voltage into the 0–5 V range acceptable by the microcontroller and fed to the analog input port of the microcontroller.

At a cost, according to the authors, lower than $35, a portable potentiostat capable of performing cyclic voltammetry was developed by Huang *et al.* [172]. This three-electrode system was combined with an operational amplifier and a negative feedback loop to accomplish the potentiostatic control. As the follow-up of previous work [162], the system includes a mixed-signal microprocessor (C8051F020), an instrumental amplifier for current-to-voltage conversion, three voltage level shifters, an operational amplifier, and an RS232 serial data transfer interface. The microprocessor contains an 8051-based CPU, 12-bit DAC, programmable gain amplifier (PGA), 12-bit 8-channel ADC, and digital peripherals. Since the C8051F020 chip can only process positive potential, the level shifters are indispensable for processing the scan voltages from negative potential to positive potential in cyclic voltammetry. Another example of a potentiostat developed under open-source licensing is the CheapStat [173]. According to the authors, this handheld potentiostat,which can be constructed for less than $80 by anyone who is proficient at assembling circuits, supports a number of potential waveforms necessary to perform cyclic, square-wave, linear sweep, and anodic stripping voltammetry. As demonstrated, it is suitable for a wide range of applications ranging from food- and drug-quality testing to environmental monitoring, rapid DNA detection, and educational exercises. A three-channel microcontroller-unit (MCU)-based potentiostat capable of performing chronoamperometry was developed by Friedman *et al.* [174] This potentiostat (shown in Figure 16.9) was accurate within $0.95 \pm 0.6\%$ (95% confidence limit) of a commercial potentiostat (VSP, Biologic; http://www.bio-logic.info/) when used to measure microbial respiration that occurs through exocellular electron transfer. According to the report, the system can be manufactured at relatively low cost ($600) and integrated with a solar-panel-based power system for remote analysis.

Although not necessarily involving a formal potentiostat, a broadly applicable and inexpensive, on-board power source was recently developed by Liu and Crooks [46]. As a very clever alternative, a metal/air battery that drives electrolytic reactions was integrated with a paper-based sensor, defined by wax printing. Electrical contact was made to the

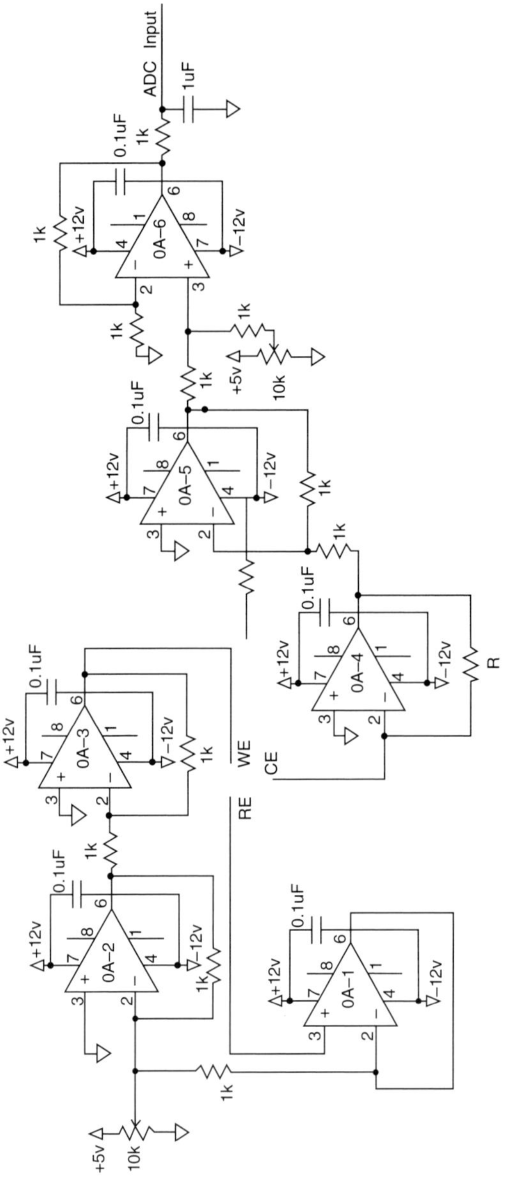

Figure 16.9 *Op amps (OA) that provide the core potentiostat circuitry. The working-electrode (WE) potential is poised with respect to the reference electrode (RE). The current necessary to maintain this potential is measured at the counter electrode (CE), and processed through a series of op amps before being read via analog-to-digital conversion. OA-3 inverts the applied potential to allow the application of a negative potential, and can be bypassed when a positive potential is required. A current-to-voltage conversion is performed at OA-4 through a precision resistor that sets the measurement range of the channel. The feedback resistor (R) for current-to-voltage conversion can be changed through a series of jumpers. Reproduced from Ref. [174] with permission from Elsevier*

paper via a pair of transparent indium-doped tin oxide (ITO) electrodes. The chip, which was preloaded with glucose oxidase, $Fe(CN)_6^{3-}$, and Prussian blue was used to measure glucose at concentrations above 0.10 mM.

16.3.6 Remotely Controlled Instruments

Aiming to provide connectivity, Desmond *et al.* developed a wireless electrochemical detector composed of a Xilinx field programmable gate array, an analog voltammetry subsystem (containing converters and operational amplifiers) and a voltammetry sensor subsystem [175]. The digital-to-analog (DA) converter forces arbitrary voltage waveforms on to the voltammetry electrode system while the analog-to-digital (AD) converter measures the resulting current flow in the sensor, which is stored in a portable computer. A custom potentiostat (including readout circuitry, ADC, and microcontroller) and a radiofrequency (RF) module for sending detected signals to a base station through wireless communication were also presented by Yun *et al.* [176].

Due to their increased availability and computing capabilities, different groups have presented potentiostats that can be controlled with a handheld device. These integrated devices have the potential to be combined with cloud computing services and provide information on scales ranging from microscopic to global. Although a number of these sensors have been integrated with agriculture and food industries [177], the needs for real-time environmental monitoring, precision agriculture, cold chain control, and traceability are largely unmet. Along these lines, our group developed a series of instruments that integrate a microchip-CE device into a mobile platform to perform remote wireless analysis of volatile compounds [178, 179].

Although the setup of each system is very different, both instruments (shown in Figure 16.10) are capable of collecting a gas sample, perform the subsequent analysis (injection, separation, and ECD), and send the data to a distant control unit. In the first case [178], a potentiostat, an analog signal conditioning circuitry, an analog/digital interface (A/D), and digital hardware were arranged. The potentiostat incorporates three operational amplifiers (TLC2202, Texas Instrument, Dallas, TX) as its main components and a 1 MΩ feedback resistor which allowed the offset to be adjusted in the signal conditioning

Figure 16.10 *Pictures of two prototypes of robotic platforms to perform remotely controlled analyses using electrochemical detection. Details about these systems can be found in Refs. [178, 179]. Reproduced with permission from [178]. Copyright © 2008 WILEY-VCH Verlag GmbH & Co. KGaA. Reproduced with permission from [179. Copyright © 2012 WILEY-VCH Verlag GmbH & Co. KGaA, Weinheim*

circuitry, rather than through the potentiostat, and did not significantly affect the sensitivity of the system (noise level lower than 1 nA). DACs (DAC121S101, National Semiconductor; Santa Clara, CA) as well as ADCs(ADCS7476, National Semiconductor) were used to design the interface between the analog and the digital circuitry. Both converters have an analog range of 0–3.3 V and are digitally controlled by digital hardware via a field-programmable gate array. The voltage from the DAC used to control the auxiliary voltage of the potentiostat was shifted by −1.65 V and magnified by a factor of 1.515 using analog signal conditioning. This arrangement allowed an auxiliary voltage range from −2.5 to 2.5 V, controlled with a Moore type state machine implemented in the digital hardware. The electrochemical detector is capable of performing amperometry and PAD. The second prototype was built on an all-terrain vehicle (four stroke engines) and included a series of subsystems that include gas sensors, drivers, communication, computers, and a CE module. This instrument allowed air sampling using a polypropylene porous tube, coupled to a flow system that transports the sample to the inlet of a fused silica capillary. The proposed system was used to analyze a mixture of formic, acetic, and propionic acids [179]. Both prototypes have the potential to be applied in the field of agriculture (e.g., to remotely assess the decomposition of organic matter).

Recently, Gunasekara *et al.* [180] reported some instrumental advances based on the use of a wireless-isolated potentiostat from Pinnacle Technology (Lawrence, KS, USA). The authors described that the in-channel detection mode using this insolated potentiostat can be performed with a peak-to-peak noise level slightly lower than that for the end-channel configuration. Furthermore, the wireless-insolated potentiostat has provided better sensitivity, higher efficiency, and better LOD values.

Aiming to address the demand for wireless sensors, the first ultra-low-power radio-frequency identification (RFID) wireless sensor tag with potentiometric input for use with pH and ion-selective electrodes was developed [181]. The proposed tag is able to autonomously measure and store electrode potential in its internal memory, and then transfer logged data wirelessly by RFID to a nearby reader or by near-field communication (NFC) to a smart phone.

16.3.7 Electrochemical Detectors Coupled to Microchip Capillary Electrophoresis

When analyzing complex samples, like the ones in food industry, a separation method is needed to identify and quantify the components of the sample. Among all the commonly used separation methods, CE and microchip-CE display integration compatibility with ECD systems. In addition, microchip-CE offers short analysis time, high separation efficiency, low cost, and low consumption of sample and reagents, and miniaturization. These techniques are particularly well suited for portable instruments and field applications. The combinations of microfluidics, CE, and ECD methods have led to the development of successful portable systems with applications in environmental analysis, food quality control, and medical testing [55, 56, 182–184].

One of the first portable instruments for CE coupled to ECD was developed in 1998 by Kappes and Hauser [185]. The instrument included a 30-W HVPS module capable of providing 30 kV DC, a signal amplifier and an ADC. Potentiometric detection was carried out using a coated-wire-ion-selective electrode. The system was powered by two 12 V lead-acid batteries and was contained in a PVC case 340 × 175 × 175 mm^3 ($l \times w \times h$) size with a total weight of 7.5 kg. Amperometric [38] and conductivity detection [186] were later added to the instrument. In these systems, the injection was limited to electrokinetic mode as no vacuum or pressure system for hydrodynamic injection was implemented. Recently, the same group reported portable instrument featuring an automated valve-based injection system [187]. The system relies on pressurized air for solution delivery and a micromembrane pump for sample aspiration. C^4D system was employed due to its versatility and low power requirement. The instrument has a compact design, with all components arranged in a briefcase with dimensions of $450 \times 350 \times 150\,mm^3$ ($w \times d \times h$) and a weight of about 8 kg. It can operate continuously for 9 h in the battery-powered mode. A picture of the system is shown in Figure 16.11a. The authors demonstrated simultaneous separation of 11 slowly migrating organic anions, namely oxalate, malonate, citrate, succinate, phthalate, acetate, lactate, benzoate, vanillate, ascorbate, and gluconate. These compounds are found in various beverages either as major constituents or as additives. Figure 16.11b shows the analysis of cola beverage, orange juice, and lactate-containing soft drinks. The system showed good performance with high reproducibility.

Scott Martin *et al.* [188] reported in 2002 the development of a battery-powered electrically insulated potentiostat which allowed positioning the WE directly inside the separation channel of a microfluidic device, without the use of a decoupler. The resulting system displayed analytical performance that was comparable to fluorescence detection while ensuring reproducibility, sensitivity, and high separation efficiency. In 2003, Jackson *et al.* [189] described a compact instrument based on the combination of electrophoresis-based system and amperometric detection. In their work, the authors reported the development of a portable battery-powered dual source HVPS and a battery-powered potentiostat for potential control and current monitoring for amperometric detection on chip-based systems. The circuit was designed using the small and lightweight EMCO

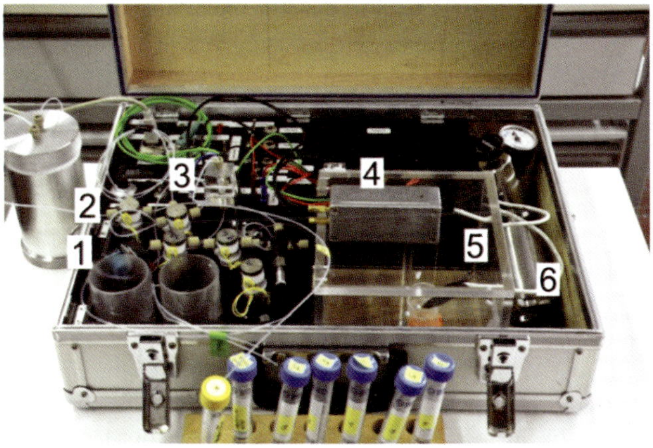

Figure 16.11 *(A) Photograph of the instrument showing the membrane pump (1), valves (2), splitter (3), detector (4), safety cage for application of highvoltage (5), and pressurized air (6). (B) High efficiency separations of 11 organic compounds found in beverages and carbonate (a), lactate-containing soft drinks (b), cola (c), and orange juice (d). For peak identification and experimental conditions. Reproduced with permission from Ref. [187]. Copyright © 2013, American Chemical Society*

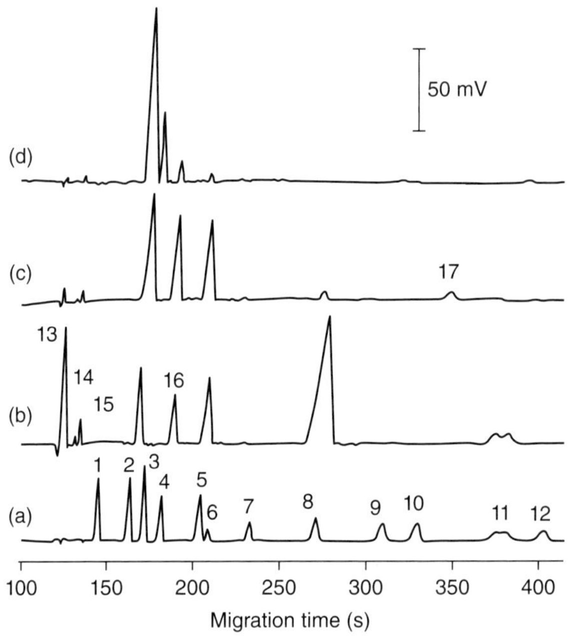

Figure 16.11 *(continued)*

(http://www.emcohighvoltage.com/) Q12-5 and Q12-5N modules, which are rated at ±1.2 kV, 400 μA.

Commercial sources have also proposed solutions providing portability. For example, MicruX Technologies, Spain, has developed two generations of portable instruments for microchip electrophoresis with amperometric detection. The first generation included a HVPS for electrokinetic control (floating injection and separation steps), a bipotentiostat for amperometric detection, and a communication system for control and data acquisition [190]. The HVPS has four outputs with a maximum voltage of 3.0 kV and it can be selected with positive or negative polarity. The dimensions of this instrument are $150 \times 165 \times 70\,\text{mm}^3$ ($w \times d \times h$) and it weighs approximately 2 kg. This system is powered by a lithium-ion polymer battery (3300 mAh) capable of operating a full working day. Besides the HVPS and the potentiostat, the instrument contains a chip holder for housing microchip-CE devices. A user-friendly interface, written in LabVIEW for control and data acquisition, has contributed to the popularity of this instrument which can also communicate with a computer (PC, laptop, iPad) by Bluetooth. Pyrex and SU-8/Pyrex microchips with WE located at end-channel configuration were used to demonstrate the feasibility of this portable instrument. The second generation of a battery-powered portable instrument from MicruX (shown in Figure 16.12) included a new reusable microfluidic platform to incorporate microchip-CE devices inside the instrument [191]. The portable instruments from MicruX have exhibited similar analytical performance compared to other commercial potentiostats. A remarkable feature achieved in their equipment is the higher peak current due to, probably, the implementation of additional filters in the hardware and software to isolate the electrical noise produced by the HVPS. As pointed out by the

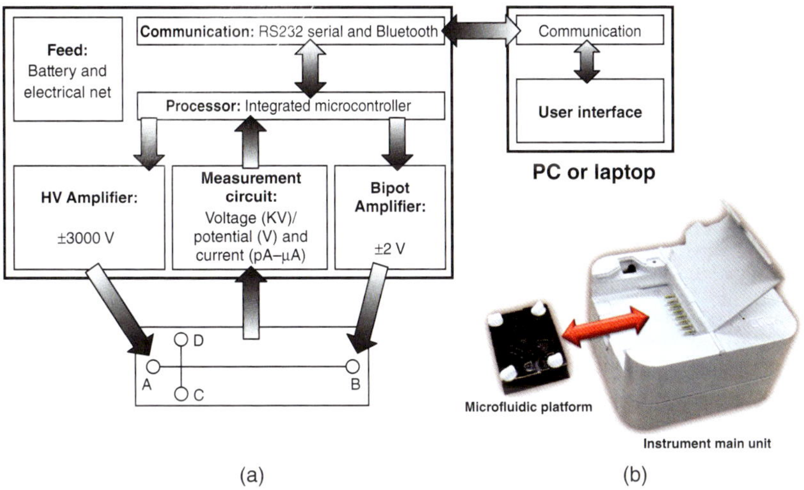

Figure 16.12 Representation of the (a) scheme for instrumental controlling. Reproduced with permission from [190]. Copyright © 2010 WILEY-VCH Verlag GmbH & Co. KGaA, Weinheim and [187] © 2012 WILEY-VCH Verlag GmbH & Co.; (b) the main unit of a portable chip-based system coupled with amperometric detection. Reproduced with permission from Ref. [191]. Copyright © 2010 WILEY-VCH Verlag GmbH & Co. KGaA, Weinheim

authors, the use of a fully portable instrument for microchip-CE-ECD can represent a true lab-on-a-chip for "*in situ*" real sample analysis.

Microfluidic ChipShop (http://www.microfluidic-chipshop.com/) has also developed portable chip-based electrophoresis systems coupled with C^4D for on-site food analysis (ChipGenie edition E) [192]. Because contactless conductivity allows label-free detection, it is quite attractive to monitor the stability or the degradation products related to foodstuff. The portable instrument developed by microfluidic ChipShop comprises a box with dimensions of $190 \times 120 \times 80\,mm^3$ $(w \times d \times h)$. Basically, it consists of two parts, as shown in Figure 16.13, which are closed together and locked before running

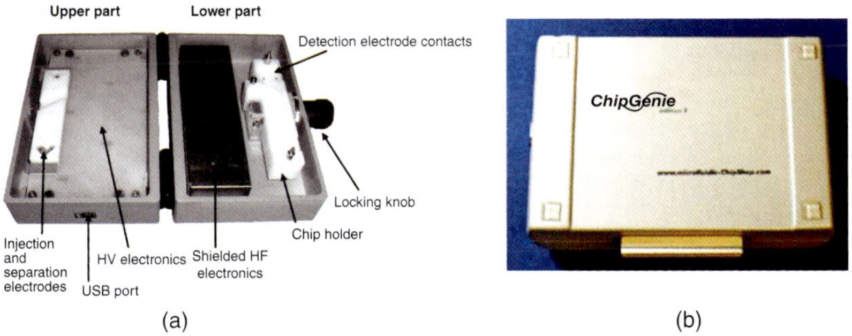

Figure 16.13 Representation of the ChipGenie® instrument from microfluidic ChipShop for electrophoresis microchip coupled with contactless conductivity detection. Images (a) and (b) depict the assembly of the instrumentation inside a box and a photo of the commercially available system, respectively. Reproduced with permission from Ref. [192]. Copyright © 2008 SPIE

experiments. The lower piece contains the high-frequency (HF) waveform and a chip holder with electrical contacts to assemble the microchip-CE device. To minimize the noise generation, the HF part is housed in a shielded box. C^4D is performed by using a frequency of 4 MHz. The upper piece houses two miniature regulated power supply modules from EMCO HighVoltage Corporation with capability of delivering a maximum voltage output of 4 kV (maximum current limited in 250 μA). Data acquisition is carried out through a user-interface software at a rate of 10 Hz with a USB-adapter plugged into a computer. ChipGenie edition E has been designed to operate with commercial microfluidic devices fabricated in poly(methyl-methacrylate). Commercial chips are integrated with detection electrodes, which are fabricated on the thin cover lid. The company has provided devices with different dimensions. This instrument has been successfully applied to the detection of inorganic ions and carboxylic acids in different food samples including mineral water, wine, and milk.

16.4 Conclusions

In summary, this chapter highlights the instrumental aspects related to ECD methods applied in food analysis. Complete and portable systems dedicated to food testing can be customized according to the requirements of the final user. Based on the advances presented here, it can be concluded that ECD is and will continue to be a powerful tool for agriculture and food analysis. Although the selection of specific instruments is mainly limited by the number and type of electrochemical techniques to be performed, the performance of current systems is certainly sufficient for most applications related to this book. Portable instruments, while not yet offering the performance of lab-grade instruments, can provide users very competitive tools for handheld or remotely controlled tests. We believe that future developments in this area will place ECD in a leading position for food-related analysis. We also envision that as more scientists become familiar with the instrumental improvements linked to electrochemical analysis and troubleshooting procedures, these techniques will become standard methods accepted by official agencies like the US Food and Drug Administration (FDA, http://www.fda.gov/).

Acknowledgments

Authors would like to thank Eric Costa (USP) for reviewing the manuscript and providing helpful suggestions. CDG would also like to thank the financial support provided, in part, by Clemson University, UT San Antonio, the National Aeronautics and Space Administration (NNX13CP49C), the National Institutes of Health through the National Institute of General Medical Sciences (2SC3GM081085) and the Research Centers at Minority Institutions (G12MD007591).

References

1. Tsarouhas, P. (2012) Reliability, availability and maintainability analysis in food production lines: a review. *Int. J. Food Sci. Technol.*, **47** (11), 2243–2251.

2. Ruiz-Capillas, C. and Jiménez-Colmenero, F. (2009) Application of flow injection analysis for determining sulphites in food and beverages: a review. *Food Chem.*, **112** (2), 487–493.
3. Pérez-Olmos, R., Soto, J.C., Zárate, N., Araújo, A.N., Lima, J.L.F.C., and Saraiva, M.L.M.F.S. (2005) Application of sequential injection analysis (SIA) to food analysis. *Food Chem.*, **90** (3), 471–490.
4. Chan-Eam, S., Teerasong, S., Damwan, K., Nacapricha, D., and Chaisuksant, R. (2011) Sequential injection analysis with electrochemical detection as a tool for economic and rapid evaluation of total antioxidant capacity. *Talanta*, **84** (5), 1350–1354.
5. Liu, M., Lin, Z., and Lin, J.-M. (2010) A review on applications of chemiluminescence detection in food analysis. *Anal. Chim. Acta*, **670** (1–2), 1–10.
6. Awad, T.S., Moharram, H.A., Shaltout, O.E., Asker, D., and Youssef, M.M. (2012) Applications of ultrasound in analysis, processing and quality control of food: a review. *Food Res. Int.*, **48** (2), 410–427.
7. Escuder-Gilabert, L. and Peris, M. (2010) Review: highlights in recent applications of electronic tongues in food analysis. *Anal. Chim. Acta*, **665** (1), 15–25.
8. Tiwari, K., Tudu, B., Bandyopadhyay, R., and Chatterjee, A. (2013) Identification of monofloral honey using voltammetric electronic tongue. *J. Food Eng.*, **117** (2), 205–210.
9. Frazier, R.A. (2001) Recent advances in capillary electrophoresis methods for food analysis. *Electrophoresis*, **22** (19), 4197–4206.
10. Martín, A., Vilela, D., and Escarpa, A. (2012) Food analysis on microchip electrophoresis: an updated review. *Electrophoresis*, **33** (15), 2212–2227.
11. Davies, A.M.C. and Grant, A. (1987) Review: near infra-red analysis of food. *Int. J. Food Sci. Technol.*, **22** (3), 191–207.
12. Bernal, J., Nozal, M.J., Jiménez, J.J., Martín, M.T., and Sanz, E. (2009) A new and simple method to determine trace levels of sulfonamides in honey by high performance liquid chromatography with fluorescence detection. *J. Chromatogr. A*, **1216** (43), 7275–7280.
13. Mannina, L., Sobolev, A.P., and Viel, S. (2012) Liquid state ^{1}H high field NMR in food analysis. *Prog. Nucl. Magn. Reson. Spectrosc.*, **66**, 1–39.
14. Marcone, M.F., Wang, S., Albabish, W., Nie, S., Somnarain, D., and Hill, A. (2013) Diverse food-based applications of nuclear magnetic resonance (NMR) technology. *Food Res. Int.*, **51** (2), 729–747.
15. Situ, C., Mooney, M.H., Elliott, C.T., and Buijs, J. (2010) Advances in surface plasmon resonance biosensor technology towards high-throughput, food-safety analysis. *TrAC, Trends Anal. Chem.*, **29** (11), 1305–1315.
16. Nielen, M.W.F., Hooijerink, H., Zomer, P., and Mol, J.G.J. (2011) Desorption electrospray ionization mass spectrometry in the analysis of chemical food contaminants in food. *TrAC, Trends Anal. Chem.*, **30** (2), 165–180.
17. Kuhnert, N., Jaiswal, R., Yassin, G., and Golon, A. (2013) What is under the hump? Mass spectrometry based analysis of complex mixtures in processed food – Lessons from the characterisation of black tea thearubigins, coffee melanoidines and caramel. *Food Funct.*, **4**, 1130–1147.
18. Herrero, M., Ibáñez, E., Cifuentes, A., and Bernal, J. (2009) Multidimensional chromatography in food analysis. *J. Chromatogr. A*, **1216** (43), 7110–7129.

19. Núñez, O., Gallart-Ayala, H., Martins, C.P.B., and Lucci, P. (2012) New trends in fast liquid chromatography for food and environmental analysis. *J. Chromatogr. A*, **1228**, 298–323.
20. Di Stefano, V., Avellone, G., Bongiorno, D., Cunsolo, V., Muccilli, V., Sforza, S., Dossena, A., Drahos, L., and Vékey, K. (2012) Applications of liquid chromatography–mass spectrometry for food analysis. *J. Chromatogr. A*, **1259**, 74–85.
21. Jeleń, H.H., Majcher, M., and Dziadas, M. (2012) Microextraction techniques in the analysis of food flavor compounds: a review. *Anal. Chim. Acta*, **738**, 13–26.
22. Altundag, H. and Tuzen, M. (2011) Comparison of dry, wet and microwave digestion methods for the multi element determination in some dried fruit samples by ICP-OES. *Food Chem. Toxicol.*, **49** (11), 2800–2807.
23. Mamone, G., Picariello, G., Caira, S., Addeo, F., and Ferranti, P. (2009) Analysis of food proteins and peptides by mass spectrometry-based techniques. *J. Chromatogr. A*, **1216** (43), 7130–7142.
24. Corradini, C., Lantano, C., and Cavazza, A. (2013) Innovative analytical tools to characterize prebiotic carbohydrates of functional food interest. *Anal. Bioanal. Chem.*, **405** (13), 4591–4605.
25. Cevallos-Cevallos, J.M., Reyes-De-Corcuera, J.I., Etxeberria, E., Danyluk, M.D., and Rodrick, G.E. (2009) Metabolomic analysis in food science: a review. *Trends Food Sci. Technol.*, **20** (11–12), 557–566.
26. Ridgway, K., Lalljie, S.P.D., and Smith, R.M. (2009) Analysis of food taints and off-flavours: a review. *Food Addit. Contam. Part A Chem. Anal. Control Expo. Risk Assess.*, **27** (2), 146–168.
27. Arapitsas, P. (2012) Hydrolyzable tannin analysis in food. *Food Chem.*, **135** (3), 1708–1717.
28. Farré, M. and Barceló, D. (2013) Analysis of emerging contaminants in food. *TrAC, Trends Anal. Chem.*, **43**, 240–253.
29. Gallart-Ayala, H., Núñez, O., and Lucci, P. (2013) Recent advances in LC-MS analysis of food-packaging contaminants. *TrAC, Trends Anal. Chem.*, **42**, 99–124.
30. Mezcua, M., Martínez-Uroz, M.A., Gómez-Ramos, M.M., Gómez, M.J., Navas, J.M., and Fernández-Alba, A.R. (2012) Analysis of synthetic endocrine-disrupting chemicals in food: a review. *Talanta*, **100**, 90–106.
31. Felhofer, J., Hanrahan, G., and García, C.D. (2009) Multivariate versus univariate optimization of separation conditions in micellar electrokinetic chromatography. *Talanta*, **77**, 1172–1178.
32. Motilva, M.-J., Serra, A., and Macià, A. (2013) Analysis of food polyphenols by ultra high-performance liquid chromatography coupled to mass spectrometry: an overview. *J. Chromatogr. A*, **1292**, 66–82.
33. Mejia, E., Ding, Y., Mora, M.F., and Garcia, C.D. (2007) Determination of banned sudan dyes in chili powder by capillary electrophoresis. *Food Chem.*, **102**, 1027–1033.
34. Chailapakul, O., Wonsawat, W., Siangproh, W., Grudpan, K., Zhao, Y., and Zhu, Z. (2008) Analysis of sudan I, sudan II, sudan III, and sudan IV in food by HPLC with electrochemical detection: comparison of glassy carbon electrode with carbon nanotube-ionic liquid gel modified electrode. *Food Chem.*, **109** (4), 876–882.

35. Rodriguez-Amaya, D.B. (2010) Quantitative analysis, in vitro assessment of bioavailability and antioxidant activity of food carotenoids – A review. *J. Food Compos. Anal.*, **23** (7), 726–740.
36. Martín-Calero, A., Pino, V., and Afonso, A.M. (2011) Ionic liquids as a tool for determination of metals and organic compounds in food analysis. *TrAC, Trends Anal. Chem.*, **30** (10), 1598–1619.
37. Herrero, M., Simó, C., García-Cañas, V., Fanali, S., and Cifuentes, A. (2010) Chiral capillary electrophoresis in food analysis. *Electrophoresis*, **31** (13), 2106–2114.
38. Kappes, T. and Hauser, P.C. (1999) Electrochemical detection methods in capillary electrophoresis and applications to inorganic species. *J. Chromatogr. A*, **834**, 89–101.
39. Trouillon, R., Svensson, M.I., Berglund, E.C., Cans, A.-S., and Ewing, A.G. (2012) Highlights of selected recent electrochemical measurements in living systems. *Electrochim. Acta*, **84**, 84–95.
40. Hickling, A. (1942) Studies in electrode polarisation. Part IV. The automatic control of the potential of a working electrode. *Trans. Faraday Soc.*, **38**, 27–33.
41. Stock, J.T. (1968) A simple low-current potentiostat coulometric analysis. *J. Chem. Educ.*, **45** (11), 736.
42. Goldsworthy, W.W. and Clem, R.G. (1971) Digital potentiostat. *Anal. Chem.*, **43** (12), 1718–1720.
43. Schwabe, K., Suschke, H.D., and Wachler, G. (1980) Electrochemical instrumentation. *Electrochim. Acta*, **25** (1), 59–76.
44. Smith, J.P. and Hinson-Smith, V. (2002) Flow injection analysis: quietly pushing ahead. *Anal. Chem.*, **74** (13), 385A–388A.
45. Cheng, K.Y., Cord-Ruwisch, R., and Ho, G. (2009) A new approach for in situ cyclic voltammetry of a microbial fuel cell biofilm without using a potentiostat. *Bioelectrochemistry*, **74** (2), 227–231.
46. Liu, H. and Crooks, R.M. (2012) Paper-based electrochemical sensing platform with integral battery and electrochromic read-out. *Anal. Chem.*, **84** (5), 2528–2532.
47. Brett, C.M.A. and Brett, A.M.O. (2005) *Electrochemistry: Principles, Methods, and Applications*, Oxford University Press.
48. Privett, B.J., Shin, J.H., and Schoenfisch, M.H. (2010) Electrochemical sensors. *Anal. Chem.*, **82** (12), 4723–4741.
49. Svorc, L. (2013) Determination of caffeine: a comprehensive review on electrochemical methods. *Int. J. Electrochem. Sci.*, **8**, 5755–5773.
50. Gulaboski, R. and Pereira, C.M. (2008) Electroanalytical techniques and instrumentation in food analysis, in *Handbook of Food Analysis Instruments* (ed S. Otles), CRC Press.
51. Fanjul-Bolado, P., Hernández-Santos, D., Lamas-Ardisana, P.J., Martín-Pernía, A., and Costa-García, A. (2008) Electrochemical characterization of screen-printed and conventional carbon paste electrodes. *Electrochim. Acta*, **53** (10), 3635–3642.
52. Khan, R., Gorski, W., and Garcia, C.D. (2011) Nanomolar detection of L-glutamate using screen-printed electrodes containing multi-wall carbon nanotubes. *Electroanalysis*, **23**, 2357–2363.
53. Grieshaber, D., MacKenzie, R., Vörös, J., and Reimhult, E. (2008) Electrochemical biosensors – Sensor principles and architectures. *Sensors*, **8** (3), 1400–1458.

54. Pacáková, V. and Štulík, K. (1985) Electrochemical detection in HPLC part 2. Applications. *Mol. Nutr. Food Res.*, **29** (7), 651–664.
55. Vandaveer, W.R., Pasas-Farmer, S.A., Fischer, D.J., Frankenfeld, C.N., and Lunte, S.M. (2004) Recent developments in electrochemical detection for microchip capillary electrophoresis. *Electrophoresis*, **25** (21-22), 3528–3549.
56. Escarpa, A., González, M.C., Crevillén, A.G., and Blasco, A.J. (2007) CE microchips: an opened gate to food analysis. *Electrophoresis*, **28** (6), 1002–1011.
57. Mark, J.J.P., Scholz, R., and Matysik, F.-M. (2012) Electrochemical methods in conjunction with capillary and microchip electrophoresis. *J. Chromatogr. A*, **1267**, 45–64.
58. Laude, N.D., Atcherley, C.W., and Heien, M.L. (2012) Rethinking data collection and signal processing. 1. Real-time oversampling filter for chemical measurements. *Anal. Chem.*, **84** (19), 8422–8426.
59. Chailapakul, O., Korsrisakul, S., Siangproh, W., and Grudpan, K. (2008) Fast and simultaneous detection of heavy metals using a simple and reliable microchip-electrochemistry route: an alternative approach to food analysis. *Talanta*, **74** (4), 683–689.
60. Romanazzo, D., Ricci, F., Volpe, G., Elliott, C.T., Vesco, S., Kroeger, K., Moscone, D., Stroka, J., Van Egmond, H., Vehniäinen, M., and Palleschi, G. (2010) Development of a recombinant Fab-fragment based electrochemical immunosensor for deoxynivalenol detection in food samples. *Biosens. Bioelectron.*, **25** (12), 2615–2621.
61. Viswanathan, S., Rani, C., and Ho, J.-A.A. (2012) Electrochemical immunosensor for multiplexed detection of food-borne pathogens using nanocrystal bioconjugates and MWCNT screen-printed electrode. *Talanta*, **94**, 315–319.
62. Yin, H., Meng, X., Su, H., Xu, M., and Ai, S. (2012) Electrochemical determination of theophylline in foodstuff, tea and soft drinks based on urchin-like CdSe microparticles modified glassy carbon electrode. *Food Chem.*, **134** (2), 1225–1230.
63. Kergaravat, S.V., Gómez, G.A., Fabiano, S.N., Laube Chávez, T.I., Pividori, M.I., and Hernández, S.R. (2012) Biotin determination in food supplements by an electrochemical magneto biosensor. *Talanta*, **97**, 484–490.
64. del Torno-de Román, L., Alonso-Lomillo, M.A., Domínguez-Renedo, O., and Arcos-Martínez, M.J. (2013) Gluconic acid determination in wine by electrochemical biosensing. *Sens. Actuators B*, **176**, 858–862.
65. Guyenet, S.J., Matsen, M.E., Morton, G.J., Kaiyala, K.J., and Schwartz, M.W. (2013) Rapid glutamate release in the mediobasal hypothalamus accompanies feeding and is exaggerated by an obesogenic food. *Mol. Metab.*, **2** (2), 116–122.
66. Yoneyama, Y., Yonemori, Y., Murata, M., Ohnuki, H., Hibi, K., Hayashi, T., Ren, H., and Endo, H. (2009) Wireless biosensor system for real-time cholesterol monitoring in fish "Nile tilapia". *Talanta*, **80** (2), 909–915.
67. Lee, K.H., Kristic, K., van Hoff, R., Hitti, F.L., Blaha, C., Harris, B., Roberts, D.W., and Leiter, J.C. (2007) High-frequency stimulation of the subthalamic nucleus increases glutamate in the subthalamic nucleus of rats as demonstrated by in vivo enzyme-linked glutamate sensor. *Brain Res.*, **1162**, 121–129.

68. Veloso, A.J., Nagy, P.M., Zhang, B., Dhar, D., Liang, A., Ibrahim, T., Mikhaylichenko, S., Aubert, I., and Kerman, K. (2013) Miniaturized electrochemical system for cholinesterase inhibitor detection. *Anal. Chim. Acta*, **774**, 73–78.
69. Martín-Yerga, D., González-García, M.B., and Costa-García, A. (2013) Biosensor array based on the in situ detection of quantum dots as electrochemical label. *Sens. Actuators B*, **182**, 184–189.
70. Golitsch, F., Bücking, C., and Gescher, J. (2013) Proof of principle for an engineered microbial biosensor based on *Shewanella oneidensis* outer membrane protein complexes. *Biosens. Bioelectron.*, **47**, 285–291.
71. Stefan-van Staden, R.-I., Nejem, R.A.M., van Staden, J.F., and Aboul-Enein, H.Y. (2012) Amperometric biosensor based on diamond paste for the enantioanalysis of l-lysine. *Biosens. Bioelectron.*, **35** (1), 439–442.
72. Salinas, E., Rivero, V., Torriero, A.A.J., Benuzzi, D., Sanz, M.I., and Raba, J. (2006) Multienzymatic-rotating biosensor for total cholesterol determination in a FIA system. *Talanta*, **70** (2), 244–250.
73. Marszałł, M.L., Lebiedzińska, A., Czarnowski, W., Makarowski, R., Kłos, M., and Szefer, P. (2009) Application of the high-performance liquid chromatography method with coulometric detection for determination of vitamin B6 in human plasma and serum. *J. Chromatogr. B*, **877** (27), 3151–3158.
74. Webster, R.D. (2012) Voltammetry of the liposoluble vitamins (A, D, E and K) in organic solvents. *Chem. Rec.*, **12** (1), 188–200.
75. Baldwin, R.P. (1999) Electrochemical determination of carbohydrates: enzyme electrodes and amperometric detection in liquid chromatography and capillary electrophoresis. *J. Pharm. Biomed. Anal.*, **19** (1–2), 69–81.
76. Jin, P., Yamaguchi, A., Oi, F.A., Matsuo, S., Tan, J., and Misawa, H. (2001) Glucose sensing based on interdigitated array microelectrode. *Anal. Sci.*, **17** (7), 841–846.
77. García, C.D. and Ortiz, P.I. (1998) Determination of *tert*-butylhydroxytoluene by flow injection analysis at polymer modified glassy carbon electrodes. *Electroanalysis*, **10** (12), 832–835.
78. Blasco, A.J., Rogerio, M.C., Gonzalez, M.C., and Escarpa, A. (2005) "Electrochemical Index" as a screening method to determine "total polyphenolics" in foods: a proposal. *Anal. Chim. Acta*, **539** (1–2), 237–244.
79. Guan, Y., Chu, Q., Fu, L., Wu, T., and Ye, J. (2006) Determination of phenolic antioxidants by micellar electrokinetic capillary chromatography with electrochemical detection. *Food Chem.*, **94** (1), 157–162.
80. Palchetti, I., Laschi, S., and Mascini, M. (2009) Electrochemical biosensor technology: application to pesticide detection, in *Biosensors and Biodetection*, vol. **504** (eds A. Rasooly and K. Herold), Humana Press, pp. 115–126.
81. Llorent-Martínez, E.J., Ortega-Barrales, P., Fernández-de Córdova, M.L., and Ruiz-Medina, A. (2011) Trends in flow-based analytical methods applied to pesticide detection: a review. *Anal. Chim. Acta*, **684** (1–2), 30–39.
82. Vidal, J.C., Bonel, L., Ezquerra, A., Hernández, S., Bertolín, J.R., Cubel, C., and Castillo, J.R. (2013) Electrochemical affinity biosensors for detection of mycotoxins: a review. *Biosens. Bioelectron.*, **49**, 146–158.
83. Mello, L.D. and Kubota, L.T. (2002) Review of the use of biosensors as analytical tools in the food and drink industries. *Food Chem.*, **77** (2), 237–256.

84. Ricci, F., Volpe, G., Micheli, L., and Palleschi, G. (2007) A review on novel developments and applications of immunosensors in food analysis. *Anal. Chim. Acta*, **605** (2), 111–129.
85. Pilolli, R., Monaci, L., and Visconti, A. (2013) Advances in biosensor development based on integrating nanotechnology and applied to food-allergen management. *TrAC, Trends Anal. Chem.*, **47**, 12–26.
86. Lotierzo, M., Abuknesha, R., Davis, F., and Tothill, I.E. (2012) A membrane-based ELISA assay and electrochemical immunosensor for microcystin-LR in water samples. *Environ. Sci. Technol.*, **46** (10), 5504–5510.
87. Croci, L., Delibato, E., Volpe, G., Medici, D.D., and Palleschi, G. (2004) Comparison of PCR, electrochemical enzyme-linked immunosorbent assays, and the standard culture method for detecting salmonella in meat products. *Appl. Environ. Microbiol.*, **70** (3), 1393–1396.
88. Li, X.J.,Nie, Z.H.,and Cheng, C.-M. (2010) Paper-based electrochemical ELISA. 14th International Conference on Miniaturized Systems for Chemistry and Life Sciences, Groningen, The Netherlands, 2010.
89. Chatrathi, M.P., Wang, J., and Collins, G.E. (2007) Sandwich electrochemical immunoassay for the detection of Staphylococcal enterotoxin B based on immobilized thiolated antibodies. *Biosens. Bioelectron.*, **22** (12), 2932.
90. Scampicchio, M., Ballabio, D., Arecchi, A., Cosio, S.M., and Mannino, S. (2008) Amperometric electronic tongue for food analysis. *Microchim. Acta*, **163**, 11–21.
91. Gutierrez, M., Llobera, A., Vila-Planas, J., Capdevila, F., Demming, S., Buttgenbach, S., Minguez, S., and Jimenez-Jorquera, C. (2010) Hybrid electronic tongue based on optical and electrochemical microsensors for quality control of wine. *Analyst*, **135** (7), 1718–1725.
92. Wang, J., Jiang, M., and Lu, F. (1998) Electrochemical quartz crystal microbalance investigation of surface fouling due to phenol oxidation. *J. Electroanal. Chem.*, **444** (1), 127–132.
93. García, G., García, C.D., Ortiz, P.I., and De Pauli, C.P. (2002) Reflectometry applied to electrochemically generated phenoxy radical adsorption monitoring. *J. Electroanal. Chem.*, **519** (1–2), 53–59.
94. Ferreira, M., Varela, H., Torresi, R.M., and Tremiliosi-Filho, G. (2006) Electrode passivation caused by polymerization of different phenolic compounds. *Electrochim. Acta*, **52** (2), 434–442.
95. Yang, X., Kirsch, J., Fergus, J., and Simonian, A. (2013) Modeling analysis of electrode fouling during electrolysis of phenolic compounds. *Electrochim. Acta*, **94**, 259–268.
96. Tanimoto, S. and Ichimura, A. (2013) Discrimination of inner- and outer-sphere electrode reactions by cyclic voltammetry experiments. *J. Chem. Educ.*, **90** (6), 778–781.
97. LaCourse, W.R. and Modi, S.J. (2005) Microelectrode applications of pulsed electrochemical detection. *Electroanalysis*, **17** (13), 1141–1152.
98. Garcia, C.D. and Henry, C.S. (2003) Direct determination of carbohydrates, amino acids and antibiotics by microchip electrophoresis with pulsed amperometric detection. *Anal. Chem.*, **75** (18), 4778–4783.

99. Hompesch, R.W., Garcia, C.D., Weiss, D.J., Vivanco, J.M., and Henry, C.S. (2005) Analysis of natural flavonoids by microchip-micellar electrokinetic chromatography with pulsed amperometric detection. *Analyst*, **130** (5), 694–700.
100. Garcia, C.D. and Henry, C.S. (2005) Comparison of pulsed electrochemical detection modes coupled with microchip capillary electrophoresis. *Electroanalysis*, **17** (3), 223–229.
101. Liang, L., Cai, Y., Mou, S., and Cheng, J. (2005) Comparisons of disposable and conventional silver working electrode for the determination of iodide using high-performance anion-exchange chromatography with pulsed amperometric detection. *J. Chromatogr. A*, **1085** (1), 37–41.
102. Mercier, G. and Campargue, C. (2012) Partially hydrolyzed guar gum characterization and sensitive quantification in food matrices by high performance anion exchange chromatography with pulsed amperometric detection – Validation using accuracy profile. *J. Chromatogr. A*, **1262**, 180–187.
103. Sim, H.-J., Moon, E., Kim, S.Y., and Hong, S.-P. (2013) Determination of proline–hydroxyproline dipeptide in rat urine by high-performance anion-exchange chromatography coupled with pulsed amperometric detection. *J. Chromatogr. B*, **930**, 70–74.
104. Coutinho, C.F.B., Coutinho, L.F.M., Lanças, F.M., Câmara, C.A.P., Nixdorf, S.L., and Mazo, L.H. (2008) Development of instrumentation for amperometric and coulometric detection using ultramicroelectrodes. *J. Braz. Chem. Soc.*, **19**, 131–139.
105. Mann, C.K. (1965) An instrument for cyclic voltammetry. *Anal. Chem.*, **37** (3), 326–329.
106. Kissinger, P.T. and Heineman, W.R. (1983) Cyclic voltammetry. *J. Chem. Educ.*, **60** (9), 702.
107. Alghamdi, A.H. (2010) Applications of stripping voltammetric techniques in food analysis. *Arabian J. Chem.*, **3** (1), 1–7.
108. Herzog, G. and Beni, V. (2013) Stripping voltammetry at micro-interface arrays: a review. *Anal. Chim. Acta*, **769**, 10–21.
109. Heien, M.L.A.V., Johnson, M.A., and Wightman, R.M. (2004) Resolving neurotransmitters detected by fast-scan cyclic voltammetry. *Anal. Chem.*, **76** (19), 5697–5704.
110. Chen, A. and Shah, B. (2013) Electrochemical sensing and biosensing based on square wave voltammetry. *Anal. Methods*, **5** (9), 2158–2173.
111. Brazill, S.A., Bender, S.E., Hebert, N.E., Cullison, J.K., Kristensen, E.W., and Kuhr, W.G. (2002) Sinusoidal voltammetry: a frequency based electrochemical detection technique. *J. Electroanal. Chem.*, **531** (2), 119–132.
112. Gholivand, M.B., Torkashvand, M., and Malekzadeh, G. (2012) Fabrication of an electrochemical sensor based on computationally designed molecularly imprinted polymers for determination of cyanazine in food samples. *Anal. Chim. Acta*, **713**, 36–44.
113. Gholivand, M.B., Karimian, N., and Malekzadeh, G. (2012) Computational design and synthesis of a high selective molecularly imprinted polymer for voltammetric sensing of propazine in food samples. *Talanta*, **89**, 513–520.
114. Abbasi, S., Bahiraei, A., and Abbasai, F. (2011) A highly sensitive method for simultaneous determination of ultra trace levels of copper and cadmium in food and water

samples with luminol as a chelating agent by adsorptive stripping voltammetry. *Food Chem.*, **129** (3), 1274–1280.
115. Abbasi, S. and Bahiraei, A. (2012) Ultra trace quantification of chromium(VI) in food and water samples by highly sensitive catalytic adsorptive stripping voltammetry with rubeanic acid. *Food Chem.*, **133** (3), 1075–1080.
116. Chaiyo, S., Chailapakul, O., Sakai, T., Teshima, N., and Siangproh, W. (2013) Highly sensitive determination of trace copper in food by adsorptive stripping voltammetry in the presence of 1,10-phenanthroline. *Talanta*, **108**, 1–6.
117. Feier, B., Floner, D., Cristea, C., Bodoki, E., Sandulescu, R., and Geneste, F. (2012) Flow electrochemical analyses of zinc by stripping voltammetry on graphite felt electrode. *Talanta*, **98**, 152–156.
118. Yardım, Y., Gülcan, M., and Şentürk, Z. (2013) Determination of vanillin in commercial food product by adsorptive stripping voltammetry using a boron-doped diamond electrode. *Food Chem.*, **141**, 1821–1827.
119. Souza, D.D., Machado, S.A.S., and Pires, R.C. (2006) Multiple square wave voltammetry for analytical determination of paraquat in natural water, food, and beverages using microelectrodes. *Talanta*, **69** (5), 1200–1207.
120. Chen, J. and Chen, C. (2013) A new data analysis method to determine pesticide concentrations by cyclic voltammetry. *Measurement*, **46** (6), 1828–1833.
121. Lin, X., Ni, Y., and Kokot, S. (2013) Glassy carbon electrodes modified with gold nanoparticles for the simultaneous determination of three food antioxidants. *Anal. Chim. Acta*, **765**, 54–62.
122. Yin, H., Zhou, Y., Meng, X., Tang, T., Ai, S., and Zhu, L. (2011) Electrochemical behaviour of Sudan I at Fe_3O_4 nanoparticles modified glassy carbon electrode and its determination in food samples. *Food Chem.*, **127** (3), 1348–1353.
123. Medeiros, R.A., Lourencao, B.C., Rocha-Filho, R.C., and Fatibello-Filho, O. (2012) Simultaneous voltammetric determination of synthetic colorants in food using a cathodically pretreated boron-doped diamond electrode. *Talanta*, **97**, 291–297.
124. Thomsen, K.N., Skov, H.J., and Dam, M.E.R. (1994) A flexible instrument for voltammetry, amperometry and stripping potentiometry. *Anal. Chim. Acta*, **293** (1-2), 1–17.
125. Nawghare, P.M., Singh, K., and Limaye, S.S. (1996) PC based instrument for impedance analysis, cyclic voltammetry and transient analysis. *Solid State Ion.*, **90** (1-4), 295–301.
126. Huang, H. and Dasgupta, P.K. (1999) A field-deployable instrument for the measurement and speciation of arsenic in potable water. *Anal. Chim. Acta*, **380** (1), 27–37.
127. Cox, J.T., Guerrette, J.P., and Zhang, B. (2012) Steady-state voltammetry of a microelectrode in a closed bipolar cell. *Anal. Chem.*, **84** (20), 8797–8804.
128. Takmakov, P., McKinney, C.J., Carelli, R.M., and Wightman, R.M. (2011) Instrumentation for fast-scan cyclic voltammetry combined with electrophysiology for behavioral experiments in freely moving animals. *Rev. Sci. Instrum.*, **82** (7), 074302.
129. Brand, M.J.D. and Rechnitz, G.A. (1970) Differential potentiometry with ion-selective electrodes. New instrumental approach. *Anal. Chem.*, **42** (6), 616–622.
130. DuPre, D.B. and Just, W.E. (1994) A digital pH-mV meter for overhead projection. *J. Chem. Educ.*, **71** (12), 1081.

131. da Rocha, R.T., Gutz, I.G.R., and do Lago, C.L. (1995) From christmas ornament to glass electrode. *J. Chem. Educ.*, **72** (12), 1135.
132. Chow, C.W.K., Davey, D.E., and Mulcahy, D.E. (1998) Comparison of detector systems in oxidative stripping potentiometry. *Lab. Autom. Inf. Manage.*, **33** (3), 207–215.
133. Santini, A.O., Pezza, H.R., Carloni-Filho, J., Sequinel, R., and Pezza, L. (2009) Potentiometric sensor for sorbic acid determination in food products. *Food Chem.*, **115** (4), 1563–1567.
134. Santini, A.O., Pezza, H.R., and Pezza, L. (2012) Development of a sensitive potentiometric sensor for determination of fumaric acid in powdered food products. *Food Chem.*, **134** (1), 483–487.
135. Amorim, C.G., Souza, R.C., Araújo, A.N., Montenegro, M.C.B.S.M., and Silva, V.L. (2010) SI lab-on-valve analysis of histamine using potentiometric detection for food quality control. *Food Chem.*, **122** (3), 871–876.
136. Švancara, I., Ogorevc, B., Nović, M., and Vytřas, K. (2002) Simple and rapid determination of iodide in table salt by stripping potentiometry at a carbon-paste electrode. *Anal. Bioanal. Chem.*, **372** (7–8), 795–800.
137. Saei-Dehkordi, S.S. and Fallah, A.A. (2011) Determination of copper, lead, cadmium and zinc content in commercially valuable fish species from the Persian Gulf using derivative potentiometric stripping analysis. *Microchem. J.*, **98** (1), 156–162.
138. Giacomo, D., Pera Lara, L., Daniele, G., Francesco, S., and Turco Vincenzo, L. (2004) Influence of the olive variety and the zone of provenience on selenium content determined by cathodic stripping potentiometry (CSP) in virgin olive oils. *Food Chem.*, **88** (1), 135–140.
139. Estela, J.M., Tomás, C., Cladera, A., and Cerdà, V. (1995) Potentiometric stripping analysis: areview. *Crit. Rev. Anal. Chem.*, **25** (2), 91–141.
140. Coltro, W.K.T., Lima, R.S., Segato, T.P., Carrilho, E., de Jesus, D.P., do Lago, C.L., and da Silva, J.A.F. (2012) Capacitively coupled contactless conductivity detection on microfluidic systems – ten years of development. *Anal. Methods*, **4** (1), 25–33.
141. Fracassi da Silva, J.A. and do Lago, C.L. (1998) An oscillometric detector for capillary electrophoresis. *Anal. Chem.*, **70** (20), 4339–4343.
142. Zemann, A.J., Schnell, E., Volgger, D., and Bonn, G.K. (1998) Contactless conductivity detection for capillary electrophoresis. *Anal. Chem.*, **70** (3), 563–567.
143. Brito-Neto, J.G.A., Fracassi da Silva, J.A., Blanes, L., and do Lago, C.L. (2005) Understanding capacitively coupled contactless conductivity detection in capillary and microchip electrophoresis. Part 1. Fundamentals. *Electroanalysis*, **17** (13), 1198–1206.
144. Brito-Neto, J.G.A., Fracassi da Silva, J.A., Blanes, L., and do Lago, C.L. (2005) Understanding capacitively coupled contactless conductivity detection in capillary and microchip electrophoresis. Part 2. Peak shape, stray capacitance, noise, and actual electronics. *Electroanalysis*, **17** (13), 1207–1214.
145. Pumera, M. (2007) Contactless conductivity detection for microfluidics: designs and applications. *Talanta*, **74** (3), 358–364.
146. Kubán, P. and Hauser, P.C. (2008) A review of the recent achievements in capacitively coupled contactless conductivity detection. *Anal. Chim. Acta*, **607** (1), 15–29.

147. Elbashir, A.A. and Aboul-Enein, H.Y. (2012) Recent advances in applications of capillary electrophoresis with capacitively coupled contactless conductivity detection (CE-C4D): an update. *Biomed. Chromatogr.*, **26**, 990–1000.
148. Mori, M., Kaseda, M., Yamamoto, T., Yamada, S., and Itabashi, H. (2012) Capillary ion electrophoresis–capacitively coupled contactless conductivity detection of inorganic cations in human saliva on a polyvinyl alcohol-coated capillary. *Anal. Bioanal. Chem.*, **402** (7), 2425–2430.
149. Segato, T.P., Bhakta, S.A., Gordon, M.T., Carrilho, E., Willis, P.A., Jiao, H., and Garcia, C.D. (2013) Microfab-less microfluidic capillary electrophoresis devices. *Anal. Methods*, **5** (7), 1652–1657.
150. Francisco, K.J.M. and do Lago, C.L. (2009) *Electrophoresis*, **30**, 3458–3464.
151. Rossi, M.R., Vidal, D.T.R., and do Lago, C.L. (2012) Monoalkyl carbonates in carbonated alcoholic beverages. *Food Chem.*, **133** (2), 352–357.
152. da Costa, E.T., Mora, M.F., Willis, P.A., do Lago, C.L., Jiao, H., and Garcia, C.D. (2014) Getting started with open-hardware: development and control of microfluidic devices. *Electrophoresis*, **35** (16), 2370–2377.
153. Talamond, P., Doulbeau, S., Rochette, I., Guyot, J.-P., and Treche, S. (2000) Anion-exchange high-performance liquid chromatography with conductivity detection for the analysis of phytic acid in food. *J. Chromatogr. A*, **871** (1–2), 7–12.
154. Niemann, R.A., Krynitsky, A.J., and Nortrup, D.A. (2006) Ion chromatographic determination of perchlorate in foods by on-line enrichment and suppressed conductivity detection. *J. Agric. Food Chem.*, **54** (4), 1137–1143.
155. Pormsila, W., Krähenbühl, S., and Hauser, P.C. (2010) Determination of carnitine in food and food supplements by capillary electrophoresis with contactless conductivity detection. *Electrophoresis*, **31** (13), 2186–2191.
156. Cinquina, A.L., Calı, A., Longo, F., Santis, L.D., Severoni, A., and Abballe, F. (2004) Determination of biogenic amines in fish tissues by ion-exchange chromatography with conductivity detection. *J. Chromatogr. A*, **1032** (1–2), 73–77.
157. Palermo, C., Muscarella, M., Nardiello, D., Iammarino, M., and Centonze, D. (2013) A multiresidual method based on ion-exchange chromatography with conductivity detection for the determination of biogenic amines in food and beverages. *Anal. Bioanal. Chem.*, **405** (2–3), 1015–1023.
158. Nogueira, T. and do Lago, C.L. (2007) Determination of caffeine in coffee products by dynamic complexation with 3,4-dimethoxycinnamate and separation by CZE. *Electrophoresis*, **28** (19), 3570–3574.
159. Tuma, P., Málková, K., Samcová, E., and Štulík, K. (2011) Rapid monitoring of mono- and disaccharides in drinks, foodstuffs and foodstuff additives by capillary electrophoresis with contactless conductivity detection. *Anal. Chim. Acta*, **698** (1-2), 1–5.
160. Mark, J.J.P., Kumar, A., Demattio, H., Hoffmann, W., Malik, A., and Matysik, F.-M. (2011) Combination of headspace single-drop microextraction, microchip electrophoresis and contactless conductivity detection for the determination of aliphatic amines in the biodegradation process of seafood samples. *Electroanalysis*, **23** (1), 161–168.

161. Masot, R., Alcañiz, M., Garcia-Breijo, E., Olguin, C., Ibáñez, J., and Gil-Sánchez, L. (2012) Portable measurement system for voltammetry and impedance spectroscopy. Application for TNT detection. *Procedia Eng.*, **47**, 1129–1132.
162. Huang, C.-Y., Syu, M.-J., Chang, Y.-S., Chang, C.-H., Chou, T.-C., and Liu, B.-D. (2007) A portable potentiostat for the bilirubin-specific sensor prepared from molecular imprinting. *Biosens. Bioelectron.*, **22** (8), 1694–1699.
163. Jiang, L., Jiang, X., Lu, Y., Dai, Z., Xie, M., Qin, J., and Lin, B. (2007) Development of a universal serial bus-powered mini-high-voltage power supply for microchip electrophoresis. *Electrophoresis*, **28** (8), 1259–1264.
164. Chan, F.-L., Chang, W.-Y., Kuo, L.-M., Lin, C.-H., Wang, S.-W., Yang, Y.-S., and Lu, M.S.-C. (2008) An electrochemical dopamine sensor with a CMOS detection circuit. *J. Micromech. Microeng.*, **18** (7), 075028.
165. Kim, B.N., Herbst, A.D., Kim, S.J., Minch, B.A., and Lindau, M. (2013) Parallel recording of neurotransmitters release from chromaffin cells using a 10 × 10 CMOS IC potentiostat array with on-chip working electrodes. *Biosens. Bioelectron.*, **41**, 736–744.
166. Jin, Y., Wang, H., Lv, Z., Yang, S., Cai, H., and Jiang, J. (2009) Reliable remote-monitoring electrochemical potentiostat for glucose measurements. *Tsinghua Sci. Technol.*, **14** (5), 593–600.
167. Ramfos, I., Vassiliadis, N., Blionas, S., Efstathiou, K., Fragoso, A., O'Sullivan, C.K., and Birbas, A. (2013) A compact hybrid-multiplexed potentiostat for real-time electrochemical biosensing applications. *Biosens. Bioelectron.*, **47**, 482–489.
168. Lillehoj, P.B., Huang, M.-C., Truong, N., and Ho, C.-M. (2013) Rapid electrochemical detection on a mobile phone. *Lab Chip*, **13** (15), 2950–2955.
169. Delaney, J.L., Doeven, E.H., Harsant, A.J., and Hogan, C.F. (2013) Use of a mobile phone for potentiostatic control with low cost paper-based microfluidic sensors. *Anal. Chim. Acta*, **790**, 56–60.
170. Nemiroski, A., Christodouleas, D.C., Hennek, J.W., Kumar, A.A., Maxwell, E.J., Fernández-Abedul, M.T., and Whitesides, G.M. (2014) Universal mobile electrochemical detector designed for use in resource-limited applications. *Proc. Natl. Acad. Sci. U.S.A.*, **111** (33), 11984–11989.
171. Gopinath, A.V. and Russell, D. (2006) An inexpensive field-portable programmable potentiostat. *Chem. Educ.*, **11** (1), 23–28.
172. Huang, C.-Y., Tsai, T.-C., Thomas, J.L., Lee, M.-H., Liu, B.-D., and Lin, H.-Y. (2009) Urinalysis with molecularly imprinted poly(ethylene-*co*-vinyl alcohol) potentiostat sensors. *Biosens. Bioelectron.*, **24** (8), 2611–2617.
173. Rowe, A.A., Bonham, A.J., White, R.J., Zimmer, M.P., Yadgar, R.J., Hobza, T.M., Honea, J.W., Ben-Yaacov, I., and Plaxco, K.W. (2011) CheapStat: an open-source, "do-it-yourself" potentiostat for analytical and educational applications. *PLoS One*, **6** (9), e23783.
174. Friedman, E.S., Rosenbaum, M.A., Lee, A.W., Lipson, D.A., Land, B.R., and Angenent, L.T. (2012) A cost-effective and field-ready potentiostat that poises subsurface electrodes to monitor bacterial respiration. *Biosens. Bioelectron.*, **32** (1), 309–313.

175. Desmond, D., Lane, B., Alderman, J., Hill, M., Arrigan, D.W.M., and Glennon, J.D. (1998) An environmental monitoring system for trace metals using stripping voltammetry. *Sens. Actuators B*, **48** (1-3), 409–414.
176. Yun, K.-S., Gil, J., Kim, J., Kim, H.-J., Kim, K., Park, D., Kim, M.S., Shin, H., Lee, K., Kwak, J., and Yoon, E. (2004) A miniaturized low-power wireless remote environmental monitoring system based on electrochemical analysis. *Sens. Actuators B*, **102** (1), 27–34.
177. Ruiz-Garcia, L., Lunadei, L., Barreiro, P., and Robla, I. (2009) A review of wireless sensor technologies and applications in agriculture and food industry: state of the art and current trends. *Sensors*, **9** (6), 4728–4750.
178. Berg, C., Valdez, D.C., Bergeron, P., Mora, M.F., Garcia, C.D., and Ayon, A. (2008) Lab-on-a-robot: integrated microchip – Capillary electrophoresis, power supply, potentiostat, wireless unit, and controller circuitry. *Electrophoresis*, **29**, 4914–4921.
179. da Costa, E.T., Neves, C.A., Hotta, G.M., Vidal, D.T.R., Barros, M.F., Ayon, A.A., Garcia, C.D., and do Lago, C.L. (2012) Unmanned platform for long-range remote analysis of volatile compounds in air samples. *Electrophoresis*, **33** (17), 2650–2659.
180. Gunasekara, D.B., Hulvey, M.K., and Lunte, S.M. (2011) In-channel amperometric detection for microchip electrophoresis using a wireless isolated potentiostat. *Electrophoresis*, **32** (8), 832–837.
181. Kassal, P., Steinberg, I.M., and Steinberg, M.D. (2013) Wireless smart tag with potentiometric input for ultra low-power chemical sensing. *Sens. Actuators B*, **184**, 254–259.
182. Rossier, J., Reymond, F., and Michel, P.E. (2002) Polymer microfluidic chips for electrochemical and biochemical analyses. *Electrophoresis*, **23** (6), 858–867.
183. Uchiyama, K., Nakajima, H., and Hobo, T. (2004) Detection method for microchip separations. *Anal. Bioanal. Chem.*, **379** (3), 375–382.
184. Xu, J.-J., Wang, A.-J., and Chen, H.-Y. (2007) Electrochemical detection modes for microchip capillary electrophoresis. *TrAC, Trends Anal. Chem.*, **26** (2), 125–132.
185. Kappes, T. and Hauser, P.C. (1998) Portable capillary electrophoresis instrument with potentiometric detection. *Anal. Commun.*, **35** (10), 325–329.
186. Kappes, T., Galliker, B., Schwarz, M.A., and Hauser, P.C. (2001) Portable capillary electrophoresis instrument with amperometric, potentiometric and conductometric detection. *TrAC, Trends Anal. Chem.*, **20** (3), 133–139.
187. Mai, T.D., Pham, T.T.T., Pham, H.V., Sáiz, J., Ruiz, C.G., and Hauser, P.C. (2013) Portable capillary electrophoresis instrument with automated injector and contactless conductivity detection. *Anal. Chem.*, **85** (4), 2333–2339.
188. Martin, R.S., Ratzlaff, K.L., Huynh, B.H., and Lunte, S.M. (2002) In-channel electrochemical detection for microchip capillary electrophoresis using an electrically isolated potentiostat. *Anal. Chem.*, **74** (5), 1136–1143.
189. Jackson, D.J., Naber, J.F., Roussel, T.J. Jr.,, Crain, M.M., Walsh, K.M., Keynton, R.S., and Baldwin, R.P. (2003) Portable high-voltage power supply and electrochemical detection circuits for microchip capillary electrophoresis. *Anal. Chem.*, **75**, 3311–3317.
190. Fernández-la-Villa, A., Pozo-Ayuso, D.F., and Castaño-Álvarez, M. (2010) New analytical portable instrument for microchip electrophoresis with electrochemical detection. *Electrophoresis*, **31** (15), 2641–2649.

191. Fernández-la-Villa, A., Sánchez-Barragán, D., Pozo-Ayuso, D.F., and Castaño-Álvarez, M. (2012) Smart portable electrophoresis instrument based on multipurpose microfluidic chips with electrochemical detection. *Electrophoresis*, **33** (17), 2733–2742.
192. Becker, H., Mühlberger, H., Hoffmann, W. *et al.* (2008) Portable CE system with contactless conductivity detection in an injection molded polymer chip for on-site food analysis. *Proc.SPIE*, Microfluidics, BioMEMS, and Medical Microsystems VI, **6886**, 68860C.

Index

Note: Page references in *italic* refers to figures and those are in **bold** refers to tables.

Agricultural and Food Electroanalysis, First Edition.
Edited by Alberto Escarpa, María Cristina González and Miguel Ángel López.